This book presents a comprehensive and coherent account of the theory of quantum fields on a lattice, an essential technique for the study of the strong and electroweak nuclear interactions.

Quantum field theory describes basic physical phenomena over an extremely wide range of length or energy scales. Quantum fields exist in space and time, which can be approximated by a set of lattice points. This approximation allows the application of powerful analytical and numerical techniques, and has provided a powerful tool for the study of both the strong and the electroweak interaction. After introductory chapters on scalar fields, gauge fields and fermion fields, the book studies quarks and gluons in QCD and fermions and bosons in the electroweak theory. The last chapter is devoted to numerical simulation algorithms which have been used in recent large-scale numerical simulations.

This book will be valuable for graduate students and researchers in theoretical physics, elementary particle physics, and field theory, interested in non-perturbative approximations and numerical simulations of quantum field phenomena.

CAMBRIDGE MONOGRAPHS ON MATHEMATICAL PHYSICS

General Editors: P. V. Landshoff, D. R. Nelson, D. W. Sciama, S. Weinberg

QUANTUM FIELDS ON A LATTICE

CAMBRIDGE MONOGRAPHS ON
MATHEMATICAL PHYSICS

A. M. Anile   *Relativistic Fluids and Magneto-Fluids*
J. Bernstein   *Kinetic Theory in the Early Universe*
N. D. Birrell and P. C. W. Davies   *Quantum Fields in Curved Space*[†]
D. M. Brink   *Semiclassical Methods in Nucleus–Nucleus Scattering*
J. C. Collins   *Renormalization*[†]
P. D. B. Collins   *An Introduction to Regge Theory and High Energy Physics*
M. Creutz   *Quarks, Gluons and Lattices*[†]
F. de Felice and C. J. S. Clarke   *Relativity on Curved Manifolds*[†]
B. DeWittt   *Supermanifolds, second edition*[†]
P. G. O. Freund   *Introduction to Supersymmetry*[†]
F. G. Friedlander   *The Wave Equation on a Curved Space-Time*
J. Fuchs   *Affine Lie Algebras and Quantum Groups*
J. A. H. Futterman, F. A. Handler and R. A. Matzner   *Scattering from Black Holes*
M. Göckeler and T. Schücker   *Differential Geometry, Gauge Theories and Gravity*[†]
M. B. Green, J. H. Schwarz and E. Witten   *Superstring Theory, volume 1: Introduction*[†]
M. B. Green, J. H. Schwarz and E. Witten   *Superstring Theory, volume 2: Loop Amplitudes, Anomalies and Phenomenology*[†]
S. W. Hawking and G. F. R. Ellis   *The Large Scale Structure of Space-Time*[†]
F. Iachello and A. Arima   *The Interacting Boson Model*
F. Iachello and P. van Isacker   *The Interacting Boson–Fermion Model*
C. Itzkyson and J.-M. Drouffe   *Statistical Field Theory, volume 1: From Brownian Motion to Renormalization and Lattice Gauge Theory*[†]
C. Itzkyson and J.-M. Drouffe   *Statistical Field Theory, volume 2: Strong Coupling, Monte Carlo Methods, Conformal Field Theory, and Random Systems*[†]
J. I. Kapusta   *Finite-Temperature Field Theory*[†]
D. Kramer, H. Stephani, M. A. H. MacCallum and E. Herit   *Exact solutions of Einstein's Field Equations*
N. H. March   *Liquid Metals: Concepts and Theory*
I. Montvay and G. Münster   *Quantum Fields on a Lattice*
L. O'Raifeartaigh   *Group Structure of Gauge Theories*[†]
A. Ozorio de Almeida   *Hamiltonian Systems: Chaos and Quantization*[†]
R. Penrose and W. Rindler   *Spinors and Space-Time, volume 1: Two-Spinor Calculus and Relativistic Fields*[†]
R. Penrose and W. Rindler   *Spinors and Space-Time, volume 2: Spinor and Twistor Methods in Space-Time Geometry*[†]
S. Pokorski   *Gauge Field Theories*[†]
V. N. Popov   *Functional Integrals and Collective Excitations*[†]
R. Rivers   *Path Integral Methods in Quantum Field Theory*[†]
R. G. Roberts   *The Structure of the Proton*[†]
W. C. Saslaw   *Gravitational Physics of Stellar and Galactic Systems*[†]
J. M. Stewart   *Advanced General Relativity*[†]
R. S. Ward and R. O. Wells Jr   *Twistor Geometry and Field Theories*[†]

[†] Issued as a paperback

QUANTUM FIELDS
ON A LATTICE

ISTVÁN MONTVAY

Deutsches Elektronen Synchrotron, DESY

GERNOT MÜNSTER

Westfälische Wilhelms-Universität, Münster

CAMBRIDGE
UNIVERSITY PRESS

PUBLISHED BY THE PRESS SYNDICATE OF THE UNIVERSITY OF CAMBRIDGE
The Pitt Building, Trumpington Street, Cambridge CB2 1RP, United Kingdom

CAMBRIDGE UNIVERSITY PRESS
The Edinburgh Building, Cambridge CB2 2RU, United Kingdom
40 West 20th Street, New York, NY 10011–4211, USA
10 Stamford Road, Oakleigh, Melbourne 3166, Australia

First published 1994
First paperback edition 1997

Typeset in Latex Times 11/13pt

A catalogue record for this book is available from the British Library

Library of Congress Cataloguing in Publication data
Montvay, I.
Quantum fields on a lattice / István Montvay, Gernot Münster.
p. cm.
Includes bibliographical references and index
ISBN 0 521 40432 0
1. Lattice field theory. 2. Quantum field theory. 3. Electroweak interactions.
4. Gauge fields (Physics). I. Münster, Gernot. II. Title.
QC793.3.F5M66 1993
530.1′43–dc20 93-1026 CIP

ISBN 0 521 40432 0 hardback
ISBN 0 521 59917 2 paperback

Transferred to digital printing 2003

Contents

Contents   xi

Preface

Since K. Wilson's seminal work on lattice gauge theory in 1974 the regularization of quantum field theories by a space–time lattice has become one of the basic methods for non-perturbative studies in field theory. It has been applied to Quantum Chromodynamics, to the electroweak theory, and to various other model field theories. Many techniques have been developed to deal with quantum fields on a lattice, and with their help many insights into non-perturbative phenomena have been gained.

During a workshop in Cargèse in 1989 Peter Landshoff explained to us his impression that the development of this subject in recent years justified a new book after the one by Michael Creutz on lattice gauge theory, which was published by Cambridge University Press in 1983, and he invited us to undertake the task of writing a book about lattice field theory. As usual the writing took more time than anticipated by the authors and was only finished three years later.

The resulting book is by no means homogeneous, which corresponds to our intentions and also reflects the continuous but irregular development of the field. Some parts are of a more introductory nature and may be useful for graduate students who would like to enter the field. Other parts explain specific techniques in some detail and may serve as a reference for active workers on the subject. Still other parts summarize some of the results obtained about particular questions, or can be used as a guide to the literature. It is, however, not possible to review the present status of the whole subject in such a book, and we did not try to achieve this. For the most recent results we refer the reader to the proceedings of the annual lattice conferences. Similarly, we did not attempt to include an exhaustive list of references, because the vast amount of relevant literature would not allow this. If any colleague feels that important work has been overlooked, we would like to apologize for the omission.

We would like to thank all those from whom we learned about physics, and Tünde and Ulrike for their patience and help.

István Montvay and Gernot Münster

1

Introduction

1.1 Historical remarks

The theory of quantized fields has its origin in the early days of quantum
mechanics. For accounts on the early history of quantum field theory
see [1.1, 1.2, 1.3, 1.4, 1.5]. After the seminal work of Heisenberg [1.6]
the subsequent article of Born and Jordan [1.7] already contains a section in which the quantization of the electromagnetic field is sketched.
In the following 'Drei-Männer-Arbeit' [1.8] the quantization of an arbitrary number of degrees of freedom is worked out in more detail and is
applied to a calculation of the thermal and quantal fluctuations of the
electromagnetic field. These parts were entirely due to P. Jordan.

In the following years the idea of quantizing classical fields was pursued
further by him. But he did not only think of the quantization of the
electromagnetic field, which was, apart from the gravitational field, the
only known fundamental classical field at that time. He also aimed at a
representation of matter by quantized fields. In this case the Schrödinger
waves were considered as classical fields. Therefore he called this procedure
'second quantization'. In this way he attempted a unified description of
matter and radiation, in which both were treated on equal footing. Such
a formulation should form the framework for a mathematical realization
of the duality between waves and particles. Despite the criticisms of
many colleagues progress was made along these lines. The canonical
quantization of bosons [1.9] and of fermions [1.10] indeed revealed the
equivalence of field quantization and many-particle quantum mechanics.

On the other side of the channel, quantum electrodynamics had been
founded in the same years by Dirac [1.11]. In his ingenious work the
emission and absorption of radiation was treated through an application
of quantum theory to the radiation field. Matter was, however, dealt
with in the particle picture. Dirac maintained the distinction between

1

particles and fields and rejected matter-wave quantization for many years (see [1.1]).

But history was in favour of field quantization. The formal development of quantum field theory made progress through the contributions of many physicists. The paper [1.12] marks the beginning of relativistic quantum field theory, and in [1.13, 1.14] quantum electrodynamics was formulated in terms of interacting Dirac and Maxwell fields, quantized in a way compatible with the requirements of special relativity. Fermi achieved essential simplifications of the quantization procedure for QED [1.15].

Another landmark of quantum field theory is the article by Pauli and Weisskopf [1.16], which reveals that the quantization of the Klein–Gordon field solves the problems associated with a one-particle interpretation.

A source of big concern was the appearance of divergencies in the perturbative treatment of interacting electron and photon fields. This problem led to the procedure of renormalization, which has its origin in the work of Dirac, Heisenberg, Weisskopf, Pauli, Fierz and Kramers, and was fully developed in the well-known work of Tomonaga, Schwinger, Feynman and Dyson after the second world war (see [1.2]). Although renormalization paved the way for the calculation of higher orders of perturbation theory, a deep-seated uneasiness remained with many physicists, who considered it as a dubious way to sweep infinities under the rug (see [1.5, 1.17]).

A deeper conceptual understanding of renormalization was advanced in modern quantum field theory. We would like to characterize modern quantum field theory as the non-perturbative approach to quantum field theory through regularized Euclidean functional integrals. This already names its essential ingredients.

The first one is Feynman's path integral or functional integral formalism for quantum theory [1.18, 1.19]. The second one is the Euclidean formulation, which is obtained by continuing the time variable to imaginary values. It goes back to Dyson, Wick, Schwinger and Symanzik [1.20, 1.21, 1.22, 1.23]. Taken together, Euclidean functional integrals provide an equivalence of quantum field theory with statistical mechanics. This is particularly evident after introducing a space–time lattice as a regularization for the infinities. (For early uses of the lattice regularization see [1.24, 1.25].) With this third ingredient the effective application of non-perturbative methods has become possible.

The relation of quantum field theory to statistical mechanics has turned out fruitful for both sides. The development of the renormalization group by Wilson and others (see [1.26]) was an essential step in the theoretical development of both fields. In particular it allowed new insights into the essence of renormalization.

Euclidean quantum field theory regularized on a lattice is the subject of our book. In this first chapter the three ingredients of modern quantum field theory specified above will be introduced.

1.2 Path integral in quantum mechanics

The functional integral formalism lies at the heart of modern quantum field theory. It has been introduced in the framework of quantum mechanics by Feynman [1.18, 1.19] based on earlier considerations by Dirac [1.27]. The term 'path integral' properly applies only to the case of quantum mechanics but is commonly used also in quantum field theory. As a preparation for the latter case we introduce the quantum mechanical path integral in this section.

1.2.1 Feynman path integral

In quantum mechanics the probability amplitude for a particle to move from point y to point x within the time interval t is

$$\langle x|e^{-iHt}|y\rangle, \tag{1.1}$$

where $|y\rangle$ denotes an improper eigenstate of the position operator, H is the Hamiltonian, and we have set $\hbar = 1$. In the case of a free particle,

$$H \equiv H_0 = \frac{\mathbf{p}^2}{2m}, \tag{1.2}$$

the amplitude can be obtained in closed form as follows:

$$\langle x|\exp\left\{-i\frac{\mathbf{p}^2}{2m}t\right\}|y\rangle = \int dp \,\langle x|p\rangle \exp\left\{-i\frac{p^2}{2m}t\right\}\langle p|y\rangle$$

$$= \int \frac{dp}{2\pi}\, e^{ip(x-y)} \exp\left\{-i\frac{p^2}{2m}t\right\}$$

$$= \left(\frac{m}{2\pi it}\right)^{1/2} \exp\left\{i\frac{m}{2t}(x-y)^2\right\}. \tag{1.3}$$

If the particle moves in a potential,

$$H = H_0 + V(\mathbf{x}), \tag{1.4}$$

the amplitude can in general not be written down explicitly. For small times ϵ, however, the time evolution operator

$$U_\epsilon = \exp\left(-iH\epsilon\right) \tag{1.5}$$

can be approximated by the operator

$$W_\epsilon = \exp\left(-iV\frac{\epsilon}{2}\right)\exp\left(-iH_0\epsilon\right)\exp\left(-iV\frac{\epsilon}{2}\right), \tag{1.6}$$

because

$$U_\epsilon = W_\epsilon + O(\epsilon^3), \tag{1.7}$$

which can be checked by e.g. comparing the power series expansions. W_ϵ has the advantage that its matrix elements are explicitly known:

$$\langle x|W_\epsilon|y\rangle = \left(\frac{m}{2\pi i\epsilon}\right)^{1/2}\exp\left\{i\frac{m}{2\epsilon}(x-y)^2 - i\frac{\epsilon}{2}[V(x)+V(y)]\right\}. \tag{1.8}$$

We can use this, when the time interval t is divided into small elements

$$\epsilon = \frac{t}{N}. \tag{1.9}$$

The precise formulation of this statement is given by a slight modification of the Lie–Kato–Trotter product formula [1.28]. Let H_0 and V be self-adjoint operators such that $H_0 + V$ is essentially self-adjoint on their common domain. Then

$$\exp\{-i(H_0+V)t\} = \lim_{N\to\infty} W_\epsilon^N, \tag{1.10}$$

with W_ϵ defined as above. The proof is given here only for the case of finite-dimensional matrices, which is the Lie product formula. We have

$$U_\epsilon^N - W_\epsilon^N = \sum_{k=0}^{N-1} U_\epsilon^k(U_\epsilon - W_\epsilon)W_\epsilon^{N-1-k}, \tag{1.11}$$

which can be checked by directly expanding the sum and cancelling terms. Taking the matrix norm of both sides and using

$$\|F\cdot G\| \le \|F\|\cdot\|G\|, \qquad \|F+G\| \le \|F\|+\|G\|, \tag{1.12}$$

yields

$$\|U_\epsilon^N - W_\epsilon^N\| \le N\|U_\epsilon - W_\epsilon\|. \tag{1.13}$$

Together with

$$\|U_\epsilon - W_\epsilon\| \le \frac{\text{const.}}{N^3} \tag{1.14}$$

the desired result

$$\lim_{N\to\infty} \|U_\epsilon^N - W_\epsilon^N\| = 0 \tag{1.15}$$

is obtained.

Now we insert $N - 1$ complete sets of position eigenstates:

$$\langle x|e^{-iHt}|y\rangle = \lim_{N\to\infty} \int dx_1 \cdots dx_{N-1} \langle x|W_\epsilon|x_1\rangle \cdots \langle x_{N-1}|W_\epsilon|y\rangle$$

$$= \lim_{N\to\infty} \left(\frac{m}{2\pi i\epsilon}\right)^{N/2} \int dx_1 \cdots dx_{N-1}$$

$$\exp\left\{i\frac{m}{2\epsilon}[(x - x_1)^2 + \ldots + (x_{N-1} - y)^2]\right.$$

$$\left. -i\epsilon\left[\frac{1}{2}V(x) + V(x_1) + \ldots + V(x_{N-1}) + \frac{1}{2}V(y)\right]\right\}. \qquad (1.16)$$

The quantity in the exponent, which we may call iS_ϵ, is an approximation to the classical action S of a particle moving from y to x along a path $x(t)$ with $x_k = x(k\epsilon)$:

$$S = \int_0^t dt' \left[\frac{m}{2}\dot{x}^2 - V(x)\right] = S_\epsilon + O(\epsilon^2). \qquad (1.17)$$

Therefore we may denote the limit by

$$\langle x|e^{-iHt}|y\rangle = \int Dx\, e^{iS}, \qquad (1.18)$$

where the symbol Dx stands for

$$Dx = \lim_{N\to\infty} \left(\frac{m}{2\pi i\epsilon}\right)^{N/2} dx_1 \cdots dx_{N-1}. \qquad (1.19)$$

The expression (1.18) represents the quantum mechanical amplitude as an integral over all classical paths weighted by the exponential of i times the classical action. Quantum mechanical operators have been eliminated in favour of an infinite-dimensional integral, the so-called *path integral* or *functional integral*. It is difficult to give to the expression above a satisfactory mathematical meaning as an integral over some space of functions. This is due to the fact that the integral is complex and strongly oscillating. On the other hand, the functional integral formalism has been proved very useful, in particular in field theory, because many results can be derived in a compact and easy way through formal manipulations of functional integrals. This status makes it highly desirable to provide functional integrals with a more sound basis. As we shall discuss below, this can be achieved by introducing imaginary times, which leads to the Feynman–Kac formula [1.29].

1.2.2 Euclidean path integral

If instead of a real time t we choose to take

$$t = -i\tau, \qquad \tau > 0, \tag{1.20}$$

the time evolution operator is replaced by $\exp(-H\tau)$, which is a well-defined positive bounded operator, if V is bounded from below. The variable τ is called 'Euclidean time', the origin of which name comes from the field theoretic context. Repeating the same steps as above, the path integral for the corresponding amplitude results as

$$\langle x|e^{-H\tau}|y\rangle = \lim_{N\to\infty} \left(\frac{m}{2\pi\epsilon}\right)^{N/2} \int dx_1 \cdots dx_{N-1}$$

$$\exp\left\{-\frac{m}{2\epsilon}[(x-x_1)^2 + \ldots + (x_{N-1}-y)^2]\right.$$

$$\left. -\epsilon\left[\frac{1}{2}V(x) + V(x_1) + \ldots + V(x_{N-1}) + \frac{1}{2}V(y)\right]\right\}. \tag{1.21}$$

We abbreviate this expression by

$$\langle x|e^{-H\tau}|y\rangle = \int Dx\, e^{-S_E}, \tag{1.22}$$

where Dx differs from the earlier one by the absence of the factors of i, and the 'Euclidean' action is defined by

$$S_E = \int_0^\tau d\tau' \left[\frac{m}{2}\dot{x}^2 + V(x)\right]. \tag{1.23}$$

It is related to the action S through

$$S = iS_E, \tag{1.24}$$

if t is substituted by τ. The path integral (1.22) now is manifestly real and has an integrand, which is damped for wildly oscillating paths with a large Euclidean action. In fact, these properties have allowed such integrals a well-defined meaning as measures over certain function-spaces [1.30, 1.31]. For $V = 0$ it is called the Wiener measure [1.32]. As can be seen from (1.21), the measure is concentrated on paths for which

$$|x_{k+1} - x_k| \sim \epsilon^{1/2} \tag{1.25}$$

in the limit $\epsilon \to 0$. These paths are continuous everywhere, but not necessarily differentiable. In a more fashionable language they can be characterized by having a fractal dimension of 2.

We shall not dwell further on the mathematical details of the functional integral, but consider it here to be defined by the limit procedure in (1.21). Before the limit is taken, the formula defines a discretization of the quantum mechanical amplitude on a Euclidean time-lattice with lattice spacing ϵ. On this lattice the Euclidean time evolution operator, which shifts by one ϵ-unit in τ is the *transfer matrix*. It is given by

$$\mathbf{T} = \exp\left(-V\frac{\epsilon}{2}\right) \exp\left(-H_0\epsilon\right) \exp\left(-V\frac{\epsilon}{2}\right). \qquad (1.26)$$

Alternatively it may be defined through its matrix elements

$$\langle x|\mathbf{T}|y\rangle = \left(\frac{m}{2\pi\epsilon}\right)^{1/2} \exp\left\{-\frac{m}{2\epsilon}(x-y)^2 - \frac{\epsilon}{2}[V(x)+V(y)]\right\}. \qquad (1.27)$$

The corresponding Hamiltonian H_ϵ is defined by

$$\mathbf{T} = \exp\left(-\epsilon H_\epsilon\right). \qquad (1.28)$$

In the limit $\epsilon \to 0$ it becomes equal to the Hamiltonian H.

Let $E_0, E_1, \dots$ and $|0\rangle, |1\rangle, \dots$ be the eigenvalues and eigenvectors respectively of H_ϵ in ascending order. Very often one is interested in ground state expectation values $\langle 0|A|0\rangle$ of operators A. The Euclidean formulation provides a convenient way to express these as path integrals. Consider

$$\text{Tr}\left(e^{-H\tau}A\right) = \sum_{n=0}^{\infty} e^{-E_n\tau} \langle n|A|n\rangle \qquad (1.29)$$

and

$$Z(\tau) = \text{Tr}\left(e^{-H\tau}\right) = \sum_{n=0}^{\infty} e^{-E_n\tau}. \qquad (1.30)$$

For large τ the term with $n=0$ dominates the sums and therefore the projection onto the ground state is achieved by taking the limit

$$\langle 0|A|0\rangle = \lim_{\tau \to \infty} \frac{1}{Z(\tau)} \text{Tr}\left(e^{-H\tau}A\right). \qquad (1.31)$$

In particular let us consider the case of correlation functions. For real times they are given by

$$\langle x(t_1)\cdots x(t_n)\rangle \equiv \langle 0|\mathbf{x}(t_1)\cdots\mathbf{x}(t_n)|0\rangle$$

$$= \langle 0|e^{iE_0 t_1}\,\mathbf{x}\,e^{-iH(t_1-t_2)}\,\mathbf{x}\cdots\mathbf{x}\,e^{-iH(t_{n-1}-t_n)}\,\mathbf{x}\,e^{-iE_0 t_n}|0\rangle, \qquad (1.32)$$

with

$$\mathbf{x}(t) = e^{iHt}\,\mathbf{x}\,e^{-iHt}. \qquad (1.33)$$

The correlation functions can be continued analytically to Euclidean times $\tau_k = it_k$:

$$\langle x(\tau_1)\cdots x(\tau_n)\rangle = \langle 0|e^{E_0 \tau_1}\,\mathbf{x}\,e^{-H(\tau_1-\tau_2)}\,\mathbf{x}\cdots\mathbf{x}\,e^{-H(\tau_{n-1}-\tau_n)}\,\mathbf{x}\,e^{-E_0 \tau_n}|0\rangle, \quad (1.34)$$

if the times τ_k are ordered:

$$\tau_1 > \tau_2 > \ldots > \tau_n.$$

Using (1.31) we write (1.34) as

$$\lim_{\tau \to \infty} \frac{1}{Z(\tau)} \mathrm{Tr}\,(e^{-H(\tau/2-\tau_1)}\,\mathbf{x}\,e^{-H(\tau_1-\tau_2)}\,\mathbf{x}\cdots\mathbf{x}\,e^{-H(\tau_n+\tau/2)})$$

$$= \lim_{\tau \to \infty} \frac{1}{Z(\tau)} \int \mathrm{d}x\,\langle x|e^{-H(\tau/2-\tau_1)}\,\mathbf{x}\,e^{-H(\tau_1-\tau_2)}\,\mathbf{x}\cdots\mathbf{x}\,e^{-H(\tau_n+\tau/2)}|x\rangle$$

$$= \lim_{\tau \to \infty} \frac{1}{Z(\tau)} \int Dx\,x(\tau_1)\cdots x(\tau_n)\,\exp\{-S_E[x(\tau)]\}, \qquad (1.35)$$

where the paths in the path integral obey periodic boundary conditions

$$x\left(-\frac{\tau}{2}\right) = x\left(\frac{\tau}{2}\right). \qquad (1.36)$$

Similarly

$$Z(\tau) = \int Dx\,\exp\{-S_E[x(\tau)]\}. \qquad (1.37)$$

Our aim in the next sections will be to obtain the corresponding representations of field theoretic correlation functions in terms of functional integrals.

Let us mention here that the Schrödinger equation for wave functions

$$i\frac{\partial}{\partial t}\psi(x,t) = H\psi(x,t) \qquad (1.38)$$

via continuation to Euclidean time goes over into

$$\frac{\partial}{\partial \tau}\psi_E(x,\tau) + H\psi_E(x,\tau) = 0. \qquad (1.39)$$

In the case of the free Hamiltonian $H = H_0$ this is the equation for heat conduction or diffusion. The Euclidean path integral can then be considered as describing an average over random paths occurring in Brownian motion, which was the starting point of the work of Wiener mentioned above.

1.3   Euclidean quantum field theory

The use of imaginary times in quantum field theory has a long history, going back to Dyson, Wick, Schwinger, Symanzik [1.20, 1.21, 1.22, 1.23] and others. If the time coordinate is purely imaginary

$$x^0 = -ix^4, \qquad x^4 \in \mathbf{R}, \qquad (1.40)$$

the space–time metric for the coordinates $x^1, \ldots, x^4$ is a Euclidean one, whence the name 'Euclidean quantum field theory'. The Euclidean formulation is taken as the starting point in lattice field theory for all kinds of fields. In this section we introduce the continuation from Minkowski space to Euclidean space in the framework of scalar field theory.

In Minkowski space the scalar product between vectors is defined by

$$x * y \equiv g_{\mu\nu} x^\mu y^\nu = x^0 y^0 - x^1 y^1 - x^2 y^2 - x^3 y^3 . \tag{1.41}$$

It is denoted by an asterisk in order to distinguish it from the scalar product in Euclidean space, which is introduced later and which is used most in the book.

1.3.1 Scalar fields

To start with let us recapitulate the main ingredients of quantum field theory for a real scalar field in Minkowski space [1.33].

A1: There is a Hilbert space $\mathcal{H}$ of physical states, containing a vacuum $|0\rangle$.

A2: On $\mathcal{H}$ we have a unitary representation $U(a, \Lambda)$ of the Poincaré group, where Λ is a Lorentz transformation and a is a translation vector. The vacuum is invariant under these transformations.

A3 (spectrum condition): The generators P^μ of translations,

$$U(a, \mathbf{1}) = \exp(iP_\mu a^\mu), \tag{1.42}$$

have a spectrum, which is contained in the forward light cone

$$\overline{V}_+ = \left\{ q \in \mathbf{R}^4 \mid q^0 \geq 0, \ q^\mu q_\mu \geq 0 \right\} . \tag{1.43}$$

$P^0 \equiv H$ is the Hamiltonian.

A4: The vacuum is the only vector invariant under $U(a, \Lambda)$.

F1: We have a field $\phi(x)$ acting as an operator on $\mathcal{H}$. More precisely $\phi(x)$ is an operator-valued distribution.

F2: The field transforms covariantly:

$$U(a, \Lambda)\phi(x)U^{-1}(a, \Lambda) = \phi(\Lambda x + a) . \tag{1.44}$$

F3 (locality): The field commutes for space-like separations:

$$[\phi(x), \phi(y)] = 0 \qquad \text{for} \quad (x - y)^2 \leq 0 . \tag{1.45}$$

The list above is not complete but contains those properties which are most relevant to us. The n-point correlation functions of the scalar field

$$\mathscr{W}(x_1,\ldots,x_n) = \langle 0|\phi(x_1)\cdots\phi(x_n)|0\rangle \tag{1.46}$$

are also called *Wightman functions*. They contain all the important information about the theory in the sense that from them the Hilbert space and the field can be reconstructed [1.33].

1.3.2 Schwinger functions

Writing

$$\phi(x) = e^{iP*x}\,\phi(0)\,e^{-iP*x}, \tag{1.47}$$

the Wightman functions are

$$\mathscr{W}(x_1,\ldots,x_n)$$

$$= \langle 0|\phi(0)\,\exp\{-iP*(x_1-x_2)\}\,\phi(0)\cdots\exp\{-iP*(x_{n-1}-x_n)\}\,\phi(0)|0\rangle. \tag{1.48}$$

If the x_k are continued to complex values,

$$x_k = u_k - iy_k, \qquad u_k, y_k \in \mathbf{R}^4, \tag{1.49}$$

the exponentials become

$$\exp\{-iP*(u_k-u_{k+1})\}\exp\{-P*(y_k-y_{k+1})\}. \tag{1.50}$$

As long as all

$$y_k - y_{k+1} \in \overline{V}_+ \tag{1.51}$$

the terms $P*(y_k-y_{k+1})$ are positive due to the spectrum condition. Therefore the second factor in (1.50) is bounded and the Wightman functions can be continued analytically into this region. In particular it contains the *Euclidean points*

$$x_k = (-ix_k^4, \mathbf{x}_k), \qquad x_k^4 \in \mathbf{R}, \quad \mathbf{x}_k \in \mathbf{R}^3 \tag{1.52}$$

with

$$x_k^4 - x_{k+1}^4 > 0. \tag{1.53}$$

Consequently the *Schwinger functions* or *Euclidean Green functions*

$$\mathscr{S}(\ldots;\mathbf{x}_k, x_k^4;\ldots) \equiv \mathscr{W}(\ldots;-ix_k^4, \mathbf{x}_k;\ldots) \tag{1.54}$$

are for the time being only defined for

$$x_1^4 > x_2^4 > \ldots > x_n^4. \tag{1.55}$$

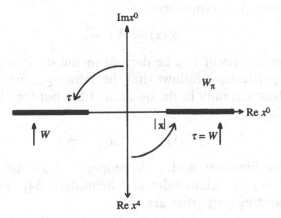

Fig. 1.1.   The analyticity region of the two-point function $W(x)$ in the complex x^0-plane. The Wightman function W is obtained by approaching the real line from below. The Schwinger function is defined on the imaginary x^0-axis. The time-ordered τ-function is obtained from the Schwinger function by means of a Wick rotation, indicated by curved arrows.

In analogy to the Euclidean correlation functions of quantum mechanics they can be represented as

$$\mathscr{S}(x_1,\ldots,x_n) = \langle 0|\phi(0,\mathbf{x}_1)\,\exp\{-H(x_1^4 - x_2^4)\}\,\phi(0,\mathbf{x}_2)\cdots|0\rangle\,. \qquad (1.56)$$

As an example consider the two-point function, which, due to translation invariance, is a function of the difference of its arguments:

$$\mathscr{W}(x_1,x_2) \equiv W(x_1 - x_2)\,. \qquad (1.57)$$

As a function of complex x^0, $W(x)$ is analytic in the lower half plane.

Now the complex region, where W is defined and analytic, will be extended. Permuting the arguments of the two-point function yields

$$W_\pi(x) \equiv W(-x)\,, \qquad (1.58)$$

which, as a function of x^0, is analytic in the upper half plane. The crucial fact is now that in Minkowski space, i.e. on the real x^0-axis, the functions W and W_π are equal for space-like x since

$$\mathscr{W}(x_1,x_2) = \mathscr{W}(x_2,x_1) \qquad (1.59)$$

for space-like $x_1 - x_2$ due to locality. See fig. 1.1. So on their boundaries W and W_π have an open region in common where they coincide. This implies (according to the edge-of-the-wedge theorem [1.33]) that they form a single analytic function in the union of their domains of definition. What has been achieved is an extension of W into the whole complex x^0-plane with the exception of those parts of the real line where $|x^0| > |\mathbf{x}|$ (see fig. 1.1). Moreover it is symmetric:

$$W(x) = W(-x).\tag{1.60}$$

The corresponding result can be derived in the same way for all n-point functions. In particular it follows that the Schwinger functions are defined and real analytic not only in the domain (1.55) but for all non-coinciding Euclidean points:

$$x_j \neq x_k \qquad \text{for} \quad j \neq k.\tag{1.61}$$

The Schwinger functions and their properties have been studied in an axiomatic setting by Osterwalder and Schrader[1.34]. For our purposes the most interesting properties are

    E1 (Euclidean covariance): The Schwinger functions are covariant under Euclidean transformations

$$\mathscr{S}(x_1,\ldots,x_n) = \mathscr{S}(\Lambda x_1 + a,\ldots,\Lambda x_n + a),\tag{1.62}$$

where now $\Lambda \in SO(4)$ is a rotation.

    E2 (reflection positivity): is discussed below.

    E3 (symmetry): Schwinger functions are symmetric in their arguments.

This follows from the considerations above.
Symmetry rests on the fact that Euclidean points are always space-like relative to each other.

    From the Schwinger functions, the Wightman functions in Minkowski space, and consequently the whole quantum field theory, can be reconstructed [1.34]. The advantage of the Euclidean formulation, however, is that the Schwinger functions obey simpler properties and are easier to handle than Wightman functions or field operators. In particular their symmetry is the crucial property which opens the way to a representation in terms of functional integrals.

    Nevertheless we should keep in our minds that we live in Minkowski space and therefore the Minkowskian Green functions are the physical ones.

1.3.3   Wick rotation

According to the considerations above the Wightman functions in Minkowski space are boundary values of the analytic functions $\mathscr{S}(x_1,\ldots,x_n)$. In the case of the two-point function $W(x)$ the real line in the complex

x^0-plane has to be approached from the lower half plane (see fig. 1.1). The generalization to the n-point function is

$$\mathscr{W}(x_1,\ldots,x_n) = \lim_{\substack{\epsilon_k \to 0 \\ \epsilon_k - \epsilon_{k+1} > 0}} \mathscr{S}(\ldots; \mathbf{x}_k, ix_k^0 + \epsilon_k; \ldots), \qquad x_k \in \mathbf{R}^4. \quad (1.63)$$

Another class of correlation functions which play an important rôle in quantum field theory are the time-ordered Green functions

$$\tau(x_1,\ldots,x_n) = \langle 0|T\phi(x_1)\cdots\phi(x_n)|0\rangle, \qquad (1.64)$$

where the time ordering T orders the operators such that fields with larger time x^0 stand to the left of fields with smaller time, e.g.

$$T\phi(x_1)\phi(x_2) = \theta(x_1^0 - x_2^0)\phi(x_1)\phi(x_2) + \theta(x_2^0 - x_1^0)\phi(x_2)\phi(x_1). \quad (1.65)$$

By definition the τ-functions are symmetric in their arguments. They too are boundary values of the Schwinger functions. In our example of the two-point function we have

$$\tau(x) \equiv \tau(x,0) = \begin{cases} W(x), & \text{for} \quad x^0 > 0 \\ W(-x), & \text{for} \quad x^0 < 0. \end{cases} \quad (1.66)$$

Consequently the τ-functions are obtained by approaching the real line in the complex x^0-plane through a counter-clockwise rotation (see fig. 1.1). In the same way this prescription holds for all τ-functions:

$$\tau(x_1,\ldots,x_n) = \lim_{\varphi \to \pi/2} \mathscr{S}(\ldots; \mathbf{x}_k, e^{i\varphi}x_k^0; \ldots). \quad (1.67)$$

This rule, which obviously is simpler than the one for the Wightman functions, is called *Wick rotation*.

Here a paragraph about the notation is in order. In Minkowski space the metric $g_{\mu\nu}$, defined in (1.41), is used to build scalar products and to raise or lower indices. The Euclidean components of a contravariant vector are defined, e.g. for positions and momenta, through

$$x^4 = ix^0, \qquad x^k = x^k, \quad k = 1, 2, 3,$$

$$p^4 = ip^0, \qquad p^k = p^k, \quad k = 1, 2, 3. \quad (1.68)$$

In Euclidean space the metric $\delta_{\mu\nu}$ is used. Thus the Euclidean scalar product is

$$x \cdot y \equiv \delta_{\mu\nu}x^\mu y^\nu = x^1 y^1 + x^2 y^2 + x^3 y^3 + x^4 y^4 = -x * y. \quad (1.69)$$

The covariant and contravariant components of a Euclidean vector are identical:

$$x_\mu^{(E)} = x^{(E)\mu}. \quad (1.70)$$

Thus for lower indices the space components differ in sign from the corresponding Minkowskian ones:

$$x_k^{(E)} = -x_k^{(M)}. \tag{1.71}$$

Since in later chapters only Euclidean vectors will occur we omit the marks (E) and write them in terms of lower indices in general.

1.3.4   Free field

We would like to illustrate the analytic continuations discussed above by the case of a free field. The two-point functions of the free scalar field are [A2]

$$W(x) = \langle 0|\phi(x)\phi(0)|0\rangle = \Delta_+(x) = \int \frac{\mathrm{d}^4 k}{(2\pi)^3}\, \delta(k^2 - m^2)\, \theta(k^0)\, \mathrm{e}^{-ik*x} \tag{1.72}$$

$$\tau(x) = \theta(x^0)\langle 0|\phi(x)\phi(0)|0\rangle + \theta(-x^0)\langle 0|\phi(0)\phi(x)|0\rangle$$

$$= \mathrm{i}\Delta_F(x) = \mathrm{i}\int \frac{\mathrm{d}^4 k}{(2\pi)^4}\, \frac{\mathrm{e}^{-ik*x}}{k^2 - m^2 + \mathrm{i}\epsilon}. \tag{1.73}$$

Performing the analytic continuation of $W(x)$ yields

$$\mathscr{S}(x,0) = \int \frac{\mathrm{d}^4 k}{(2\pi)^4}\, \frac{\mathrm{e}^{ik\cdot x}}{k^2 + m^2}. \tag{1.74}$$

$\tau(x)$ is obtained from $\mathscr{S}(x,0)$ by means of a Wick rotation of $x^4 = ix^0$ in a counter-clockwise sense. In the integral we have to rotate at the same time $k^4 = ik^0$ in a clockwise sense. The resulting integration path in the complex k^0-plane bypasses the poles at $\pm(m^2 + \mathbf{k}^2)^{1/2}$ in the way indicated by the $i\epsilon$ prescription. It goes from $+\infty$ to $-\infty$ and an additional sign change has to be introduced to bring it into the form above.

1.3.5   Reflection positivity

In order that Euclidean correlation functions can be continued back to Minkowski space they have to obey a positivity condition, which has been discussed first in [1.34]. It is called *Osterwalder–Schrader positivity* or *reflection positivity*. Let the Euclidean field be

$$\phi(x) = \mathrm{e}^{Hx^4}\, \phi(\mathbf{x},0)\, \mathrm{e}^{-Hx^4}. \tag{1.75}$$

As an operator this expression is not well defined since H is not bounded, but it makes sense inside vacuum expectation values, where it is defined through the correlation functions. Define the Euclidean time reflection by

$$\theta(\mathbf{x}, x^4) = (\mathbf{x}, -x^4), \tag{1.76}$$

$$\Theta\phi(x) = \overline{\phi(\theta x)}. \tag{1.77}$$

In the case of a real scalar field, which we consider presently, the complex conjugation can of course be omitted. The action of Θ is extended to arbitrary functions F of the field by means of the requirement of antilinearity

$$\Theta(\lambda F) = \overline{\lambda}\Theta F, \tag{1.78}$$

and of

$$\Theta(FG) = \Theta F\Theta G. \tag{1.79}$$

Roughly speaking, Θ is the Euclidean equivalent of Hermitean conjugation in Minkowski space. The fact that for the time evolution operator $\exp(-iHt)$ Hermitean conjugation amounts to changing the sign of t explains why a time reflection is needed in Euclidean space.

Now let F be a function of the fields at positive times, e.g.

$$F = \sum_j \int dx_1 \cdots dx_j \, f_j(x_1, \ldots, x_j)\, \phi(x_1) \cdots \phi(x_j), \tag{1.80}$$

with some functions f_j which decrease sufficiently fast at infinity and have support only at positive times: $x_1^4 \geq 0, \cdots, x_j^4 \geq 0$. Then reflection positivity states that

$$\langle (\Theta F)\, F \rangle \geq 0. \tag{1.81}$$

This is meant as a statement about Schwinger functions, which reads in full:

$$\sum_{j,k} \int dx_1 \cdots dx_j \, dy_1 \cdots dy_k \, \overline{f_j(x_1, \ldots, x_j)}\, f_k(y_1, \ldots, y_k)$$

$$\mathscr{S}(\theta x_1, \ldots, \theta x_j, y_1, \ldots, y_k) \geq 0. \tag{1.82}$$

It can be derived by writing the left hand side as the norm squared of the vector $\Theta F|0\rangle$, and thus originates in the positivity of the scalar product in Hilbert space. As the analysis shows [1.34] reflection positivity replaces Hilbert space positivity and the spectral condition of the Minkowskian formulation.

Given some Euclidean correlation functions, it is important to check for reflection positivity in order to ensure that a Hilbert space formulation with a well-defined Hamiltonian can be set up. Reflection positivity for theories regularized on a lattice will be discussed later.

1.4 Euclidean functional integrals

The symmetry of Euclidean correlation functions means that the Euclidean fields commute. In this respect they behave like classical fields. It was Symanzik's [1.23] idea to consider the Euclidean fields not as operators but as random variables, whose expectation values yield the correlation functions. The problem is then to find an appropriate probability distribution or 'measure' $d\mu$ for the $\phi(x)$ such that

$$\langle F[\phi] \rangle = \int d\mu \, F[\phi] \,. \tag{1.83}$$

The Schwinger functions

$$\mathscr{S}(x_1,\ldots,x_n) = \langle \phi(x_1) \cdots \phi(x_n) \rangle \tag{1.84}$$

are then called the *moments of the measure* $d\mu$ and should fix the measure uniquely. This idea has been pursued by Nelson [1.35] and many others.

Formally one would write the measure as

$$d\mu = \frac{1}{Z} \, e^{-S[\phi]} \prod_x d\phi(x) \tag{1.85}$$

with a Euclidean action $S[\phi]$. This is the Euclidean functional integral. The main task is to give a meaning to such an expression. In this section we will discuss the functional integral in a more formal way for scalar field theory in the continuum.

1.4.1 Gaussian integrals

The main tools in dealing with functional integrals are Gaussian integrals. The simplest case is the well-known integral

$$\int_{-\infty}^{\infty} d\phi \, \exp\left\{-\frac{1}{2}A\phi^2\right\} = (2\pi/A)^{1/2} \,, \quad \text{for } \operatorname{Re} A > 0. \tag{1.86}$$

Proof :

$$\left(\int_{-\infty}^{\infty} d\phi \, \exp\left\{-\frac{1}{2}A\phi^2\right\}\right)^2 = \int d\phi_1 \, d\phi_2 \, \exp\left\{-\frac{1}{2}A(\phi_1^2 + \phi_2^2)\right\}$$

$$= \int_0^{\infty} dr \int_0^{2\pi} d\theta \, r \exp\left\{-\frac{1}{2}Ar^2\right\} = \pi \int_0^{\infty} d(r^2) \, \exp\left\{-\frac{1}{2}Ar^2\right\} = \frac{2\pi}{A} \tag{1.87}$$

$\blacksquare$

Now let

$$\phi = (\phi_1,\ldots,\phi_k) \in \mathbf{R}^k \tag{1.88}$$

and

$$A = (A_{ij}), \qquad i, j = 1, \ldots, k \qquad (1.89)$$

be a real, symmetric and positive matrix, i.e. all its eigenvalues are positive. With

$$(\phi, A\phi) = \phi_i A_{ij} \phi_j \qquad (1.90)$$

we have

$$Z_0 \equiv \int d^k \phi \, \exp \left\{ -\frac{1}{2}(\phi, A\phi) \right\} = (2\pi)^{k/2} (\det A)^{-1/2}. \qquad (1.91)$$

Proof: A can be diagonalized:

$$S A S^t = D = \text{diag}\,(\lambda_1, \ldots, \lambda_k), \qquad \text{all } \lambda_i > 0. \qquad (1.92)$$

Changing variables to

$$y = S\phi, \qquad d^k \phi = |\det S|^{-1} d^k y = d^k y, \qquad (1.93)$$

yields

$$\int d^k \phi \, \exp \left\{ -\frac{1}{2}(\phi, A\phi) \right\} = \int d^k y \, \exp \left\{ -\frac{1}{2} \sum_i \lambda_i y_i^2 \right\}$$

$$= \prod_{i=1}^{k} \left(\frac{2\pi}{\lambda_i} \right)^{1/2} = (2\pi)^{k/2} (\det A)^{-1/2} \qquad (1.94)$$

∎

For an arbitrary vector J we get, under the same conditions as before,

$$Z_0(J) \equiv \frac{1}{Z_0} \int d^k \phi \, \exp \left\{ -\frac{1}{2}(\phi, A\phi) + (J, \phi) \right\}$$

$$= \exp \frac{1}{2}(J, A^{-1}J). \qquad (1.95)$$

Proof:

$$-\frac{1}{2}(\phi, A\phi) + (J, \phi) = -\frac{1}{2}(\phi + A^{-1}J, A(\phi + A^{-1}J)) + \frac{1}{2}(J, A^{-1}J) \quad (1.96)$$

With

$$y = \phi + A^{-1}J, \qquad d^k \phi = d^k y \qquad (1.97)$$

we obtain

$$\int d^k \phi \, \exp \left\{ -\frac{1}{2}(\phi, A\phi) + (J, \phi) \right\}$$

$$= \int \mathrm{d}^k y \, \exp\left\{-\frac{1}{2}(y, Ay)\right\} \exp\frac{1}{2}(J, A^{-1}J),\qquad (1.98)$$

from which the result follows.   ∎

The moments of the Gaussian weight function $\exp-\frac{1}{2}(\phi, A\phi)$ are now defined by

$$\langle \phi_{i_1} \cdots \phi_{i_n}\rangle \equiv \frac{1}{Z_0} \int \mathrm{d}^k\phi \, \phi_{i_1} \cdots \phi_{i_n} \exp-\frac{1}{2}(\phi, A\phi).\qquad (1.99)$$

They can be obtained by differentiating $Z_0(J)$:

$$\langle \phi_{i_1} \cdots \phi_{i_n}\rangle = \frac{\partial}{\partial J_{i_1}} \cdots \frac{\partial}{\partial J_{i_n}} Z_0(J)\bigg|_{J=0}$$

$$= \frac{\partial}{\partial J_{i_1}} \cdots \frac{\partial}{\partial J_{i_n}} \exp\frac{1}{2}(J, A^{-1}J)\bigg|_{J=0}.\qquad (1.100)$$

In particular

$$\langle \phi_i \rangle = 0,\qquad (1.101)$$

$$\langle \phi_i \phi_j \rangle = A_{ij}^{-1},\qquad (1.102)$$

$$\langle \phi_i \phi_j \phi_l \rangle = 0,\qquad (1.103)$$

$$\langle \phi_i \phi_j \phi_l \phi_m \rangle = A_{ij}^{-1} A_{lm}^{-1} + A_{il}^{-1} A_{jm}^{-1} + A_{im}^{-1} A_{jl}^{-1}.\qquad (1.104)$$

In general the odd moments are zero whereas for the even moments we find

$$\langle \phi_{i_1} \cdots \phi_{i_{2n}}\rangle = \sum_{\text{pairings}} A_{j_1 k_1}^{-1} A_{j_2 k_2}^{-1} \cdots A_{j_n k_n}^{-1}.\qquad (1.105)$$

1.4.2 Euclidean free field

Before the functional integral formalism is applied to Euclidean field theory we review the free field in this section. We consider a real one-component scalar field $\phi(x)$ with Euclidean action

$$S_0[\phi] = \int \mathrm{d}^4 x \left\{\frac{1}{2}(\partial_\mu \phi)^2 + \frac{m^2}{2}\phi^2\right\}$$

$$= \int \mathrm{d}^4 x \, \frac{1}{2}\phi(x)(\Box + m^2)\phi(x),\qquad (1.106)$$

where

$$\Box = -\partial_\mu \partial^\mu.\qquad (1.107)$$

The propagator

$$G(x, y) = \langle \phi(x)\phi(y) \rangle \tag{1.108}$$

satisfies

$$(\Box + m^2)G(x, y) = \delta(x - y) \tag{1.109}$$

and can be represented as

$$G(x, y) = \int \frac{d^4 p}{(2\pi)^4} \, e^{ip \cdot (x-y)} \frac{1}{p^2 + m^2}. \tag{1.110}$$

For the Schwinger functions or Euclidean Green functions we have according to Wick's theorem

$$\mathcal{S}(x_1, \ldots, x_{2n}) \equiv G(x_1, \ldots, x_{2n}) \equiv \langle \phi(x_1) \cdots \phi(x_{2n}) \rangle$$

$$= \sum_{\text{pairings}} G(x_{j_1}, x_{k_1}) \cdots G(x_{j_n}, x_{k_n}). \tag{1.111}$$

The connected Green functions $G_c(x_1, \ldots, x_n)$ are defined in any field theory through

$$G(x_1, \ldots, x_n) = \sum_{\mathcal{P}} G_c(x_i, \ldots, x_j) \cdots G_c(x_k, \ldots, x_l), \tag{1.112}$$

where $\mathcal{P}$ denotes a partition of the indices $\{1, \ldots, n\}$ into non-empty subsets. Explicitly

$$G(x) = G_c(x), \tag{1.113}$$

$$G(x, y) = G_c(x, y) + G_c(x)G_c(y), \tag{1.114}$$

$$G(x, y, z) = G_c(x, y, z) + G_c(x)G_c(y, z) + G_c(y)G_c(x, z) + G_c(z)G_c(x, y)$$

$$+ G_c(x)G_c(y)G_c(z), \tag{1.115}$$

etc. The inverse relation is

$$G_c(x_1, \ldots, x_n) = \sum_{\mathcal{P}} (-1)^{(k-1)}(k - 1)! \, G(x_i, \ldots, x_j) \cdots G(x_l, \ldots, x_m), \tag{1.116}$$

where k is the number of factors in the summand. The first examples are

$$G_c(x) = G(x), \tag{1.117}$$

$$G_c(x, y) = G(x, y) - G(x)G(y), \tag{1.118}$$

$$G_c(x, y, z) = G(x, y, z) - G(x)G(y, z) - G(y)G(x, z) - G(z)G(x, y)$$

$$+ 2G(x)G(y)G(z). \tag{1.119}$$

For the free field we obtain

$$G_c(x) = 0, \tag{1.120}$$

$$G_c(x, y) = G(x, y) \tag{1.121}$$

$$G_c(x_1, \ldots, x_n) = 0 \qquad \text{for} \quad n \geq 3. \tag{1.122}$$

For a classical source $J(x)$ let

$$(J, \phi) = \int d^4x \, J(x)\phi(x). \tag{1.123}$$

The generating functional of Green's functions is defined by

$$Z[J] \equiv \langle e^{(J,\phi)} \rangle$$

$$= \sum_{n=0}^{\infty} \frac{1}{n!} \int d^4x_1 \ldots d^4x_n \, J(x_1) \cdots J(x_n) \, G(x_1, \ldots, x_n). \tag{1.124}$$

It is normalized to 1:

$$Z[0] = 1. \tag{1.125}$$

From $Z[J]$ the Green functions can be obtained by means of functional derivation. Let us briefly recall what functional derivatives are. The *functional derivative*

$$\frac{\delta}{\delta J(x)} F[J]$$

of a functional $F[J]$ is defined through the formula

$$\lim_{\varepsilon \to 0} \frac{1}{\varepsilon} \{F[J + \varepsilon h] - F[J]\} = \int d^4x \, h(x) \frac{\delta}{\delta J(x)} F[J]. \tag{1.126}$$

Examples:

$$F_1[J] = \int d^4x \, J(x)f(x), \qquad \frac{\delta}{\delta J(x)} F_1[J] = f(x), \tag{1.127}$$

$$F_2[J] = \int d^4x \, d^4y \, J(x)J(y)f(x, y),$$

$$\frac{\delta}{\delta J(x)} F_2[J] = \int d^4y \, J(y)\{f(x, y) + f(y, x)\}, \tag{1.128}$$

$$F_3[J] = J(z), \qquad \frac{\delta}{\delta J(x)} F_3[J] = \delta(x - z). \tag{1.129}$$

Now from the definition of the generating functional it follows that

$$G(x_1,\ldots,x_n) = \frac{\delta^n Z[J]}{\delta J(x_1)\cdots\delta J(x_n)}\bigg|_{J=0}. \tag{1.130}$$

The generating functional for the connected Green functions, $W[J]$,

$$G_c(x_1,\ldots,x_n) = \frac{\delta^n W[J]}{\delta J(x_1)\cdots\delta J(x_n)}\bigg|_{J=0}, \tag{1.131}$$

is related to $Z[J]$ via

$$W[J] = \log Z[J], \tag{1.132}$$

as can be derived from (1.112).

These relations hold in general. For the free field we have in particular

$$W[J] \equiv W_0[J] = \frac{1}{2}(J, GJ) \equiv \frac{1}{2}\int d^4x\, d^4y\; J(x)G(x,y)J(y), \tag{1.133}$$

$$Z[J] \equiv Z_0[J] = \exp\left\{\frac{1}{2}\int d^4x\, d^4y\; J(x)G(x,y)J(y)\right\}. \tag{1.134}$$

1.4.3 Functional integral for the Euclidean free field

Comparing the formulae of the two previous sections it appears that the Gaussian integral (1.95) corresponds to the generating functional $Z_0[J]$ of the Green functions, if we make the replacements

$$\phi_i \to \phi(x), \qquad \frac{\partial}{\partial J_i} \to \frac{\delta}{\delta J(x)} \tag{1.135}$$

and correspondingly

$$\frac{1}{2}(J, A^{-1}J) \to \frac{1}{2}(J, GJ). \tag{1.136}$$

Thus the matrix A^{-1} corresponds to the operator G, which has kernel $G(x,y)$, i.e.

$$(GJ)(x) = \int d^4y\; G(x,y)J(y). \tag{1.137}$$

The matrix A, which appears in the exponential of the Gaussian integrals, then corresponds to the operator G^{-1}, which is equal to the Klein–Gordon operator

$$G^{-1} = \Box + m^2, \tag{1.138}$$

because G satisfies

$$(\Box + m^2)G = 1, \tag{1.139}$$

or equivalently

$$(\Box + m^2)G(x, y) = \delta(x - y). \tag{1.140}$$

So we have for the argument of the exponential

$$\frac{1}{2}(\phi, A\phi) \rightarrow \frac{1}{2}(\phi, G^{-1}\phi) = \frac{1}{2}(\phi, (\Box + m^2)\phi)$$

$$= \frac{1}{2}\int d^4x \ \phi(x) (\Box + m^2)\phi(x) = S_0, \tag{1.141}$$

which is the Euclidean action. Using these correspondences we now write $Z_0[J]$ according to (1.95) in the form of a Gaussian integral:

$$Z_0[J] = \exp\frac{1}{2}(J, GJ)$$

$$= \frac{1}{Z_0}\int \prod_x d\phi(x) \ \exp\left\{-\frac{1}{2}(\phi, (\Box + m^2)\phi) + (J, \phi)\right\}, \tag{1.142}$$

where

$$Z_0 = \int \prod_x d\phi(x) \ \exp-\frac{1}{2}(\phi, (\Box + m^2)\phi). \tag{1.143}$$

The infinite-dimensional integral looks quite ill-defined, but in this case we can turn the argument around and consider (1.142) as a definition of the measure

$$d\mu_0(\phi) = \frac{1}{Z_0}\prod_x d\phi(x) \ \exp-\frac{1}{2}(\phi, (\Box + m^2)\phi) \equiv \frac{1}{Z_0}D[\phi] \ e^{-S_0(\phi)}. \tag{1.144}$$

A definition of this expression which uses a lattice discretization will be discussed in section 1.5.

For the free field the aim is now achieved. We have a functional integral over classical fields $\phi(x)$ such that the Euclidean correlation functions are equal to the moments of the Gaussian distribution defined above:

$$\langle\phi(x_1)\cdots\phi(x_n)\rangle = \frac{1}{Z_0}\int \prod_x d\phi(x) \ e^{-S_0[\phi]} \phi(x_1)\cdots\phi(x_n). \tag{1.145}$$

1.4.4 *Functional integral for the interacting field*

Whereas in the case of a free field the representation of the generating functional $Z[J]$ as a functional integral can be defined unambiguously, the case of an interacting field is much more problematic. The problems show up in the form of divergencies when perturbation theory is applied to the interaction. Therefore a more careful path has to be followed in defining the functional integral here.

In this section, however, we will temporarily neglect the problems associated with divergencies and renormalization and introduce a formal expression for the functional integral for an interacting scalar field. The derivation follows Coleman [1.36].

Again we consider a real scalar field. Let the Euclidean action be

$$S[\phi] = S_0[\phi] + S_I[\phi] = \frac{1}{2}(\phi, (\Box + m^2)\phi) + S_I[\phi], \qquad (1.146)$$

where S_I is the interaction component of the action. In general it is a polynomial of the field and of its derivatives. The standard example would be

$$S_I[\phi] = \frac{g}{4!} \int d^4x \, \phi(x)^4. \qquad (1.147)$$

Dyson's formula for the correlation functions reads in Euclidean space:

$$\langle \phi(x_1) \cdots \phi(x_n) \rangle = \frac{\langle 0 | \phi_{in}(x_1) \cdots \phi_{in}(x_n) \exp\{-S_I[\phi_{in}]\} | 0 \rangle_{in}}{\langle 0 | \exp\{-S_I[\phi_{in}]\} | 0 \rangle_{in}}. \qquad (1.148)$$

Here ϕ_{in} is a free field of mass m and the vacuum expectation values on the right hand side are to be taken with respect to the vacuum of the free field theory. In terms of the generating functional the same state of affairs is expressed as

$$Z[J] = \frac{\langle 0 | \exp\{-S_I[\phi_{in}] + (J, \phi_{in})\} | 0 \rangle_{in}}{\langle 0 | \exp\{-S_I[\phi_{in}]\} | 0 \rangle_{in}}. \qquad (1.149)$$

Now we use for the free field expectation values the Gaussian functional integral of the previous section to obtain

$$Z[J] = \frac{1}{Z} \int \prod_x d\phi(x) \, e^{-S_0[\phi] - S_I[\phi] + (J, \phi)}$$

$$= \frac{1}{Z} \int \prod_x d\phi(x) \, e^{-S[\phi] + (J, \phi)} \qquad (1.150)$$

with

$$Z = \int \prod_x d\phi(x) \, e^{-S[\phi]}, \qquad (1.151)$$

where the label '*in*' on the field was omitted. Returning to the correlation functions, this translates to

$$\langle \phi(x_1) \cdots \phi(x_n) \rangle = \frac{1}{Z} \int \prod_x d\phi(x) \, e^{-S[\phi]} \phi(x_1) \cdots \phi(x_n). \qquad (1.152)$$

This very formula is the functional integral for the interacting scalar field theory in Euclidean space. From it all formal properties of expectation values can be derived without using operators on Hilbert space. Many manipulations are made more lucid than in the operator formalism by using rules from calculus carried over to the infinite-dimensional case. Moreover, as is obvious from the expression above, the covariance is manifest, since the action and not the Hamiltonian enters it. These advantages are particularly important when gauge theories or non-perturbative problems are treated.

1.4.5 *Perturbation theory*

The rules of perturbation theory, i.e. the Feynman rules, can be derived from the functional integral by applying the techniques for dealing with Gaussian integrals. As an example we consider the scalar field theory with a quartic self-interaction. The action is the one defined in (1.146),(1.147). Expanding the exponential of the interaction we get for the n-point function

$$G(x_1,\ldots,x_n)$$

$$= \frac{1}{Z} \int \prod_x \mathrm{d}\phi(x)\; \phi(x_1)\cdots\phi(x_n) \sum_{n=0}^{\infty} \frac{1}{n!} \left[-\frac{g}{4!} \int \mathrm{d}^4x\; \phi(x)^4 \right]^n e^{-S_0[\phi]} .$$

$$(1.153)$$

This is a Gaussian integral which we treat with the help of the generating functional

$$Z[J] = \frac{1}{Z} \sum_{n=0}^{\infty} \frac{1}{n!} \int \prod_x \mathrm{d}\phi(x)\; \left[-\frac{g}{4!} \int \mathrm{d}^4x\; \phi(x)^4 \right]^n e^{-S_0[\phi]+(J,\phi)}$$

$$= \frac{1}{Z} \sum_{n=0}^{\infty} \frac{1}{n!} \left[-\frac{g}{4!} \int \mathrm{d}^4x\; \left(\frac{\delta}{\delta J(x)}\right)^4 \right]^n \int \prod_x \mathrm{d}\phi(x)\; e^{-S_0[\phi]+(J,\phi)} . \quad (1.154)$$

In the last expression the integral equals $Z_0[J]\, Z_0$ and we write more compactly

$$Z[J] = \frac{Z_0}{Z} \exp\left\{ -\frac{g}{4!} \int \mathrm{d}^4x\; \left(\frac{\delta}{\delta J(x)}\right)^4 \right\} Z_0[J]$$

$$= \frac{Z_0}{Z} \exp\left\{ -S_I\left[\frac{\delta}{\delta J(x)}\right] \right\} \exp \frac{1}{2}(J, G_0 J), \quad (1.155)$$

where the free propagator is denoted by $G_0(x,y)$. Similarly the Green functions are

$$G(x_1,\ldots,x_n)$$

$$= \frac{Z_0}{Z} \frac{\delta}{\delta J(x_1)} \cdots \frac{\delta}{\delta J(x_n)} \exp\left\{ -S_I\left[\frac{\delta}{\delta J(x)}\right] \right\} \exp \frac{1}{2}(J, G_0 J) \bigg|_{J=0} . \quad (1.156)$$

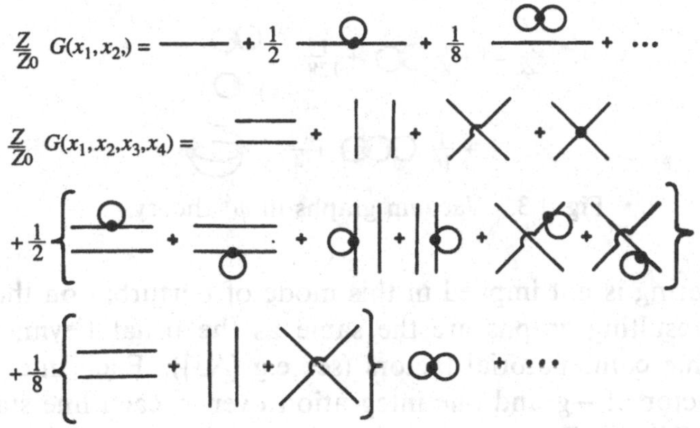

$$\frac{Z}{Z_0} G(x_1, x_2,) = \text{———} + \frac{1}{2}\text{——O——} + \frac{1}{8}\text{——OO——} + \cdots$$

$$\frac{Z}{Z_0} G(x_1, x_2, x_3, x_4) = \equiv + \| \| + \times + \times$$

Fig. 1.2. Graphs appearing in the perturbative expansion of the two- and four-point functions in ϕ^4 theory.

Expanding the exponential of the interaction term again yields terms containing products of functional derivatives, which are represented graphically as follows: each derivative $\delta/\delta J(x_i)$ is indicated by an external point x_i from which a line emerges, each factor of $-g \int d^4x\, (\delta/\delta J(x))^4$ is indicated by an internal vertex, from which four lines emerge. On the other hand the expansion of the second exponential yields products of terms $\int d^4x\, d^4y\, J(x) G_0(x, y) J(y)$, which are represented by internal lines. Since the derivatives have to be evaluated at $J = 0$, each factor $\delta/\delta J$ has to be paired with some factor J. This results in graphs where the external points $x_1, \ldots, x_n$ and a number of internal vertices are linked by lines. The graphs may also include disconnected pieces and 'tadpoles' (connected subgraphs with one external line). See fig. 1.2 for the cases $n = 2$ and $n = 4$.

Now we turn to the remaining factor Z_0/Z. The normalization of the generating functional,

$$Z(0) = 1, \tag{1.157}$$

implies that

$$\frac{Z}{Z_0} = \exp\left\{-S_I\left[\frac{\delta}{\delta J(x)}\right]\right\} \exp \frac{1}{2}(J, G_0 J)\bigg|_{J=0}. \tag{1.158}$$

The graphical representation of this expression is analogous to the case above but without external points, see fig. 1.3. These graphs are called 'vacuum graphs'. Since in (1.156) the sum of all possible vacuum graphs appears as an overall factor, the vacuum graphs just cancel out in the graphical representation of Green's functions. We remain with graphs which do not contain vacuum graphs as subgraphs, but may be disconnected or contain tadpoles. The appearance of tadpoles means that

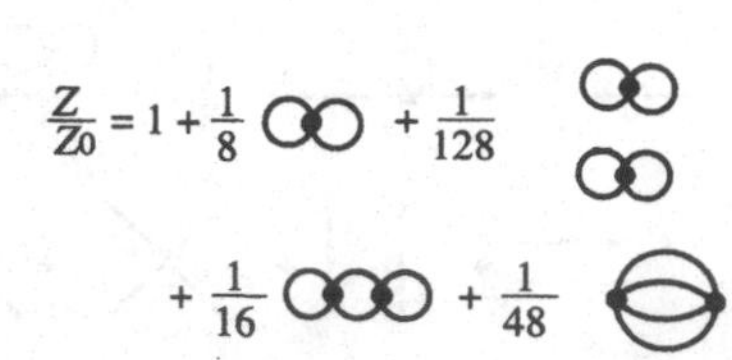

$$\frac{Z}{Z_0} = 1 + \frac{1}{8}\ \text{OO}\ + \frac{1}{128}\ \text{OO}$$

$$+ \frac{1}{16}\ \text{OOO}\ + \frac{1}{48}\ \text{⊖}$$

Fig. 1.3. Vacuum graphs in ϕ^4 theory.

normal ordering is not implied in this mode of perturbation theory. Otherwise the resulting graphs are the same as the usual Feynman graphs with the same combinatorial factors (see e.g. [A3]). Each internal vertex implies a factor of $-g$ and one integration over x, each line stands for a propagator $G_0(x, y)$. Equivalently, the Feynman rules can be written down in momentum space.

As is well known, divergencies appear when loop integrals, corresponding to graphs containing closed loops, are evaluated. This necessitates renormalization. In particular it means that the functional integral for $Z[J]$, as it stands above, is not well defined. The definition is purely formal and has to be made meaningful by regularizing it in such a way that no infinities appear. In perturbation theory, different regularizations are possible. In order to define the functional integral also outside perturbation theory, however, a non-perturbative regularization has to be called upon. In this book the discretization of space–time by means of a lattice is invoked.

For early approaches towards a rigorous definition of Euclidean functional integrals via space–time discretization, in the case of two-dimensional scalar field theory, see [1.37].

1.5   Quantum field theory on a lattice

1.5.1   Lattice regularization

In the quantum mechanical case the path integral was defined as a limit of a finite-dimensional integral which resulted from a discretization of time. This prescription will now be carried over to field theory by considering the functional integral as a limit of a well-defined integral over discretized Euclidean space–time. To this end we introduce a hypercubical lattice

$$\Lambda = a\mathbf{Z}^4 = \{x \mid x_\mu/a \in \mathbf{Z}\}, \tag{1.159}$$

where a is the lattice constant. The scalar field $\phi(x)$ is defined on the points $x \in \Lambda$. In analogy to the continuous case we set

$$(f, g) = \sum_x a^4 f(x)\, g(x) \tag{1.160}$$

and define the lattice forward derivative by the finite difference operation

$$\Delta_\mu^f f(x) = \frac{1}{a} \left(f(x + a\hat{\mu}) - f(x) \right) , \tag{1.161}$$

where $\hat{\mu}$ is the unit vector in the direction indicated by μ. We have

$$(\Delta_\mu^f f, g) = -(f, \Delta_\mu^b g) \tag{1.162}$$

or equivalently

$$(\Delta_\mu^f)^+ = -\Delta_\mu^b \tag{1.163}$$

with the backward derivative

$$\Delta_\mu^b g(x) = \frac{1}{a} \left(g(x) - g(x - a\hat{\mu}) \right) . \tag{1.164}$$

This implies

$$(\Delta_\mu^f f, \Delta_\mu^f f) = -(f, \Delta_\mu^b \Delta_\mu^f f) \equiv (f, \Box f), \tag{1.165}$$

where the lattice d'Alembert operator is

$$\Box = -\Delta_\mu^b \Delta_\mu^f . \tag{1.166}$$

It acts on functions as

$$\Box f(x) = \sum_{\mu=1}^{4} \frac{1}{a^2} \left(2f(x) - f(x + a\hat{\mu}) - f(x - a\hat{\mu}) \right) . \tag{1.167}$$

With the help of these elements the lattice action of the free field is defined to be

$$S_0[\phi, a] = \frac{1}{2} (\phi, (\Box + m^2)\phi)$$

$$= \frac{1}{2} \sum_{x,y} a^8 \, \phi(x)(\Box + m^2)_{x,y} \phi(y), \tag{1.168}$$

where

$$(\Box + m^2)_{x,y} = a^{-4}(\Box + m^2)\delta_{x,y} \tag{1.169}$$

is a matrix with indices labelled by points on the lattice.

The calculation of the lattice propagator proceeds as follows. $G(x, y; a)$ is defined to be the matrix inverse of $(\Box + m^2)$, i.e.

$$\sum_y a^4 \, (\Box + m^2)_{x,y} G(y, z; a) = a^{-4}\delta_{x,z} . \tag{1.170}$$

The solution of this equation is obtained through Fourier transformation:

$$G(x, y; a) = \int_{-\pi/a}^{\pi/a} \frac{d^4 p}{(2\pi)^4} \, e^{ip \cdot (x-y)} \, \tilde{G}(p; a), \tag{1.171}$$

which leads to

$$\left\{ \sum_{\mu=1}^{4} a^{-2}\, 2(1 - \cos a p_\mu) + m^2 \right\} \widetilde{G}(p;a) = 1 \,. \tag{1.172}$$

Therefore

$$\widetilde{G}(p;a) = \left\{ \sum_{\mu=1}^{4} a^{-2}\, 2(1 - \cos a p_\mu) + m^2 \right\}^{-1}$$

$$= \left\{ \sum_{\mu=1}^{4} a^{-2}\, 4\sin^2 \frac{a p_\mu}{2} + m^2 \right\}^{-1} \,. \tag{1.173}$$

$G(x, y; a)$ is a well-defined matrix for all x, y.

The integration over all scalar field configurations is defined by

$$D[\phi] = \prod_x d\phi(x) \,, \tag{1.174}$$

which is a discrete product. Even better would be to start with a finite volume, such that the number of integration variables is finite, and take the infinite-volume limit afterwards. But in the present case the infinite-volume limit is trivial and we ignore these subtleties. Thus the generating functional for the discretized theory is

$$Z_0[J,a] = \frac{1}{Z_0(a)} \int \prod_x d\phi(x) \, \exp\left\{ -\frac{1}{2}(\phi, (\square + m^2)\phi) + (J, \phi) \right\} \tag{1.175}$$

with

$$Z_0(a) = \int \prod_x d\phi(x) \, \exp\left\{ -\frac{1}{2}(\phi, (\square + m^2)\phi) \right\} \,. \tag{1.176}$$

Being a discrete Gaussian integral, it is amenable to our tools and we obtain

$$Z_0[J,a] = \exp \frac{1}{2}(J, GJ) = \exp\left\{ \frac{1}{2} \sum_{x,y} a^8\, J(x)\, G(x, y; a)\, J(y) \right\} \,. \tag{1.177}$$

Now let us consider the continuum limit where $a \to 0$. For the propagator in momentum space we get, in accordance with (1.110)

$$\lim_{a \to 0} \widetilde{G}(p;a) = \left(\sum_{\mu=1}^{4} p_\mu^2 + m^2 \right)^{-1} = (p^2 + m^2)^{-1} = \widetilde{G}(p) \,. \tag{1.178}$$

Let us consider a source $J(x)$ which is defined for all $x \in \mathbf{R}^4$ and has a Fourier transform

$$J(x) = \int \frac{\mathrm{d}^4 k}{(2\pi)^4} \, \tilde{J}(k) \, \mathrm{e}^{ik \cdot x} . \qquad (1.179)$$

This implies

$$\lim_{a \to 0} \sum_{x,y} a^8 \, J(x) \, G(x, y; a) \, J(y) = \int \frac{\mathrm{d}^4 k}{(2\pi)^4} \, \tilde{J}(k) \, \tilde{J}(-k) \, \tilde{G}(k)$$

$$= \int \mathrm{d}^4 x \, \mathrm{d}^4 y \, J(x) \, G(x, y) \, J(y) , \qquad (1.180)$$

and consequently

$$\lim_{a \to 0} Z_0[J, a] = Z_0[J] = \exp \frac{1}{2} \int \mathrm{d}^4 x \, \mathrm{d}^4 y \, J(x) \, G(x, y) \, J(y) , \qquad (1.181)$$

which is the generating functional in the continuum. Thus we have confirmed the definition of the functional integral for the free scalar field theory by means of a limit procedure.

Encouraged by this result one defines the generating functional for the interacting theory on the lattice in the same way:

$$Z[J, a] = \frac{1}{Z(a)} \int \prod_x \mathrm{d}\phi(x) \, \exp\{-S[\phi, a] + (J, \phi)\} , \qquad (1.182)$$

where the action on the lattice is

$$S[\phi, a] = \frac{1}{2}(\phi, (\Box + m^2)\phi) + S_I[\phi]$$

$$= \frac{1}{2} \sum_x a^4 \, \{\Delta_\mu^f \phi(x) \, \Delta_\mu^f \phi(x) + m^2 \, \phi(x)^2\} + \frac{g}{4!} \sum_x a^4 \, \phi(x)^4 . \qquad (1.183)$$

The problem is now to find a limit $\lim_{a \to 0} Z[J, a]$ which produces well-defined Green's functions.

1.5.2 Transfer matrix

Before the problem of a continuum limit is addressed we consider some aspects of the theory at finite lattice spacing. In order to simplify the notation we set $a = 1$ in most formulae.

In the same way as was done for the discretized quantum mechanical path integral, a transfer matrix can be defined for field theory on a lattice. Let

$$\Phi_t = \{\phi(x) \mid x_4 = t\} \qquad (1.184)$$

be the field configuration on a Euclidean timeslice $x_4 = t$. We decompose the lattice action as

$$S[\phi] = \sum_t L[\Phi_{t+1}, \Phi_t], \tag{1.185}$$

$$L[\Phi_{t+1}, \Phi_t] = \sum_{\mathbf{x}} \frac{1}{2} \left(\phi(\mathbf{x}, t+1) - \phi(\mathbf{x}, t) \right)^2 + \frac{1}{2} L_1[\Phi_t] + \frac{1}{2} L_1[\Phi_{t+1}], \tag{1.186}$$

with

$$L_1[\Phi_t] = \sum_{\mathbf{x}} \left\{ \frac{1}{2} \sum_{k=1}^{3} (\phi(\mathbf{x}, t) - \phi(\mathbf{x} + \hat{k}, t))^2 + \frac{m^2}{2} \phi(\mathbf{x}, t)^2 + \frac{g}{4!} \phi(\mathbf{x}, t)^4 \right\}. \tag{1.187}$$

The transfer matrix is defined by

$$\mathbf{T}[\Phi_{t+1}, \Phi_t] = \exp -L[\Phi_{t+1}, \Phi_t]. \tag{1.188}$$

The timeslice fields serve as indices for the transfer matrix. The transfer matrix can be considered as the kernel of an operator $\mathbf{T}$, which is usually also called a transfer matrix. It acts on wave functions $\Psi_t(\Phi)$, depending on the fields $\Phi = \{\phi(\mathbf{x}) \mid \mathbf{x} \in \mathbf{Z}^3\}$ at fixed times, by means of

$$|\Psi_{t+1}\rangle = \mathbf{T}|\Psi_t\rangle, \tag{1.189}$$

or, more explicitly,

$$\Psi_{t+1}(\Phi) = \int \prod_{\mathbf{x}} d\phi'(\mathbf{x}) \ \mathbf{T}[\Phi, \Phi'] \ \Psi_t(\Phi'). \tag{1.190}$$

The square integrable wave functions form the Hilbert space of the theory. The action of $\mathbf{T}$ on $|\Psi_t\rangle$ can also be expressed in terms of differential operators as

$$\mathbf{T} = \text{const.} \cdot \exp\left(-L_1 \frac{a}{2}\right) \exp\left(-H_0 a\right) \exp\left(-L_1 \frac{a}{2}\right) \tag{1.191}$$

with

$$H_0 = -\sum_{\mathbf{x}} \frac{1}{2} \frac{\partial^2}{\partial(\phi(\mathbf{x}))^2}. \tag{1.192}$$

Let us introduce the multiplication operators $\hat{\phi}(\mathbf{x})$ via

$$\hat{\phi}(\mathbf{x})\Psi(\Phi) = \phi(\mathbf{x})\Psi(\Phi). \tag{1.193}$$

The functional integral formula for the two-point function

$$\langle \phi(x_1) \phi(x_2) \rangle = \frac{1}{Z} \int \prod_x d\phi(x) \ e^{-S[\phi]} \ \phi(x_1) \ \phi(x_2) \tag{1.194}$$

reads in terms of the transfer matrix

$$\langle \phi(x_1)\,\phi(x_2) \rangle = \frac{1}{Z} \int \prod_x d\phi(x)\ \phi(x_1)\,\phi(x_2)\ \prod_t \mathbf{T}[\Phi_{t+1}, \Phi_t]$$

$$= \lim_{N \to \infty} \frac{\mathrm{Tr}\left(\mathbf{T}^{N-k}\,\hat{\phi}(\mathbf{x_1})\,\mathbf{T}^k\,\hat{\phi}(\mathbf{x_2})\right)}{\mathrm{Tr}\,(\mathbf{T}^N)}, \tag{1.195}$$

where $k = t_1 - t_2 > 0$ is assumed. Corresponding formulae hold for the other n-point functions. For finite N the expectation values are those of a theory with finite Euclidean time extent Na and periodic boundary conditions in x_4. The normalization factor

$$Z = \mathrm{Tr}\,(\mathbf{T}^N) \tag{1.196}$$

is called the *partition function.*

The lattice Hamiltonian H is defined by

$$\mathbf{T} = e^{-Ha}, \tag{1.197}$$

and the eigenvector of H with lowest eigenvalue E_0 is the vacuum $|0\rangle$. Inserting the complete set of eigenvectors of H into (1.195) yields

$$\langle \phi(x_1)\,\phi(x_2) \rangle = \sum_{n \geq 0} |\langle 0|\hat{\phi}(\mathbf{0})|n\rangle|^2\ e^{-(E_n - E_0)k}, \tag{1.198}$$

where we have assumed $\mathbf{x_1} = \mathbf{x_2} = \mathbf{0}$ and a discrete spectrum E_n for notational simplicity. In fact, as long as the lattice has a finite spatial extent the spectrum is always discrete. The connected two-point function

$$\langle \phi(x_1)\,\phi(x_2) \rangle_c = \sum_{n > 0} |\langle 0|\hat{\phi}(\mathbf{0})|n\rangle|^2\ e^{-(E_n - E_0)k} \tag{1.199}$$

consequently decays exponentially for large distances like

$$\langle \phi(x_1)\,\phi(x_2) \rangle_c \propto e^{-(E_1 - E_0)|x_1 - x_2|}. \tag{1.200}$$

The *correlation length* ξ, which determines the rate of the exponential decay, is therefore given by the inverse of the *mass gap*:

$$\xi = (E_1 - E_0)^{-1}. \tag{1.201}$$

1.5.3 *Reflection positivity*

From the explicit representation of the transfer matrix discussed above one can infer that $\mathbf{T}$ is a bounded, symmetric, and positive operator. The positivity is essential for the existence of a self-adjoint Hamiltonian H. For other field theoretical models an explicit representation of the transfer matrix, which would allow investigation of the above mentioned properties directly, cannot always be given. Therefore it is desirable to have a characterization in terms of expectation values. This leads us to consider reflection positivity again. As we shall see, the lattice adds some subtleties when a model is tested for reflection positivity. Reflection positivity for lattice models has been investigated first in [1.38, 1.39, 1.40].

On a lattice there are two possible types of time reflections. First of all we may reflect with respect to a hyperplane which contains sites of the lattice, e.g. the plane $x_4 = 0$:

$$\theta_s : \quad x_4 \rightarrow -x_4. \qquad (1.202)$$

We call this *site-reflection*. Secondly we may reflect with respect to a hyperplane lying in between the lattice planes, e.g. the plane $x_4 = 1/2$:

$$\theta_l : \quad x_4 \rightarrow 1 - x_4. \qquad (1.203)$$

Since such planes are pierced by lattice links, this kind of reflection is called *link-reflection*.

Let us consider site-reflection first. Suppose F is any function of the fields at non-negative times, which means that F depends only on $\phi(x)$ with $x_4 \geq 0$. ΘF is defined as in section (1.3.5) with the reflection θ_s. It depends on the fields at negative times. Site-reflection positivity states that

$$\langle (\Theta F) F \rangle \geq 0. \qquad (1.204)$$

With regard to the general relation between Euclidean expectation values and the Hilbert space formalism this property amounts to the positivity of scalar products

$$\langle \Psi | \Psi \rangle \geq 0, \qquad (1.205)$$

where $|\Psi\rangle$ represents a wave function depending on the fields at time $x_4 = 0$.

If we now shift F by n lattice-spacings in the positive time direction, the resulting function F' still depends on fields at positive times only and is separated from $\Theta F'$ by $2n$ lattice spacings. The positivity of $\langle (\Theta F') F' \rangle$ implies the positivity of $\langle \Psi | \mathbf{T}^{2n} | \Psi \rangle$ from which it follows that $\mathbf{T}^2$ is positive. Although from this result one can not be sure about the positivity of $\mathbf{T}$, it is sufficient for the definition of a Hamiltonian by means of

$$\mathbf{T}^2 = e^{-2Ha}. \qquad (1.206)$$

In general this Hamiltonian may contain terms with third or fourth (lattice) derivatives. To summarize, site-reflection positivity guarantees the positivity of the scalar product in Hilbert space, the positivity of $\mathbf{T}^2$, which generates time translations by 2 units, and the existence of a self-adjoint Hamiltonian. As our discussion was more informal we point out that a rigorous construction of the Hilbert space and Hamiltonian from

Euclidean expectation values satisfying reflection positivity is contained in [1.34, 1.40].

Let us see how site-reflection positivity is confirmed for the case of scalar field theory, which we discussed before. We split the action according to

$$S = S_+ + S_- + S_0, \tag{1.207}$$

where

$$S_0 = L_1[\Phi_0] \tag{1.208}$$

depends on the fields at $x_4 = 0$ only, S_+ depends on the fields at positive times and $S_- = \Theta S_+$. The numerator of the functional integral for $\langle (\Theta F) F \rangle$ is

$$\int \prod_x d\phi(x) \, e^{-S_0} \, \Theta \left(F e^{-S_+} \right) \, F e^{-S_+}. \tag{1.209}$$

With

$$\mathscr{F}[\Phi_0] = \int \prod_{x_4>0} d\phi(x) \, F e^{-S_+} \tag{1.210}$$

it can be written as

$$\int \prod_{x_4=0} d\phi(x) \, e^{-S_0} |\mathscr{F}[\Phi_0]|^2. \tag{1.211}$$

The integrand is manifestly positive and the desired result follows.

Now we come to the consideration of link-reflection. The operator Θ is defined as above but with θ_l instead of θ_s. Link-reflection positivity requires that $\langle (\Theta F) F \rangle \geq 0$ for all F depending on the fields at strictly positive times, i.e. $x_4 \geq 1$. In contrast to the case before, F and ΘF are now separated by an odd number of lattice spacings in the time direction. Shifting F one lattice spacing in the positive time direction increases the distance by 2. Therefore reflection positivity implies again that the operator $\mathbf{T}^2$, which generates time shifts of two units, is positive and a Hamiltonian can be defined. A Hilbert space with positive scalar product can also be constructed [1.40].

Let us verify link-reflection positivity for scalar field theory. This time the action is split as

$$S = S_c + S'_+ + S'_-, \tag{1.212}$$

where S'_+ only depends on the fields with $x_4 \geq 1$, $S'_- = \Theta S'_+$, and

$$S_c = - \sum_{\mathbf{x}} \phi(\mathbf{x}, 1) \, \phi(\mathbf{x}, 0) \tag{1.213}$$

is the component which connects the time-layers $x_4 = 0$ and $x_4 = 1$. In view of later applications we generalize the action such that

$$S_c = -2\kappa \sum_{\mathbf{x}} \phi(\mathbf{x}, 1)\, \phi(\mathbf{x}, 0), \tag{1.214}$$

where a *hopping parameter* κ has been introduced. The functional integral then reads

$$\int \prod_x d\phi(x)\, e^{-S_c}\, \Theta\left(F e^{-S'_+}\right)\, F e^{-S'_+}. \tag{1.215}$$

$\exp - S_c$ can be expanded in the form

$$e^{-S_c} = \sum_{n=0}^{\infty} \kappa^n \sum_i c_{ni}(\Theta S_i) S_i, \tag{1.216}$$

with positive coefficients c_{ni}, and we obtain for the functional integral

$$\sum_{n=0}^{\infty} \kappa^n \sum_i c_{ni} \left| \int \prod_{x_4 > 0} d\phi(x)\, S_i\, F e^{-S'_+} \right|^2. \tag{1.217}$$

For positive κ this expression is positive and we find link-reflection positivity. If κ is negative, however, link-reflection positivity does not hold. For example the correlation function $\langle \phi(\mathbf{0}, 0)\, \phi(\mathbf{0}, 1) \rangle$ turns out to be negative in this case. This is also intuitively clear since for negative κ the action is lower if neighbouring fields have opposite signs. Thus it prefers configurations with antiferromagnetic ordering.

As we have seen, it is sufficient to have either site-reflection positivity or link-reflection positivity in order that a Hamiltonian can be defined and a Hilbert space formalism exists. If one is interested mainly in the continuum limit this serves the needs. At finite lattice spacings, however, the best thing is to have reflection positivity of both kinds, because then the transfer matrix $\mathbf{T}$ itself is positive and one can consider correlations at arbitrary time distances for the determination of the spectrum. In our example this is the case for $\kappa > 0$.

1.6   Continuum limit and critical behaviour

1.6.1   *Lattice field theory and statistical systems*

The lattice supplies the field theory with a cut-off. The Fourier transforms of functions on the lattice are periodic in momentum space with periodicity $2\pi/a$, since $\exp(2\pi i x_\mu / a) = 1$ for all lattice points x. Therefore all momenta can be restricted to the first *Brillouin zone* (cf. (1.171))

$$\mathscr{B} = \left\{ p \,\middle|\, -\frac{\pi}{a} < p_\mu \leq \frac{\pi}{a} \right\}, \tag{1.218}$$

and the momentum cut-off is π/a. As a result, loop integrations in perturbation theory yield finite results. In order to obtain the theory in the continuum the lattice spacing has to be sent to zero, as has been done for free field theory in section 1.5.1. At the same time the cut-off goes to infinity and renormalization is necessary to obtain finite results for physical quantities.

Renormalization introduces renormalized fields, couplings etc. which have to be distinguished from the corresponding bare quantities. In the following we denote the bare quantities by the subscript 0 and write the lattice action for scalar field theory in (1.183) as

$$S = \frac{1}{2} \sum_x a^4 \{\Delta_\mu^f \phi_0(x) \Delta_\mu^f \phi_0(x) + m_0^2 \phi_0(x)^2\} + \frac{g_0}{4!} \sum_x a^4 \phi_0(x)^4. \quad (1.219)$$

Another way of writing the action, which is useful for numerical investigations, is

$$S = \sum_x \left\{ -2\kappa \sum_{\mu=1}^4 \phi(x)\phi(x + a\hat{\mu}) + \phi(x)^2 + \lambda[\phi(x)^2 - 1]^2 - \lambda \right\}. \quad (1.220)$$

This form only contains dimensionless quantities. The field ϕ and the parameters κ and λ are related to the bare field, bare mass and bare coupling through

$$a\phi_0(x) = (2\kappa)^{1/2} \phi(x), \qquad a^2 m_0^2 = \frac{1 - 2\lambda}{\kappa} - 8, \qquad g_0 = \frac{6\lambda}{\kappa^2}. \quad (1.221)$$

If we use this action the functional integral

$$Z = \int \prod_x d\phi(x) \, e^{-S} \quad (1.222)$$

has the form of a partition function for a model of statistical mechanics. The field variables $\phi(x)$ play the rôle of one-component real spins attached to the lattice points. The integrand $\exp(-S)$ corresponds to the Boltzmann factor and the action S corresponds to $\beta\mathcal{H}$, where $\mathcal{H}$ is the classical Hamiltonian of a four-dimensional spin model. (It should not be confused with the quantum Hamiltonian H, which generates time translations in the operator formulation of field theory.) For positive κ the coupling is of ferromagnetic type and κ is analogous to the inverse temperature β. (This 'analogue temperature' should not be confused with the physical temperature considered e.g. in section 1.8.) The limit $\lambda \to \infty$ represents a special case, namely the four-dimensional Ising model. In this limit all configurations with $\phi(x)^2 \neq 1$ are suppressed and the field variables only assume the two values $\phi(x) = \pm 1$.

The analogy between Euclidean quantum field theory on a lattice and statistical mechanics has turned out to be very useful. Many concepts and methods of statistical mechanics have been applied to field theory and conversely the field theoretic renormalization group is an important tool in statistical mechanics. These topics will be discussed in more detail in later chapters.

Let us extend the relation between field theory and statistical mechanics further. The vacuum expectation value of the field

$$v = \langle \phi(x) \rangle \tag{1.223}$$

corresponds to the mean magnetization M per site of a ferromagnet. The propagator $\langle \phi(x)\phi(y) \rangle_c$ is equal to the spin–spin correlation function. The correlation length, which governs the exponential decay of the correlation function, is related to the mass gap m by

$$\xi = 1/ma, \tag{1.224}$$

as we have seen in section 1.5.2. m is the energy of the lowest state above the vacuum and thus equals the mass of the lightest particle in the theory.

As is well known in statistical mechanics the magnetic *susceptibility* is related to the correlation function by

$$\chi = \sum_x \langle \phi(x)\phi(0) \rangle_c \tag{1.225}$$

and therefore equals the propagator at zero momentum.

1.6.2   *Renormalization and critical behaviour*

Now we turn to the question of a continuum limit. If a continuum limit with a finite physical mass exists it means that by a suitable choice of the bare parameters we can approach a limit where a goes to zero while m remains finite. According to (1.224) the correlation length then has to diverge. A point in the space of parameters (κ, λ), where ξ diverges and where the first derivatives of the relevant thermodynamic potential, say the Gibbs potential, exist, is called a *critical point* or a *second order phase transition*. For most systems the behaviour of many quantities near a critical point is governed by simple power laws [1.41, 1.42]. As κ approaches its critical value κ_c the correlation length and susceptibility, for example, diverge according to

$$\xi \sim |\kappa - \kappa_c|^{-\nu}, \qquad \chi \sim |\kappa - \kappa_c|^{-\gamma}, \tag{1.226}$$

with certain *critical exponents* ν and γ. The relation $\sim$ is to be understood here in the sense that

$$f(s) \sim s^\sigma \tag{1.227}$$

is equivalent to

$$\lim_{s \to 0} \frac{\log f(s)}{\log s} = \sigma.$$ (1.228)

The magnetization in the low temperature phase ($\kappa > \kappa_c$) vanishes like

$$M \sim |\kappa - \kappa_c|^{\beta'}.$$ (1.229)

In four dimensions these laws are modified by logarithmic corrections [1.43].

On the other hand, in field theory the focus is on the behaviour of correlation functions. Consider the two-point function

$$G(x, 0) = \langle \phi(x)\phi(0) \rangle_c$$ (1.230)

and its Fourier transform $\tilde{G}(p)$. For the free theory at finite lattice spacing we have (see section 1.5.1)

$$\tilde{G}(p)^{-1} = 2\kappa a^{-2}(m_0^2 + \hat{p}^2)$$ (1.231)

with

$$\hat{p}_\mu = \frac{2}{a} \sin \frac{ap_\mu}{2}.$$ (1.232)

(The argument a in $\tilde{G}(p, a)$ is omitted from now on.) In general the physical particle mass m is given by the pole of the propagator in momentum space closest to the origin:

$$\tilde{G}(p)^{-1} = 0, \qquad p = (im, 0, 0, 0).$$ (1.233)

For practical purposes it is convenient to introduce another mass, the *renormalized mass* m_R, by means of the small momentum behaviour of the propagator:

$$\tilde{G}(p)^{-1} = (2\kappa/a^2 Z_R)\{m_R^2 + p^2 + \mathcal{O}(p^4)\}.$$ (1.234)

In the vicinity of the critical point m and m_R turn out to be nearly equal. Equation (1.234) also defines the *wave function renormalization factor* Z_R. With the help of this factor a *renormalized field*

$$\phi_R(x) = Z_R^{-1/2} \phi_0(x) = (2\kappa/a^2 Z_R)^{1/2} \phi(x)$$ (1.235)

can be defined, whose inverse propagator is normalized to

$$\tilde{G}_R(p)^{-1} = m_R^2 + p^2 + \mathcal{O}(p^4).$$ (1.236)

In the continuum limit the thus normalized propagator is expected to remain finite. More generally, the consistency of the continuum limit requires that all n-point correlation functions of the renormalized field ϕ_R possess finite limits.

As long as the lattice spacing is finite the correlation functions are symmetric under reflections in lattice planes and under rotations by an angle of $\pi/2$, but not under the full $O(4)$ rotation group. In the continuum limit they should approach functions which obey the full $O(4)$ rotational invariance, restoring the Euclidean invariance which we require for the resulting field theory in the continuum. This is a further condition which we impose on a meaningful continuum limit.

The bare coupling g_0 has a renormalized counterpart too. It is defined in terms of the four-point vertex function at zero momentum. *Vertex functions* are defined completely analogous to the case of field theory in the continuum (see e.g. [A3]). The n-point vertex function of the field ϕ is denoted by $\Gamma^{(n)}(p_1, \ldots, p_n)$ in momentum space. By definition it is equal to the amputated, one-particle-irreducible part of the connected n-point correlation function $G_c(p_1, \ldots, p_n)$. It will be discussed in more detail in the next chapter.

The renormalized n-point vertex function $\Gamma_R^{(n)}(p_1, \ldots, p_n)$ is the vertex function of the renormalized field $\phi_R(x)$. It is related to the corresponding vertex function of $\phi(x)$ through

$$\Gamma_R^{(n)} = (2\kappa/a^2 Z_R)^{-n/2} \Gamma^{(n)}. \tag{1.237}$$

The *renormalized coupling* g_R is defined by

$$g_R = -\Gamma_R^{(4)}(0,0,0,0). \tag{1.238}$$

It is a dimensionless quantity (in four space–time dimensions). The definition above ensures that in perturbation theory

$$g_R = g_0 + \mathcal{O}(g_0^2). \tag{1.239}$$

As has been said before, in the continuum limit we require m_R and g_R to remain finite. In perturbation theory one finds that this can be achieved by a suitable tuning of the bare parameters m_0 and g_0 as the cut-off goes to infinity. A priori this does not, however, imply that all correlation functions have a good continuum limit also at non-zero momenta. The central statement of renormalization theory is that nevertheless this property holds for a certain class of theories to which ϕ^4 theory belongs. These are called *perturbatively renormalizable* theories. The complete theory of perturbative renormalization will not be dealt with in this book and we refer to [1.44, 1.45, 1.46, 1.47],[A8] and in particular to the work of Reisz [1.48], where the perturbative renormalization of lattice theories is treated.

The quantities introduced above can be related to the ones used in the context of statistical mechanics. The susceptibility for example is given by

$$\chi = \frac{Z_R}{2\kappa a^2 m_R^2}, \tag{1.240}$$

which implies that the critical behaviour of Z_R is

$$Z_R \sim |\kappa - \kappa_c|^{-\gamma+2\nu}.$$

(1.241)

1.6.3 *Universality*

The investigation of various systems of statistical mechanics near their critical points has revealed the property of universality [1.41, 1.42]. To be precise, the systems fall into a relatively small number of *universality classes*. The members of a class show identical critical behaviour in the sense that their critical exponents as well as certain other universal quantities are equal. The universality classes are distinguished by

1. the number of degrees of freedom of the microscopic field and the symmetry of the system,

2. the number of dimensions of space (or Euclidean space–time).

Universality means that the long-range properties of a critical system do not depend on the details of the microscopic interaction. In particular also the size of the lattice spacing becomes unimportant for the large-distance behaviour of correlation functions if the correlation length is large. According to the *scaling hypothesis* the correlation length is the only relevant length scale for the system near criticality. This hypothesis leads to various relations between critical exponents, such that only two independent exponents remain.

Scaling theory and universality have found a theoretical basis in the Kadanoff–Wilson *renormalization group* [1.49, 1.50, 1.51]. Here we shall try to give only a brief sketch of the basic ideas of the Kadanoff–Wilson renormalization group. The field-theoretical renormalization group, which is used in other chapters, will be introduced in the next section.

The original action S with cut-off Λ is considered to be embedded in an infinite-dimensional space of actions

$$S = \sum_i K_i S_i,$$

(1.242)

where the coefficients K_i are called couplings and

$$S_i = \sum_x \mathscr{L}_i$$

(1.243)

with so-called *local operators* $\mathscr{L}_i$. They are functions depending on the fields at point x, and a finite number of points near x, and having a certain engineering dimension, e.g. $\phi(x)^n$ has dimension n. A renormalization group transformation R_λ is a mapping

$$S \to S^{(\lambda)}$$

(1.244)

in this space such that both S and $S^{(\lambda)}$ describe the same physics at large distances but the cut-off Λ gets lowered by a factor $\lambda > 1$:

$$\Lambda \to \frac{1}{\lambda}\Lambda. \qquad (1.245)$$

In other words, $S^{(\lambda)}$ is obtained from S by integrating out degrees of freedom with high momenta near the cut-off. R_λ can be described in terms of the changes of coefficients

$$K_i \to K_i^{(\lambda)}. \qquad (1.246)$$

The most important points in the space of actions are the *fixed points* S^*:

$$R_\lambda S^* = S^*, \qquad (1.247)$$

in particular those where the correlation length is infinite. Near a fixed point the action of R_λ can in general be linearized and diagonalized such that in a suitable basis

$$K_\alpha = K_\alpha^* + \delta K_\alpha \qquad (1.248)$$

it reads

$$K_\alpha^{(\lambda)} = K_\alpha^* + \lambda^{d_\alpha}\delta K_\alpha. \qquad (1.249)$$

Those terms with negative *scaling dimension* d_α die out after repeated application of the renormalization group transformation and are called *irrelevant* since their presence does not affect the long-distance physics. The terms with positive d_α, which are a few in general, are *relevant*. The values of the corresponding coefficients K_α are decisive for the long-distance physics. Terms with $d_\alpha = 0$ are called *marginal*.

In this picture universality emerges in the following way. Two original actions S' and S'', which belong to the *domain* of the same fixed point, are mapped under the action of the renormalization group into the neighbourhood of the same low-dimensional manifold

$$S = S^* + \sum_{\text{relevant } \alpha} K_\alpha S_\alpha, \qquad (1.250)$$

where for simplicity we assume that no marginal operators are present. The critical behaviour is then determined only by the few relevant operators in the vicinity of the fixed point. In particular it can be shown that the critical indices are simple algebraic combinations of the dimensions d_α belonging to them. Thus the fixed points of the renormalization group determine the universality classes of actions.

In four dimensions perturbative calculations indicate that for the scalar field theory under consideration there are only two relevant operators, which are essentially the mass term $\phi(x)^2$ and the linear term $\phi(x)$, which appears when an external 'magnetic' field is present. The quartic self-interaction $\phi(x)^4$ is marginal, but its coupling g decreases under the renormalization group transformation. The associated fixed point therefore has $g = 0$ and belongs to a free field theory. It is called the *Gaussian fixed point*. As will be discussed in chapter 2, non-perturbative methods lead to the result that the Gaussian fixed point is with high certainty the only fixed point for this theory.

1.7 Renormalization group equations

In the framework of field theory the popular versions of the renormalization group are different from the Kadanoff–Wilson one. They come in various forms and in the following we shall be concerned with those which play a rôle for the continuum limit of lattice field theories.

The concepts will again be illustrated by the case of scalar field theory on a lattice. We use the renormalization prescription introduced in the previous section. For convenience we collect the formulae below. The wave function renormalization factor Z_R relates the renormalized field and vertex functions to the unrenormalized ones by

$$\phi_R(x) = Z_R^{-1/2}\phi_0(x) \tag{1.251}$$

$$\Gamma_R^{(n)} = Z_R^{n/2}\Gamma_0^{(n)}. \tag{1.252}$$

It is fixed through the renormalization condition

$$\frac{\partial}{\partial p^2}\Gamma_R^{(2)}(0) = -1. \tag{1.253}$$

The renormalized mass and coupling are defined by

$$m_R^2 = -\Gamma_R^{(2)}(0), \qquad g_R = -\Gamma_R^{(4)}(0,0,0,0). \tag{1.254}$$

1.7.1 Renormalization group equations for the bare theory

We start with a discussion of those renormalization group equations which come closest in spirit to the Kadanoff–Wilson approach. The essential difference is that instead of the infinite-dimensional space of actions only the two-dimensional subspace parametrized by $m_0 a$ and g_0 is taken into consideration. Therefore it is clearly not possible to leave the long-distance behaviour of all renormalized vertex functions unaffected by a change in the cut-off. One has to stick to two physical, renormalized quantities, which we take to be m_R and g_R. m_R just fixes the scale for all dimensionful quantities. In particular the cut-off is characterized by the dimensionless quantity $1/(m_R a)$, and by a change of the cut-off we mean a change in

$m_R a$. The renormalized coupling g_R is dimensionless and depends on the bare parameters $m_0 a$ and g_0.

Now we consider those bare theories which have the same value of g_R. They define a one-parameter subspace, which is the analogue of the renormalization group trajectory in the Kadanoff–Wilson approach. Every point on it has a particular value of $m_R a$. As we change the cut-off while holding g_R fixed we move along the line. Locally the dependence of $m_R a$ and g_R on the bare parameters $m_0 a$ and g_0 can be inverted and we might consider $m_0 a$ and g_0 to be functions of $m_R a$ and g_R. The question of the global invertibility of the relation between bare and renormalized parameters will not be addressed here.

Now we ask for the change of renormalized vertex functions under a change of the cut-off with fixed g_R. A detailed analysis in perturbation theory [1.52] reveals that the renormalized vertex functions differ from their continuum limits by terms of the order a^2:

$$\Gamma_R^{(n)}(p_i; g_R, m_R, m_R a) = \Gamma_R^{(n)}(p_i; g_R, m_R, 0) + \mathcal{O}\left(a^2 (\ln a)^k\right) \qquad (1.255)$$

in k-loop order. Consequently

$$a \frac{\mathrm{d}}{\mathrm{d}a} \Gamma_R^{(n)}(p_i; g_R, m_R, m_R a) = \mathcal{O}\left(a^2 (\ln a)^k\right). \qquad (1.256)$$

The terms on the right hand side are called *scaling violations*. Since they are small near the continuum limit we neglect them in the following and write 0 instead. Now we express $\Gamma_R^{(n)}$ through $\Gamma_0^{(n)}$ in the form

$$\Gamma_R^{(n)} = Z_R^{n/2}(g_0, m_R a)\, \Gamma_0^{(n)}(g_0, m_R, m_R a), \qquad (1.257)$$

where $m_0 a$ has been eliminated in favour of $m_R a$ on the right hand side. Applying the chain rule we obtain from (1.256) an equation for the bare vertex functions:

$$\left\{ a \frac{\partial}{\partial a} - \beta_{\mathrm{LAT}} \frac{\partial}{\partial g_0} + n \gamma_{\mathrm{LAT}} \right\} \Gamma_0^{(n)} = 0. \qquad (1.258)$$

A similar equation for the massless theory is discussed in [1.43].

The functions β_{LAT} and γ_{LAT} are defined by

$$\beta_{\mathrm{LAT}}(g_0, m_R a) = -\, a \left. \frac{\partial g_0}{\partial a} \right|_{g_R} \qquad (1.259)$$

$$\gamma_{\mathrm{LAT}}(g_0, m_R a) = \frac{1}{2}\, a \left. \frac{\partial \ln Z_R}{\partial a} \right|_{g_R}. \qquad (1.260)$$

Equation (1.258) describes how the bare vertex functions change under a variation of the cut-off keeping g_R fixed. According to this equation the change of a can be compensated by changing g_0 and the wave function renormalization factor at the same time. These changes are globally valid in the sense that β_{LAT} and γ_{LAT} do not depend on the momenta or on the index n of the vertex function considered.

The functions β_{LAT} and γ_{LAT} actually depend only on g_0 up to scaling violations, which we do not prove here. This crucial property allows one to solve the renormalization group equations and to extract information about the behaviour of the bare vertex functions near the continuum limit. On the other hand, if β_{LAT} is known, the relation

$$\beta_{LAT}(g_0) = -a \left. \frac{\partial g_0}{\partial a} \right|_{g_R} \tag{1.261}$$

determines the change of g_0 with the cut-off. Therefore β_{LAT} is of central importance if one asks about the behaviour of the bare coupling when approaching the continuum limit. We shall come back to this point later in the discussion of fixed points.

1.7.2 Callan–Symanzik equations

The philosophy behind the Callan–Symanzik equations [1.53, 1.54, 1.43] is opposite to the one of the previous section. Now one considers the bare coupling g_0 to be fixed and looks for the changes of the renormalized vertex functions under a change of the cut-off. The renormalized vertex functions $\Gamma_R^{(n)}(p_i; g_R, m_R, m_R a)$ depend on the renormalized coupling g_R. Therefore we have according to the chain rule

$$m_R \left. \frac{d}{dm_R} \Gamma_R^{(n)} \right|_{g_0} = m_R \frac{\partial}{\partial m_R} \Gamma_R^{(n)} + m_R \left. \frac{\partial g_R}{\partial m_R} \right|_{g_0} \frac{\partial}{\partial g_R} \Gamma_R^{(n)}. \tag{1.262}$$

On the other hand

$$m_R \left. \frac{d}{dm_R} \Gamma_R^{(n)} \right|_{g_0} = m_R \left. \frac{d}{dm_R} \left(Z_R^{n/2} \Gamma_0^{(n)} \right) \right|_{g_0}$$

$$= \frac{n}{2} m_R \left. \frac{\partial \ln Z_R}{\partial m_R} \right|_{g_0} \Gamma_R^{(n)} + Z_R^{n/2} m_R \left. \frac{\partial}{\partial m_R} \Gamma_0^{(n)} \right|_{g_0}. \tag{1.263}$$

Combining these equations we get the Callan–Symanzik equations:

$$\left\{ m_R \frac{\partial}{\partial m_R} + \beta \frac{\partial}{\partial g_R} - n\gamma \right\} \Gamma_R^{(n)} = \Delta \Gamma_R^{(n)}, \tag{1.264}$$

where the functions β and γ are defined by

$$\beta(g_R, m_R a) = m_R \left. \frac{\partial g_R}{\partial m_R} \right|_{g_0} \tag{1.265}$$

$$\gamma(g_R, m_R a) = \frac{1}{2} m_R \left. \frac{\partial \ln Z_R}{\partial m_R} \right|_{g_0}. \tag{1.266}$$

The right hand side of the Callan–Symanzik equation is defined by

$$\Delta\Gamma_R^{(n)} = Z_R^{n/2}\, m_R \frac{\partial}{\partial m_R}\Gamma_0^{(n)}\Big|_{g_0}. \tag{1.267}$$

It can be shown, e.g. by an analysis of Feynman graphs, that this quantity equals the n-point vertex function with an additional insertion of the operator $m_R^2\phi_R^2$ at zero momentum. Thus it also has a finite continuum limit from which it differs by scaling violations like the other vertex functions. Since (1.264) involves renormalized vertex functions on both sides, the coefficient functions β and γ on the left hand side must scale in the same way as the vertex functions for small $m_R a$. That means they possess continuum limits too and being dimensionless they are independent of $m_R a$ up to scaling violations.

The equations (1.264) have the same structure as (1.258) in the previous section. The difference is that the bare vertex functions and coupling have interchanged their rôle with the renormalized ones now. In particular the function

$$\beta(g_R) = m_R \frac{\partial g_R}{\partial m_R}\Big|_{g_0} = m_R a \frac{\partial g_R}{\partial m_R a}\Big|_{g_0} \tag{1.268}$$

describes the variation of the renormalized coupling g_R with the cut-off for fixed g_0, whereas for $\beta_{\text{LAT}}(g_0)$ it was the other way around.

Although β and β_{LAT} are different functions they are related to each other in a particular way. To see this let us consider the situation of the previous section, where g_R was constant. This implies

$$a\frac{\mathrm{d}}{\mathrm{d}a}g_R = \left\{ a\frac{\partial}{\partial a} - \beta_{\text{LAT}}(g_0)\frac{\partial}{\partial g_0}\right\} g_R(g_0, m_R a) = 0. \tag{1.269}$$

On the other hand

$$a\frac{\partial}{\partial a}g_R(g_0, m_R a) = m_R\frac{\partial}{\partial m_R}g_R(g_0, m_R a) = \beta(g_R), \tag{1.270}$$

and we obtain

$$\beta_{\text{LAT}}(g_0)\frac{\partial g_R(g_0, m_R a)}{\partial g_0} = \beta(g_R). \tag{1.271}$$

The beta-function $\beta(g_R)$ can be calculated in renormalized perturbation theory directly in the continuum and the result is [1.55] :

$$\beta(g_R) = \beta_1\, g_R^2 + \beta_2\, g_R^3 + \cdots \tag{1.272}$$

with

$$\beta_1 = \frac{3}{16\pi^2}, \qquad \beta_2 = -\frac{17}{768\pi^4}. \tag{1.273}$$

Furthermore we know that to lowest order in perturbation theory both couplings are equal to each other:

$$g_R = g_0 + \mathcal{O}(g_0^2).$$

(1.274)

Inserting these relations into (1.271) yields

$$\beta_{\mathrm{LAT}}(g_0) = \hat{\beta}_1 g_0^2 + \hat{\beta}_2 g_0^3 + \ldots$$

(1.275)

with

$$\hat{\beta}_1 = \beta_1, \qquad \hat{\beta}_2 = \beta_2, \qquad \hat{\beta}_3 \neq \beta_3.$$

(1.276)

The first two coefficients of the beta-functions are therefore universal.

1.7.3 Renormalization group equations for a massless theory

In the case of a massless field theory the renormalization conditions at zero momenta (1.253,1.254) cannot be used since these equations contain infrared divergencies in the limit $m_R \to 0$. Instead one introduces an arbitrary mass μ as a *renormalization scale* and fixes the renormalization scheme through the conditions

$$\Gamma_R^{(2)}(0; g_R, \mu) = 0,$$

(1.277)

$$\left. \frac{\partial}{\partial p^2} \Gamma_R^{(2)}(p; g_R, \mu) \right|_{p^2 = \mu^2} = -1,$$

(1.278)

$$\left. \Gamma_R^{(4)}(p_i; g_R, \mu) \right|_{S.P.(\mu)} = -g_R.$$

(1.279)

The last equation defines the renormalized coupling at the *symmetry point* S.P.(μ):

$$p_i \cdot p_j = \frac{1}{3}(4\delta_{ij} - 1)\mu^2.$$

(1.280)

As μ is arbitrary we may consider changes in μ for a fixed bare theory [1.43]. Since the bare vertex functions are independent of μ we have

$$\left. \mu \frac{\mathrm{d}}{\mathrm{d}\mu} \Gamma_0^{(n)} \right|_{g_0} = 0.$$

(1.281)

With

$$\Gamma_0^{(n)} = Z_R^{-n/2} \Gamma_R^{(n)}$$

(1.282)

we obtain in the same way as before the renormalization group equations

$$\left\{ \mu \frac{\partial}{\partial \mu} + \beta \frac{\partial}{\partial g_R} - n\gamma \right\} \Gamma_R^{(n)} = 0$$

(1.283)

with

$$\beta(g_R) = \mu \frac{\partial g_R}{\partial \mu}\bigg|_{g_0} \tag{1.284}$$

$$\gamma(g_R) = \frac{1}{2}\mu \frac{\partial \ln Z_R}{\partial \mu}\bigg|_{g_0}. \tag{1.285}$$

The β- and γ-functions appearing here are different from the previous ones. Their first two coefficients, however, are universal and coincide with the corresponding coefficients discussed in the previous section, as can be shown in a similar way.

Being homogeneous the equations above are easier to solve than the Callan–Symanzik equations for the massive theory. In fact in the latter case one usually concentrates on the so-called asymptotic parts of the vertex functions, which are solutions of the corresponding homogeneous partial differential equations. They differ from the vertex functions of the massless theory by a finite renormalization.

Let us mention here that for the vertex functions of the massive theory homogeneous renormalization group equations, the 't Hooft–Weinberg equations [1.56, 1.57] also exist. Their β- and γ-functions in general depend on μ/m_R, which would render a solution difficult. In the MS (minimal subtraction) scheme, however, they are independent of μ/m_R, which makes them a useful alternative to the Callan–Symanzik equations.

1.7.4   Fixed points

The various renormalization group equations discussed above can be utilized to gain knowledge about the behaviour of vertex functions. The method of solution of the equations is explained in many textbooks and will not be considered here. Instead we shall concentrate on the behaviour of the bare and renormalized couplings respectively near a continuum limit. The relevant functions which give us information about this question are the β-functions $\beta(g_R)$ and $\beta_{\text{LAT}}(g_0)$. Let us start with the equation for the renormalized coupling, which we write in the form

$$m_R a \frac{\partial g_R}{\partial m_R a}\bigg|_{g_0} = \frac{\partial g_R}{\partial \ln m_R a}\bigg|_{g_0} = \beta(g_R). \tag{1.286}$$

Its solution, which describes the dependence of g_R on $m_R a$, is given by

$$m_R a = C \exp \int^{g_R} \frac{dg}{\beta(g)}, \tag{1.287}$$

where C is an integration constant. The situation under study is that we move in the space of bare parameters $(m_0 a, g_0)$ along a line $g_0 = \text{const.}$ towards the continuum limit $m_R a \to 0$. If we start out in a region where

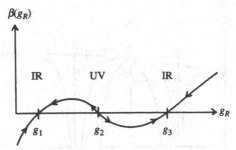

Fig. 1.4.   A model β-function with one ultraviolet and two infrared fixed points. The arrows indicate the flow of g_R for decreasing $m_R a$.

$\beta(g_R)$ is positive, the renormalized coupling will decrease until a zero of the β-function is hit. Such a zero would be reached asymptotically in the limit. On the other hand, if we begin in a region with negative $\beta(g_R)$ the renormalized coupling will increase and approach the next zero of the β-function from below.

We see that the zeroes of the β-function play a prominent rôle. If g_R would assume precisely the value of a zero of $\beta(g_R)$, it would not change under variations of the cut-off. Therefore the zeroes of the β-function are called fixed points. Let us consider simple zeroes for simplicity. According to the discussion above we have to distinguish two kinds of fixed points.

If the slope of $\beta(g_R)$ at a zero is positive the renormalized coupling is driven towards the fixed point when we approach the continuum limit. These fixed points are called *infrared (IR) fixed points* since they are attraction points for g_R, which is defined in terms of vertex functions at small momenta, i.e. in the infrared limit.

If on the other hand the slope is negative, the renormalized coupling g_R would be driven away from the fixed point. Such points are called *ultraviolet (UV) fixed points*. In fig. 1.4 a model β-function is shown with three fixed points, an ultraviolet one at g_2 and two infrared ones at g_1 and g_3. The arrows on the curve indicate in which direction g_R moves for decreasing $m_R a$. Depending on where one starts the renormalized coupling is driven to either g_1 or g_3. The regions to the left and to the right of g_2 respectively are the corresponding domains of attraction.

Now let us consider the equation for the bare coupling g_0:

$$\left. \frac{\partial g_0}{\partial \ln m_R a} \right|_{g_R} = -\beta_{\text{LAT}}(g_0, m_R a) . \tag{1.288}$$

It applies to movements in the space of parameters along lines of constant g_R. Owing to the minus sign the situation is opposite to that considered before. The zeroes of β_{LAT} with a positive slope, i.e. the infrared fixed

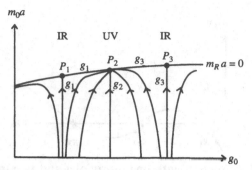

Fig. 1.5.   A model phase diagram of scalar field theory in the plane of bare parameters $m_0 a$ and g_0. On the critical line $m_R a = 0$ one ultraviolet and two infrared fixed points are drawn. In the region below the critical line some curves of constant g_R are displayed. The arrows on these curves indicate the flow for decreasing $m_R a$. The lines labelled g_1, g_2 and g_3 correspond to the fixed point values of the coupling.

points, repel the bare coupling as we move towards the continuum limit. On the other hand, the ultraviolet fixed points of β_{LAT} attract g_0. This explains their name since the bare coupling is a quantity defined at the cut-off scale, i.e. in the UV limit.

If we combine the information about the behaviour of g_R and g_0, the following picture emerges. Again we take the example of three fixed points, which was used above. In fig. 1.5 the critical line $m_R a = 0$ as well as curves of constant g_R are drawn in the space of bare parameters. For lucidity only the region below the critical line is considered. In the region above it the state of affairs would be similar but upside down.

On the critical line are the infrared fixed points P_1 and P_3 and the ultraviolet fixed point P_2. The bare coupling moves towards P_2 for fixed g_R which means that the curves of constant g_R converge to P_2 in the domain of attraction. Therefore, approaching the UV fixed point in a suitable way, continuum limits with different values of g_R would be possible. The allowed values of g_R, however, are bounded by the fixed point values of g_R at P_1 and P_3:

$$g_1 < g_R < g_3 . \tag{1.289}$$

For the particular values $g_R = g_1$ and $g_R = g_3$ the lines run into the IR fixed points P_1 and P_3 respectively.

If the critical line is approached at some point different from the fixed points, between P_1 and P_2 say, the renormalized coupling approaches the value g_1, as follows from our discussion of $\beta(g_R)$ above and from the figure. This fact is a reflection of universality, which was addressed in the section about the Kadanoff–Wilson renormalization group. The solution of the Callan–Symanzik equations, which we did not consider here, reveals that the infrared fixed points not only fix the renormalized coupling but also the long-distance behaviour of correlation functions, which is therefore universal in the domain of attraction of the fixed point.

To summarize: UV fixed points yield the possibility of continuum limits with a variety of renormalized couplings within certain bounds. IR fixed points determine these bounds on g_R. Furthermore, defining a continuum limit away from an UV fixed point, the value of the renormalized coupling approaches an IR fixed point.

The perturbative expansion of the β-functions of ϕ^4-theory (1.272) shows that the origin $g_R = 0$ or $g_0 = 0$ respectively is an infrared fixed point. For small g_R the solution (1.287) reads

$$m_R a = C \exp\left(-\frac{1}{\beta_1 g_R}\right)(\beta_1 g_R)^{-\beta_2/\beta_1^2}\{1 + \mathcal{O}(g_R)\}, \qquad (1.290)$$

or

$$g_R = \frac{1}{\beta_1 \ln\left(C/m_R a\right)} + \dots . \qquad (1.291)$$

In the domain of this fixed point any continuum limit would have $g_R = 0$, and would thus lead to a non-interacting field theory. This fixed point is called a *Gaussian fixed point* or *trivial fixed point*. The possibility of a non-trivial continuum limit would require the existence of a UV fixed point. This question will be addressed in the next chapter.

1.8 Thermodynamics of quantum fields

1.8.1 *Field theory at finite physical temperature*

We have seen that in the infinite-volume limit the functional integral yields the vacuum expectation values of observables. This corresponds to a situation where the physical temperature of the system is zero. So far a non-zero physical temperature has not been introduced.

On the other hand, it has been pointed out that Euclidean quantum field theory in the functional integral formalism, in particular in its lattice regularized version, is formally identical to a system of classical statistical mechanics in four dimensions. This is, however, only a mathematical correspondence and has to be distinguished strictly from a quantum system at finite physical temperature.

For various applications, e.g. the physics of hot hadronic matter or of the early universe, one would like to study quantum field theory at finite physical temperatures T_{ph} also. Let H be the Hamiltonian, k the

Boltzmann constant, and

$$\beta = \frac{1}{kT_{ph}}. \tag{1.292}$$

(This should not be confused with the betas which appear in other sections of this book.) Then in quantum statistical mechanics the partition function of the finite-temperature system is defined by

$$Z(\beta) = \text{Tr}\, e^{-\beta H}, \tag{1.293}$$

and the thermal expectation value of any observable A is given by

$$\langle A \rangle = \frac{1}{Z(\beta)}\, \text{Tr}\,(e^{-\beta H} A). \tag{1.294}$$

Comparing these expressions with the ones considered in sections 1.2.2 and 1.5, we recognize that they can be represented in terms of functional integrals over a finite Euclidean time extent β. In particular

$$Z(\beta) = \int D[\phi]\, e^{-S} \tag{1.295}$$

with

$$S = \int_0^\beta \mathrm{d}x_4 \int \mathrm{d}^3x\, \mathscr{L}, \tag{1.296}$$

where $\mathscr{L}$ is the Lagrangian, and the fields obey periodic boundary conditions in the fourth direction:

$$\phi(x + \beta\hat{4}) = \phi(x). \tag{1.297}$$

This remarkable correspondence between field theory at finite temperature and field theory in a finite Euclidean time interval has been known for a long time [1.58].

Introducing the lattice regularization now, we have to consider a lattice which extends over a finite number $L_t \equiv L_4$ of lattice spacings in the fourth direction, but is infinite in the other three directions: $L_1 = L_2 = L_3 \equiv L_s = \infty$. The fields have to obey periodic boundary conditions as specified above. The inverse temperature β is related to L_t by

$$\beta = L_t a, \tag{1.298}$$

where a is the lattice spacing. In numerical simulations the spatial extensions L_s would of course also be finite, but in order to approximate the thermodynamic limit one should respect the relation

$$L_s \gg L_t. \tag{1.299}$$

In the limit $L_t \to \infty$ we recover the zero-temperature field theory considered before.

In order to obtain the temperature in physical units one has to calculate some dimensionful quantity like a mass m. The lattice calculation would yield the pure number am. Then from the product

$$amL_t = \frac{m}{kT_{ph}} \qquad (1.300)$$

the physical temperature is known in units of m.

1.8.2 Anisotropic lattice regularization

If one wants to study lattice field theory at various temperatures, but at fixed lattice spacing, one has to change L_t. In this way only discrete changes of the temperature are possible. There is, however, a possibility of circumventing this restriction. It is achieved through the introduction of an anisotropic lattice regularization.

For definiteness let us consider scalar field theory again, although the same idea can be applied to any other field theory. In lattice notations consider the action

$$S = \sum_x \left\{ -2\kappa_s \sum_{l=1}^{3} \phi(x)\phi(x + a_s\hat{l}) - 2\kappa_t \phi(x)\phi(x + a_t\hat{4}) \right.$$

$$\left. + \phi(x)^2 + \lambda[\phi(x)^2 - 1]^2 - \lambda \right\}, \qquad (1.301)$$

in which we have introduced separate lattice spacings in the spatial directions (a_s) and in the temporal direction (a_t), and where the hopping parameters for space-like and time-like links

$$\kappa_s = \frac{1}{\gamma}\kappa, \qquad \kappa_t = \gamma\kappa \qquad (1.302)$$

are different from each other. γ is the *coupling-anisotropy parameter*. The action is no longer symmetric under interchanges of the time axis with any of the space axes. Therefore the correlation functions will also be anisotropic with respect to the lattice axes. In particular, the correlation length in lattice units in the time direction, ξ_t, will be different from the space-like one, ξ_s.

In view of the field theoretic context we may reinterpret this fact in another way, where we consider the correlation lengths in physical units to be equal, but where the lattice spacing for the time direction, a_t, and the lattice spacing for spatial directions, a_s, differ from each other. This means that

$$a_s\xi_s \equiv a_t\xi_t . \qquad (1.303)$$

This introduces another *correlation-anisotropy parameter*

$$\xi = \frac{a_s}{a_t},$$ (1.304)

which will depend on γ.

By varying γ it is now possible to change a_t, while keeping a_s fixed. Consequently the temperature

$$kT_{ph} = \frac{1}{L_t a_t}$$ (1.305)

can be changed continuously at fixed a_s.

Let us transform the lattice action into the continuum notation. The lattice derivative in the fourth direction is defined by

$$\Delta_4^f f(x) = \frac{1}{a_t} \left(f(x + a_t \hat{4}) - f(x) \right).$$ (1.306)

Introducing the bare field

$$\phi_0(x) = \left(\frac{2\kappa_s}{a_s a_t} \right)^{1/2} \phi(x)$$ (1.307)

one obtains

$$S = \frac{1}{2} \sum_x a_s^3 a_t \left\{ \sum_{i=1}^{3} \Delta_i^f \phi_0(x) \, \Delta_i^f \phi_0(x) + \frac{\gamma^2}{\xi^2} \Delta_4^f \phi_0(x) \, \Delta_4^f \phi_0(x) \right.$$

$$\left. + m_0^2 \, \phi_0(x)^2 + \frac{g_0}{12} \, \phi_0(x)^4 \right\}.$$ (1.308)

The bare mass m_0 and bare coupling g_0 are given by

$$a_s^2 m_0^2 = \frac{1 - 2\lambda}{\kappa} \gamma - 6 - 2\gamma^2, \qquad g_0 = \frac{6\lambda\gamma^2}{\kappa^2 \xi}.$$ (1.309)

Most remarkably, the time-component of the kinetic term has an additional factor γ^2/ξ^2, which generally is not equal to 1. This factor is sometimes called *renormalization of the speed of light*.

From the action we can read off the bare propagator, whose inverse in momentum space is

$$\tilde{\Delta}(p)^{-1} = m_0^2 + \sum_{i=1}^{3} \hat{p}_i^2 + \frac{\gamma^2}{\xi^2} \hat{p}_4^2,$$ (1.310)

where

$$\hat{p}_i = \frac{2}{a_s} \sin \frac{a_s p_i}{2}, \qquad \hat{p}_4 = \frac{2}{a_t} \sin \frac{a_t p_4}{2}.$$ (1.311)

If we define the correlation lengths in terms of the second moments of the propagator, we get from the bare propagator

$$a_s \xi_s = \frac{1}{m_0}, \qquad a_t \xi_t = \frac{\gamma}{m_0 \xi}.$$
(1.312)

Since these quantities are by definition equal, we obtain

$$\xi = \gamma$$
(1.313)

for the free field theory. For the interacting theory this equality only holds in the lowest order of perturbation theory. The corrections to ξ/γ can be calculated in higher orders of perturbation theory, but must be determined non-perturbatively in general.

2

Scalar fields

2.1 ϕ^4 model on the lattice

The ϕ^4 theory is one of the simplest field theories with interaction terms. Therefore we have used it in the introduction to illustrate various concepts of field theory. But ϕ^4 theory is not only useful for pedagogical purposes. It is also an important part of the standard model of electromagnetic and weak interactions (see e.g. [A4]). The Higgs particle is described by a four-component scalar field interacting with other fields and with itself. Switching off the other interactions yields the four-component ϕ^4 theory.

In this chapter, however, we shall stick to the scalar field theory with one real component only in order to pursue the simplest case in some more detail. This is sufficient for a discussion of the most important methods and properties.

In general we consider a lattice which extends over $L = L_1 = L_2 = L_3$ lattice spacings in the spatial directions and $T = L_4$ lattice spacings in the temporal direction:

$$\Lambda = \{x = (\mathbf{x}, x_4) \in a\mathbf{Z}^4 \mid 0 \leq x_\mu \leq a(L_\mu - 1)\}. \tag{2.1}$$

On the fields periodic boundary conditions are imposed:

$$\phi(x + a\hat{\mu}L_\mu) = \phi(x). \tag{2.2}$$

The number of lattice points is denoted by

$$\Omega \equiv V \cdot T = L^3 \cdot T = L_1 L_2 L_3 L_4. \tag{2.3}$$

The Fourier transform of a function $f(x)$ is defined by

$$\tilde{f}(p) \equiv \sum_x a^4 \, e^{-ip \cdot x} f(x), \tag{2.4}$$

54

where the allowed lattice momenta are given by

$$p_\mu = \frac{2\pi}{aL_\mu} n_\mu, \qquad n_\mu = 0, 1, \ldots, L_\mu - 1. \tag{2.5}$$

These momentum values form the Brillouin zone $\mathscr{B}$. The inverse transformation is

$$f(x) = \frac{1}{a^4\Omega} \sum_p e^{ip\cdot x} \tilde{f}(p). \tag{2.6}$$

We denote these lattice summations by

$$\int_p \equiv \frac{1}{a^4\Omega} \sum_{p\in\mathscr{B}}. \tag{2.7}$$

In the limit $L_1, L_2, L_3, L_4 \to \infty$ they go over into the integral

$$\int_p = \frac{1}{(2\pi)^4} \int_{-\pi/a}^{\pi/a} d^4p. \tag{2.8}$$

2.1.1 Green's functions

From now on we set $a = 1$ in order to simplify the notation. The action $S[\phi]$ of Euclidean ϕ^4 theory on a hypercubic lattice in four dimensions has been defined in (1.219),(1.220). In the lattice notation it is

$$S = \sum_x \left\{ -2\kappa \sum_{\mu=1}^4 \phi(x)\phi(x+\hat{\mu}) + \phi(x)^2 + \lambda[\phi(x)^2 - 1]^2 - \lambda \right\}. \tag{2.9}$$

The Green functions

$$G(x_1, \ldots, x_n) = \langle \phi(x_1) \cdots \phi(x_n) \rangle \tag{2.10}$$

are generated by the functional

$$Z[J] = \frac{1}{Z} \int \prod_x d\phi(x) \, e^{-S[\phi]+(J,\phi)} \tag{2.11}$$

with

$$Z = \int \prod_x d\phi(x) \, e^{-S[\phi]}. \tag{2.12}$$

The generating functional of connected Green's functions $G_c(x_1, \ldots, x_n)$ is given by

$$W[J] = \log Z[J]. \tag{2.13}$$

Taking into account the translation invariance of the Green functions, their Fourier transforms are defined by

$$\tilde{G}(p_1, \ldots, p_n) \equiv \sum_{x_1, \ldots, x_{n-1}} e^{-i(p_1\cdot x_1 + \cdots + p_{n-1}\cdot x_{n-1})} G(x_1, \ldots, x_{n-1}, 0), \tag{2.14}$$

where

$$p_n \equiv -(p_1 + \ldots + p_{n-1}). \tag{2.15}$$

The inverse transformation is

$$G(x_1, \ldots, x_n) = \int_{p_1, \ldots, p_{n-1}} e^{i(p_1 \cdot x_1 + \cdots + p_n \cdot x_n)} \, \widetilde{G}(p_1, \ldots, p_n). \tag{2.16}$$

Corresponding relations hold for the connected Green functions and vertex functions. For the propagator the particular notation

$$\widetilde{G}(p) \equiv \widetilde{G}_c(p, -p) \tag{2.17}$$

is also used.

Sometimes it is useful to consider the partial Fourier transform

$$C(t, \mathbf{p}) \equiv \sum_{\mathbf{x}} e^{-i\mathbf{p} \cdot \mathbf{x}} \, G_c((\mathbf{x}, t), 0) = \int \frac{dp_4}{2\pi} \, e^{ip_4 t} \, \widetilde{G}_c(\mathbf{p}, p_4). \tag{2.18}$$

It is related to the correlation function of *timeslices*

$$S(t, \mathbf{p}) \equiv \frac{1}{L^{3/2}} \sum_{\mathbf{x}} e^{-i\mathbf{p} \cdot \mathbf{x}} \, \phi(\mathbf{x}, t). \tag{2.19}$$

Namely we have

$$\langle S(t_1, \mathbf{p}) \, S(t_2, -\mathbf{p}) \rangle_c = C(t_1 - t_2, \mathbf{p}). \tag{2.20}$$

Let us consider $L = T = \infty$ in the following. The behaviour of $C(t, \mathbf{p})$ for large t is intimately connected to the complex singularities of the propagator in momentum space. Consider $\widetilde{G}(\mathbf{p}, p_4)$ as a function of p_4 for fixed spatial momentum $\mathbf{p}$. Let us assume that its singularities closest to the origin are simple poles at

$$p_4 = \pm iE(\mathbf{p}). \tag{2.21}$$

This is what is found in perturbation theory, and corresponds to the presence of a particle with energy–momentum relation $E(\mathbf{p})$. Let us concentrate on $\mathbf{p} = 0$ and define the *physical mass* by

$$m \equiv E(\mathbf{0}). \tag{2.22}$$

If we denote the residuum of the pole by $Z_3/4m\kappa$, the inverse propagator behaves near the poles as

$$[2\kappa\widetilde{G}(\mathbf{0}, p_4)]^{-1} = \frac{1}{Z_3}(p_4^2 + m^2) + \mathcal{O}\left((p_4^2 + m^2)^2\right). \tag{2.23}$$

The additional factor of 2κ, which may appear unusual here, results from the fact that the normalization of the lattice field ϕ differs from the one in (1.183) or (1.219) by a factor $(2\kappa)^{-1/2}$. Application of the residuum theorem yields

$$C(t, \mathbf{0}) = \frac{Z_3}{4m\kappa} e^{-m|t|} + \dots, \tag{2.24}$$

where the dots indicate terms which fall off faster at large t. Thus the asymptotic exponential decay of $C(t, \mathbf{0})$ is governed by the physical mass.

Let us furthermore assume that for small momenta the energy E behaves as

$$E(\mathbf{p}) = m + \frac{\mathbf{p}^2}{2m_*} + \mathcal{O}(\mathbf{p}^4). \tag{2.25}$$

The coefficient m_* is called *kinetic mass*. In order to estimate the real space propagator along the time axis,

$$G_c((\mathbf{0}, t), 0) = \int_{-\pi}^{\pi} \frac{d^3 p}{(2\pi)^3} \, C(t, \mathbf{p}), \tag{2.26}$$

for large t, we may apply the saddle-point method. Using (2.24) and (2.25) the Gaussian integration yields

$$G_c((\mathbf{0}, t), 0) = \frac{Z_3}{4m\kappa} \left(\frac{m_*}{2\pi t} \right)^{3/2} e^{-m|t|} + \dots, \tag{2.27}$$

where the dots indicate terms which are subdominant for large t. Again the exponential decay is determined by the physical mass, but an additional power of t is present, which is inconvenient in numerical investigations.

In other directions off axis the propagator will decay exponentially too, but the coefficient in front of $|t|$ will in general differ from m, manifesting the breaking of rotational invariance. The coefficient is known to be not smaller than m [2.1].

Now we turn to the *vertex functions*. The n-point vertex function $\Gamma^{(n)}(x_1, \dots, x_n)$ for $n \geq 2$ is equal to the amputated, one-particle-irreducible part of the connected n-point Green function $G_c(x_1, \dots, x_n)$. Its generating functional

$$\Gamma[\Phi] = \sum_{n=2}^{\infty} \frac{1}{n!} \sum_{x_1, \dots, x_n} \Phi(x_1) \cdots \Phi(x_n) \, \Gamma^{(n)}(x_1, \dots, x_n) \tag{2.28}$$

is related to $W[J]$ through a Legendre transformation:

$$\Gamma[\Phi] = \left\{ W[J] - \sum_x \Phi(x) J(x) \right\} \Bigg|_{J = J[\Phi]}, \tag{2.29}$$

where

$$\Phi(x) = \frac{\delta W}{\delta J(x)} \tag{2.30}$$

is the expectation value of the scalar field in the presence of the external source $J(x)$, and $J(x)$ is supposed to be expressed in terms of $\Phi(x)$ by solving (2.30). We denote vertex functions in momentum space by $\Gamma^{(n)}(p_1, \ldots, p_n)$, omitting the tilde, whenever there is no danger of confusion. The vertex functions for $n = 0$ and 1 vanish. For the two-point vertex function one finds that it is equal to the negative inverse propagator:

$$\Gamma^{(2)}(p, -p) = -\tilde{G}(p)^{-1}. \tag{2.31}$$

The three-point vertex function is given by

$$\Gamma^{(3)}(p_1, p_2, p_3) = \tilde{G}_c(p_1, p_2, p_3)\tilde{G}(p_1)^{-1}\tilde{G}(p_2)^{-1}\tilde{G}(p_3)^{-1}. \tag{2.32}$$

As a final example we write down the explicit expression for $\Gamma^{(4)}$ for later purposes:

$$\Gamma^{(4)}(p_1, p_2, p_3, p_4) = \tilde{G}(p_1)^{-1}\tilde{G}(p_2)^{-1}\tilde{G}(p_3)^{-1}\tilde{G}(p_4)^{-1}$$

$$\times \Big\{ \tilde{G}_c(p_1, p_2, p_3, p_4)$$

$$-\tilde{G}_c(p_1, p_2, -p_1 - p_2)\, \tilde{G}_c(p_3, p_4, -p_3 - p_4)\, \tilde{G}(p_1 + p_2)^{-1}$$

$$-\tilde{G}_c(p_1, p_3, -p_1 - p_3)\, \tilde{G}_c(p_2, p_4, -p_2 - p_4)\, \tilde{G}(p_1 + p_3)^{-1}$$

$$-\tilde{G}_c(p_1, p_4, -p_1 - p_4)\, \tilde{G}_c(p_2, p_3, -p_2 - p_3)\, \tilde{G}(p_1 + p_4)^{-1} \Big\}. \tag{2.33}$$

2.1.2 Particle states

In the introductory chapter the transfer matrix $\mathbf{T}$ and the Hilbert space of scalar lattice field theory have been introduced. Transformations of the field, which leave the action invariant, commute with the transfer matrix and are symmetries of the theory. Of particular importance is the Z_2-symmetry $\phi(x) \rightarrow -\phi(x)$. For values of κ below the critical κ_c Green's functions respect this symmetry. This is called the *symmetric phase*. In the symmetric phase Green's functions with an odd number of arguments vanish, and the Hamiltonian has a unique ground state $|0\rangle$ which is symmetric with respect to the Z_2-symmetry. By adding a suitable constant its energy is defined to be zero.

For $\kappa > \kappa_c$ the Z_2-symmetry is broken spontaneously, if the volume is infinite, and the field accquires a non-zero vacuum expectation value:

$$\langle\phi(x)\rangle = \pm v, \qquad v > 0. \tag{2.34}$$

In the phase with broken symmetry Green's functions or vertex functions $\Gamma^{(n)}$ with an odd number n of fields do not vanish in general. The spectrum of the transfer matrix is doubly degenerate in this phase. There are two ground states $|0_\pm\rangle$ with

$$\langle 0_+|\phi(x)|0_+\rangle = v, \qquad \langle 0_-|\phi(x)|0_-\rangle = -v. \qquad (2.35)$$

They yield two sectors of the system such that matrix elements of local operators between different sectors vanish. The reflection $\phi(x) \to -\phi(x)$ transforms the sectors into each other. We choose the ground state $|0_+\rangle$ to be the vacuum vector of the Hilbert space.

The transfer matrix $\mathbf{T}$ generates translations in the Euclidean time direction by an amount of one lattice spacing. There is a corresponding operator, acting on the wave functions $\Psi(\Phi)$, which generates translations along the spatial directions. Let $\mathbf{b} = (b_1, b_2, b_3)$ be a spatial lattice vector. Translating the time-zero field $\Phi = \{\phi(\mathbf{y})\}$ by

$$\phi(\mathbf{y}) \to \phi(\mathbf{y} - \mathbf{b}) \qquad (2.36)$$

induces a transformation of wave functions by means of

$$\Psi'[\phi(\mathbf{y})] \equiv \Psi[\phi(\mathbf{y} - \mathbf{b})]. \qquad (2.37)$$

This defines a unitary operator $\exp i\mathbf{b} \cdot \mathbf{P}$ through

$$\Psi' \equiv e^{i\mathbf{b}\cdot\mathbf{P}}\Psi. \qquad (2.38)$$

The operator $\mathbf{P}$ is the *momentum operator*. Its components are only defined modulo 2π on the lattice. (Remember $a = 1$.)

For the multiplication operator $\hat{\phi}(\mathbf{x})$, introduced in section 1.5.2, the spatial translation is defined by

$$\hat{\phi}'(\mathbf{x}) = \hat{\phi}(\mathbf{x} - \mathbf{b}). \qquad (2.39)$$

It satisfies

$$\hat{\phi}'(\mathbf{x})\Psi' = \left(\hat{\phi}(\mathbf{x})\Psi\right)' \qquad (2.40)$$

and consequently

$$\hat{\phi}'(\mathbf{x}) = e^{i\mathbf{b}\cdot\mathbf{P}}\,\hat{\phi}(\mathbf{x})\,e^{-i\mathbf{b}\cdot\mathbf{P}}. \qquad (2.41)$$

Let us consider eigenstates of $\mathbf{P}$. First of all the vacuum obeys

$$\mathbf{P}|0\rangle = \mathbf{0}. \qquad (2.42)$$

More interesting momentum eigenstates are constructed as follows. Define

$$|\mathbf{k}\rangle \equiv \frac{1}{L^{3/2}} \sum_{\mathbf{x}} e^{i\mathbf{k}\cdot\mathbf{x}} \left(\hat{\phi}(\mathbf{x}) - \langle\phi(x)\rangle\right) |0\rangle. \qquad (2.43)$$

Using (2.41) one finds

$$e^{i\mathbf{b}\cdot\mathbf{P}}|\mathbf{k}\rangle = e^{i\mathbf{b}\cdot\mathbf{k}}|\mathbf{k}\rangle, \qquad (2.44)$$

which means that $|\mathbf{k}\rangle$ is a state with momentum $\mathbf{k}$:

$$\mathbf{P}|\mathbf{k}\rangle = \mathbf{k}|\mathbf{k}\rangle. \tag{2.45}$$

The transfer matrix commutes with $\mathbf{P}$. Therefore the Hamiltonian can be diagonalized simultaneously with the momentum operator. Denote the resulting eigenvectors by $|\mathbf{k}, \alpha\rangle$, such that

$$\mathbf{P}|\mathbf{k}, \alpha\rangle = \mathbf{k}|\mathbf{k}, \alpha\rangle, \tag{2.46}$$

$$H|\mathbf{k}, \alpha\rangle = E(\mathbf{k}, \alpha)|\mathbf{k}, \alpha\rangle, \tag{2.47}$$

where α labels the states of different energies. Usually the states of lowest energy, say $E(\mathbf{k}, 0)$, will be one-particle states.

The momentum eigenstates $|\mathbf{k}\rangle$ are decomposed as

$$|\mathbf{k}\rangle = \sum_\alpha c_\alpha |\mathbf{k}, \alpha\rangle. \tag{2.48}$$

Then we get

$$C(t, \mathbf{k}) = \langle S(t, \mathbf{k}) \, S(0, -\mathbf{k})\rangle = \langle \mathbf{k}| \, \mathrm{e}^{-tH} \, |\mathbf{k}\rangle = \sum_\alpha |c_\alpha|^2 \, \mathrm{e}^{-tE(\mathbf{k}, \alpha)}. \tag{2.49}$$

For large t the lowest energy $E(\mathbf{k}, 0)$ dominates this timeslice correlation function. Comparing with (2.24) one recognizes

$$E(\mathbf{k}) = E(\mathbf{k}, 0) \tag{2.50}$$

to be identical with the energy of a one-particle state. This justifies the denotation for the physical mass

$$m = E(\mathbf{0}, 0). \tag{2.51}$$

2.1.3 Renormalized quantities

The physical mass m and the factor Z_3 are given by a pole of the propagator and its residuum. Using Z_3 as a wave function renormalization factor one could define a renormalized field and corresponding renormalized Green functions.

For practical purposes, however, it is much more convenient to work with the wave function renormalization factor Z_R and the renormalized mass m_R, which have been defined in subsection 1.6.2 in terms of the behaviour of the propagator at small momenta:

$$\widetilde{G}(p)^{-1} = (2\kappa/Z_R)\{m_R^2 + p^2 + \mathcal{O}(p^4)\}. \tag{2.52}$$

The renormalized Green and vertex functions are related to the unrenormalized ones by

$$G_R(x_1, \ldots, x_n) \equiv (Z_R/2\kappa)^{-n/2} \, G_c(x_1, \ldots, x_n), \tag{2.53}$$

$$\Gamma_R^{(n)} = (Z_R/2\kappa)^{n/2} \Gamma^{(n)}. \tag{2.54}$$

The *renormalized field expectation value* in the phase with broken symmetry is defined by

$$v_R \equiv (Z_R/2\kappa)^{-1/2} v. \tag{2.55}$$

The renormalized mass m_R, the wave function renormalization factor Z_R, and the renormalized coupling

$$g_R = -\Gamma_R^{(4)}(0,0,0,0) \tag{2.56}$$

can be expressed in terms of quantities which are primarily obtained in numerical simulations or series expansions. These are the *susceptibilities*

$$\chi_n \equiv \sum_{x_1,\dots,x_{n-1}} G_c(x_1,\dots,x_{n-1},0) \tag{2.57}$$

and the second moment

$$\mu_2 \equiv \sum_x x^2 \, G_c(x,0). \tag{2.58}$$

Equally well they may be defined in terms of timeslice correlations:

$$\chi_n = L^{3n/2} \sum_{t_1,\dots,t_{n-1}} \langle S(t_1,\mathbf{0}) \cdots S(t_{n-1},\mathbf{0}) \, S(0,\mathbf{0}) \rangle_c, \tag{2.59}$$

$$\mu_2 = 4L^3 \sum_t t^2 \, \langle S(t,\mathbf{0}) \, S(0,\mathbf{0}) \rangle_c. \tag{2.60}$$

If the symmetry

$$\phi(x) \rightarrow -\phi(x) \tag{2.61}$$

is unbroken, the odd susceptibilities χ_{2n+1} vanish.

For $n = 2$ we have

$$\chi_2 = \tilde{G}(0) = \frac{Z_R}{2\kappa m_R^2}, \tag{2.62}$$

$$\mu_2 = -\frac{\partial}{\partial p_\nu} \frac{\partial}{\partial p_\nu} \tilde{G}(p) \Big|_{p=0} = \frac{8 Z_R}{2\kappa m_R^4}, \tag{2.63}$$

which yields

$$m_R^2 = \frac{8\chi_2}{\mu_2}, \tag{2.64}$$

$$Z_R = 2\kappa \frac{8\chi_2^2}{\mu_2}. \tag{2.65}$$

Next we get for $n = 3, 4$

$$\chi_3 = \tilde{G}(0,0,0) = \left(\frac{Z_R}{2\kappa}\right)^{3/2} m_R^{-6}\, \Gamma_R^{(3)}(0,0,0)\,, \qquad (2.66)$$

$$\chi_4 = \tilde{G}(0,0,0,0) = \left(\frac{Z_R}{2\kappa}\right)^2 m_R^{-8}\left\{\Gamma_R^{(4)}(0,0,0,0) + \frac{3}{m_R^2}\, \Gamma_R^{(3)}(0,0,0)^2\right\}\,,$$
$$(2.67)$$

from which follows

$$g_R = -\frac{64}{\mu_2^2}\left\{\chi_4 - 3\frac{\chi_3^2}{\chi_2}\right\}\,. \qquad (2.68)$$

These formulae are employed in the analysis of results from numerical simulations or from high temperature series expansions.

2.2   Perturbation theory

The cut-off which is provided by a space–time lattice is not particularly convenient for perturbative calculations. The main purpose of the lattice is to provide us with a regularization which allows the application of various non-perturbative methods.

Nevertheless, it is sometimes necessary to perform perturbative calculations with a lattice cut-off. In particular this is so, if quantities evaluated by non-perturbative methods are to be related to quantities calculated perturbatively. Furthermore some quantities, like finite-volume effects, which are of interest in numerical simulations, can be estimated in lattice perturbation theory.

In the framework of lattice gauge theory, lattice perturbation theory has been applied to various questions. Some of these will be dealt with in later chapters. For pointing out characteristic features of perturbation theory on a lattice, however, it is best to consider ϕ^4 theory, as will be done in this section.

2.2.1   Free field theory

Before we discuss perturbative corrections let us return to the free scalar field theory for a moment. In the first chapter the propagator in momentum space has been found to be (see (1.231))

$$\tilde{G}(p) = (2\kappa)^{-1}\tilde{\Delta}(p)\,, \qquad (2.69)$$

where

$$\tilde{\Delta}(p) \equiv (m_0^2 + \hat{p}^2)^{-1} \qquad (2.70)$$

and

$$\hat{p}_\mu = 2 \sin \frac{p_\mu}{2}. \tag{2.71}$$

The poles of the propagator, which fix the physical mass, are therefore given by

$$\hat{p}_4 = \pm i (m_0^2 + \hat{\mathbf{p}}^2)^{1/2}. \tag{2.72}$$

This implies

$$p_4 = \pm i \omega_0(\mathbf{p}), \tag{2.73}$$

where $\omega_0(\mathbf{p})$ satisfies

$$2 \sinh \frac{\omega_0(\mathbf{p})}{2} = (m_0^2 + \hat{\mathbf{p}}^2)^{1/2} \tag{2.74}$$

or, equivalently,

$$\cosh \omega_0(\mathbf{p}) = 1 + \frac{1}{2}(m_0^2 + \hat{\mathbf{p}}^2). \tag{2.75}$$

Therefore we obtain the energy–momentum relation

$$\omega_0(\mathbf{p}) = \text{Ar} \cosh \left(1 + \frac{1}{2}(m_0^2 + \hat{\mathbf{p}}^2) \right)$$

$$= 2 \log \left\{ \left(1 + \frac{1}{4}(m_0^2 + \hat{\mathbf{p}}^2) \right)^{1/2} + \frac{1}{2}(m_0^2 + \hat{\mathbf{p}}^2)^{1/2} \right\}. \tag{2.76}$$

For small $\mathbf{p}$ this is expanded as

$$\omega_0(\mathbf{p}) = \overline{m}_0 + \frac{1}{2 m_{0*}} \mathbf{p}^2 + \cdots \tag{2.77}$$

with the physical mass

$$m = \overline{m}_0 \equiv 2 \log \left((1 + m_0^2/4)^{1/2} + \frac{m_0}{2} \right) \tag{2.78}$$

and the kinetic mass

$$m_{0*} = m_0 (1 + m_0^2/4)^{1/2} = \sinh \overline{m}_0. \tag{2.79}$$

The renormalized mass m_R, on the other hand, coincides with the bare mass m_0 in the case of free field theory.

Calculating the residuum of the pole, the factor Z_3 is obtained to be

$$Z_3 = \frac{\overline{m}_0}{\sinh \overline{m}_0}, \tag{2.80}$$

whereas the wave function renormalization factor Z_R is

$$Z_R = 1. \tag{2.81}$$

If one likes to take the continuum limit, the lattice spacing a has to be reintroduced. Then we get, for example,

$$\overline{m}_0 = \frac{2}{a} \log \left((1 + a^2 m_0^2/4)^{1/2} + \frac{am_0}{2} \right) = m_0 \{ 1 + \mathcal{O}(a^2 m_0^2) \} \qquad (2.82)$$

and

$$Z_3 = \frac{a\overline{m}_0}{\sinh a\overline{m}_0} = 1 + \mathcal{O}(a^2 m_0^2) . \qquad (2.83)$$

2.2.2 *Perturbation theory in the symmetric phase*

In order to obtain the expansion of Green's functions in powers of the bare coupling g_0, the exponentiated interaction part of the Euclidean action has to be expanded inside the functional integral. This leads to the Feynman rules, as indicated in section 1.4.5. In the case of field theory in the continuum this exercise is performed in many textbooks. On a lattice the derivation of Feynman rules proceeds in much the same way. Therefore, instead of presenting it in detail here, we restrict ourselves to emphasizing the differences in the presence of a lattice cut-off. The form of the action most convenient for the purpose of perturbation theory in the symmetric phase is (1.219):

$$S = \sum_x \left\{ \frac{1}{2} \Delta_\mu^f \phi_0(x) \, \Delta_\mu^f \phi_0(x) + \frac{m_0^2}{2} \, \phi_0(x)^2 + \frac{g_0}{4!} \, \phi_0(x)^4 \right\} , \qquad (2.84)$$

which can be obtained by a suitable field normalization according to (1.221). In momentum space the Feynman rules for a Feynman diagram with n external lines, contributing to the Green function $\tilde{G}(p_1, \ldots, p_n)$, are the following:

1. each line is associated with a propagator $\tilde{\Delta}(q)$,

2. each vertex is an end point of four lines and is associated with a factor $-g_0$,

3. at inner vertices momentum conservation holds modulo 2π,

4. loop momenta are to be integrated over the first Brillouin zone $\mathscr{B}$ with the integration measure $\int_q$,

5. finally there is an overall factor $(2\kappa)^{-n/2}$, resulting from our normalization of the lattice scalar field.

The additional symmetry factors are as in the continuum. In the case of connected Green's functions only connected diagrams contribute, as is well known.

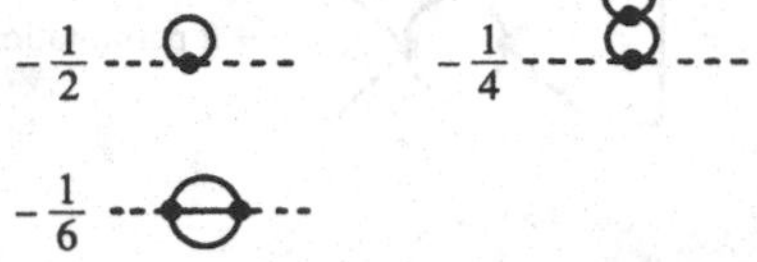

Fig. 2.1. Perturbative diagrams up to two loops contributing to the two-point vertex function in ϕ^4 theory in the symmetric phase.

If vertex functions $\Gamma^{(n)}(p_1,\ldots,p_n)$ are being considered, only connected one-particle irreducible diagrams are to be taken, and there are no propagators associated with external lines as usual. Furthermore the overall factor is $(2\kappa)^{n/2}$ in this case, as can easily be checked.

For $n = 2$ the diagrams of fig. 2.1 yield for the negative inverse propagator up to two loops

$$-\frac{1}{2\kappa}\widetilde{G}(p)^{-1} = -(\hat{p}^2 + m_0^2) - \frac{g_0}{2}J_1(m_0)$$

$$+ \frac{g_0^2}{4}J_1(m_0)J_2(m_0) + \frac{g_0^2}{6}I_3(m_0, p) + \mathcal{O}(g_0^3), \qquad (2.85)$$

where the functions

$$J_n(m_0) \equiv \int_q \widetilde{\Delta}(q)^n, \qquad (2.86)$$

$$I_3(m_0, p) \equiv \int_{q_1}\int_{q_2} \widetilde{\Delta}(q_1)\widetilde{\Delta}(q_2)\widetilde{\Delta}(p - q_1 - q_2) \qquad (2.87)$$

have been introduced.

Up to one-loop order the four-point and six-point vertex functions are given by the diagrams of fig. 2.2. The four-point function is

$$\frac{1}{(2\kappa)^2}\Gamma^{(4)}(p_1, p_2, p_3, p_4) = -g_0$$

$$+ \frac{g_0^2}{2}[I_2(m_0, p_1 + p_2) + I_2(m_0, p_1 + p_3) + I_2(m_0, p_1 + p_4)] + \mathcal{O}(g_0^3), \quad (2.88)$$

where

$$I_2(m_0, p) \equiv \int_q \widetilde{\Delta}(q)\widetilde{\Delta}(p - q). \qquad (2.89)$$

At zero momentum it simplifies to

$$\frac{1}{(2\kappa)^2}\Gamma^{(4)}(0, 0, 0, 0) = -g_0 + \frac{3}{2}g_0^2 J_2(m_0) + \mathcal{O}(g_0^3). \qquad (2.90)$$

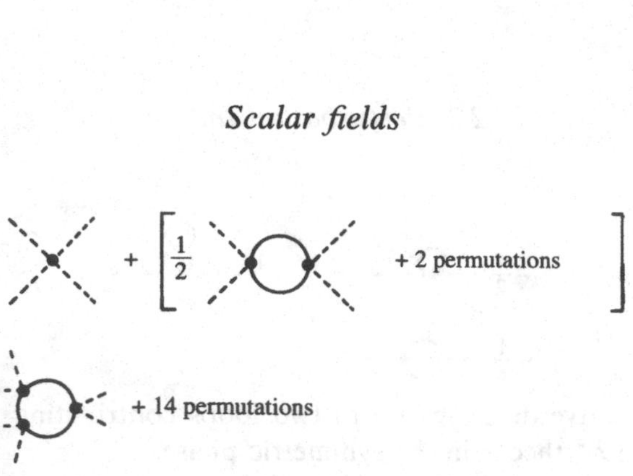

Fig. 2.2. Perturbative diagrams up to one loop contributing to the four-point and six-point vertex functions in ϕ^4 theory in the symmetric phase.

For the six-point vertex function at zero momentum one obtains

$$\frac{1}{(2\kappa)^3}\Gamma^{(6)}(0,\ldots,0) = -15g_0^3 J_3(m_0) + \mathcal{O}(g_0^4). \tag{2.91}$$

Let us consider renormalization in one-loop order. The inverse propagator differs from the bare one only by the addition of a momentum-independent term. According to the definitions of the wave function renormalization factor and the renormalized mass, see (2.52), we obtain

$$Z_R = 1 + \mathcal{O}(g_0^2), \tag{2.92}$$

$$m_R^2 = m_0^2 + \frac{g_0}{2}J_1(m_0) + \mathcal{O}(g_0^2). \tag{2.93}$$

Consequently the renormalized inverse propagator to this order is just

$$\tilde{G}_R(p)^{-1} = -\Gamma_R^{(2)}(p) = m_R^2 + \hat{p}^2. \tag{2.94}$$

From this expression the physical mass is obtained as in the case of the free propagator with the result

$$m = 2\log\left((1 + m_R^2/4)^{1/2} + \frac{m_R}{2}\right) + \mathcal{O}(g_0^2)$$

$$= \bar{m}_0 + \frac{1}{2m_{0*}}\frac{g_0}{2}J_1(m_0) + \mathcal{O}(g_0^2). \tag{2.95}$$

The renormalized four-point vertex function at zero momentum determines the renormalized coupling to be

$$g_R = -\Gamma_R^{(4)}(0,0,0,0) = g_0 - \frac{3}{2}g_0^2 J_2(m_0) + \mathcal{O}(g_0^3). \tag{2.96}$$

The relation between the bare parameters m_0, g_0 and the renormalized ones m_R, g_R involves the loop integrals J_1 and J_2. For finite lattice spacings these integrals converge. On the other hand, if one tries to take the continuum limit at fixed m_0 and g_0, the integrals diverge in the limit

$a \to 0$. In order to study these divergencies we introduce the lattice spacing again and consider

$$m_R^2 = m_0^2 + \frac{g_0}{2} \frac{1}{a^2} J_1(am_0) + \dots . \tag{2.97}$$

We need to know $J_1(am_0)$ for small am_0. At $am_0 = 0$ the integral converges:

$$J_1(0) = \int_{-\pi}^{\pi} \frac{d^4q}{(2\pi)^4} \left(4 \sum_{\mu=1}^{4} \sin^2 q_\mu/2 \right)^{-1} \equiv r_0 = 0.154\,933\,390\dots . \tag{2.98}$$

Near this point the corrections are obtained with the help of the decomposition

$$\frac{1}{m_0^2 + \hat{q}^2} = \frac{1}{\hat{q}^2} - \frac{m_0^2}{\hat{q}^2(m_0^2 + \hat{q}^2)}, \tag{2.99}$$

which leads to

$$J_1(y) = r_0 - y^2 \int_{-\pi}^{\pi} \frac{d^4q}{(2\pi)^4} \left[\hat{q}^2(y^2 + \hat{q}^2)\right]^{-1}. \tag{2.100}$$

For $y = 0$ this integral develops a logarithmic singularity near the origin. A straightforward estimation shows that

$$J_1(y) = r_0 + y^2 \left\{ \frac{1}{16\pi^2} \ln y^2 + r_1 + \mathcal{O}(y^2) \right\} \tag{2.101}$$

with

$$r_1 = -0.030\,345\,755\dots . \tag{2.102}$$

Therefore in the limit $a \to 0$ the renormalized mass contains a quadratic and a logarithmic divergence according to

$$m_R^2 = m_0^2 + \frac{g_0}{2} \frac{r_0}{a^2} + \frac{g_0}{32\pi^2} m_0^2 \ln(a^2 m_0^2) + \frac{g_0}{2} r_1 m_0^2 + \dots . \tag{2.103}$$

In lattice units, however, the equation

$$(am_R)^2 = (am_0)^2 + \frac{g_0}{2} r_0 + \frac{g_0}{32\pi^2} (am_0)^2 \ln(a^2 m_0^2) + \frac{g_0}{2} r_1 (am_0)^2 + \dots \tag{2.104}$$

does not contain any divergent terms, and just says that the in the continuum limit, $am_R \to 0$, the numerical parameter am_0 has to be tuned appropriately:

$$(am_0)^2 \to -\frac{g_0}{2} r_0 + \mathcal{O}(g_0^2). \tag{2.105}$$

Expressed in terms of the hopping parameter κ, this is equivalent to

$$\kappa \to \kappa_c \equiv \frac{1}{8} + \left(3r_0 - \frac{1}{4} \right) \lambda + \mathcal{O}(\lambda^2). \tag{2.106}$$

The behaviour of $J_2(y)$ for small y is obtained with the help of the recursion

$$J_{n+1}(y) = -\frac{1}{n}\frac{d}{d(y^2)}J_n(y),$$ (2.107)

resulting in

$$J_2(y) = -r_1 - \frac{1}{16\pi^2}(1 + \ln y^2) + \mathcal{O}(y^2).$$ (2.108)

According to (2.96) this implies that the renormalized coupling contains a logarithmic divergence:

$$g_R = g_0 + \frac{3}{32\pi^2}g_0^2 \ln(a^2 m_0^2) + \frac{3}{2}g_0^2\left(\frac{1}{16\pi^2} + r_1\right) + \dots.$$ (2.109)

The functional relation between m_0, g_0 on the one hand and m_R, g_R on the other hand can be used to express renormalized quantities in terms of the renormalized parameters m_R and g_R order by order in perturbation theory. The resulting expressions, which are power series in g_R, go under the name of *renormalized perturbation theory*. To one-loop order the substitution is done with

$$m_0^2 = m_R^2 - \frac{g_R}{2}J_1(m_R) + \mathcal{O}(g_R^2),$$ (2.110)

$$g_0 = g_R + \frac{3}{2}g_R^2 J_2(m_R) + \mathcal{O}(g_R^3).$$ (2.111)

For the renormalized propagator we obtain nothing else but (2.94). In the case of the four-point vertex function the substitution yields

$$\Gamma_R^{(4)}(p_1, p_2, p_3, p_4) = -g_R$$

$$+ \frac{1}{2}g_R^2[I_2'(m_R, p_1 + p_2) + I_2'(m_R, p_1 + p_3) + I_2'(m_R, p_1 + p_4)] + \mathcal{O}(g_R^3),$$ (2.112)

where

$$I_2'(m_R, p) \equiv I_2(m_R, p) - J_2(m_R, p) = \int_q \left(\tilde{\Delta}(q)\tilde{\Delta}(p - q) - \tilde{\Delta}(q)^2\right).$$ (2.113)

In this integral the propagators $\tilde{\Delta}$ are understood to contain the renormalized mass m_R in place of m_0, as indicated by the first argument of I_2'. This function satisfies

$$I_2'(m_R, 0) = 0$$ (2.114)

and is finite in the continuum limit.

As a further example the renormalized six-point vertex at zero momentum equals

$$\Gamma_R^{(6)}(0, \dots, 0) = -15 g_R^3 J_3(m_R) + \mathcal{O}(g_R^4)$$ (2.115)

in renormalized perturbation theory. This is finite in the continuum limit too:

$$\Gamma_R^{(6)}(0,\ldots,0) \to -\frac{15}{32\pi^2 m_R^2}g_R^3 + \mathcal{O}(g_R^4).\qquad(2.116)$$

The physical mass is given by

$$m = 2\log\left((1+m_R^2/4)^{1/2} + \frac{m_R}{2}\right) + \mathcal{O}(g_R^2).\qquad(2.117)$$

Up to one-loop order the renormalized quantities as well as the physical mass possess a finite continuum limit in renormalized perturbation theory. This means that for fixed m_R and g_R the cut-off can be removed in the perturbation series of renormalized Green's functions.

A theory is defined to be perturbatively renormalizable, if renormalized perturbation theory to all orders remains finite in the continuum limit.

Theorem: ϕ^4 theory in four dimensions is perturbatively renormalizable.

This fact is known since long for perturbation theory with various regularizations in the continuum. In the framework of lattice regularization it has been shown by Reisz [2.2].

Let us finally come to the renormalization group β-function

$$\beta(g_R, m_R) = m_R\frac{\partial g_R}{\partial m_R}\bigg|_{g_0}\qquad(2.118)$$

defined in section 1.7.2. From the perturbative expansions discussed above we obtain

$$m_0\frac{\partial g_R}{\partial m_0}\bigg|_{g_0} = -\frac{3}{2}g_0^2 m_0\frac{\partial}{\partial m_0}J_2(m_0) + \mathcal{O}(g_0^3)$$

$$= 6g_0^2 m_0^2 J_3(m_0) + \mathcal{O}(g_0^3).\qquad(2.119)$$

On the other hand

$$\frac{m_0}{m_R}\frac{\partial m_R}{\partial m_0}\bigg|_{g_0} = 1 + \mathcal{O}(g_0).\qquad(2.120)$$

Combining these equations one gets

$$\beta(g_R, m_R) = 6g_0^2 m_0^2 J_3(m_0) + \mathcal{O}(g_0^3) = 6g_R^2 m_R^2 J_3(m_R) + \mathcal{O}(g_R^3).\qquad(2.121)$$

This is the lattice β-function in the one-loop approximation. It depends on m_R (i.e. on am_R) explicitly, which is a scaling violation. In the continuum limit, however, it must be a function of g_R alone, as has been discussed earlier. In fact, using (2.107) and (2.108) we obtain

$$\beta(g_R, m_R) \to \beta(g_R) = \frac{3}{16\pi^2}g_R^2 + \mathcal{O}(g_R^3),\qquad \text{as } m_R \to 0.\qquad(2.122)$$

The two-loop term, which is universal too, has been given in section 1.7.2.

2.2.3 Perturbation theory in the phase with broken symmetry

In the phase with broken symmetry perturbation theory is based on
an expansion around a non-trivial minimum of the bare action. If the
coefficient of $\phi(x)^2$ in the bare potential is negative, the classical potential
has the shape of a double well with minima at

$$\phi(x) = \pm s_0 \,, \tag{2.123}$$

where

$$s_0^2 = \frac{1}{2\lambda}(8\kappa - 1 + 2\lambda) \,. \tag{2.124}$$

In order to expand around the minimum at s_0 we define the bare field by

$$\varphi_0(x) = (2\kappa)^{1/2}(\phi(x) - s_0) \,, \tag{2.125}$$

in contrast to the symmetric phase, where the s_0-term is absent. The bare
action can then be rewritten in the form

$$S = \sum_x \left\{ \frac{1}{2}\Delta_\mu^f \varphi_0(x)\, \Delta_\mu^f \varphi_0(x) + \frac{m_0^2}{2}\, \varphi_0(x)^2 \right.$$

$$\left. + \frac{1}{3!}(3g_0)^{1/2}\, m_0 \varphi_0(x)^3 + \frac{g_0}{4!}\, \varphi_0(x)^4 \right\} \tag{2.126}$$

where

$$m_0^2 = 16 - \frac{2}{\kappa}(1 - 2\lambda) \,, \qquad g_0 = \frac{6\lambda}{\kappa^2} \,, \tag{2.127}$$

and a constant has been added, such that the minimum of the action is
zero. Note that m_0^2, which is equal to the second derivative of the potential
at its minimum, differs from the corresponding m_0^2 in the symmetric phase.
This can be seen by writing the action in terms of

$$\phi_0(x) = (2\kappa)^{1/2}\phi(x) \,, \tag{2.128}$$

which leads to

$$S = \sum_x \left\{ \frac{1}{2}\Delta_\mu^f \phi_0(x)\, \Delta_\mu^f \phi_0(x) - \frac{m_0^2}{4}\, \phi_0(x)^2 + \frac{g_0}{4!}\, \phi_0(x)^4 + \frac{3m_0^4}{8g_0} \right\}$$

$$= \sum_x \left\{ \frac{1}{2}\Delta_\mu^f \phi_0(x)\, \Delta_\mu^f \phi_0(x) + \frac{g_0}{4!}\left(\phi_0(x)^2 - v_0^2 \right)^2 \right\} \,, \tag{2.129}$$

where

$$v_0^2 \equiv 2\kappa\, s_0^2 = \frac{3m_0^2}{g_0} \,. \tag{2.130}$$

As a consequence of the interactions the field $\varphi_0(x)$ will develop a vacuum expectation value

$$\langle \varphi_0(x) \rangle \equiv f_0 \,. \tag{2.131}$$

Correspondingly the vacuum expectation value of $\phi(x)$ is

$$\langle \phi(x) \rangle = v = \frac{1}{(2\kappa)^{1/2}} \, (f_0 + v_0) \,. \tag{2.132}$$

The wave function renormalization factor Z_R and the renormalized mass m_R are defined through the propagator in the same way as in the symmetric phase. The vacuum expectation value of the renormalized field

$$\phi_R(x) = (Z_R/2\kappa)^{-1/2} \phi(x) \tag{2.133}$$

is equal to the renormalized field expectation value

$$v_R = (Z_R/2\kappa)^{-1/2} v \,. \tag{2.134}$$

Owing to the relation

$$g_0 = 3 \, \frac{m_0^2}{v_0^2} \tag{2.135}$$

the perturbative expansion of m_R^2/v_R^2 must be of the form

$$3 \, \frac{m_R^2}{v_R^2} = g_0 + \mathcal{O}(g_0^2) \,. \tag{2.136}$$

Therefore in the phase with broken symmetry, besides a definition similar to (2.56), there is an alternative way to define a renormalized coupling g_R, which equals g_0 to lowest order in perturbation theory, namely

$$g_R \equiv 3 \, \frac{m_R^2}{v_R^2} \,. \tag{2.137}$$

This definition has been employed for example in the analysis of Lüscher and Weisz [2.3]. It is also preferably used in numerical investigations, because it is easier to obtain m_R and v_R than the four-point vertex [2.4]. In order to distinguish this new definition of the renormalized coupling from the one used in the symmetric phase, we denote the latter by

$$g_R^{(4)} = -\Gamma_R^{(4)}(0,0,0,0) \,. \tag{2.138}$$

From the action (2.126) the Feynman rules are derived as in the symmetric phase. In addition to the rules discussed there, a new vertex is present, in which three lines end. So we have one more Feynman rule:

6. The triple vertex is associated with a factor $-(3g_0)^{1/2} m_0$.

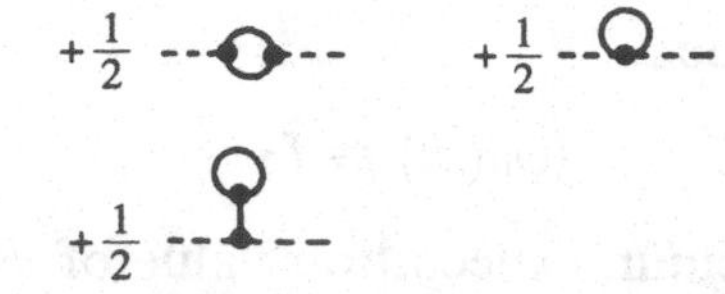

Fig. 2.3. Perturbative diagrams up to one loop contributing to the two-point vertex function in ϕ^4 theory in the phase with broken symmetry.

In the one-loop approximation the diagrams of fig. 2.3 contribute to the two-point vertex function. They give

$$-\frac{1}{2\kappa}\tilde{G}(p)^{-1} = -(\hat{p}^2 + m_0^2) + \frac{3}{2}g_0 m_0^2 I_2(m_0, p) + g_0 J_1(m_0). \qquad (2.139)$$

For small p the expansion

$$I_2(m_0, p) = J_2(m_0) + I_2''(m_0)p^2 + \mathcal{O}(p^4) \qquad (2.140)$$

with

$$I_2''(m_0) = \frac{1}{24}\left(J_2(m_0) - (m_0^2 + 8)J_3(m_0)\right) \qquad (2.141)$$

yields

$$Z_R = 1 + \frac{3}{2}g_0 m_0^2 I_2''(m_0) + \mathcal{O}(g_0^2) \qquad (2.142)$$

and

$$m_R^2 = m_0^2 - g_0\left[J_1(m_0) + \frac{3}{2}m_0^2 J_2(m_0) - \frac{3}{2}m_0^4 I_2''(m_0)\right] + \mathcal{O}(g_0^2). \qquad (2.143)$$

With the help of this relation it is possible to express m_0^2 in terms of m_R^2 and g_R in the one-loop approximation and to write down the renormalized inverse propagator:

$$\tilde{G}_R(p)^{-1} = m_R^2 + \hat{p}^2 + \frac{3}{2}g_R m_R^2[\hat{p}^2 I_2''(m_R) - I_2'(m_R, p)] + \mathcal{O}(g_R^2). \qquad (2.144)$$

Calculating its pole one obtains the following result for the physical mass:

$$m = \bar{m} - \frac{3g_R m_R}{4(1 + m_R^2/4)^{1/2}}[m_R^2 I_2''(m_R) + I_2'(m_R, q)] + \mathcal{O}(g_R^2), \qquad (2.145)$$

where

$$\bar{m} = 2\log\left((1 + m_R^2/4)^{1/2} + \frac{m_R}{2}\right), \qquad q = (i\bar{m}, \mathbf{0}). \qquad (2.146)$$

The continuum limit can be taken as discussed in the context of the symmetric phase. After evaluating the numerical integrals one finds

$$m = m_R\left\{1 - \frac{g_R}{16\pi^2}\left(\frac{11}{8} - \frac{\sqrt{3}}{4}\pi\right) + \mathcal{O}(g_R^2)\right\}. \qquad (2.147)$$

Fig. 2.4. Perturbative diagram up to one loop contributing to the vacuum expectation value of $\varphi(x)$ in ϕ^4 theory in the phase with broken symmetry.

Let us turn to the vacuum expectation value of $\varphi(x)$. At one-loop order only the tadpole graph of fig. 2.4 contributes, which leads to

$$(2\kappa)^{1/2} v = (3/g_0)^{1/2} m_0 \left(1 - \frac{g_0}{2m_0^2} J_1(m_0) + \mathcal{O}(g_0^2) \right), \qquad (2.148)$$

and

$$v_R^2 = \frac{3m_0^2}{g_0} \left(1 - \frac{g_0}{m_0^2} J_1(m_0) - \frac{3}{2} g_0 m_0^2 I_2''(m_0) + \mathcal{O}(g_0^2) \right). \qquad (2.149)$$

The definition of the renormalized coupling (2.137) in the phase with broken symmetry then gives

$$g_R = g_0 \left\{ 1 - \frac{3}{2} g_0 J_2(m_0) + 3g_0 m_0^2 I_2''(m_0) + \mathcal{O}(g_0^2) \right\}. \qquad (2.150)$$

The calculation of renormalized Green's and vertex functions proceeds in a standard fashion now. Let us just quote the following results for zero-momentum vertex functions [2.4]:

$$\Gamma_R^{(3)}(0,0,0) = -(3g_R)^{1/2} m_R [1 + 3g_R m_R^2 J_3(m_R) + \mathcal{O}(g_R^2)], \qquad (2.151)$$

$$\Gamma_R^{(4)}(0,\ldots,0) = -g_R [1 + 18g_R m_R^2 J_3(m_R) - 27g_R m_R^4 J_4(m_R) \\ + \mathcal{O}(g_R^2)],$$

$$\Gamma_R^{(5)}(0,\ldots,0) = -g_R^2 (3g_R)^{1/2} m_R [15 J_3(m_R) - 90 m_R^2 J_4(m_R) \\ + 108 m_R^4 J_5(m_R) + \mathcal{O}(g_R)],$$

$$\Gamma_R^{(6)}(0,\ldots,0) = -g_R^3 [15 J_3(m_R) - 405 m_R^2 J_4(m_R) + 1620 m_R^4 J_5(m_R) \\ - 1620 m_R^6 J_6(m_R) + \mathcal{O}(g_R)].$$

In the continuum limit these expressions go over into

$$\Gamma_R^{(3)}(0,0,0) = -(3g_R)^{1/2} m_R [1 + 3g_R/32\pi^2 + \mathcal{O}(g_R^2)],$$

$$\Gamma_R^{(4)}(0,\ldots,0) = -g_R [1 + 9g_R/32\pi^2 + \mathcal{O}(g_R^2)],$$

$$\Gamma_R^{(5)}(0,\ldots,0) = -g_R^2 (3g_R)^{1/2}/m_R [3/32\pi^2 + \mathcal{O}(g_R)],$$

$$\Gamma_R^{(6)}(0,\ldots,0) = g_R^3/m_R^2 [3/8\pi^2 + \mathcal{O}(g_R)]. \qquad (2.152)$$

With some more effort the complete vertex functions at non-vanishing momenta can also be calculated in renormalized perturbation theory, but the resulting expressions are too lengthy to be included here.

2.3 Hopping parameter expansions

Perturbation theory is based on an expansion of ϕ^4 theory around a soluble limit, namely free field theory. There is another limiting case, where ϕ^4 theory can be solved exactly too. This is the limit $\kappa = 0$, in which the action

$$S = -2\kappa \sum_x \sum_{\mu=1}^{4} \phi(x)\phi(x+\hat{\mu}) + \sum_x u(\phi(x)) \tag{2.153}$$

reduces to a sum of non-interacting pieces

$$S = \sum_x u(\phi(x)), \tag{2.154}$$

where

$$u(\phi) = \phi^2 + \lambda(\phi^2 - 1)^2 - \lambda. \tag{2.155}$$

Therefore the partition function factorizes into a product of one-site partition functions:

$$Z = \prod_x Z_1 = Z_1^\Omega, \tag{2.156}$$

$$Z_1 = \int d\phi \; e^{-u(\phi)}. \tag{2.157}$$

The Green functions factorize in a similar way.

The hopping parameter expansion is an expansion around $\kappa = 0$. In the literature it is often called *high temperature expansion* due to the statistical mechanics analogy discussed in chapter 1. In this section we shall explain the basic strategy of the hopping parameter expansion for the example of the partition function.

In order to reduce the notation we rewrite the part of the action proportional to κ, the *hopping term*, in the form

$$-2\kappa \sum_{<xy>} \phi(x)\phi(y), \tag{2.158}$$

where $<xy>$ denotes a pair of nearest neighbour points on the lattice. The partition function then reads

$$Z = \int \prod_z \left\{ d\phi(z) \; e^{-u(\phi(z))} \right\} \prod_{<xy>} e^{2\kappa\phi(x)\phi(y)}. \tag{2.159}$$

The second factor is expanded into powers of κ:

$$\prod_{<xy>} e^{2\kappa\phi(x)\phi(y)} = \prod_{<xy>} \left\{ 1 + 2\kappa\phi(x)\phi(y) + \frac{1}{2!}(2\kappa)^2(\phi(x)\phi(y))^2 + \ldots \right\}$$

$$= \sum_{\mathcal{G}} (2\kappa)^{L(\mathcal{G})} c(\mathcal{G}) \prod_{b \in \mathcal{G}} [\phi(i(b)) \, \phi(f(b))] \,. \tag{2.160}$$

The notation is as follows: $\mathcal{G}$ is a graph consisting of vertices v and bonds b, such that every bond b connects two vertices $i(b)$ and $f(b)$, the 'initial' and 'final' endpoints of b. The number of bonds, which end in a particular vertex v, is denoted $N(v)$. Every vertex is associated with a lattice point in such a way that different vertices correspond to different lattice points. A pair of vertices connected by a bond has to be a nearest neighbour pair $< xy >$ on the lattice. There may be several distinct bonds connecting the same pair of vertices $< xy >$. Their number is called multiplicity $m(x, y)$. The total number of bonds in the graph is $L(\mathcal{G})$. Finally the coefficient $c(\mathcal{G})$ is given by

$$c(\mathcal{G}) = \prod_{<xy>} \frac{1}{m(x, y)!} \,. \tag{2.161}$$

Inserting the expansion into the integral the integration over any variable $\phi(z)$ is of the form

$$\int \mathrm{d}\phi \; \phi^k \, \mathrm{e}^{-u(\phi)} \,. \tag{2.162}$$

Therefore we introduce the one-point expectation values

$$\gamma_k \equiv \langle \phi^k \rangle_1 = \frac{1}{Z_1} \int \mathrm{d}\phi \; \phi^k \, \mathrm{e}^{-u(\phi)} \,. \tag{2.163}$$

Only the even ones are non-zero:

$$\gamma_{2n+1} = 0 \,. \tag{2.164}$$

The generating function of these coefficients,

$$z(j) \equiv \sum_{k=0}^{\infty} \frac{1}{k!} \gamma_k \, j^k \,, \tag{2.165}$$

is given by

$$z(j) = \frac{1}{Z_1} \int \mathrm{d}\phi \; \mathrm{e}^{-u(\phi)+j\phi} \,. \tag{2.166}$$

With the help of these numbers the partition function can be represented as

$$Z = Z_1^{\Omega} \sum_{\mathcal{G}} (2\kappa)^{L(\mathcal{G})} c(\mathcal{G}) \prod_{v \in \mathcal{G}} \gamma_{N(v)} \,. \tag{2.167}$$

From the graphs displayed in fig. 2.5 we obtain in d dimensions

$$Z/Z_1^{\Omega} = 1 + (2\kappa)^2 \Omega d \frac{1}{2} \gamma_2^2 + (2\kappa)^4 \Omega d \frac{1}{4!} \gamma_4^2 + (2\kappa)^4 \Omega d (2d-1) \frac{1}{(2!)^2} \gamma_4 \gamma_2^2$$

Fig. 2.5. Graphs in the hopping parameter expansion of the partition function up to order κ^4 in ϕ^4 theory.

$$+ (2\kappa)^4 \frac{1}{2} \Omega d(d-1)\gamma_2^4 + (2\kappa)^4 \frac{1}{2} \Omega d(\Omega d - 4d + 1)\frac{1}{(2!)^2}\gamma_2^4 + \mathcal{O}(\kappa^6). \quad (2.168)$$

In higher orders of the expansion higher powers of the space–time volume Ω will appear.

For the *free energy*, however, we get

$$F \equiv -\frac{1}{\Omega}\ln Z$$

$$= -\ln Z_1 - (2\kappa)^2 \frac{d}{2}\gamma_2^2 - (2\kappa)^4 \left\{ \frac{d}{24}\gamma_4^2 + \frac{1}{4}d(2d-1)\gamma_4\gamma_2^2 - \frac{3d}{8}\gamma_2^4 \right\} + \mathcal{O}(\kappa^6),$$
$$(2.169)$$

which is independent of the volume. This property persists to all orders of the expansion, which amounts to the fact that the hopping parameter expansion for Z exponentiates.

In order to prove this one may derive graphical rules for the expansion of the free energy [2.5]:

$$-F = \ln Z_1 + \frac{1}{\Omega}\sum_{\mathscr{G}'}(2\kappa)^{L(\mathscr{G}')}c'(\mathscr{G})\prod_{v \in \mathscr{G}'}m_{N(v)}. \quad (2.170)$$

This expansion is called *linked cluster expansion*. We shall not discuss this topic in detail, but only state the main results. The graphs $\mathscr{G}'$ contributing to $-F$ are similar to the ones considered above, but now different vertices are allowed to cover the same point on the lattice. Furthermore the graphs have to be connected, which is the crucial property responsible for the absence of any volume dependence in F. The coefficients m_k are related to the γ_k by means of

$$\ln z(j) \equiv \sum_{k=0}^{\infty}\frac{1}{k!}m_k j^k. \quad (2.171)$$

The coefficients $c'(\mathscr{G}')$ involve certain additional symmetry factors, which can be found in [2.5].

The linked cluster expansion can be extended to the case of connected Green's functions. The relevant graphs do then contain external points. For details we refer to the literature cited above.

With the help of refined graphical techniques various quantities, in particular the susceptibilities χ_2, χ_4 and the second moment μ_2, have been calculated to 14th order in κ by Lüscher and Weisz [2.6]. Previous results went up to tenth order [2.7].

With the help of the resulting power series in κ the quantities mentioned before and physical quantities derived from them can be evaluated numerically in the symmetric phase.

2.4 Lüscher–Weisz solution and triviality of the continuum limit

2.4.1 Triviality of four-dimensional ϕ^4 theory

Lattice ϕ^4 theory has two bare parameters, the hopping parameter κ and the quartic coupling λ. The symmetric phase at small values of κ is separated from the phase with broken symmetry at large κ by a line of second order phase transitions, where κ assumes its critical values $\kappa = \kappa_c(\lambda)$. Any continuum limit must be taken by approaching this critical line. The nature of possible continuum limits is related to the fixed points of the renormalization group, as discussed in section 1.7.4.

At $\lambda = 0$ we have the Gaussian infrared fixed point, corresponding to free field theory. Any continuum limit in the vicinity of this fixed point is trivial in the sense that it has a vanishing renormalized coupling $g_R = 0$.

In order to find a continuum limit representing an interacting field theory with $g_R \neq 0$ one has to look for a non-trivial ultraviolet fixed point. The question whether such a fixed point exists and where it is located could be answered once the renormalized mass m_R and the renormalized coupling g_R were known as functions of κ and λ.

The task of determining these functions in the whole symmetric phase of one-component ϕ^4 theory has been tackled by Lüscher and Weisz [2.8] by means of a combination of the hopping parameter expansion and renormalization group equations. With the help of perturbation theory they extended their solution through the critical line into the scaling region of the phase with broken symmetry [2.3]. Furthermore they also treated the case of an n-component model with O(n) symmetry [2.9]. In the following we shall review their work on ϕ^4 theory with a one-component field. For related earlier work of other authors we refer to the literature cited in [2.3].

In a first step the critical line $\kappa_c(\lambda)$ was determined from the hopping parameter expansion of the susceptibility χ_2, using series analysis techniques. Starting from $\lambda = 0$, where the critical value is $\kappa_c(0) = 1/8$, it increases smoothly until it reaches a maximum near $\kappa_c(0.5) = 0.14$, and then decreases to the limiting value $\kappa(\infty) = 0.07475$.

Next the hopping parameter expansions of m_R, g_R and Z_R were derived from those of χ_2, χ_4 and μ_2. The latter were taken from Baker and Kincaid [2.7] up to tenth order in κ. The series have been extended to 14th order by Lüscher and Weisz later [2.6].

The analysis showed that the expansions yield accurate values for the quantities under consideration as long as $\kappa < 0.95\,\kappa_c$ which corresponds to $m_R \simeq 0.5$. It turned out that the curves of constant m_R are to a good approximation given by $\kappa/\kappa_c = $ constant. The renormalized coupling g_R decreases continuously when the critical line is approached at fixed values of λ. For fixed m_R, on the other hand, g_R increases with increasing λ, but the limit which is attained for $\lambda = \infty$ is finite. In particular, for $\kappa = 0.95\,\kappa_c$ the maximal value found for the renormalized coupling is $g_R = 41 \pm 6$. Furthermore the wave function renormalization factor Z_R was always close to 1.

The values of g_R might look large on first sight, but in fact they are not. If one considers e.g. the β-function $\beta(g_R)$ in renormalized perturbation theory, it turns out that the natural expansion parameter is $g_R/16\pi^2$, which is rather small then.

One can therefore employ the renormalization group equations, in particular

$$m_R \frac{\partial g_R}{\partial m_R}\bigg|_{g_0} = \beta(g_R), \qquad (2.172)$$

to cover the region closer to the critical line, i.e. $0.95\,\kappa_c < \kappa < \kappa_c$. Approaching the critical line the renormalized coupling even decreases further according to

$$m_R = C \exp\left(-\frac{1}{\beta_1 g_R}\right)(\beta_1 g_R)^{-\beta_2/\beta_1^2}\{1 + \mathcal{O}(g_R)\}, \qquad (2.173)$$

or

$$g_R = \frac{1}{\beta_1 \ln(C/m_R)} + \dots, \qquad (2.174)$$

see section 1.7.4. Therefore we have good reason to trust the results from the renormalization group equations in this region, which is called *scaling region*.

Lüscher and Weisz integrated the renormalization group equations numerically, using the results from the hopping parameter expansion at $\kappa = 0.95\,\kappa_c$ as initial values. In this way it was possible to obtain m_R, g_R and Z_R in the whole symmetric phase as functions of the bare parameters. Also, as a consequence of (2.174), the renormalized coupling vanishes in the continuum limit, no matter how large we choose the bare coupling λ.

In order to arrive at this conclusion one must assume the validity of the renormalization group equations with the perturbative β-functions in the

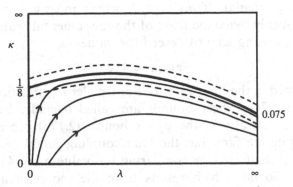

Fig. 2.6. Schematic phase diagram of four-dimensional ϕ^4 theory in the plane of bare parameters κ and λ. The bold line indicates the critical line $\kappa_c(\lambda)$. The dashed lines represent the points with $m_R a = 1/2$. In the symmetric phase below the critical line three curves of constant g_R are drawn. The arrows on them point towards decreasing $m_R a$. They end on the boundary of the phase diagram at $\lambda = \infty$.

scaling region. However, due to the smallness of $g_R/16\pi^2$ at the boundary of the scaling region there remains little doubt that the analysis reflects the true behaviour of the theory. A schematic plot of the lines of constant g_R is shown in fig. 2.6.

In the phase with spontaneously broken symmetry there is no analogue of the hopping parameter expansion available to repeat the calculation, except in the limit $\lambda = \infty$ [2.12]. Nevertheless information about the scaling region in this phase can be gained too. In the broken symmetry phase the scaling behaviour (2.173) also holds for small enough g_R, but with a different constant C'. This constant is related to the one of the symmetric phase, C, through the *scaling connection* [2.3]:

$$C' = e^{1/6}C, \tag{2.175}$$

which is obtained from perturbation theory in the critical theory at $\kappa = \kappa_c(\lambda)$. With the help of this connection the renormalization group equations can be integrated upwards from the critical line until g_R gets too large for the perturbative β-function to be trusted any longer. The region covered in this way includes all points above $\kappa_c(\lambda)$ with $m_R < 0.5$.

The results of this analysis imply the absence of any non-trivial fixed point for lattice ϕ^4 theory in four dimensions with the standard action. Consequently the continuum limit would be trivial, as had already been conjectured by Wilson [2.10] and others. A rigorous proof of triviality is, however, still missing despite serious attempts, see e.g. [2.11].

For finite lattice spacings the theory may involve interaction, i.e. $g_R \neq 0$, but the triviality of the continuum limit places upper bounds on g_R for given m_R. In particular, if $m_R < 1/2$, which means that the cut-off is higher than roughly twice the mass of the fundamental scalar particle, the renormalized coupling cannot exceed the value

$$g_{Rmax} \approx 41, \qquad (2.176)$$

which is assumed in the limit $\lambda \to \infty$. For smaller m_R this bound decreases according to (2.174). These bounds are called *triviality bounds*. As has already been noted above, the upper bound (2.176) represents a weak coupling, despite the fact that the bare coupling might be infinite. This statement is substantiated by comparing the value $g_R = 41$ with the tree level unitarity bound, which results from the requirement of unitarity of the scattering matrix in lowest order of perturbation theory, or by studying the rate of convergence of the perturbation series for various physical quantities [2.8, 2.3].

Even if the strict continuum limit leads to a trivial theory, ϕ^4 theory may be used as an *effective field theory*. This means that the cut-off can be made so large that cut-off effects on Green's functions $\widetilde{G}(p_i)$, which are typically of order $p_i^2 a^2$, are completely negligible, while the renormalized coupling, which only vanishes logarithmically, is still noticeable.

The numbers cited above apply to the ϕ^4 theory regularized on a hypercubic lattice with the standard action. By choosing other lattices or other actions, including other coupling terms, the triviality upper bounds can be changed quantitatively [2.13], but the qualitative conclusions are the same.

2.4.2   Infinite bare coupling limit

In the last section the limit $\lambda \to \infty$ of ϕ^4 theory has been mentioned in connection with the behaviour of the renormalized coupling. The Boltzmann factor of the functional integral involves a factor

$$\exp\left(-\lambda[\phi(x)^2 - 1]^2\right)$$

for every lattice point. Taking into account the normalization of the functional integral these factors go over into δ-functions

$$\delta(\phi(x)^2 - 1)$$

in the limit of infinite λ. Consequently the field is restricted to the values

$$\phi(x) = \pm 1. \qquad (2.177)$$

The action reduces to

$$S = -2\kappa \sum_{<xy>} \phi(x)\phi(y), \qquad (2.178)$$

which defines the *Ising model* in four dimensions.

As has been discussed in the previous section the renormalized coupling g_R assumes its finite maximum in the Ising limit. Therefore this is the limit relevant for a determination of upper bounds on g_R. On the other hand the simplicity of the field variables allows one to simulate the Ising model numerically much more efficiently than the general ϕ^4 theory with scalar fields of variable length.

Owing to this fact, the Ising limit has been used to perform high precision Monte Carlo calculations for comparison with the results from analytical work. In the symmetric phase various points corresponding to values of m_R between 0.5 and 0.2 have been investigated in [2.14, 2.15]. In [2.14] the emphasis was on the study of finite-size effects. The main concern of [2.15] was the scaling behaviour of g_R by comparing the obtained values at three points $m_R \approx 0.5$, 0.4 and 0.3 with the predictions of the renormalization group and of [2.8]. The agreement between the numerical data and the analytical calculations turned out to be excellent, giving more support to the triviality of four-dimensional ϕ^4 theory.

Analogous calculations for the broken symmetry phase of the four-dimensional Ising model have been done in [2.16, 2.4]. The agreement between the Monte Carlo results and those of [2.3] is good for m_R and g_R. For Z_R somewhat larger deviations are visible. They might be due to the fact that g_R is not really small enough in the points considered in the numerical work.

The phase with broken symmetry has also been the subject of analytical work based on series expansions. 'Low temperature series', which are series in powers of $\exp -8\kappa$, for the quantities under consideration have been derived in [2.12] and compared with the other methods. Again good agreement both with the results of [2.4] as well as with [2.3] was found in a range $m_R \simeq 0.3 - 0.6$.

2.5   Finite-volume effects

In numerical simulations of quantum field theories on a lattice the number of lattice points is restricted by the available computer resources. In order to have a situation which is relevant for the physics in the continuum, one tries to make the lattice spacing so small relative to the correlation length that scaling behaviour is approximately realized. On the other hand the correlation length should not be too large, since otherwise the finite size of the lattice has strong effects on measurable quantities such as masses or couplings. These are called *finite-volume effects*. The two important geometrical characteristics referred to are the ratio of the correlation length ξ to the lattice spacing a, and the ratio of the linear size of the lattice, say L, to the correlation length ξ, and the requirements stated above can be symbolized by

$$a \ll \xi \ll L. \tag{2.179}$$

The nature of finite-volume effects of course depends very much on the theory under consideration. In the case of asymptotically free theories in a finite volume, like pure non-Abelian gauge theories in four dimensions, it has been realized [2.17] that for small volumes the relevant parameter is a running coupling $g(1/L)$ associated with the scale $1/L$. Owing to asymptotic freedom, this coupling becomes small for very small volumes, which makes it possible to calculate volume dependent quantities by means of perturbation theory in the small-volume limit [2.18, 2.19].

On the other hand, in the limit of large volumes, finite-volume effects can be estimated by perturbative methods in both asymptotically free and non-free theories. In this section the influence of the finiteness of the lattice size L on different quantities will be discussed in the framework of one-component ϕ^4 theory on a four-dimensional lattice.

2.5.1   *Perturbative finite-volume effects*

A typical situation found in numerical calculations is a cylindrical geometry, which means that the spatial extensions of the lattice are equal

$$L_1 = L_2 = L_3 \equiv L, \tag{2.180}$$

and that the elongation in the Euclidean time direction is much larger than L:

$$L_4 \equiv T \gg L. \tag{2.181}$$

The largeness of T is necessary for a precise determination of masses from the asymptotic decay of timeslice correlations. For simplicity let us assume that T is infinite, whereas the spatial volume $V = L^3$ may be finite.

Perturbation theory can be carried through in a finite volume in much the same way as has been discussed before for the case of an infinite volume. The essential difference is that in loop 'integrals' the internal lattice momenta are restricted to discrete values in the Brillouin zone specified by (2.5). Consider for example the renormalized mass in the symmetric phase. Its value in a finite volume is denoted $m_R(L)$, whereas in the infinite volume it is also just called m_R. According to (2.93) the renormalized mass squared is given in bare perturbation theory to one-loop order by

$$m_R(L)^2 = m_0^2 + \frac{g_0}{2} J_1(m_0, L) + \mathcal{O}(g_0^2). \tag{2.182}$$

The loop integrals $J_n(m, L)$ in a finite volume are defined by

$$J_n(m, L) = \frac{1}{L^3} \sum_{\mathbf{k}} \int_{-\pi}^{\pi} \frac{dk_4}{(2\pi)} (\hat{k}^2 + m^2)^{-n}, \qquad (2.183)$$

where the sum over $\mathbf{k}$ is over the Brillouin zone. For fixed values of the bare parameters m_0 and g_0, the deviation from the infinite-volume expression is thus

$$m_R(L)^2 - m_R^2 = \frac{g_0}{2} \delta J_1(m_0, L) + \mathcal{O}(g_0^2), \qquad (2.184)$$

with

$$\delta J_n(m, L) \equiv J_n(m, L) - J_n(m, \infty). \qquad (2.185)$$

In general the deviation of any quantity X from its infinite-volume limit is denoted by

$$\delta X(L) \equiv X(L) - X(\infty). \qquad (2.186)$$

Let

$$g_R \equiv g_R(\infty) \qquad (2.187)$$

be the renormalized coupling in the infinite volume. Owing to

$$g_R = g_0 + \mathcal{O}(g_0^2) \qquad (2.188)$$

we can express $\delta m_R(L)^2$ in terms of the renormalized parameters g_R and m_R, namely

$$\delta(m_R(L)^2) = \frac{g_R}{2} \delta J_1(m_R, L) + \mathcal{O}(g_R^2). \qquad (2.189)$$

This is the finite-volume effect in renormalized perturbation theory to one-loop order. Equivalently one may write

$$\delta m_R(L) = \frac{g_R}{4m_R} \delta J_1(m_R, L) + \mathcal{O}(g_R^2). \qquad (2.190)$$

It is straightforward to obtain the corresponding expressions for the perturbative finite-volume effects on other quantities. In the symmetric phase one gets for example

$$\delta m(L) = \frac{g_R}{4m_*} \delta J_1(m_R, L) + \mathcal{O}(g_R^2), \qquad (2.191)$$

$$\delta g_R(L) = -\frac{3}{2} g_R^2 \delta J_2(m_R, L) + \mathcal{O}(g_R^3). \qquad (2.192)$$

The asymptotic behaviour for large L of these perturbative finite-volume effects can be approximated with the help of the relations [2.14]

$$\delta J_1(m_R, L) = 6m_*^2 (2\pi m_* L)^{-3/2} e^{-mL} (1 + \mathcal{O}(L^{-1})), \qquad (2.193)$$

$$\delta J_2(m_R, L) = \frac{3}{2} m_* L (2\pi m_* L)^{-3/2} e^{-mL} (1 + \mathcal{O}(L^{-1})). \qquad (2.194)$$

We see that for the masses the infinite-volume limit is approached from above with exponentially vanishing corrections, whereas for the renormalized coupling it is approached from below.

The finite-volume effects in the symmetric phase have been investigated in detail and compared to results of Monte Carlo computations in [2.14]. This study had been motivated in turn by work of Lüscher [2.20], in which the asymptotic large L behaviour of finite-size effects on masses was considered to all orders in perturbation theory in the continuum, and was related to elastic scattering amplitudes. Corresponding relations have been obtained for field theories on the lattice in [2.1].

Let us display the perturbative results in the phase with broken symmetry too:

$$\delta m_R(L) = -\frac{g_R}{2m_R} \left[\delta J_1(m_R, L) + \frac{3}{2} m_R^2 \delta J_2(m_R, L) - \frac{3}{2} m_R^4 \delta I_2''(m_R, L) \right.$$

$$\left. + \mathcal{O}(g_R) \right], \qquad (2.195)$$

$$\delta m(L) = -\frac{g_R}{2m_R(1 + m_R^2/4)^{1/2}} \left[\delta J_1(m_R, L) + \frac{3}{2} m_R^2 \delta I_2(m_R, L, q) \right.$$

$$\left. + \mathcal{O}(g_R) \right], \qquad (2.196)$$

$$\delta g_R = -g_R^2 \left[\frac{3}{2} \delta J_2(m_R, L) - 3 m_R^2 \delta I_2''(m_R, L) + \mathcal{O}(g_R) \right], \qquad (2.197)$$

$$\delta v_R = -\frac{(3 g_R)^{1/2}}{2 m_R} \left[\delta J_1(m_R, L) + \frac{3}{2} m_R^4 \delta I_2''(m_R, L) + \mathcal{O}(g_R) \right], \qquad (2.198)$$

$$\Delta Z_R \equiv \frac{Z_R(L)}{Z_R(\infty)} - 1 = \frac{3}{2} g_R m_R^2 \delta I_2''(m_R, L) + \mathcal{O}(g_R^2), \qquad (2.199)$$

where I_2 and I_2'' are defined according to (2.89), (2.141). For the physical mass the asymptotic behaviour is [2.20, 2.1]

$$\delta m(L) \sim -\frac{C}{L} e^{-\alpha m L}, \qquad (2.200)$$

with certain coefficients C and α, which in the scaling region are approximately given by

$$C = \frac{9}{16\pi^2} g_R, \qquad (2.201)$$

$$\alpha = \frac{\sqrt{3}}{2}. \qquad (2.202)$$

The infinite-volume limit of the mass is now approached from below.

In this phase, however, a different type of finite-volume effect also appears, which is discussed in the following.

2.5.2 *Tunneling*

In an infinite volume the Z_2-symmetry of the action is broken spontaneously for $\kappa > \kappa_c$ (see section 2.1.2). There are two ground states $|0_\pm\rangle$ with

$$\langle 0_+ | \phi(x) | 0_+ \rangle = v, \qquad \langle 0_- | \phi(x) | 0_- \rangle = -v, \qquad (2.203)$$

and the spectrum of the transfer matrix is doubly degenerate.

On the other hand it is well known in statistical mechanics that spontaneous symmetry breaking does not occur in a finite volume. As a consequence of the Frobenius–Perron theorem applied to the transfer matrix [2.21] there is a unique ground state $|0_s\rangle$ symmetric under the reflection $\phi \rightarrow -\phi$ and the vacuum expectation value of the field vanishes. This means that the degeneracy of the infinite-volume ground states $|0_\pm\rangle$ is lifted. Separated from the ground state $|0_s\rangle$ by a small *energy splitting* E_{0a} there is an antisymmetric state $|0_a\rangle$ and if one decomposes these states as

$$|0_s\rangle \equiv \frac{1}{\sqrt{2}} (|0_+\rangle + |0_-\rangle) \qquad |0_a\rangle \equiv \frac{1}{\sqrt{2}} (|0_+\rangle - |0_-\rangle), \qquad (2.204)$$

then $|0_+\rangle$ and $|0_-\rangle$ are states which go over into the degenerate vacua in the infinite-volume limit. For $\kappa > \kappa_c$, transitions between the two sectors mentioned in the previous section occur in a finite volume.

The energy splitting E_{0a} is due to tunneling between $|0_+\rangle$ and $|0_-\rangle$ in a finite volume. Its volume dependence was studied in [2.22, 2.23]. Their analysis is based on a picture of domains which extend over the spatial volume and cover certain intervals in time. Neighbouring domains with a different sign of the field are separated by domain walls, which can be considered as tunneling events. From this picture a prediction about the energy splitting of the form

$$E_{0a} \sim \exp \left\{ -\sigma L^3 \right\} \qquad (2.205)$$

is obtained, where σ is the *interface tension* associated with the domain walls. One sees that tunneling effects vanish very rapidly with increasing volume.

For a quantitative analysis in connection with Monte Carlo calculations it is important to have a more precise formula and to have an expression for σ in terms of the parameters of the theory. The energy splitting has been obtained from a semiclassical calculation in the continuum including one-loop effects in [2.24]. The result is

$$E_{0a} = C \cdot L^{1/2} \exp\left\{-\sigma L^3\right\}, \tag{2.206}$$

where the prefactor C and the surface tension σ are given in equations (2.215),(2.216),(2.217) below. The factor $L^{1/2}$ has also been observed by Brézin and Zinn-Justin [2.25] in the context of a one-loop calculation. The formula (2.206) can be applied to lattice ϕ^4 theory as long as it is in the scaling region, where finite lattice spacing effects are negligible. A corresponding calculation of E_{0a} for three-dimensional theories has been made in [2.26].

The semiclassical calculation is based on an instanton-like saddle point approximation to the Euclidean path integral as introduced in [2.27] and beautifully explained in [2.28]. To this end the tunneling amplitudes

$$\langle 0_+ | e^{-TH} | 0_\pm \rangle = \frac{1}{2}(e^{-TE_{0s}} \pm e^{-TE_{0a}}), \tag{2.207}$$

where H is the Hamiltonian, are expressed as path integrals with boundary conditions

$$\phi_0(x) \longrightarrow \begin{cases} v_0, & T \to \infty \\ \pm v_0, & T \to -\infty \end{cases} \tag{2.208}$$

where v_0 is the value of ϕ_0 at the minimum of the classical action. In the case where $|0_-\rangle$ appears in (2.207) the path integral is dominated by a classical solution, the so-called 'kink':

$$\phi_c(x) = \left(\frac{3m_0^2}{g_0}\right)^{1/2} \tanh \frac{m_0}{2}(x_4 - \tau) \tag{2.209}$$

with classical action

$$S_c = 2\frac{m_0^3}{g_0}L^3, \tag{2.210}$$

where τ is a free parameter specifying the location of the kink. For fluctuations around the classical solution

$$\phi = \phi_c + \eta$$

the quadratic part of the action is given by

$$S = S_c + \frac{1}{2}\int d^4x\, \eta(x) M \eta(x) + \mathcal{O}(\eta^3) \tag{2.211}$$

with the fluctuation operator

$$M = -\partial_\mu \partial^\mu + m_0^2 - \frac{3}{2}m_0^2 \cosh^{-2}\left(\frac{m_0}{2}(x^4 - \tau)\right). \tag{2.212}$$

The saddle-point approximation to the path integral amounts to integrating these Gaussian fluctuations. The operator M has a zero-mode corresponding to translations of the kink or shifts of the parameter τ. This zero-mode has to be treated separately by the method of collective coordinates. Taking into account also all contributions from non-interacting multi-kink configurations, which exponentiate, the result for the energy splitting is

$$E_{0a} - E_{0s} = 2\,e^{-S_c} \left(\frac{S_c}{2\pi}\right)^{1/2} \left|\frac{\det' M}{\det M_0}\right|^{-1/2}, \qquad (2.213)$$

where $\det'$ is the determinant without zero-modes and

$$M_0 = -\partial_\mu \partial^\mu + m_0^2. \qquad (2.214)$$

The factor $S_c^{1/2} \sim L^{3/2}$ is due to the zero-mode. The determinant, which represents a one-loop effect, leads to the following three types of contributions. First of all it produces precisely those counterterms which are required to convert the unrenormalized parameters appearing in (2.213) into the renormalized ones. Moreover it yields an additional factor L^{-1}. Finally it gives a one-loop correction to the term proportional to L^3 in the exponential. The final result is of the form (2.206) with

$$C = 1.65058 \left(2\,\frac{m_R^3}{g_R}\right)^{1/2} \qquad (2.215)$$

and an L-dependent interface tension

$$\sigma(L) = \sigma_\infty \left[1 - \frac{g_R}{16\pi^2}\frac{3\sqrt{3}\pi}{(m_R L)^2}\exp\left(-\frac{\sqrt{3}}{2}m_R L\right) + \mathcal{O}(e^{-m_R L}) + \mathcal{O}(g_R^2)\right] \qquad (2.216)$$

$$\sigma_\infty = 2\,\frac{m_R^3}{g_R}\left[1 - \frac{g_R}{16\pi^2}\left(\frac{1}{8} + \frac{\pi}{4\sqrt{3}}\right) + \mathcal{O}(g_R^2)\right]. \qquad (2.217)$$

A comparison of these one-loop formulae with the results of a Monte Carlo calculation has been made in [2.4] and a remarkably good agreement was found. The vacuum tunneling has been studied also in [2.16] and has been identified as the main source of finite-volume effects in the phase with broken symmetry. The formulae above can be employed for an estimation of how large L has to be in order that finite-volume effects of this type are negligible.

2.6 N-component model

The standard model of electromagnetic and weak interactions contains a scalar field with four components as an essential piece (see chapter 6).

This field is associated with the predicted Higgs particle. The spontaneous breaking of the O(4)-symmetry of the action gives rise to the generation of masses for fermions and vector bosons. Therefore the theory with a four-component scalar field in the phase with spontaneously broken O(4)-symmetry has an immediate physical relevance. Assuming that the influence of the Yukawa-couplings to the fermions is small and that the gauge interactions can be dealt with in perturbation theory, one can obtain a non-perturbative upper bound on the mass of the physical Higgs boson.

Owing to the prominent rôle of four-component scalar field theory we would not like to close this chapter without introducing ϕ^4 theory with more than one component. In this section we consider the general case of the theory of a real N-component scalar field $\phi^\alpha(x)$, $\alpha = 1,\ldots,N$. On a hypercubic lattice in four dimensions the O(N)-symmetric action of ϕ^4 theory is parametrized as

$$S = \sum_x \left\{ -2\kappa \sum_{\mu=1}^4 \phi(x) \cdot \phi(x + \hat{\mu}) + \phi(x) \cdot \phi(x) + \lambda(\phi(x) \cdot \phi(x) - 1)^2 \right\}.$$
(2.218)

The O(N)-symmetric non-linear sigma model is characterized by the additional restriction of a fixed length for the field

$$\phi(x) \cdot \phi(x) = 1,$$
(2.219)

and is equivalent to the ϕ^4 theory in the limit of an infinite bare quartic self-coupling $\lambda = \infty$.

For values of κ below a certain critical $\kappa_c(\lambda)$ the O(N)-symmetry is unbroken and the Green functions

$$G_{\alpha_1,\ldots,\alpha_n}(x_1,\ldots,x_n) = \langle \phi^{\alpha_1}(x_1) \cdots \phi^{\alpha_n}(x_n) \rangle$$

$$= \frac{1}{Z} \int \prod_{x,\alpha} \mathrm{d}\phi^\alpha(x) \; \mathrm{e}^{-S} \; \phi^{\alpha_1}(x_1) \cdots \phi^{\alpha_n}(x_n)$$
(2.220)

are O(N)-symmetric. The spectrum has a gap m corresponding to the mass of an O(N)-vector multiplet of particles. This mass is given by the pole of the propagator closest to the origin. Let $\tilde{G}_{\alpha\beta}(p)$ be the propagator in momentum space. Then

$$\tilde{G}(p)^{-1}_{\alpha\beta} = 0, \qquad p = (im, 0, 0, 0).$$
(2.221)

The renormalized mass m_R and the wave function renormalization factor Z_R on the other hand are defined through the small momentum behaviour of the propagator:

$$\tilde{G}(p)^{-1}_{\alpha\beta} = 2\kappa Z_R^{-1} \delta_{\alpha\beta} \{ m_R^2 + p^2 + \mathcal{O}(p^4) \}.$$
(2.222)

The renormalized and the unrenormalized vertex functions are related through

$$\Gamma_R^{(n)}(p_1, \ldots, p_n)_{\alpha_1, \ldots, \alpha_n} = (2\kappa Z_R^{-1})^{-n/2} \, \Gamma^{(n)}(p_1, \ldots, p_n)_{\alpha_1, \ldots, \alpha_n} \,. \tag{2.223}$$

The renormalized coupling g_R is defined in terms of the renormalized four-point vertex function by

$$\Gamma_R^{(4)}(0,0,0,0)_{\alpha\beta\gamma\delta} = -g_R \, S_{\alpha\beta\gamma\delta} \,, \tag{2.224}$$

where

$$S_{\alpha\beta\gamma\delta} = \frac{1}{3}(\delta_{\alpha\beta}\delta_{\gamma\delta} + \delta_{\alpha\gamma}\delta_{\beta\delta} + \delta_{\alpha\delta}\delta_{\beta\gamma}) \,. \tag{2.225}$$

For $\kappa > \kappa_c(\lambda)$ the $O(N)$-symmetry is broken spontaneously. The field has a non-vanishing vacuum expectation value, which can be chosen to point into the Nth direction:

$$\langle \phi^\alpha(x) \rangle = v \, \delta_{\alpha N} \,. \tag{2.226}$$

From Goldstone's theorem (see [A3]) it follows that the spectrum contains $N - 1$ massless Goldstone bosons. Furthermore one massive particle, corresponding to the Higgs boson, is present, which is unstable in the pure scalar theory without gauge fields and can decay into Goldstone bosons.

Perturbation theory can be applied as in the case of one component. In the phase with spontaneously broken symmetry, however, one has to take special care about infrared divergencies, which are produced by the massless particles. For details the reader should consult [2.9].

Perturbation theory is also applicable in the limit $\lambda \to \infty$, which yields the non-linear sigma model. The four-dimensional non-linear sigma model is a perturbatively non-renormalizable field theory. Therefore the reader might wonder how perturbation theory can be applied. This is possible in the following way. For finite bare coupling λ renormalized perturbation theory can be applied to the N-component ϕ^4 theory. This means that physical quantities are expanded in powers of the renormalized coupling g_R, the coefficients depending on the renormalized mass m_R. Renormalized perturbation theory can be successfully used near the critical point where the renormalized coupling becomes small. Outside this scaling region non-perturbative effects dominate.

The bare coupling and the renormalized coupling are numerically quite different. In particular g_R remains finite even in the limit where λ goes to infinity. This gives us the possibility to apply renormalized perturbation theory in this case too. In this limit the coupling g_R and the mass m_R are related in a certain way, which cannot be calculated within perturbation theory. If, however, for a given g_R one knows the value of m_R from other sources one may use this as an input to the perturbative formulae.

The analysis of ϕ^4 theory by Lüscher and Weisz, which was discussed in section 2.4, has been carried out by them also for the case of the N-component model [2.9]. The results do not differ qualitatively from the case $N = 1$. For the upper bound on the Higgs boson mass in the O(4)-symmetric theory they obtain [2.29]

$$m_R < 630 \text{ GeV} \qquad \text{if } \Lambda > 2m_R, \tag{2.227}$$

where the phenomenological value

$$v_R = 250 \text{ GeV} \tag{2.228}$$

has been used.

Numerical simulations of the four-component theory in the sigma-model limit have been performed in the symmetric phase in [2.30], the results being in agreement with the analytical ones. In the phase with broken symmetry, investigations by means of the Monte Carlo method have been done in [2.31, 2.32]. They yield an upper bound of (640 ± 40) GeV. Yet another non-perturbative method has been applied by Hasenfratz and Nager [2.33], who employed truncated renormalization group equations. More details about these studies can be found in the review [2.34].

Finally we would like to mention the $1/N$ expansion, which allows one to calculate various quantities in the O(N)-symmetric ϕ^4 theory non-perturbatively on the lattice as well as in the continuum (see [A5] and references cited there). For recent work on this subject the reader is referred to [2.35].

3
Gauge fields

3.1 Continuum gauge fields

3.1.1 SU(N) gauge fields

Consider N complex scalar fields $\phi^i(x)$, $i = 1, \ldots, N$ on the Euclidean continuum, which form the components of a complex vector field $\phi(x)$. For each space–time point x the vector $\phi(x)$ is an element of a vector space V_x. In our case every V_x is isomorphic to $\mathbf{C}^N$. On V_x a scalar product is defined through

$$\phi \cdot \phi' = \sum_i \overline{\phi}^i \phi'^i. \tag{3.1}$$

The action

$$S = \int d^4x \, \{\phi(x) \cdot (\Box + m^2)\phi(x) + U(\phi(x) \cdot \phi(x))\} \tag{3.2}$$

is invariant under transformations of the form

$$\phi(x) \rightarrow \phi'(x) = \Lambda^{-1}\phi(x), \qquad \Lambda \in \mathrm{SU}(N), \tag{3.3}$$

i.e. Λ is a $N \times N$ matrix, independent of x, satisfying

$$\Lambda^+\Lambda = \mathbf{1}, \qquad \det \Lambda = 1. \tag{3.4}$$

The first condition defines the group $\mathrm{U}(N)$. The second condition means restricting ourselves to $\mathrm{SU}(N)$, which we do for simplicity here, although there is invariance also with respect to the whole $\mathrm{U}(N)$ group. The transformations above are called *global gauge transformations* or *gauge transformations of the first kind*. The appearance of Λ^{-1} instead of Λ in (3.3) is just a matter of convention.

A much larger class of transformations consists of those gauge transformations

$$\phi(x) \rightarrow \phi'(x) = \Lambda^{-1}(x)\phi(x), \tag{3.5}$$

91

where $\Lambda(x)$ varies with x. These are called *local gauge transformations* or *gauge transformations of the second kind*. They can be considered as passive transformations resulting from an x-dependent change of basis. The action above is not invariant under local gauge transformations.

A generalization of the 'Nahewirkungsprinzip' is the 'Naheinformationsprinzip' [3.1], which forbids parallelism at a distance and requires the physics to be independent of the local choice of basis. Applied to the situation above, it would demand modification of the action such as to make it invariant under local gauge transformations. By analogy with General Relativity, this can be achieved by introducing a *covariant differentiation*. Covariant differentiation in turn results from a concept of parallel transport [3.2]. Parallel transport of a vector along some curve is defined in the following way. Let $\mathscr{C}_{yx}$ be some curve in space–time from x to y. It may be parametrized by means of

$$c^\mu(s), \qquad s \in [0,1], \qquad c^\mu(0) = x^\mu, \qquad c^\mu(1) = y^\mu. \qquad (3.6)$$

With $\mathscr{C}_{yx}$ we associate an SU(N) matrix

$$U(\mathscr{C}_{yx}) : V_x \to V_y, \qquad (3.7)$$

which defines a mapping from V_x into V_y. Then the vector

$$U(\mathscr{C}_{yx})\phi(x) \in V_y \qquad (3.8)$$

is defined to be the vector $\phi(x)$ parallel transported along $\mathscr{C}_{yx}$ to the point y. $U(\mathscr{C}_{yx})$ is called the *parallel transporter*. In general, to every curve in space–time we associate a parallel transporter in a continuous and differentiable way. The parallel transporters have to satisfy the following compatibility conditions:

1. $U(\emptyset) = 1,$ (3.9)

    where $\emptyset$ denotes a curve of zero length.

2. $U(\mathscr{C}_2 \circ \mathscr{C}_1) = U(\mathscr{C}_2)U(\mathscr{C}_1),$ (3.10)

    where $\mathscr{C}_2 \circ \mathscr{C}_1$ denotes the path which is composed from $\mathscr{C}_1$ followed by $\mathscr{C}_2$ (see fig. 3.1).

3. $U(-\mathscr{C}) = U(\mathscr{C})^{-1},$ (3.11)

    where $-\mathscr{C}$ denotes the path $\mathscr{C}$ traversed in the opposite way (see fig. 3.1).

Under a local gauge transformation

$$
\begin{aligned}
\phi(x) \to \phi'(x) &= \Lambda^{-1}(x)\phi(x) \\
\phi(y) \to \phi'(y) &= \Lambda^{-1}(y)\phi(y)
\end{aligned}
\qquad (3.12)
$$

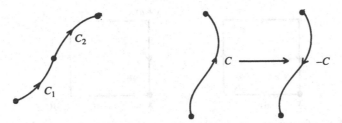

Fig. 3.1.   Composition and reversion of paths.

a parallel transporter transforms as

$$U(\mathscr{C}_{yx}) \to U'(\mathscr{C}_{yx}) = \Lambda^{-1}(y)U(\mathscr{C}_{yx})\Lambda(x). \qquad (3.13)$$

In order to define covariant differentiation we have to subtract vectors at infinitesimal neighbouring points x and $y = x + \mathrm{d}x$. As the basis in different points may be chosen arbitrarily, we can compare the vectors only by parallel transporting them to the same point. Let us consider the straight curve from x to $x + \mathrm{d}x$ and the corresponding parallel transporter. Since it deviates from the unit matrix only infinitesimally we write

$$U(\mathscr{C}_{x+dx,x}) = 1 - A_\mu(x)\,\mathrm{d}x^\mu, \qquad (3.14)$$

where

$$A_\mu(x) \in su(N) \qquad (3.15)$$

is an element of the Lie algebra of SU(N), i.e. it is a traceless anti-Hermitean $N \times N$ matrix. If we define the covariant differential of $\phi(x)$ by

$$D\phi(x) = U^{-1}(\mathscr{C}_{x+dx,x})\phi(x + \mathrm{d}x) - \phi(x), \qquad (3.16)$$

we obtain

$$D\phi(x) = D_\mu\phi(x)\,\mathrm{d}x^\mu \qquad (3.17)$$

with the covariant derivative

$$D_\mu\phi(x) = (\partial_\mu + A_\mu(x))\phi(x). \qquad (3.18)$$

$A_\mu(x)$ is called *gauge field*. From the transformation law of parallel transporters under local gauge transformations we obtain the transformation law of the gauge field:

$$\begin{aligned} A'_\mu(x) &= \Lambda^{-1}(x)A_\mu(x)\Lambda(x) - (\partial_\mu\Lambda^{-1}(x))\Lambda(x) \\ &= \Lambda^{-1}(x)(\partial_\mu + A_\mu(x))\Lambda(x). \end{aligned} \qquad (3.19)$$

The covariant derivative then transforms covariantly in the following

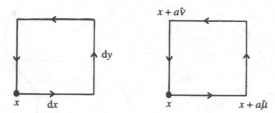

Fig. 3.2. An infinitesimal parallelogram in the continuum and an elementary plaquette on a hypercubic lattice

sense:

$$D'_\mu \phi'(x) = \Lambda^{-1}(x) D_\mu \phi(x) . \tag{3.20}$$

The object corresponding to the curvature tensor of General Relativity is the *field strength*, which can be defined through the commutator of two covariant derivatives:

$$F_{\mu\nu}(x) = [D_\mu, D_\nu] = \partial_\mu A_\nu(x) - \partial_\nu A_\mu(x) + [A_\mu(x), A_\nu(x)] . \tag{3.21}$$

Its geometrical meaning can be gathered from considering parallel transport around an infinitesimal parallelogram spanned by dx and dy (see fig. 3.2). The corresponding parallel transporter is given by

$$U(\mathscr{C}_{xx}) = 1 - F_{\mu\nu}(x)\, dx^\mu\, dy^\nu . \tag{3.22}$$

Under a local gauge transformation the field strength transforms as

$$F'_{\mu\nu}(x) = \Lambda^{-1}(x) F_{\mu\nu}(x) \Lambda(x) . \tag{3.23}$$

Let us now introduce the component notation. Every element A of the real Lie algebra su(N) can be written as a linear combination of generators iT_a :

$$A = \sum_{a=1}^{N^2-1} i\omega^a T_a, \qquad \omega^a \in \mathbf{R} . \tag{3.24}$$

The matrices T_a, which are also called generators, are traceless and Hermitean and are usually normalized as

$$\mathrm{Tr}\,(T_a T_b) = \frac{1}{2}\, \delta_{ab} . \tag{3.25}$$

They satisfy the commutation relations

$$[T_a, T_b] = i f_{abc} T_c , \tag{3.26}$$

where the *structure constants* f_{abc} are completely antisymmetric and real.
For SU(2) we have

$$T_a = \frac{\tau_a}{2} , \qquad a = 1, 2, 3 \tag{3.27}$$

with the Pauli matrices τ_a and

$$f_{abc} = \epsilon_{abc}, \tag{3.28}$$

and for SU(3)

$$T_a = \frac{\lambda_a}{2}, \qquad a = 1, \ldots, 8, \tag{3.29}$$

where λ_a are the Gell-Mann matrices (see the appendix).

Every element of SU(N) can be represented in the form

$$\Lambda = e^{iT_a\omega^a}, \tag{3.30}$$

where the parametrization in terms of the ω^a is unique in a sufficiently small neighbourhood of unity.

The component fields of A_μ and $F_{\mu\nu}$ are defined by

$$A_\mu(x) = -igA_\mu^a(x)T_a \tag{3.31}$$

$$F_{\mu\nu}(x) = -igF_{\mu\nu}^a(x)T_a, \tag{3.32}$$

and they are related via

$$F_{\mu\nu}^a = \partial_\mu A_\nu^a - \partial_\nu A_\mu^a + gf_{abc}A_\mu^b A_\nu^c, \tag{3.33}$$

where a coupling constant g has been introduced conventionally.

Now we return to the problem of formulating physics in a way invariant under local gauge transformations. With the help of the covariant derivative, a gauge invariant action for the field $\phi(x)$ can be written down as

$$S_\phi = \int d^4x \, \{D_\mu\phi(x) \cdot D_\mu\phi(x) + m^2\phi(x) \cdot \phi(x) + U(\phi(x) \cdot \phi(x))\}. \tag{3.34}$$

A dynamics for the gauge field itself is introduced by means of the *Yang–Mills action* [3.3]

$$S_{YM} = -\frac{1}{2g^2} \int d^4x \, \mathrm{Tr} \, F_{\mu\nu}F_{\mu\nu} = \frac{1}{4} \int d^4x \, F_{\mu\nu}^a F_{\mu\nu}^a, \tag{3.35}$$

which contains cubic and quartic self-interaction terms.

The field $\phi(x)$ has served as a starting point for introducing parallel transporters and gauge fields. The gauge field with its Yang–Mills action, however, may also be considered on its own without being coupled to other fields. This is then called the *pure Yang–Mills theory*.

Let us return to the parallel transporters again. They determine the gauge field uniquely as we have seen above. On the other hand, given the gauge field $A_\mu(x)$, the parallel transporters can be reconstructed [3.2]. Consider a curve $\mathscr{C}_{yx}$ parametrized as in (3.6) and let $\mathscr{C}_s$ be the corresponding curve with the parameter t running from 0 to s only. Varying the end-point of $\mathscr{C}_s$ one obtains the differential equation

$$\frac{\mathrm{d}}{\mathrm{d}s}U(\mathscr{C}_s) = -A_\mu(c(s))\frac{\mathrm{d}c^\mu}{\mathrm{d}s}U(\mathscr{C}_s).\tag{3.36}$$

With the initial condition

$$U(\mathscr{C}_0) = 1\tag{3.37}$$

its solution is given by *Dyson's formula*, well known from quantum mechanics:

$$U(\mathscr{C}_s) = \mathrm{P}\exp\left\{-\int_0^s A_\mu(c(s))\frac{\mathrm{d}c^\mu}{\mathrm{d}s}\mathrm{d}s\right\}$$

$$\equiv \mathrm{P}\exp\left\{-\int_{\mathscr{C}_s} A_\mu\,\mathrm{d}x^\mu\right\},\tag{3.38}$$

where the symbol P denotes path ordering with respect to the parameter s such that matrices $A_\mu(c(s))$ with larger s stand to the left of those with smaller s.

3.1.2   Abelian gauge fields

In the previous section we have restricted ourselves to the case of the gauge groups SU(N). These belong to the most important cases for physics. There is, however, no obstacle to formulation of gauge theories for any compact gauge group. The only other example which will be considered here explicitly is the one of the Abelian group U(1), which is related to electromagnetism. The group U(1) consists of the complex numbers of modulus 1:

$$\mathrm{U}(1) = \{\mathrm{e}^{-\mathrm{i}\alpha}|\ 0 \le \alpha < 2\pi\}.\tag{3.39}$$

It is a commutative or Abelian group with dimension 1. Complex charged fields $\phi(x)$ transform under local U(1) gauge transformations like

$$\phi(x) \to \phi'(x) = \mathrm{e}^{-\mathrm{i}\alpha(x)}\phi(x).\tag{3.40}$$

Since there is only one generator, the component notation is being used exclusively, and $A_\mu(x)$ denotes the real component of the gauge field. In terms of it the covariant derivative is

$$D_\mu = \partial_\mu - \mathrm{i}gA_\mu(x).\tag{3.41}$$

The transformation law reads

$$A'_\mu(x) = A_\mu(x) - \frac{1}{g}\partial_\mu\alpha(x).\tag{3.42}$$

The field strength

$$F_{\mu\nu} = \partial_\mu A_\nu - \partial_\nu A_\mu\tag{3.43}$$

is gauge invariant in contrast to the non-Abelian case. Moreover the fact that it is linear in A_μ implies that the action

$$S_A = \frac{1}{4} \int d^4x \, F_{\mu\nu} F_{\mu\nu} \tag{3.44}$$

does not contain interaction terms and describes a free field theory. The field equations are

$$\partial_\mu F^{\mu\nu} = 0. \tag{3.45}$$

Obviously A_μ is the electromagnetic vector potential and $F_{\mu\nu}$ the Maxwell field strength tensor. For Quantum Electrodynamics, where $A_\mu(x)$ couples to a Dirac field representing the electron, the coupling g should be replaced by $-e$, where e is the negative electron charge.

Since the gauge fields commute, the formula for the parallel transporters

$$U(\mathscr{C}) = \exp\left\{ ig \int_{\mathscr{C}} A_\mu(x) \, dx^\mu \right\} \tag{3.46}$$

does not require path ordering. In the particular case where $\mathscr{C}$ is a closed curve, which is the boundary of some area S,

$$\mathscr{C} = \partial S, \tag{3.47}$$

the argument of the exponential equals the magnetic flux through S:

$$U(\mathscr{C}) = \exp\left\{ ig \int_S F_{\mu\nu}(x) \, df^{\mu\nu} \right\}, \tag{3.48}$$

and we obtain the Aharonov–Bohm phase factor.

3.2 Lattice gauge fields and Wilson's action

3.2.1 Lattice gauge fields

In this section the concept of a gauge field will be transcribed to the case of a lattice regularization of the Euclidean continuum. For simplicity we consider a hypercubic lattice again. The matter field $\phi(x)$ is defined only for x being a lattice point, and correspondingly the local gauge transformations

$$\phi(x) \rightarrow \phi'(x) = \Lambda^{-1}(x)\phi(x) \tag{3.49}$$

are only defined on lattice points. The kinetic part of the action

$$\frac{1}{2} \sum_x a^4 \Delta_\mu^f \phi \cdot \Delta_\mu^f \phi = - \sum_{<xy>} a^2 \phi(x) \cdot \phi(y) + 4 \sum_x a^2 \phi(x) \cdot \phi(x), \tag{3.50}$$

where $< xy >$ are pairs of nearest neighbour points, is not invariant under local gauge transformations. In order to make it invariant, a gauge field must be introduced.

In contrast to the case of the continuum, where the gauge field is given in terms of parallel transporters along infinitesimal distances, the shortest non-zero distance on a hypercubic lattice is the lattice spacing a. Therefore the elementary parallel transporters on the lattice are those associated with the links (or bonds) b connecting nearest neighbour points. Let x be a point on the lattice and $x + a\hat{\mu}$ the neighbouring point in the direction of the lattice axis $\mu = 1, 2, 3, 4$. The corresponding directed link b is the straight path from x to $x + a\hat{\mu}$. It is identified with the ordered pair of points and denoted by

$$b = < x + a\hat{\mu}, x > \equiv (x, \mu). \tag{3.51}$$

The parallel transporter associated with a link b is denoted

$$U(b) \equiv U(x + a\hat{\mu}, x) \equiv U_{x\mu} \in G, \tag{3.52}$$

where G is the gauge group. It is called *link variable* and satisfies

$$U(y, x) = U^{-1}(x, y). \tag{3.53}$$

For an arbitrary path on the lattice

$$\mathscr{C} = b_n \circ \ldots \circ b_2 \circ b_1 \tag{3.54}$$

the parallel transporter

$$U(\mathscr{C}) = U(b_n) \ldots U(b_1) \equiv \prod_{b \in \mathscr{C}} U(b) \tag{3.55}$$

is determined by the link variables. Therefore we consider the collection of all link variables $\{U(b)\}$ as the lattice gauge field.

Under a gauge transformation the link variables transform as

$$U'(y, x) = \Lambda^{-1}(y) U(y, x) \Lambda(x). \tag{3.56}$$

Therefore a coupling term of the form

$$\sum_{<xy>} \phi(x) \cdot U(x, y) \phi(y) \tag{3.57}$$

is invariant under local gauge transformations. If we define the covariant lattice (forward) derivative through

$$D_\mu \phi(x) = \frac{1}{a} \left(U^{-1}(x, \mu) \phi(x + a\hat{\mu}) - \phi(x) \right), \tag{3.58}$$

the replacement of ordinary derivatives by covariant derivatives in the kinetic term yields

$$\frac{1}{2} \sum_x a^4 D_\mu \phi \cdot D_\mu \phi = -a^2 \sum_{<xy>} \phi(x) \cdot U(x, y) \phi(y) + 4a^2 \sum_x \phi(x)^2, \tag{3.59}$$

which contains the gauge invariant nearest neighbour coupling.

Other gauge invariant quantities can be constructed from expressions of the form

$$\phi(x) \cdot U(\mathscr{C}_{x,y})\phi(y) \tag{3.60}$$

or

$$\mathrm{Tr}\, U(\mathscr{C}_{x,x}), \tag{3.61}$$

where $\mathscr{C}_{x,x}$ is a closed curve.

3.2.2 Wilson's action

The latter type of terms are used for the construction of a gauge invariant action for the gauge field. The smallest closed loops on the lattice are the *plaquettes* consisting of four links, see fig. 3.2. A plaquette containing the points

$$x, \qquad x + a\hat{\mu}, \qquad x + a\hat{\mu} + a\hat{\nu}, \qquad x + a\hat{\nu} \tag{3.62}$$

and being oriented as in fig. 3.2, is denoted by

$$\mathrm{p} = (x; \mu, \nu). \tag{3.63}$$

The corresponding parallel transporter is abbreviated as

$$U_\mathrm{p} \equiv U_{x;\mu\nu} \equiv$$

$$U(x, x + a\hat{\nu})U(x + a\hat{\nu}, x + a\hat{\mu} + a\hat{\nu})U(x + a\hat{\mu} + a\hat{\nu}, x + a\hat{\mu})U(x + a\hat{\mu}, x) \tag{3.64}$$

and is called a *plaquette variable*. The action which has been proposed by Wilson [3.4] for the pure lattice gauge theory is defined in terms of the plaquette variables:

$$S[U] = \sum_\mathrm{p} S_\mathrm{p}(U_\mathrm{p}) \tag{3.65}$$

with the plaquette term

$$S_\mathrm{p}(U) = -\beta \left\{ \frac{1}{2\,\mathrm{Tr}\,\mathbf{1}} \left(\mathrm{Tr}\, U + \mathrm{Tr}\, U^{-1} \right) - 1 \right\} \tag{3.66}$$

$$= \beta \left\{ 1 - \frac{1}{N}\, \mathrm{Re}\,\mathrm{Tr}\, U \right\} \qquad \text{for SU}(N). \tag{3.67}$$

Here the sum over all plaquettes p is meant to include every plaquette only with one orientation:

$$\sum_\mathrm{p} \equiv \sum_x \sum_{1 \leq \mu < \nu \leq 4}. \tag{3.68}$$

The constant term in the action is physically insignificant and is often left out. The *Wilson action* is gauge invariant since

$$\text{Tr}\, U'_p = \text{Tr}\, U_p. \tag{3.69}$$

Furthermore it is real and positive. There are other possibilities for defining gauge invariant actions, which will be discussed later, but Wilson's choice appears to be the simplest one. Let us note that in an earlier work Wegner [3.5] studied models with this action for a discrete gauge group Z_2 (generalized Ising model) in the context of statistical mechanics.

Now we consider the question in which sense Wilson's action for $SU(N)$ is related to the Yang–Mills action for gauge fields on the continuum. In order to reveal their connection let

$$A_\mu(x) = -\mathrm{i}g A^b_\mu(x) T_b \tag{3.70}$$

be a Lie algebra valued vector field defined on the lattice and let

$$U(x, \mu) \equiv \mathrm{e}^{-aA_\mu(x)} = 1 - aA_\mu(x) + \frac{a^2}{2} A_\mu(x)^2 + \ldots. \tag{3.71}$$

Using

$$A_v(x + a\hat{\mu}) = A_v(x) + a\Delta^f_\mu A_v(x) \tag{3.72}$$

and the Campbell–Baker–Hausdorff formula

$$\mathrm{e}^x \mathrm{e}^y = \mathrm{e}^{x+y+(1/2)[x,y]+\ldots} \tag{3.73}$$

one finds

$$U_{x;\mu v} = \exp -a^2 G_{\mu v}(x) \tag{3.74}$$

with

$$G_{\mu v}(x) = F_{\mu v}(x) + \mathcal{O}(a) \tag{3.75}$$

$$F_{\mu v}(x) = \Delta^f_\mu A_v(x) - \Delta^f_v A_\mu(x) + [A_\mu(x), A_v(x)]. \tag{3.76}$$

This yields

$$\text{Tr}\,(U_p + U_p^{-1}) = 2\,\text{Tr}\,1 + a^4\,\text{Tr}\,(F_{\mu v}(x))^2 + \mathcal{O}(a^5) \tag{3.77}$$

since

$$\text{Tr}\, G_{\mu v}(x) = 0. \tag{3.78}$$

With

$$\sum_p \text{Tr}\,(F_{\mu v}(x))^2 = \frac{1}{2} \sum_{x,\mu,v} \text{Tr}\,(F_{\mu v}(x))^2 \tag{3.79}$$

we finally obtain for the Wilson action

$$S = -\frac{\beta}{4N} \sum_x a^4 \operatorname{Tr} F_{\mu\nu}(x) F^{\mu\nu}(x) + \mathcal{O}(a^5) \,. \tag{3.80}$$

Thus the leading term for small a coincides with the Yang–Mills action if we set

$$\beta = \frac{2N}{g^2} \,, \tag{3.81}$$

and identify g with the bare coupling constant of the lattice theory.

In the Abelian case, $G = \mathrm{U}(1)$, the same calculation gives

$$S = \frac{\beta g^2}{4} \sum_x a^4 F_{\mu\nu}(x) F^{\mu\nu}(x) + \mathcal{O}(a^5) \tag{3.82}$$

such that the relation between β and g in this case is

$$\beta = \frac{1}{g^2} \,. \tag{3.83}$$

3.2.3 Functional integral

After having defined the field variables and an action the next step in our approach to quantizing gauge fields is to specify the functional integral. In the continuum one would write down the formal expression

$$\langle \mathcal{O} \rangle = \frac{1}{Z} \int \mathcal{D}[A_\mu] \; \mathcal{O} \; \exp\left(-S[A_\mu]\right) \tag{3.84}$$

for the expectation value of some observable $\mathcal{O}$. The integral is meant to be a functional integral over all configurations of the gauge field. As it stands this expression is not well defined. However, as is well known, this expression can be taken as the starting point for the derivation of perturbation theory, if it is supplied with further *gauge fixing factors*.

Let us now consider the case of lattice gauge fields. On a lattice the gauge field is given by the configuration of link variables

$$\{U(b)\} \equiv U \,, \tag{3.85}$$

and observables are functions

$$\mathcal{O}(\{U(b)\}) \tag{3.86}$$

of the link variables. The expectation value of $\mathcal{O}$ is given by

$$\langle \mathcal{O} \rangle = \frac{1}{Z} \int \prod_b dU(b) \; \mathcal{O} \; \exp\left(-S(U)\right), \tag{3.87}$$

where

$$Z = \int \prod_b dU(b) \, \exp(-S(U)) \tag{3.88}$$

and $S(U)$ is the Wilson action. If matter fields $\phi(x)$ are present in addition, the corresponding integrals have to be included:

$$\langle \mathcal{O} \rangle = \frac{1}{Z} \int \prod_b dU(b) \prod_x d\phi(x) \, \mathcal{O} \, \exp(-S(U, \phi)). \tag{3.89}$$

In these formulae the finite-dimensional integration measure $dU(b)$ for the link variables $U(b)$ has to be specified. A good choice should respect gauge invariance. As we shall see below, this is guaranteed if dU is taken to be the *invariant group measure* or *Haar measure* on the gauge group (see e.g. [3.6]). For any compact group G the Haar measure is the unique measure dU on G which obeys

1. invariance:

$$\int_G f(U) \, dU = \int_G f(VU) \, dU = \int_G f(UV) \, dU \quad \text{for all } V \in G, \tag{3.90}$$

2. normalization:

$$\int_G dU = 1. \tag{3.91}$$

It satisfies

$$\int_G f(U) \, dU = \int_G f(U^{-1}) \, dU. \tag{3.92}$$

As an example let us take $G = \mathrm{SU}(2)$. If the group elements are parametrized as

$$U = x^0 \mathbf{1} + i \vec{x} \cdot \vec{\tau} = \begin{pmatrix} x^0 + ix^3 & x^2 + ix^1 \\ -x^2 + ix^1 & x^0 - ix^3 \end{pmatrix}, \tag{3.93}$$

the parameters x^i have to satisfy the condition

$$\det U = x^2 = (x^0)^2 + \vec{x}^2 = 1, \tag{3.94}$$

which defines the sphere S^3. The Haar measure then coincides with the uniform measure on S^3:

$$dU = \frac{1}{\pi^2} \delta(x^2 - 1) \, d^4 x. \tag{3.95}$$

Another common parametrization of SU(2) is in terms of an angle φ and a unit vector $\vec{n}$:

$$U = e^{i\varphi \vec{n} \cdot \vec{\tau}/2} = \left(\cos \frac{\varphi}{2} \right) \mathbf{1} + i \left(\sin \frac{\varphi}{2} \right) \vec{n} \cdot \vec{\tau}, \qquad 0 \le \varphi < 2\pi, \ |\vec{n}| = 1. \tag{3.96}$$

In terms of these the Haar measure reads:

$$dU = \frac{1}{4\pi^2} \, d\varphi \, d\Omega(\check{n}) \left(\sin \frac{\varphi}{2} \right)^2 , \tag{3.97}$$

where $d\Omega(\check{n})$ is the uniform measure on the unit sphere S^2.

Some properties of the functional integral which are worth noting here are the following:

1.) In a finite volume the number of variables is finite. Because the domain of integration is compact the functional integral is well defined without fixing of a gauge.

Sometimes it is said in the framework of continuum gauge theories that gauge fixing is made necessary by the infinite volume of the group of gauge transformations. This is not quite correct. On a lattice the volume of the gauge group is unity through normalization. This is independent of the lattice volume and of the lattice spacing and therefore also holds in the continuum limit. Indeed the volume of the gauge group would factor out in the expression for expectation values and is irrelevant.

Gauge fixing becomes necessary if a saddle-point approximation to the functional integral, like in perturbation theory, is attempted, because then one cannot accept zero modes in the quadratic component of the action.

2.) Gauge invariance: under gauge transformations

$$U'(x, y) = \Lambda^{-1}(x) U(x, y) \Lambda(y) \tag{3.98}$$

the measure and action are invariant:

$$dU = dU', \qquad S(U) = S(U'). \tag{3.99}$$

Let $f(\{U(b)\})$ be some function which only depends on variables inside a finite volume. The gauge transform of f is

$$f_\Lambda(\{U(x, y)\}) = f(\{\Lambda^{-1}(x) U(x, y) \Lambda(y)\}). \tag{3.100}$$

Then we have

$$\langle f \rangle = \frac{1}{Z} \int \prod_b dU'(b) \, f(\{U'(b)\}) \, e^{-S(U')}$$

$$= \frac{1}{Z} \int \prod_b dU(b) \, f_\Lambda(\{U(b)\}) \, e^{-S(U)} = \langle f_\Lambda \rangle, \tag{3.101}$$

which is the property of gauge invariance. In particular, if

$$\bar{f}(\{U(b)\}) = \int \prod_x d\Lambda(x) \, f_\Lambda(\{U(b)\}) \tag{3.102}$$

is the gauge invariant mean of f, we find

$$\langle f \rangle = \langle \bar{f} \rangle. \tag{3.103}$$

At this point we would like to emphasize that local gauge invariance cannot be broken spontaneously, a fact which is also called *Elitzur's theorem* [3.7]. This is due to the fact that, for observables of the type considered above, the expectation value can be seen to be gauge invariant by employing gauge transformations which are restricted to a compact space–time volume only. Spontaneous breaking of a symmetry, on the other hand, requires symmetry transformations acting on an infinite number of variables. Formal proofs of the absence of spontaneous breaking of local gauge symmetry can be found in [3.7, 3.8].

A particular consequence is the vanishing of the expectation value of link variables:

$$\langle U(b) \rangle = 0. \tag{3.104}$$

3.) Discrete gauge groups are possible, in contrast to the continuum case. As a popular example we mention $Z_2 = \{+1, -1\}$, in which case the link variables assume values

$$\sigma(b) = \pm 1 \tag{3.105}$$

as in the Ising model. Wilson's action is

$$S = -\beta \sum_{\mathrm{p}} \{\sigma_{\mathrm{p}} - 1\} \tag{3.106}$$

with

$$\sigma_{\mathrm{p}} = \sigma(b_1)\sigma(b_2)\sigma(b_3)\sigma(b_4) \tag{3.107}$$

for a plaquette p consisting of the links b_1, b_2, b_3, b_4. The invariant measure on Z_2 is given by

$$\int_{Z_2} f(\sigma)\,\mathrm{d}\sigma = \frac{1}{2}\{f(1) + f(-1)\}. \tag{3.108}$$

Other discrete gauge groups like Z_n, crystallographic subgroups of SO(3), etc. have been investigated.

The functional integrals of lattice gauge theory are formally identical to partition functions or expectation values of statistical systems in four dimensions, as has been noted earlier for a general Euclidean quantum field theory. The bare coupling constant squared, g^2, plays a rôle analogous to the temperature β^{-1}. Therefore methods which are used in statistical mechanics can also be employed for the investigation of lattice gauge theories. In particular the following methods are at our disposal, some of which shall be discussed in more detail in other sections:

a) High-temperature expansions,
    which are expansions in powers of $\beta \sim g^{-2}$. They are therefore also
    called *strong-coupling expansions*.

b) Low-temperature expansions,

which amount to so-called *contour expansions* in the case of discrete gauge groups, and to perturbation theory if the gauge group is continuous.

c) Mean-field approximation.

d) Monte Carlo integration, which is a numerical evaluation of the functional integral by statistical methods.

e) Exact results: correlation inequalities etc. (see e.g. [3.9]).

3.2.4   Observables

In the context of Euclidean functional integrals, observables are gauge invariant functions of the basic variables of the theory. Restricting ourselves to the pure gauge theory without matter fields, observables are gauge invariant functions of the link variables. As has been mentioned earlier, particular examples are the traces of parallel transporters around closed loops

$$\text{Tr}_j\, U(\mathscr{C}),\tag{3.109}$$

where Tr_j denotes the trace in some representation j of the gauge group. The expectation value of such a loop variable

$$W(\mathscr{C}) \equiv \langle \text{Tr}\, U(\mathscr{C}) \rangle\tag{3.110}$$

is called the *Wilson loop*. More generally one may consider

$$W(\mathscr{C}_1, \mathscr{C}_2, \ldots, \mathscr{C}_n) \equiv \langle \text{Tr}\, U(\mathscr{C}_1)\, \text{Tr}\, U(\mathscr{C}_2) \ldots \text{Tr}\, U(\mathscr{C}_n) \rangle.\tag{3.111}$$

The special rôle of these observables is revealed by the following fact [3.10]: every observable, which depends continuously on the link variables, can be approximated arbitrarily well by expressions of the form

$$\sum_{n \geq 0} \sum_{\mathscr{C}_1, \ldots, \mathscr{C}_n} a(\mathscr{C}_1, \ldots, \mathscr{C}_n)\, \text{Tr}\, U(\mathscr{C}_1) \ldots \text{Tr}\, U(\mathscr{C}_n).\tag{3.112}$$

Some observables, which are considered frequently, deserve special mention. The *internal energy* is defined as

$$E \equiv \left\langle 1 - \frac{1}{\text{Tr}\,\mathbf{1}} \text{Tr}\, U_\text{p} \right\rangle \geq 0.\tag{3.113}$$

It equals the expectation value of the contribution of a single plaquette to the action.

Next consider the connected correlation function

$$G(t) = \langle \text{Tr}\, U_{\text{p}_1}\, \text{Tr}\, U_{\text{p}_2} \rangle - \langle \text{Tr}\, U_{\text{p}_1} \rangle \langle \text{Tr}\, U_{\text{p}_2} \rangle\tag{3.114}$$

between two plaquettes p_1 and p_2, separated in time x_4 by a distance t. Its fall-off for large t determines the *correlation length* through the definition

$$\xi^{-1} \equiv -\lim_{t\to\infty} \frac{1}{t} \log G(t). \qquad (3.115)$$

Typically a behaviour like

$$G(t) \underset{t\to\infty}{\sim} c\, t^p\, e^{-t/\xi} \qquad (3.116)$$

is to be expected. Whereas the correlation length is dimensionless, the *mass gap*

$$m \equiv \frac{1}{\xi a} \qquad (3.117)$$

has dimensions of a mass. It equals the smallest mass in the particle spectrum as the transfer matrix formalism shows (see sections 1.5.2 and 3.2.6).

Next consider a rectangular loop $\mathscr{C}_{R,T}$ of side lengths R and T. The *static quark potential* $V(R)$ is defined by means of the large T behaviour of the corresponding Wilson loop in the fundamental representation:

$$V(R) \equiv -\lim_{T\to\infty} \frac{1}{T} \log W(\mathscr{C}_{R,T}), \qquad (3.118)$$

so that

$$W(\mathscr{C}_{R,T}) \underset{T\to\infty}{\sim} C e^{-T\,V(R)}. \qquad (3.119)$$

$V(R)$ gives the energy of the gauge field in the presence of two static colour sources separated by a distance R. We postpone the explanation of this fact to section 3.5.

Finally the coefficient

$$\alpha \equiv \lim_{R\to\infty} \frac{1}{R} V(R) = -\lim_{R,T\to\infty} \frac{1}{RT} \log W(\mathscr{C}_{R,T}) \qquad (3.120)$$

is called *string tension*. If the string tension does not vanish, the potential $V(R)$ asymptotically rises linearly with R:

$$V(R) \underset{R\to\infty}{\sim} \alpha R. \qquad (3.121)$$

In this case a constant force α would act between widely separated colour sources, and one speaks of *static quark confinement*. Large Wilson loops then obey the *area law* [3.4]

$$W(\mathscr{C}_{R,T}) \underset{R,T\to\infty}{\sim} C e^{-\alpha RT}. \qquad (3.122)$$

3.2.5 Gauge fixing

Although gauge fixing is not necessary, it may sometimes be advantageous to fix a gauge in lattice gauge theory. On a lattice there is a special way of fixing the gauge such that some of the link variables are set equal to prescribed values [3.11]. This is described in the following.

Let B be a set of links which does not contain any closed loop. For this set B choose some fixed configuration

$$U_0(b), \qquad b \in B \tag{3.123}$$

of link variables. It is not difficult to verify that for each lattice gauge field $\{U(b)\}$ there is a gauge transformation

$$U(x, y) \rightarrow U'(x, y) = \Lambda^{-1}(x) U(x, y) \Lambda(y) \tag{3.124}$$

which achieves

$$U'(b) = U_0(b) \qquad \text{for all } b \in B. \tag{3.125}$$

Now let f be some function of the link variables. Owing to gauge invariance of the functional integral, it is sufficient to consider f to be gauge invariant,

$$f = f_\Lambda, \tag{3.126}$$

with f_Λ defined in (3.100). The expectation value of f can be written in the form

$$\langle f \rangle = \frac{1}{Z} \int \prod_{b \in B} \mathrm{d}U(b)\, F(U|_B), \tag{3.127}$$

where

$$F(U|_B) = \int \prod_{b \notin B} \mathrm{d}U(b)\, f(U)\, \mathrm{e}^{-S(U)} \tag{3.128}$$

only depends on the variables

$$U|_B = \{U(b)|b \in B\}. \tag{3.129}$$

Since the gauge transformation Λ, defined above, only depends on $U|_B$, we can perform the change of variables $U \rightarrow U'$ in the integral (3.128). Owing to

$$f(U) = f(U'), \; S(U) = S(U'), \; \mathrm{d}U(b) = \mathrm{d}U'(b) \text{ for } b \notin B \tag{3.130}$$

we find that

$$F(U|_B) = F(U'|_B) = F(U_0) \tag{3.131}$$

is a constant. Therefore finally

$$\langle f \rangle = \frac{1}{Z} F(U_0) = \frac{1}{Z} \int \prod_{b \notin B} \mathrm{d}U(b) \, f(\tilde{U}) \, \mathrm{e}^{-S(\tilde{U})} , \qquad (3.132)$$

with

$$\tilde{U}(b) = \left\{ \begin{array}{ll} U(b), & b \notin B \\ U_0, & b \in B . \end{array} \right. \qquad (3.133)$$

This result means that in the calculation of expectation values of gauge invariant functions, in particular of observables, one may fix the link variables on B to arbitrary prescribed values. This way of gauge fixing does not introduce any non-trivial Jacobian or Faddeev–Popov ghosts.

An example of such a gauge fixing is the *temporal gauge*, where on an infinite lattice one chooses

$$B = \{ < x, y > \mid y = x \pm a\hat{4} \}, \qquad \hat{4} = (0, 0, 0, 1) , \qquad (3.134)$$

$$U_0(b) = \mathbf{1} . \qquad (3.135)$$

It corresponds to the gauge

$$A_4 = 0 \qquad (3.136)$$

in the continuum. This is not a complete gauge fixing, however, because in a given timeslice, say $x_4 = 0$, one can still fix some link variables to unity. A complete gauge fixing is achieved by fixing the link variables on a *maximal tree*, which is a maximal set of links without closed loops.

3.2.6 Transfer matrix

As has been discussed in the first chapter, the transfer matrix of a lattice model provides the relation between the functional integral and the Hamiltonian, and is of central importance for a quantum mechanical interpretation of correlation functions. In the following we shall write down the transfer matrix for lattice gauge theory with Wilson's action [3.4, 3.11, 3.12]. For this purpose some notation has to be introduced.

We denote the set of links that are completely contained in the timeslice $x_4 = t$ by B_t. Those links which connect the timeslices $x_4 = t + 1$ and $x_4 = t$ form the set $B_{t+1,t}$. The corresponding collections of link variables are denoted

$$\mathcal{U}_t \equiv \{ U(b) \mid b \in B_t \} , \qquad (3.137)$$

$$\mathcal{U}_{t+1,t} \equiv \{ U(b) \mid b \in B_{t+1,t} \} . \qquad (3.138)$$

The plaquettes are divided in a similar way, namely P_t consists of all plaquettes which are completely contained in the timeslice $x_4 = t$, and $P_{t+1,t}$ of those which connect $x_4 = t + 1$ and $x_4 = t$.

The action can be written as

$$S = \sum_t L[\mathcal{U}_{t+1}, \mathcal{U}_{t+1,t}, \mathcal{U}_t], \qquad (3.139)$$

$$L[\mathcal{U}_{t+1}, \mathcal{U}_{t+1,t}, \mathcal{U}_t] = \frac{1}{2} L_1[\mathcal{U}_{t+1}] + \frac{1}{2} L_1[\mathcal{U}_t] + L_2[\mathcal{U}_{t+1}, \mathcal{U}_{t+1,t}, \mathcal{U}_t], \quad (3.140)$$

with the following components:

$$L_1[\mathcal{U}_t] = -\frac{\beta}{N} \sum_{p \in P_t} \text{Re Tr } U_p \qquad (3.141)$$

is the contribution of space-like plaquettes on a timeslice and

$$L_2[\mathcal{U}_{t+1}, \mathcal{U}_{t+1,t}, \mathcal{U}_t] = -\frac{\beta}{N} \sum_{p \in P_{t+1,t}} \text{Re Tr } U_p \qquad (3.142)$$

is the contribution of time-like plaquettes between two timeslices.

The transfer matrix is a matrix which acts on square-integrable wave functions

$$\Psi[\mathcal{U}] \qquad (3.143)$$

depending on the link variables in a timeslice. Its matrix elements are defined by

$$\mathbf{T}[\mathcal{U}_{t+1}, \mathcal{U}_t] = \int \prod_{b \in B_{t+1,t}} dU(b) \; \exp -L[\mathcal{U}_{t+1}, \mathcal{U}_{t+1,t}, \mathcal{U}_t]. \qquad (3.144)$$

For a lattice which extends over n units in the time direction with periodic boundary conditions, it is obvious that the partition function Z defined in (3.88) is given by

$$Z = \text{Tr}(\mathbf{T}^n) = \int \prod_t \left(\mathbf{T}[\mathcal{U}_{t+1}, \mathcal{U}_t] \prod_{b \in B_t} dU(b) \right). \qquad (3.145)$$

It is easy to check that $\mathbf{T}$ is a self-adjoint, bounded operator. The positivity of $\mathbf{T}$, which is necessary for the existence of the Hamiltonian

$$H \equiv -\frac{1}{a} \log \mathbf{T}, \qquad (3.146)$$

is a consequence of reflection positivity, which will be discussed below.

A simpler expression for the transfer matrix is obtained by fixing the temporal gauge, in which

$$U(b) = \mathbf{1} \qquad \text{for } b \in B_{t+1,t}. \qquad (3.147)$$

In this case

$$\mathbf{T}[\mathscr{U}_{t+1}, \mathscr{U}_t] = \exp -L[\mathscr{U}_{t+1}, \mathbf{1}, \mathscr{U}_t]. \qquad (3.148)$$

Since the local gauge invariance is not broken spontaneously there is a unique vacuum vector $|\Omega\rangle$, which is the eigenvector belonging to the largest eigenvalue of $\mathbf{T}$. The projection onto $|\Omega\rangle$ can be represented as

$$P_0 = \lim_{n\to\infty} \frac{1}{Z} \mathbf{T}^n. \qquad (3.149)$$

Using this formula the vacuum wave function $\Omega(\mathscr{U}_0)$, written as a function of the variables at time zero, can be represented as a functional integral over the variables at negative times [3.13]:

$$\Omega(\mathscr{U}_0) = \int \prod_{b\in B_-} \mathrm{d}U(b)\; \mathrm{e}^{-S_-}, \qquad (3.150)$$

where

$$B_- = \bigcup_{t<0} (B_t \cup B_{t+1,t}), \qquad (3.151)$$

$$S_- = \sum_{t<0} L[\mathscr{U}_{t+1}, \mathscr{U}_{t+1,t}, \mathscr{U}_t]. \qquad (3.152)$$

An analogous expression in quantum gravity is called the *Hartle–Hawking wave function* [3.14].

3.2.7   Group characters

In later chapters we shall repeatedly make use of the characters of the gauge group. Therefore we insert here a section in which some group theoretical material related to characters is reviewed. More details can be found in [3.6, 3.15].

So far only the fundamental N-dimensional representation of SU(N) has been considered. More generally let r be some unitary representation of the compact Lie group G. The corresponding matrices representing the group elements U are denoted

$$D^{(r)}(U). \qquad (3.153)$$

A matter field in the representation r transforms under gauge transformations according to

$$\phi'(x) = D^{(r)}(\Lambda^{-1}(x))\phi(x) \qquad (3.154)$$

and the gauge invariant coupling term is

$$\phi(x) \cdot D^{(r)}(U(x,y))\phi(y). \qquad (3.155)$$

The *character* of r is the function

$$\chi_r(U) \equiv \mathrm{Tr}\, D^{(r)}(U). \tag{3.156}$$

Its value at unity

$$\chi_r(\mathbf{1}) = d_r \tag{3.157}$$

yields the dimension of the representation r. Characters are invariant functions or *class functions* on the group, which means that

$$\chi_r(VUV^{-1}) = \chi_r(U), \qquad U, V \in G. \tag{3.158}$$

Of particular importance are the unitary irreducible representations of G and their characters. The set of all inequivalent unitary irreducible representations of G is denoted $\hat{G}$. In the following r and s are always understood to be in $\hat{G}$. A crucial property is the orthogonality of characters:

$$\int dU\, \overline{\chi_r}(U)\chi_s(U) = \delta_{rs}. \tag{3.159}$$

Moreover the theorem of Peter and Weyl states that the χ_r form a basis of the space of square-integrable class functions on the group G. That means that every square-integrable function $f(U)$, obeying

$$f(U) = f(VUV^{-1}), \tag{3.160}$$

can be expanded into characters:

$$f(U) = \sum_{r \in \hat{G}} f_r \chi_r(U) \tag{3.161}$$

$$f_r = \int dU\, \overline{\chi_r}(U) f(U). \tag{3.162}$$

This is the Fourier analysis or *harmonic analysis* on group manifolds. The completeness relation of characters can be written as

$$\sum_{r \in \hat{G}} \chi_r(U)\overline{\chi_r}(V) = \delta(UV^{-1}). \tag{3.163}$$

The invariant δ-function on G is defined by means of

$$\int dU\, f(U)\delta(UV^{-1}) = f(V), \tag{3.164}$$

and obeys

$$\delta(U) = \delta(U^{-1}). \tag{3.165}$$

Let us take $G = \mathrm{SU}(2)$ as an example. The unitary irreducible representations of SU(2) are labelled by $j = 0, 1/2, 1, 3/2, \ldots$ and have dimensions

$d_j = 2j + 1$. In the parametrization (3.96) of group elements, class functions only depend on the invariant angle φ, which we indicate by writing them as

$$f(U) \equiv f(\varphi). \tag{3.166}$$

The characters are

$$\chi_j(U) = \frac{\sin\left(j + \frac{1}{2}\right)\varphi}{\sin\frac{1}{2}\varphi}. \tag{3.167}$$

The expression (3.162) for the Fourier coefficients explicitly reads

$$f_j = \frac{1}{2\pi} \int_0^{2\pi} d\varphi \, \sin\frac{1}{2}\varphi \, \sin\left(j + \frac{1}{2}\right)\varphi \, f(\varphi). \tag{3.168}$$

In the case of gauge group U(1) the representations are labelled by integers n and the characters are

$$\chi_n(e^{-i\varphi}) = e^{-in\varphi}, \qquad d_n = 1. \tag{3.169}$$

Harmonic analysis on U(1) is just ordinary Fourier analysis.

In lattice gauge theory with Wilson's action, each plaquette contributes a factor

$$\exp -S_p(U_p) \tag{3.170}$$

to the total Boltzmann factor in the functional integral. We write its character expansion in the form

$$\exp -S_p(U) = \sum_{r \in \hat{G}} d_r c_r(\beta) \chi_r(U). \tag{3.171}$$

Explicit expressions for the coefficients c_r are

$$c_n = I_n(\beta) \, e^{-\beta} \qquad \text{for U(1)}, \tag{3.172}$$

$$c_j = \frac{2}{\beta} I_{2j+1}(\beta) \, e^{-\beta} \qquad \text{for SU(2)}, \tag{3.173}$$

where I_n are modified Bessel functions. For other groups see [3.16].

A final topic to be treated here is differential operators on the group. A representation of G on the space of square-integrable functions on G is the (left-) regular representation. Any group element $V \in G$ is represented by an operator $R(V)$ acting on functions $f(U)$ via

$$[R(V)f](U) = f(VU). \tag{3.174}$$

For V being parametrized as

$$V = e^{-i\omega_a T^a} \tag{3.175}$$

$R(V)$ can be expressed as

$$R(V) = e^{-i\omega_a L^a} . \tag{3.176}$$

The operators L^a are the generators in the regular representation. They are defined by

$$[L^a f](U) = i\frac{d}{ds} f(e^{-isT^a}U)_{s=0} \tag{3.177}$$

and act as differential operators on functions f. Needless to say, they obey the commutation relations of the Lie algebra of G. The quadratic Casimir operator of the regular representation

$$C_R^{(2)} = \sum_a L^a L^a \tag{3.178}$$

is a second order differential operator, which commutes with the L^a. It coincides with the negative Laplace–Beltrami operator on G:

$$\sum_a L^a L^a = -\Delta_G . \tag{3.179}$$

3.2.8 Reflection positivity

With the character expansion at our disposal we can make up for showing that reflection positivity holds for pure lattice gauge theory with Wilson's action. From this follows the positivity of the transfer matrix and the existence of a Hamiltonian H, as has been discussed in section 1.3.5. In order to simplify the discussion the temporal gauge will be fixed. The Euclidean time reflection acts on link variables as

$$\Theta U(x, y) = \overline{U(\theta x, \theta y)} . \tag{3.180}$$

Site-reflection positivity is easy to check. The action is split according to

$$S = S_+ + \Theta S_+ + S_0 , \tag{3.181}$$

where

$$S_0 = L_1[\mathscr{U}_0] \tag{3.182}$$

and S_+ depends on the link variables in B_+, i.e. at positive times, only. The proof of site-reflection positivity is then completely parallel to the one in section 1.3.5.

For link-reflection positivity [3.17, 3.9] the action in temporal gauge is split as

$$S = S'_+ + \Theta S'_+ + S_c , \tag{3.183}$$

with

$$S_c = L_2[\mathcal{U}_1, \mathbf{1}, \mathcal{U}_0] = -\frac{\beta}{N} \sum_{\mathbf{x}} \sum_{i=1}^{3} \text{Re Tr } U^+_{\mathbf{x}+\hat{4},i} U_{\mathbf{x},i}, \qquad (3.184)$$

and the sum over $\mathbf{x}$ goes over all points in the timeslice $x_4 = 0$. Each term of this sum contributes a factor to the functional integral. The character expansion of such a factor is

$$\sum_{r \in \hat{G}} d_r c_r \chi_r(U^+_{\mathbf{x}+\hat{4},i} U_{\mathbf{x},i})$$

$$= \sum_{r \in \hat{G}} d_r c_r \sum_{k,l} \{(\Theta D^{(r)}_{kl}(U_{\mathbf{x},i})) D^{(r)}_{kl}(U_{\mathbf{x},i})\}. \qquad (3.185)$$

If all coefficients c_r are positive, $\exp - S_c$ is a sum of the form

$$e^{-S_c} = \sum_m b_m (\Theta S_m) S_m \qquad (3.186)$$

with positive coefficients b_m, and link-reflection positivity follows in the same way as in section 1.3.5.

The positivity of the character expansion coefficients c_r can be inferred from the integral representation

$$d_r c_r = \int dU \, \overline{\chi_r}(U) e^{-S_p(U)} \qquad (3.187)$$

and the fact that

$$\int dU \, \overline{\chi_r}(U) (\text{Tr } U)^m (\text{Tr } U^+)^n$$

is non-negative, completing the proof.

3.2.9   Other actions

Wilson's action is not the only possible choice. In the literature other gauge invariant actions for lattice gauge theory have been proposed and we would like to mention three of them.

In *Manton's action* [3.18] the plaquette term is taken to be

$$S_p(U) = \frac{1}{g^2} d^2(U, \mathbf{1}), \qquad (3.188)$$

where $d(U, \mathbf{1})$ is the distance between the group elements U and $\mathbf{1}$ with respect to the invariant metric on the group manifold. For SU(2) in the parametrization (3.96) we have

$$S_p(U) = \frac{1}{2g^2} \varphi^2. \qquad (3.189)$$

For small angles this approaches Wilson's action

$$S_\mathrm{p}(U) = \frac{4}{g^2}\left(1 - \cos\frac{\varphi}{2}\right). \tag{3.190}$$

Reflection positivity does not hold for Manton's action [3.19].

For the *heat kernel action* [3.20] the choice is

$$e^{-S_\mathrm{p}(U)} = K\left(U, \frac{N}{\beta'}\right)\bigg/ K\left(1, \frac{N}{\beta'}\right), \qquad \beta' = \frac{2N}{g^2} + \frac{N^2}{12}, \tag{3.191}$$

with the heat kernel $K(U,t)$ on the group manifold. It is equal to the matrix element

$$K(U,t) = \langle U|e^{t\Delta_G}|1\rangle, \tag{3.192}$$

where Δ_G is the Laplace–Beltrami operator on G. Alternatively it can be defined by means of the heat equation on G:

$$\frac{\mathrm{d}}{\mathrm{d}t}K(U,t) = \Delta_G K(U,t). \tag{3.193}$$

Its character expansion is explicitly known,

$$K(U,t) = \sum_{r\in\hat{G}} d_r e^{-tC_r^{(2)}}\chi_r(U), \tag{3.194}$$

as follows from the formulae of section 3.2.7, where $C_r^{(2)}$ is the value of the quadratic Casimir operator in representation r. This facilitates the derivation of strong-coupling expansions.

For the case of U(1) the heat kernel action coincides with what is called *Villain's action* [3.21], which is a periodic Gaussian:

$$e^{-S_\mathrm{p}(U)} \propto \sum_n \exp{-\frac{\beta}{2}(\varphi - 2\pi n)^2}, \qquad U = e^{-\mathrm{i}\varphi}. \tag{3.195}$$

Finally we mention the *mixed actions*

$$S_\mathrm{p}(U) = \beta_f \operatorname{Re}\operatorname{Tr}U + \beta_A\chi_A(U), \tag{3.196}$$

where A denotes the adjoint representation of the gauge group.

All these actions reproduce the Yang–Mills action in the classical continuum limit. The corresponding gauge theories have been investigated by various means [3.22, 3.23], and it is believed that they belong to the same universality class as Wilson's action. In this book we shall not be concerned further with other actions and restrict ourselves to the Wilson action in the following.

3.3 Perturbation theory

The lattice has been introduced mainly in order to provide us with a regularization of quantum field theories, which is not tied to a specific approximation method. This in contrast to other regularizations like Pauli–Villars or dimensional regularization, which are defined only in the framework of perturbation theory. The formulation of gauge theories on a lattice opens the possibility of applying a multitude of non-perturbative methods. Nevertheless, perturbation theory may be applied to lattice gauge theories too. The lattice represents a valid cut-off for perturbative loop integrals, which regularizes the divergencies that appear.

As will become clear below, perturbative calculations in lattice gauge theory are rather laborious compared to other schemes in the continuum. Therefore it would not be efficient to perform such calculations on the lattice, if it could be avoided. Lattice perturbation theory becomes necessary, however, in discussing the continuum limit. This is due to the property of asymptotic freedom, which implies that the continuum limit is to be taken at small bare couplings. In order to relate mass scales, like glueball masses, which have been evaluated on the lattice, to mass scales, which are fixed in a perturbative framework, like the so-called Λ-parameters, one has to consider lattice perturbation theory.

Another application of lattice perturbation theory concerns scaling violations near the continuum limit. These are corrections to the continuum limit of physical quantities proportional to powers of the lattice spacing a. For various quantities they can be estimated in lattice perturbation theory. Related to this is the concept of *improved actions* [3.24], which aim at a reduction of scaling violations. Improved actions for lattice gauge theory are calculable in lattice perturbation theory [3.25, 3.26, 3.27, 3.28, 3.29].

Perturbation theory amounts to an expansion of correlation functions in powers of the coupling constant. The perturbative techniques on the lattice are quite the same as in the continuum. The action is expanded around its minima in order to determine the propagator and the interaction vertices. The terms of the perturbative expansion can be represented by Feynman diagrams.

We assume that the reader is familiar with perturbation theory for gauge theories in the continuum including the gauge fixing procedure of Faddeev and Popov. Therefore only those elements of lattice perturbation theory which are characteristic for the lattice regularization will be explained in detail.

In particular it turns out that on the lattice an infinite number of vertices are present, corresponding to interaction terms with an arbitrary number of fields. Naively an action of this type would be considered to be non-renormalizable. The higher vertices, however, are accompanied by powers of the lattice spacing a, such that a cancellation of divergencies in the limit $a \to 0$ appears possible.

Perturbation theory for lattice gauge theories has been considered first in [3.30], where gauge fixing and renormalization at one-loop order is discussed for non-Abelian gauge groups. In [3.31] the question of renormalizability to all orders of perturbation theory is investigated for the case of Abelian lattice gauge theory. Hasenfratz and Hasenfratz [3.32] were the first who had the courage to undertake an explicit one-loop calculation in non-Abelian lattice gauge theory in order to relate lattice and continuum Λ parameters. The same was achieved in [3.33], where details about the calculation can be found. A proof of the renormalizability of lattice gauge theory to all orders of perturbation theory based on lattice power counting theorems was given by Reisz in [3.34].

In the presentation of lattice perturbation theory below we follow [3.33] to some extent.

3.3.1 Feynman rules

We consider perturbation theory for SU(N) lattice gauge theory with Wilson's action. For the conventions we refer to the previous sections and to the appendix.

Classical configurations with minimal action are the classical 'vacuum'

$$U_{x,\mu} = \mathbf{1} \tag{3.197}$$

and gauge transforms of it. Since we are going to fix the gauge later, we consider lattice gauge fields near the classical vacuum.

In order to be able to perform explicit calculations, a parametrization of group elements has to be chosen. We take the exponential parametrization

$$U = \exp(iT_a\omega^a), \qquad a = 1, \ldots, N^2 - 1, \tag{3.198}$$

with real parameters ω^a, which is unique near unity. The Haar measure dU on the group manifold must be expressed in terms of these parameters. This can be achieved if an invariant metric on the group manifold has been specified. On the Lie algebra su(N), which coincides with the tangent space of SU(N) at the unity element, a canonical metric or scalar product is given by the Killing form. For two elements

$$\omega = iT_a\omega^a, \qquad \xi = iT_a\xi^a \tag{3.199}$$

of the Lie algebra, the Killing form is

$$(\omega, \xi) \equiv -2\operatorname{Tr}(\omega\xi) = \omega^a\xi^a. \tag{3.200}$$

If it is translated to the tangent space at an arbitrary group element U via left multiplication with U, we obtain the metric on SU(N)

$$(ds)^2 \equiv -2\operatorname{Tr}\left\{(U^{-1}\partial_a U)(U^{-1}\partial_b U)\right\} d\omega^a \, d\omega^b$$

$$\equiv g_{ab}\, d\omega^a \, d\omega^b, \tag{3.201}$$

where

$$\partial_a U = \frac{\partial}{\partial \omega^a} U. \tag{3.202}$$

The line element above is invariant under translations

$$U \to VU. \tag{3.203}$$

Therefore the invariant group integration measure is, up to a normalization factor, given by

$$dU = (\det g)^{1/2} \prod_a d\omega^a. \tag{3.204}$$

Now there is a formula for the derivatives appearing in the metric. Let the adjoint representation of a Lie algebra element ω be defined through the commutator

$$\operatorname{Ad}\omega \cdot \xi = [\omega, \xi]. \tag{3.205}$$

It is well known that [3.35]

$$e^{-\omega}\xi e^{\omega} = \exp(-\operatorname{Ad}\omega) \cdot \xi$$

$$= \xi - [\omega, \xi] + \frac{1}{2!}[\omega, [\omega, \xi]] - \dots. \tag{3.206}$$

Define the function

$$E(x) = \frac{1 - \exp(-x)}{x} = \sum_{n=0}^{\infty} \frac{1}{(n+1)!}(-x)^n. \tag{3.207}$$

Then for $U = \exp \omega$ we have [3.35]

$$U^{-1}\partial_a U = E(\operatorname{Ad}\omega) \cdot iT_a. \tag{3.208}$$

Proof:

$$\frac{d}{ds}\left(e^{-s\omega}\partial_a e^{s\omega}\right) = e^{-s\omega}(\partial_a\omega)e^{s\omega}$$

$$= \exp(-s\operatorname{Ad}\omega) \cdot iT_a. \tag{3.209}$$

Integration with respect to s yields

$$e^{-\omega}\partial_a e^{\omega} = \int_0^1 ds \, \exp\left(-s\,\text{Ad}\,\omega\right) \cdot iT_a$$

$$= E(\text{Ad}\,\omega) \cdot iT_a \tag{3.210}$$

∎

In the basis $\{T_a\}$ the operator $\text{Ad}\,\omega$ is represented by an antisymmetric matrix

$$\overline{\omega}_{cb} = f_{cba}\omega^a \tag{3.211}$$

through

$$\text{Ad}\,\omega \cdot \xi = -i\omega^a \xi^b f_{abc} T_c \equiv iT_c \overline{\omega}_{cb}\xi^b. \tag{3.212}$$

Correspondingly we write

$$U^{-1}\partial_a U = E(\overline{\omega})_{ab} \, iT_b. \tag{3.213}$$

For the metric we obtain the expression

$$g_{ab} = 2\,\text{Tr}\left\{E(\overline{\omega})_{ac}T_c E(\overline{\omega})_{bd}T_d\right\}$$

$$= E_{ac}E_{bc} = (EE^t)_{ab} = \left(\frac{2}{\overline{\omega}}\sinh\frac{\overline{\omega}}{2}\right)^2_{ab}, \tag{3.214}$$

and this finally gives us the Haar measure up to a normalization factor:

$$dU = (\det EE^t)^{1/2} \prod_a d\omega^a. \tag{3.215}$$

Later we shall put the measure factor into the exponent, which leads to

$$\log\left(\det EE^t\right)^{1/2} = \frac{1}{2}\,\text{Tr}\log EE^t = \text{Tr}\log\left(\frac{2}{\overline{\omega}}\sinh\frac{\overline{\omega}}{2}\right)$$

$$= \text{Tr}\left(\frac{\overline{\omega}^2}{24} - \frac{\overline{\omega}^4}{2880} + \frac{\overline{\omega}^6}{181440} + \ldots\right) = -\frac{N}{24}\omega^a\omega^a + \mathcal{O}(\omega^4), \tag{3.216}$$

where the relation

$$f_{abc}f_{dbc} = N\delta_{ad} \tag{3.217}$$

has been used.

According to the considerations above and in section 3.2.2 the lattice gauge field is parametrized as

$$U_{x\mu} = \exp iag T_b A_\mu^b(x). \tag{3.218}$$

It is now straightforward but tedious to expand Wilson's action in powers of the bare coupling g:

$$S = \sum_x a^4 \{L_2 + gL_3 + g^2 L_4 + \ldots\}. \tag{3.219}$$

The leading term has already been determined in section 3.2.2:

$$L_2 = \frac{1}{4} \left(\Delta_\mu^f A_\nu^a(x) - \Delta_\nu^f A_\mu^a(x) \right)^2. \tag{3.220}$$

The next component is

$$L_3 = f_{abc} \left[A_\mu^a(x) A_\nu^b(x) \Delta_\mu^f A_\nu^c(x) + \frac{a}{2} A_\mu^a(x) \Delta_\nu^f A_\mu^b(x) \Delta_\mu^f A_\nu^c(x) \right]. \tag{3.221}$$

(Summation over μ, ν is implied here.) The first term in L_3 equals the lattice transcription of the corresponding triple gluon term of the continuum Lagrangian. The second term, however, is a typical lattice artefact. In the usual terminology of perturbation theory it corresponds to an operator of dimension five and would be called a non-renormalizable interaction. On the other hand it has an explicit factor a in front of it, which is sufficient to counteract the additional linear divergence produced by this term in Feynman diagrams [3.34]. In the language of the Kadanoff–Wilson renormalization group it is an irrelevant operator, which does not affect the scaling behaviour of the theory.

In higher orders of the expansion a proliferation of such terms takes place. Altogether the Wilson action produces an infinite number of interaction terms, leading to an infinite number of vertices.

It is in general preferable to do perturbative calculations in momentum space. Since the gauge field variable $A_\mu^b(x)$ is naturally associated with the midpoint of the link $(x, x + a\hat{\mu})$, its Fourier transform is defined by

$$\tilde{A}_\mu^b(k) = \sum_x a^4 e^{-ik \cdot (x + a\hat{\mu}/2)} A_\mu^b(x), \tag{3.222}$$

$$A_\mu^b(x) = \int_k e^{ik \cdot (x + a\hat{\mu}/2)} \tilde{A}_\mu^b(k). \tag{3.223}$$

The trilinear term is then

$$S_3 = \sum_x a^4 g L_3$$

$$= -\int_{kpq} (2\pi)^4 \delta(k + p + q) \tilde{A}_\mu^a(k) \tilde{A}_\nu^b(p) \tilde{A}_\rho^c(q) \frac{1}{3!} V_3(k, \mu, a; p, \nu, b; q, \rho, c), \tag{3.224}$$

where the triple vertex (see fig. 3.3) is

$$V_3(k, \mu, a; p, \nu, b; q, \rho, c) =$$

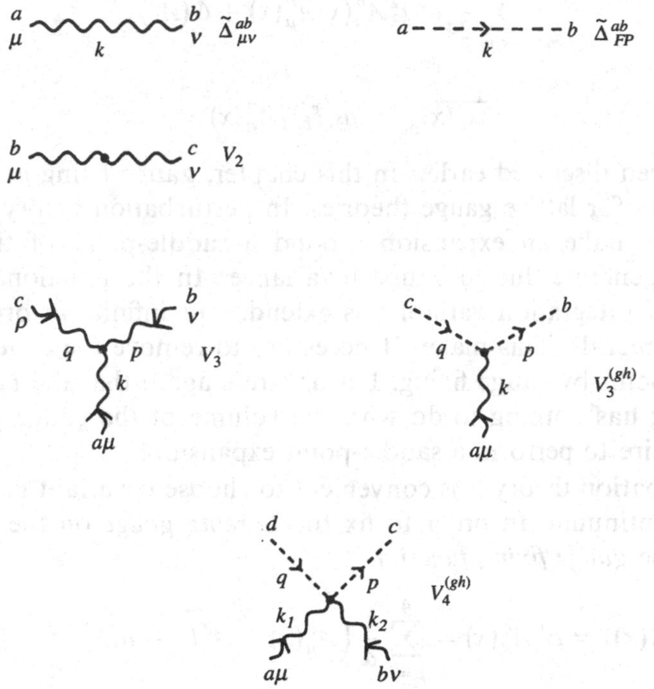

Fig. 3.3. Some Feynman rules for pure lattice gauge theory. The indices and momenta associated with the various lines represent the notation used in the text.

$$\mathrm{i}gf_{abc}\left\{\delta_{\mu\nu}(\widehat{k-p})_\rho\cos\frac{1}{2}q_\mu a + 2 \text{ cyclic permutations}\right\}. \qquad (3.225)$$

The momentum conserving delta-function is the periodic delta-function appropriate for lattice momenta. The quartic vertex can be found in [3.33] (the two sums over σ in their equation (A.7) should, however, be unrestricted [3.36]). A general algorithm for the computation of higher vertices with the help of a computer has been developed in [3.37].

In addition to the vertices from the classical action, the Haar measure also contributes to the total action. In the parametrization chosen the functional integral measure is according to (3.216):

$$\prod_{x,\mu}\mathrm{d}U_{x,\mu} = \mathrm{e}^{-S_m[A]}\prod_{x,\mu,a}\mathrm{d}A_\mu^a(x) \qquad (3.226)$$

with

$$S_m = -\sum_{x,\mu}\mathrm{Tr}\log\left(\frac{2}{A_\mu(x)}\sinh\frac{\overline{A_\mu(x)}}{2}\right)$$

$$= \sum_{x,\mu} \frac{N}{24} a^2 g^2 A_\mu^b(x) A_\mu^b(x) + \mathcal{O}(A^4), \qquad (3.227)$$

$$\overline{A_\mu(x)}_{bc} = ag f_{bcd} A_\mu^d(x). \qquad (3.228)$$

As has been discussed earlier in this chapter, gauge fixing is in general not necessary for lattice gauge theories. In perturbation theory, however, one likes to make an expansion around a saddle-point of the action, which is degenerate due to gauge invariance. In the functional integral the range of integration variables is extended to infinity in order to get Gaussian integrals. This makes it necessary to remove the degeneracy of the saddle-point by gauge fixing. Let us stress again that the necessity of gauge fixing has nothing to do with the volume of the gauge group but with the desire to perform a saddle-point expansion.

In perturbation theory it is convenient to choose covariant gauge fixing as in the continuum. In order to fix the *Lorentz gauge* on the lattice we introduce the *gauge fixing function*

$$f^c(A(x)) = \Delta_\mu^b A_\mu^c(x) = \sum_{\mu=1}^{4} \frac{1}{a}\left(A_\mu^c(x) - A_\mu^c(x - a\hat{\mu})\right). \qquad (3.229)$$

Gauge transformations on the lattice are parametrized as

$$\Lambda(x) = \exp i T_b \omega^b(x). \qquad (3.230)$$

Under an infinitesimal gauge transformation

$$\Lambda(x) = \mathbf{1} + i T_b \delta\omega^b(x) \qquad (3.231)$$

the gauge field changes according to

$$\delta A_\mu^c(x) = \frac{1}{ag}\left\{ E_{bc}^{-1}(\overline{A_\mu(x)})\delta\omega^b(x + a\hat{\mu}) - E_{cb}^{-1}(\overline{A_\mu(x)})\delta\omega^b(x) \right\}, \qquad (3.232)$$

with

$$E^{-1}(x) = \frac{x}{1 - \exp(-x)} = \sum_{n=0}^{\infty} \frac{B_n}{n!}(-x)^n, \qquad (3.233)$$

where B_n are the Bernoulli numbers. The change of the gauge fixing function is then given by

$$\delta f^c(A(x)) = \sum_y a^4 \mathcal{M}_{cx,by} \delta\omega^b(y), \qquad (3.234)$$

where we have introduced the *Faddeev–Popov operator*

$$\mathcal{M}_{cx,by}[A] =$$

$$\frac{1}{a^6 g} \sum_{\mu=1}^{4} \left\{ E_{bc}^{-1}(\overline{A_\mu(x)}) \delta_{x+a\hat{\mu},y} - E_{cb}^{-1}(\overline{A_\mu(x)}) \delta_{xy} \right.$$

$$\left. -E_{bc}^{-1}(\overline{A_\mu(x - a\hat{\mu})}) \delta_{xy} + E_{cb}^{-1}(\overline{A_\mu(x - a\hat{\mu})}) \delta_{x-a\hat{\mu},y} \right\} . \tag{3.235}$$

Denote the gauge transformed gauge field by $A^{(\Lambda)}(x)$. Since $\mathcal{M}$ is the Jacobian of the change of variables from $f^c(A^{(\Lambda)}(x))$ to $\omega^b(y)$, the integral

$$\int \prod_{x,b} d\omega^b(x) \prod_{y,c} \delta\left(f^c(A^{(\Lambda)}(y)) - g_y^c\right) \Delta_f[A^{(\Lambda)}] \tag{3.236}$$

is constant, where

$$\Delta_f[A] = \det \mathcal{M}[A] \tag{3.237}$$

is the *Faddeev–Popov determinant*. We may insert this constant into the functional integral of some gauge invariant function $\Sigma[U]$. Using gauge invariance the integral over the $\omega^b(y)$ can be performed and leads to

$$\int \prod_{x,\mu} dU_{x,\mu} \Sigma[U] = \int \prod_{x,\mu} dU_{x,\mu} \Sigma[U] \Delta_f[A] \prod_{y,c} \delta\left(f^c(A(y)) - g_y^c\right) . \tag{3.238}$$

Taking the average over g_y^c with a Gaussian weight we finally obtain the gauge fixed functional integral

$$\int \prod_{x,\mu} dU_{x,\mu} \Sigma[U] \Delta_f[A] e^{-S_{gf}[A]} \tag{3.239}$$

with the gauge fixing term

$$S_{gf}[A] = \frac{1}{2\alpha} \sum_y a^4 \left[f^c(A(y))\right]^2 . \tag{3.240}$$

In a final step the Faddeev–Popov determinant is expressed in terms of an integral over Grassmann variables η_{ax} and $\overline{\eta}_{ax}$, the so-called *Faddeev–Popov ghosts*:

$$\Delta_f[A] = \text{const.} \int \prod_{x,a} d\eta_{ax} \, d\overline{\eta}_{ax} \, e^{-S_{FP}[A,\eta,\overline{\eta}]} . \tag{3.241}$$

The Faddeev–Popov action is

$$S_{FP} = \sum_{xy} a^8 g \overline{\eta}_{cx} \mathcal{M}_{cx,by}[A] \eta_{by}$$

$$= a^2 \sum_{x,\mu} \left(\overline{\eta}_{c,x+a\hat{\mu}} - \overline{\eta}_{cx}\right) \left\{ E_{cb}^{-1}(\overline{A_\mu(x)}) \eta_{bx} - E_{bc}^{-1}(\overline{A_\mu(x)}) \eta_{b,x+a\hat{\mu}} \right\} . \tag{3.242}$$

We have anticipated the definition of Grassmann variables and Grassmann integrals here, which will be introduced in chapter 4 in full detail.

To summarize, on the lattice the gauge fixing procedure *à la* Faddeev–Popov leads to functional integrals of the type

$$\int \prod_{x,\mu,a} \mathrm{d}A^a_\mu(x)\, \mathrm{d}\eta_{ax}\, \mathrm{d}\bar{\eta}_{ax}\, \mathrm{e}^{-S_T[A]} F[U]\,, \qquad (3.243)$$

where the total action is

$$S_T = S + S_m + S_{gf} + S_{FP}\,. \qquad (3.244)$$

From it we can determine the Feynman rules. The gauge field propagator is defined through the quadratic term in $S + S_{gf}$. In momentum space it turns out to be (see fig. 3.3)

$$\tilde{\Delta}^{ab}_{\mu\nu}(k) = \delta^{ab} \frac{1}{\hat{k}^2} \left[\delta_{\mu\nu} - (1-\alpha) \frac{\hat{k}_\mu \hat{k}_\nu}{\hat{k}^2} \right]\,. \qquad (3.245)$$

This is a massless lattice propagator. At first sight one might think that the quadratic component in the measure term S_m would contribute a mass term to the propagator. Since it has a factor g^2 in it, it is, however, treated as an additional two-point interaction vertex

$$V_2(\mu,b;\nu,c) = -\frac{N}{12} g^2 \delta^{bc} \delta_{\mu\nu} \frac{1}{a^2}\,. \qquad (3.246)$$

The consistency of this procedure is guaranteed by Ward identities [3.33, 3.34], which ensure that the gauge field propagator remains massless to all orders of perturbation theory.

The propagator for the ghost fields is just

$$\tilde{\Delta}^{ab}_{FP}(k) = \delta^{ab} \frac{1}{\hat{k}^2}\,. \qquad (3.247)$$

The vertices from the classical action have been discussed before. Additional vertices come from the measure term. The lowest one is the two-point vertex mentioned above.

Further vertices are generated by the Faddeev–Popov action. There is again an infinite number of them, which couple the ghost fields to the gauge field. The lowest ones (see fig. 3.3) are [3.33]

$$V^{(gh)}_3(k,\mu,a;p,b;q,c) = \mathrm{i}g f_{abc} \hat{p}_\mu \cos \frac{1}{2} q_\mu a \qquad (3.248)$$

$$V^{(gh)}_4(k_1,\mu,a;k_2,\nu,b;p,c;q,d) = \frac{1}{12} a^2 g^2 (f_{ade} f_{bce} + f_{ace} f_{bde}) \delta_{\mu\nu} \hat{p}_\mu \hat{q}_\mu\,. \qquad (3.249)$$

3.3.2 Renormalization group

With the help of the lattice Feynman rules, correlation functions can be calculated perturbatively in much the same way as in the continuum, although the calculations are more difficult. Of particular interest are those quantities which are usually also measured in Monte Carlo simulations. As an example, we mention Wilson loops and the related static potential. They have been investigated in the framework of lattice perturbation theory by various authors [3.38, 3.25, 3.26, 3.27].

In this section we would like to consider the renormalized coupling of gauge theories, which is the basic quantity for a discussion of β-functions and the renormalization group. There are many ways to define a renormalized coupling in gauge theories in the continuum. A definition which may also be implemented in lattice perturbation theory goes under the name of *momentum space subtraction* (MOM) [3.39]. In the MOM scheme the renormalized coupling g_{MOM} is defined in terms of the triple gluon vertex function (in the Feynman gauge $\alpha = 1$). Since the bare gauge theory is massless, one cannot use renormalization conditions at zero momentum. Therefore a renormalization scale μ has to be introduced (see section 1.7.3).

The first step is to determine the wave function renormalization factor Z_3, which relates the renormalized gauge field A_{Rv} to the bare one:

$$A_{Rv}(x) = Z_3^{-1/2} A_v(x).$$
(3.250)

Z_3 is fixed by a renormalization condition on the renormalized gauge field propagator. This condition requires that the transverse part of the renormalized propagator, evaluated at $k^2 = \mu^2$, equals the transverse bare propagator:

$$Z_3^{-1} \left. \tilde{D}_{T v \rho}^{bc}(k) \right|_{k^2 = \mu^2} = \delta^{bc} \frac{1}{\mu^2} \left[\delta_{v\rho} - \frac{k_v k_\rho}{\mu^2} \right].$$
(3.251)

From the one-loop expression of the gauge field propagator $\tilde{D}_{v\rho}^{bc}(k)$ for gauge group SU(N) [3.32, 3.33] one obtains

$$Z_3 = \left\{ 1 - \frac{g^2 N}{48\pi^2} \left(5 \ln(a^2 \mu^2) + c_3 \right) + \mathcal{O}(g^4) \right\} \left(1 + \mathcal{O}(a^2 \mu^2) \right)$$
(3.252)

with some numerical constant c_3.

Next the triple gluon vertex function $\Gamma_{v\rho\sigma}^{(3)bcd}(k_1, k_2, k_3)$ is considered at the symmetry point

$$k_i \cdot k_j = \frac{1}{2}(3\delta_{ij} - 1)\mu^2.$$
(3.253)

It contains a part proportional to the bare vertex:

$$\Gamma_{v\rho\sigma}^{(3)bcd}(k_1, k_2, k_3)$$

$$= Z_1^{-1} \mathrm{i} g f_{bcd} \{ \delta_{v\rho}(k_1 - k_2)_\sigma + 2 \text{ cyclic permutations} \} + \dots,$$
(3.254)

which determines the renormalization factor Z_1. Its value turns out to be

$$Z_1 = \left\{ 1 - \frac{g^2 N}{48\pi^2} \left(2\ln\left(a^2\mu^2\right) + c_1 \right) + \mathcal{O}(g^4) \right\} \left(1 + \mathcal{O}(a^2\mu^2) \right) \qquad (3.255)$$

with some other constant c_1.

Finally the renormalized coupling is defined to be the coefficient of the corresponding part of the renormalized triple gluon vertex, i.e.

$$g_{MOM}(\mu) = g\, Z_3^{3/2} Z_1^{-1} . \qquad (3.256)$$

From the previous results one obtains

$$g_{MOM}^2(\mu)$$

$$= g^2 \left\{ 1 - \frac{g^2}{16\pi^2} \frac{11}{3} N \left(\ln\left(a^2\mu^2\right) + C \right) + \mathcal{O}(g^4) \right\} \left(1 + \mathcal{O}(a^2\mu^2) \right), \qquad (3.257)$$

with

$$C = \frac{1}{11}(3c_3 - 2c_1) = \frac{6\pi^2}{11N^2} - 9.44598 . \qquad (3.258)$$

This is the central relation concerning the discussion of the renormalization group. From it the one-loop lattice β-function is easily derived. We get

$$\beta_{LAT}(g, \mu a) = -a\left. \frac{\partial g}{\partial a} \right|_{g_{MOM}} = \left\{ -\frac{g^3}{16\pi^2} \frac{11}{3} N + \mathcal{O}(g^5) \right\} \left(1 + \mathcal{O}(a^2\mu^2) \right) . \qquad (3.259)$$

We have indicated the presence of scaling violating terms also. Neglecting them we write

$$\beta_{LAT}(g) = -\beta_0\, g^3 - \beta_1\, g^5 + \dots \qquad (3.260)$$

with

$$\beta_0 = \frac{N}{16\pi^2} \frac{11}{3} . \qquad (3.261)$$

This result of course agrees with the well-known one-loop β-function of continuum SU(N) gauge theory, since it is universal, as we have discussed earlier. Being universal too, the second coefficient can be taken over from the two-loop calculations in the continuum [3.40, 3.41]:

$$\beta_1 = \left(\frac{N}{16\pi^2} \right)^2 \frac{34}{3} . \qquad (3.262)$$

Just for the purpose of deriving these coefficients, the calculation in lattice perturbation theory would have been superfluous. The aim of the demonstration above, however, is to show that the lattice calculation is in addition able to deliver scaling violating terms, which may become important for not too small lattice spacings, or for the determination of improved actions. Even more important is the finite part C which will be considered later.

From the β-function above we see that the point $g = 0$ is an ultraviolet fixed point. This means that for fixed g_{MOM} the continuum limit is obtained by sending the bare coupling g to zero according to equation (3.259). The solution of this differential equation is given by

$$a = \exp - \int^g \frac{dg'}{\beta_{LAT}(g')}$$

$$= \Lambda_{LAT}^{-1} \exp\left(-\frac{1}{2\beta_0 g^2}\right) (\beta_0 g^2)^{-\beta_1/(2\beta_0^2)} \{1 + \mathcal{O}(g^2)\} \qquad (3.263)$$

or

$$\frac{1}{g^2} = \beta_0 \ln \frac{1}{a^2 \Lambda_{LAT}^2} + \frac{\beta_1}{\beta_0} \ln \ln \frac{1}{a^2 \Lambda_{LAT}^2} + \mathcal{O}(1/\ln a^2 \Lambda_{LAT}^2). \qquad (3.264)$$

An integration constant Λ_{LAT} with the dimensions of a mass appears here. It is called the *lattice Λ-parameter*. It may be defined by means of

$$\Lambda_{LAT} \equiv \lim_{g \to 0} \frac{1}{a} \exp\left(-\frac{1}{2\beta_0 g^2}\right) (\beta_0 g^2)^{-\beta_1/(2\beta_0^2)}, \qquad (3.265)$$

and is a constant independent of g. Being dimensionful, Λ_{LAT} provides us with a scale, which survives the continuum limit. Although classical non-Abelian gauge field theory does not contain any mass scale and is scale invariant, the process of renormalization introduces a mass scale into the quantized theory. This fact is sometimes called *dimensional transmutation*.

In pure gauge theory every physical quantity with dimensions of a mass, like a particle mass m, say, is proportional to Λ_{LAT} in the continuum limit:

$$m = C_m \Lambda_{LAT}. \qquad (3.266)$$

Therefore the mass in lattice units, am, which is the quantity being calculated in numerical simulations, varies as a function of g according to

$$am = C_m \exp\left(-\frac{1}{2\beta_0 g^2}\right) (\beta_0 g^2)^{-\beta_1/(2\beta_0^2)} \{1 + \mathcal{O}(g^2)\}. \qquad (3.267)$$

This behaviour is called *asymptotic scaling*. Equivalently it can be expressed through the renormalization group equation

$$\left\{a \frac{\partial}{\partial a} - \beta_{LAT} \frac{\partial}{\partial g}\right\} m = \mathcal{O}(a^2 m^2). \qquad (3.268)$$

If two different masses, m_1 and m_2, say, are considered, both would scale according to equation (3.267). Because m_1/m_2 depends on g only, (3.268) implies

$$\frac{\partial}{\partial g}\left(\frac{m_1}{m_2}\right) = \mathcal{O}(a^2 m^2). \tag{3.269}$$

This means that in the ratio m_1/m_2 the scaling functions would cancel out completely and one is left with

$$\frac{m_1}{m_2} = \frac{C_{m1}}{C_{m2}}\left\{1 + \mathcal{O}(a^2 m^2)\right\}. \tag{3.270}$$

If the lattice spacing a is small enough, such that this ratio appears to be nearly constant, one speaks of *scaling*. The size of the scaling violating corrections will, however, in general depend on which quantities are being considered.

3.3.3 Λ-parameters

For fixed bare coupling g the renormalized coupling g_{MOM} depends on the choice of the renormalization scale μ, and is therefore called running coupling $g_{MOM}(\mu)$. This dependence defines another β-function, see section 1.7:

$$\beta(g_{MOM}, \mu a) = \mu\frac{\partial g_{MOM}}{\partial \mu}\bigg|_g. \tag{3.271}$$

Neglecting scaling violations again, equation (3.257) yields

$$\beta(g_{MOM}) = -\beta_0 g_{MOM}^3 + \mathcal{O}(g_{MOM}^5), \tag{3.272}$$

in accordance with the universality of the lowest two coefficients. The solution of the renormalization group equation above is similar to the corresponding one for g:

$$\mu^{-1} = \Lambda_{MOM}^{-1}\exp\left(-\frac{1}{2\beta_0 g_{MOM}^2}\right)(\beta_0 g_{MOM}^2)^{-\beta_1/(2\beta_0^2)}\{1 + \mathcal{O}(g_{MOM}^2)\}, \tag{3.273}$$

and holds in the continuum limit, too. The new scale parameter Λ_{MOM} appearing here is given by

$$\Lambda_{MOM} \equiv \lim_{g_{MOM}\to 0}\mu\exp\left(-\frac{1}{2\beta_0 g_{MOM}^2}\right)(\beta_0 g_{MOM}^2)^{-\beta_1/(2\beta_0^2)}. \tag{3.274}$$

It is not specific to the lattice regularization and is also used in perturbative calculations with other regularization schemes. Λ_{MOM} can be employed to set the scale in the framework of perturbative calculations in the continuum, for example in deep inelastic scattering in QCD.

On the other hand, in non-perturbative calculations of masses in lattice gauge theory the results are naturally expressed in terms of Λ_{LAT}. In order to make contact with the scales of continuum gauge theory the ratio of Λ_{MOM} to Λ_{LAT} must be known. From equations (3.263),(3.273) we get

$$\frac{\Lambda_{MOM}}{\Lambda_{LAT}} = \lim_{g \to 0} a\mu \exp\left\{ \frac{1}{2\beta_0} \left(\frac{1}{g^2} - \frac{1}{g_{MOM}^2(\mu)} \right) \right\}. \tag{3.275}$$

Using (3.257) this reduces to

$$\frac{\Lambda_{MOM}}{\Lambda_{LAT}} = e^{-C/2}. \tag{3.276}$$

We see that the finite part C of the one-loop expression for the renormalized coupling determines the relevant ratio of Λ-parameters. Numerical values are

$$\frac{\Lambda_{MOM}}{\Lambda_{LAT}} = \begin{cases} 57.40, & \text{for} \quad N = 2 \\ 83.42, & \text{for} \quad N = 3. \end{cases} \tag{3.277}$$

They were obtained first in [3.32]. Their calculation has been extended to QCD, including quarks, in [3.33, 3.42].

In the context of phenomenological applications of perturbative QCD, other renormalization schemes with their specific Λ-parameters are utilized. A particular popular example is the $\overline{MS}$ scheme [3.43]. In order to convert to its scale the relation [3.39]

$$\frac{\Lambda_{MOM}}{\Lambda_{\overline{MS}}} = 2.895655, \tag{3.278}$$

resulting from another one-loop calculation, must be used.

The one-loop lattice calculations of the finite part C are quite tedious. It was observed in [3.44] that the result can be obtained with less labour with the help of the *background field method*. We do not want to introduce this subject here, but refer to [3.44, 3.45] for details.

The story of Λ-parameters does not end here. So far only lattice gauge theory with Wilson's action has been considered. If instead one uses some other action, for example Manton's action or the heat kernel action, the relation between the bare coupling and the renormalized coupling differs from the one above, (3.257), by finite terms. This in turn has the effect that the corresponding lambda-parameter Λ'_L differs from the previous one by a certain factor, which can be determined through a one-loop lattice calculation. For various lattice actions the corresponding calculations have been performed by means of the background field method and the results can be found in [3.46, 3.25, 3.27].

3.4 Strong-coupling expansion

Perturbation theory can be applied to gauge theories in the continuum as well as on a lattice. Several interesting aspects, however, like dynamical mass generation, confinement etc. are inaccessible to a perturbative treatment and require non-perturbative methods. One of these is the method of strong-coupling expansions, which amount to expansions in powers of the inverse coupling. Strong-coupling expansions are tied to the lattice and cannot be derived directly for a continuum theory. Therefore, in order to yield information about continuum physics, they must be supplemented with suitable procedures to extrapolate their results to the continuum limit. On the other hand, strong-coupling expansions can lead to new insights even for finite lattice spacings by revealing mechanisms and connections which are typical for strongly interacting theories.

3.4.1 *High-temperature expansions*

With regard to the analogy between Euclidean quantum field theory in terms of functional integrals and statistical mechanics we have seen that the bare coupling constant squared g^2 plays the rôle of an analogue temperature. Lattice gauge theory at strong coupling thus corresponds to a statistical system at high temperatures, i.e. small β. Therefore the well-known method of high-temperature expansions of statistical mechanics suggests itself to be applied to this situation.

In the following we consider the pure SU(N) lattice gauge theory with Wilson's action:

$$S = \sum_{\mathrm{p}} S_{\mathrm{p}}(U_{\mathrm{p}}) \tag{3.279}$$

$$S_{\mathrm{p}}(U) = -\frac{\beta}{2N} \left\{ \mathrm{Tr}\, U + \mathrm{Tr}\, U^+ - 2N \right\}. \tag{3.280}$$

The plaquette term obeys

$$0 \le S_{\mathrm{p}}(U) \le 2\beta. \tag{3.281}$$

From this follows that the single-plaquette Boltzmann factor converges to 1 uniformly in the high-temperature limit:

$$e^{-S_{\mathrm{p}}} \xrightarrow{\beta \to 0} 1. \tag{3.282}$$

The idea is then to write

$$e^{-S_{\mathrm{p}}(U)} = 1 + f_{\mathrm{p}}(U) \tag{3.283}$$

with a correction term f_p, which vanishes uniformly as β goes to zero:

$$f_p(U) \xrightarrow{\beta \to 0} 0 \qquad \text{uniformly}, \qquad (3.284)$$

and to expand

$$
\begin{aligned}
e^{-S} &= \prod_p (1 + f_p) \\
&= (1 + f_{p_1})(1 + f_{p_2}) \cdots \\
&= 1 + \sum_p f_p + \sum_{(p,p')} f_p f_{p'} + \cdots \\
&= \sum_{\mathscr{P}} \prod_{p \in \mathscr{P}} f_p, \qquad (3.285)
\end{aligned}
$$

where $\mathscr{P}$ is any set of plaquettes. For the partition function this gives

$$Z = \int \prod_b dU(b)\, e^{-S} = \sum_{\mathscr{P}} \int \prod_b dU(b) \prod_{p \in \mathscr{P}} f_p(U_p), \qquad (3.286)$$

in which the sum over sets of plaquettes can be considered as a sum over graphs on the lattice with certain weights. Pursuing this idea leads to an expansion of Z in powers of β.

3.4.2 Strong-coupling graphs

In turns out that the computations can be facilitated very much by using the character expansion

$$e^{-S_p(U)} = \sum_r d_r c_r(\beta) \chi_r(U). \qquad (3.287)$$

Factoring out the trivial character we write

$$e^{-S_p(U)} = c_0(\beta) \left\{ 1 + \sum_{r \neq 0} d_r a_r(\beta) \chi_r(U) \right\} \qquad (3.288)$$

$$\equiv c_0(\beta) \left\{ 1 + f_p(U) \right\}, \qquad (3.289)$$

with

$$a_r = \frac{c_r(\beta)}{c_0(\beta)}. \qquad (3.290)$$

These coefficients behave like powers of β for small β. For SU(2) we have

$$a_j(\beta) = \frac{I_{2j+1}(\beta)}{I_1(\beta)} = \frac{\beta^{2j}}{(2j+1)!} + \mathcal{O}(\beta^{2j+2}). \qquad (3.291)$$

Using the expansion above we obtain

$$e^{-S} = [c_0(\beta)]^{6\Omega} \prod_p \left\{ 1 + \sum_{r \neq 0} d_r a_r(\beta) \chi_r(U_p) \right\}, \qquad (3.292)$$

where Ω is the total lattice volume. Expanding the product over plaquettes one gets a sum of terms, each of which is of the form

$$d_{r_1} a_{r_1} \chi_{r_1}(U_{p_1}) \cdot d_{r_2} a_{r_2} \chi_{r_2}(U_{p_2}) \cdot \ldots \qquad (3.293)$$

Thereby to each occurring plaquette p is associated a non-trivial representation $r \neq 0$. To the plaquettes not occurring in a particular term we associate $r = 0$. Correspondingly we define a *graph* $\mathcal{G}$ to be a mapping which attaches to each plaquette p a representation r_p.
The support $|\mathcal{G}|$ of a graph is defined by

$$p \in |\mathcal{G}| \Leftrightarrow r_p \neq 0. \qquad (3.294)$$

With these definitions we write the Boltzmann factor as

$$e^{-S} = c_0^{6\Omega} \sum_{\mathcal{G}} \prod_{p \in |\mathcal{G}|} d_{r_p} a_{r_p} \chi_{r_p}(U_p). \qquad (3.295)$$

Integrating over the link variables we get

$$Z = c_0^{6\Omega} \sum_{\mathcal{G}} \Phi(\mathcal{G}) \qquad (3.296)$$

where

$$\Phi(\mathcal{G}) = \int \prod_b dU(b) \prod_{p \in |\mathcal{G}|} d_{r_p} a_{r_p} \chi_{r_p}(U_p). \qquad (3.297)$$

$\Phi(\mathcal{G})$ is called the contribution of the graph $\mathcal{G}$.

As an example consider a cube, consisting of six plaquettes, each of which has the fundamental representation $r = f$ of SU(N). Its contribution is

$$\Phi(\mathcal{G}) = a_f(\beta)^6 \int \prod dU(b) \prod_{i=1}^{6} [d_f \text{Tr}(U_{p_i})]. \qquad (3.298)$$

For the integrations the rule

$$\int dU \, d_f \text{Tr}(U V_1) \, d_f \text{Tr}(U^{-1} V_2) = d_f \text{Tr}(V_1 V_2) \qquad (3.299)$$

is used, see fig. 3.4. Successive application yields for a surface bounded by a simple curve C the value

$$d_f \text{Tr} \, U(C). \qquad (3.300)$$

In our case we take five plaquettes from the cube and get

$$d_f \text{Tr} \, U_{p_1}. \qquad (3.301)$$

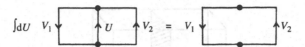

Fig. 3.4. Graphical representation of the integration rule (3.299).

The final integration is

$$\int \prod_{b \in p_1} dU(b) \, d_f \operatorname{Tr} U_{p_1} \, d_f \operatorname{Tr} U_{p_1}^+ = d_f^2 \int dU \, (1 + \chi_A(U)) = d_f^2 \,. \quad (3.302)$$

The result for the cube is thus

$$\Phi(\text{cube}, r = f) = d_f^2 a_f(\beta)^6 \,. \quad (3.303)$$

For more complicated graphs the calculation of their contributions proceeds in a similar fashion. In general, for this purpose one needs to know integrals of the type

$$\int dU \, \chi_{r_1}(V_1 U) \cdot \ldots \cdot \chi_{r_n}(V_n U) \,. \quad (3.304)$$

One way to solve this group theoretical exercise is to write the integrand in terms of the representation matrices and reduce out the resulting Kronecker product. In particular, if the resulting Clebsch–Gordan series does not contain the trivial representation, the integral vanishes, since

$$\int dU \, \chi_r(U) = \delta_{r,0} \,. \quad (3.305)$$

This leads to an important selection rule for graphs. Suppose $p_1, \ldots, p_n$ are plaquettes in $|\mathcal{G}|$ which have a link b in common, and let r_i be the representation attached to p_i. The integration over $dU(b)$ is then of the type (3.304). So that the contribution of $\mathcal{G}$ does not vanish, the Kronecker product $r_1 \otimes r_2 \otimes \ldots \otimes r_n$ must contain the trivial representation in its Clebsch–Gordan series. Two immediate consequences are:

1. the graph should not have a boundary,

2. if precisely two plaquettes with representations r and r' meet in a link, we must have $r = r'$ or $r^* = r'$, depending on how the plaquettes are oriented:

$$\int dU \, \chi_r(VU)\chi_{r'}(U^{-1}W) = \delta_{rr'} \frac{1}{d_r} \chi_r(VW) \,. \quad (3.306)$$

To summarize the selection rule: if $\Phi(\mathcal{G}) \neq 0$ the support $|\mathcal{G}|$ is a closed surface, which may have branch lines. Along a branch line the Kronecker product of the occurring representations, when reduced out, must contain the trivial representation.

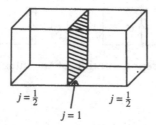

Fig. 3.5.   A double cube with an inner wall, representing a graph in the strong-coupling expansion of the partition function of lattice gauge theory. The labels j indicate representations of SU(2).

As an example consider a double cube with $j = 1/2$ on its surface and $j = 1$ on the inner partition wall (see fig. 3.5). Its contribution is not zero since

$$\frac{1}{2} \otimes \frac{1}{2} \otimes 1 = 0 \otimes 1 \otimes 1 \otimes 2. \tag{3.307}$$

In SU(2) up to β^{10} we have three contributions: the empty graph yields 1, any cube yields

$$4a_{1/2}(\beta)^6 = \mathcal{O}(\beta^6) \tag{3.308}$$

and any double cube without inner wall yields

$$4a_{1/2}(\beta)^{10} = \mathcal{O}(\beta^{10}). \tag{3.309}$$

There are 4 cubes and 36 double-cubes per lattice point and we obtain

$$Z = c_0^{6\Omega} \left\{ 1 + 16a_{1/2}(\beta)^6 \Omega + 144a_{1/2}(\beta)^{10}\Omega + \ldots \right\} \tag{3.310}$$

In higher orders we have to include graphs consisting of several disjoint components:

$$|\mathcal{G}| = |X_1| \cup \ldots \cup |X_n|, \qquad \text{with } |X_i| \cap |X_j| = 0, \text{for } i \neq j. \tag{3.311}$$

In this case the contribution factorizes obviously:

$$\Phi(\mathcal{G}) = \prod_i \Phi(X_i). \tag{3.312}$$

Even if two components touch each other at some corner only, factorization holds. Therefore we define a graph $\mathcal{G}$ as *disconnected* if $|\mathcal{G}|$ can be divided into two subsets such that the intersection of both does not contain any link. Otherwise it is called *connected*.

Given an arbitrary graph we can decompose it into connected parts.

These are called *polymers* X_i and we write

$$\mathscr{G} = \sum_{i=1}^{n} X_i, \tag{3.313}$$

$$\Phi(\mathscr{G}) = \prod_{i=1}^{n} \Phi(X_i). \tag{3.314}$$

From such a graph we may produce new graphs by translating or rotating some of the polymers in such a way that they remain disconnected. It is convenient to collect all such graphs consisting of congruent polymers. The number of ways this can be done is the *counting factor* $\kappa(\mathscr{G})$. On a finite lattice with total volume Ω and periodic boundary conditions $\kappa(\mathscr{G})$ is a polynomial in Ω of degree n. If, for example, $|\mathscr{G}|$ consists of two cubes we get

$$\kappa = 4\Omega(4\Omega - 79)\frac{1}{2} = 8\Omega^2 - 158\Omega. \tag{3.315}$$

Thus we obtain a series for Z containing higher and higher powers of Ω. Assuming that the *free energy*

$$F \equiv -\frac{1}{\Omega}\ln Z \tag{3.316}$$

exists in the thermodynamic limit $\Omega \to \infty$, as it should, the volume dependence of Z exponentiates, and we would obtain the free energy as the coefficient of the term linear in Ω:

$$Z = \exp(-\Omega F) = 1 - \Omega F + \mathcal{O}(\Omega^2). \tag{3.317}$$

The fact that the expansion for Z indeed exponentiates will be shown next.

3.4.3 Moments and cumulants

Our aim is to show the exponentiation of the strong-coupling expansion for the partition function Z. To this end we shall introduce a graphical expansion for $\ln Z$, which is valid for general lattices, i.e. also in the absence of translation invariance. In order to achieve this, the technique of moments and cumulants is needed [3.47, 3.48].

Let I be some set. A *moment* $< >$ is a sequence of symmetric, real functions over I

$$< \alpha, \ldots, \beta > \in \mathbf{R} \qquad \text{for } \alpha, \ldots, \beta \in I \tag{3.318}$$

such that

$$< \emptyset >= 0. \tag{3.319}$$

A multiplication of moments

$$<>_3 = <>_1 \circ <>_2 \qquad (3.320)$$

is defined by means of

$$< \alpha, \ldots, \beta >_3 = \sum_{P_2} < \gamma, \ldots, \delta >_1 < \epsilon, \ldots, \phi >_2, \qquad (3.321)$$

where the sum goes over all partitions P_2 of the set of variables $\{\alpha, \ldots, \beta\}$ into two subsets $\{\gamma, \ldots, \delta\}$ and $\{\epsilon, \ldots, \phi\}$. The function

$$\mathbf{1}(\alpha_1, \ldots, \alpha_n) = \begin{cases} 1, & n = 0 \\ 0, & n \neq 0 \end{cases} \qquad (3.322)$$

is the identity with respect to multiplication of moments, but is not a moment itself. The $\circ$-exponential of a moment [ ] is

$$\exp_\circ[\,] = 1 + \sum_{n=1}^{\infty} \frac{1}{n!} [\,]^n. \qquad (3.323)$$

The relation

$$\exp_\circ[\,] = 1 + < > \qquad (3.324)$$

is equivalent to

$$< \alpha, \ldots, \zeta > = \sum_{P} [\alpha, \ldots, \beta][\gamma, \ldots, \delta] \ldots [\mu, \ldots, \nu], \qquad (3.325)$$

where the sum now goes over all partitions P. In this case [ ] is called the *cumulant* of $< >$. The simplest cases of the relation between moments and cumulants are

$$< \alpha > = [\alpha] \qquad (3.326)$$

$$< \alpha, \beta > = [\alpha, \beta] + [\alpha][\beta] \qquad (3.327)$$

$$< \alpha, \beta, \gamma > = [\alpha, \beta, \gamma] + [\alpha, \beta][\gamma] + [\beta, \gamma][\alpha] + [\alpha, \gamma][\beta] + [\alpha][\beta][\gamma]. \qquad (3.328)$$

The inverse of the transformation from moments to cumulants is given by

$$[\,] = \ln_\circ(1 + < >) = \sum_{n=1}^{\infty} (-1)^{n-1} \frac{1}{n} < >^n$$

$$\Leftrightarrow \exp_\circ[\,] = 1 + < >, \qquad (3.329)$$

which amounts to

$$[\alpha, \ldots, \zeta]$$

$$= \sum_{n=1}^{\infty} (-1)^{n-1} \frac{1}{n} \sum_{P_n} n! < \alpha, \ldots, \beta > \ldots < \gamma, \ldots, \delta >$$

$$= \sum_P (-1)^{n-1} (n-1)! <\alpha, \ldots, \beta> \ldots <\gamma, \ldots, \delta>, \tag{3.330}$$

where n is the number of factors on the right hand side. For example

$$[\alpha] = <\alpha> \tag{3.331}$$

$$[\alpha, \beta] = <\alpha, \beta> - <\alpha><\beta>. \tag{3.332}$$

Now let ϕ_α, $\alpha \in I$ be complex variables indexed by elements of I, and define the generating functional of a moment $<>$ by

$$F(\{\phi_\alpha\}) = \sum_{n=1}^{\infty} \sum_{\alpha_1, \ldots, \alpha_n} \frac{1}{n!} <\alpha_1, \ldots, \alpha_n> \phi_{\alpha_1} \cdot \ldots \cdot \phi_{\alpha_n}. \tag{3.333}$$

If $<>$ is the product of two other moments,

$$<> = <>_1 \circ <>_2, \tag{3.334}$$

one can convince oneself easily that the same holds for the corresponding generating functionals:

$$F(\{\phi_\alpha\}) = F_1(\{\phi_\alpha\}) \cdot F_2(\{\phi_\alpha\}). \tag{3.335}$$

This implies the *main theorem of the moment-cumulant formalism*: let $<>$ be a moment with generating functional F and $[\ ]$ be the corresponding cumulant with generating functional f. Then

$$1 + F = \exp f. \tag{3.336}$$

The formalism introduced above is rather general and allows for various applications. In quantum field theory it can be used to show that the generating functional of connected Green's functions equals the logarithm of the generating functional of the Green functions. In statistical mechanics it is applied e.g. in the Mayer expansion for real gases. Other applications are in mathematical graph theory or combinatorics.

An important property of cumulants is the following: let A and B be subsets of I for which the moment factorizes, i.e.

$$<\alpha_1, \ldots, \alpha_n, \beta_1, \ldots, \beta_m> = <\alpha_1, \ldots, \alpha_n><\beta_1, \ldots, \beta_m>, \tag{3.337}$$

whenever $\alpha_i \in A$ and $\beta_i \in B$. Then the cumulant vanishes if it contains elements from both A and B:

$$[\alpha_1, \ldots, \alpha_j, \beta_1, \ldots, \beta_k] = 0 \qquad \text{for } j \neq 0 \text{ and } k \neq 0. \tag{3.338}$$

The proof by induction is left as an exercise.

In the field theoretic application, this property causes the connected Green function to vanish for widely separated arguments.

3.4.4   Cluster expansion for the free energy

Armed with moments and cumulants, we proceed now to the graphical expansion of the free energy. Let us recall that for the partition function Z we obtained an expansion in terms of polymers in the form

$$Z = c_0^{6\Omega} \left\{ 1 + \sum_{n=1}^{\infty} \sideset{}{'}\sum_{X_1,\ldots,X_n} \frac{1}{n!} \Phi(X_1) \cdot \ldots \cdot \Phi(X_n) \right\}, \tag{3.339}$$

where the primed sum goes over all n-tuples of polymers $X_1,\ldots,X_n$ with the property that $|X_i|$ is disconnected from $|X_j|$ for $i \neq j$. For I being the set of all polymers we define the following moment:

$$< X_1,\ldots,X_n > = \begin{cases} 1, & \text{if every pair } X_i, X_j \text{ is disconnected} \\ 0, & \text{otherwise} \end{cases} . \tag{3.340}$$

The expression for Z is then

$$Z = c_0^{6\Omega} \left\{ 1 + \sum_{n=1}^{\infty} \sum_{X_1,\ldots,X_n} \frac{1}{n!} < X_1,\ldots,X_n > \Phi(X_1) \cdot \ldots \cdot \Phi(X_n) \right\}. \tag{3.341}$$

As it stands it is ready for us to take its logarithm. The main theorem of the moment-cumulant formalism yields

$$F = -\frac{1}{\Omega} \ln Z$$

$$= -6 \ln c_0 - \frac{1}{\Omega} \left(\sum_{n=1}^{\infty} \sum_{X_1,\ldots,X_n} \frac{1}{n!} [X_1,\ldots,X_n] \Phi(X_1) \cdot \ldots \cdot \Phi(X_n) \right). \tag{3.342}$$

The crucial property of the moments appearing here is:

$$[X_1,\ldots,X_n] \neq 0 \iff |X_1| \cup |X_2| \ldots \cup |X_n| \text{ is connected.} \tag{3.343}$$

Proof: Assume that $|X_1| \cup |X_2| \ldots \cup |X_n|$ is disconnected. Then there are two subsets A and B of $\{X_1,\ldots,X_n\}$ such that elements of A are disconnected from elements of B. The definition of the moment $< >$ above implies that it factorizes in the sense

$$< A_1,\ldots,A_j, B_1,\ldots,B_k > = < A_1,\ldots,A_j > < B_1,\ldots,B_k >, \tag{3.344}$$

$$A_i \in A, \ B_i \in B.$$

According to the property of cumulants discussed in the last section the cumulant vanishes, if it contains elements from both A and B. ∎

Thus only configurations of overlapping polymers contribute to the free energy. We define a *cluster C* to be a connected collection of polymers. If a cluster C contains polymers X_i with possible multiplicities n_i we write

$$C = (X_1^{n_1}, X_2^{n_2}, \ldots). \tag{3.345}$$

The expansion of the free energy can then be rewritten in terms of clusters:

$$F = -6 \ln c_0 - \frac{1}{\Omega} \sum_{C=(X_1^{n_1}, \ldots X_k^{n_k})} a(C) \, \Phi(X_1)^{n_1} \cdot \ldots \cdot \Phi(X_k)^{n_k}, \tag{3.346}$$

where the combinatorial factor is

$$a(C) = \frac{[X_1, \ldots X_1, X_2, \ldots, X_2, \ldots, X_k]}{n_1! n_2! \ldots n_k!}. \tag{3.347}$$

This is the *cluster expansion* for the free energy. Since clusters are connected objects, the number of ways a cluster can be put onto the lattice does not grow faster than linearly with the volume. From this follows the existence of the thermodynamic limit of F.

In the presence of translation invariance, e.g. on a finite lattice with periodic boundary conditions, we can write

$$F = -6 \ln c_0 - \sum_{C=(X_i^{n_i})}' a(C) \prod_i \Phi(X_i)^{n_i}, \tag{3.348}$$

where the primed sum goes over clusters modulo translations.

In SU(2) lattice gauge theory the following clusters occur up to $\mathcal{O}(\beta^{12})$:

1. single polymers like the cube and double-cube discussed earlier, and some closed surfaces made from 12 plaquettes. For these clusters the combinatorial factor is $a(C) = 1$.

2. clusters consisting of two distinct polymers:

$$C = (X_1, X_2), \qquad a(C) = -1. \tag{3.349}$$

a) both X_1 and X_2 are cubes, but at different places, with one common plaquette. The contribution is

$$- (4 a_{1/2}(\beta)^6)^2. \tag{3.350}$$

b) X_1 and X_2 are the same cube:

$$C = (X_1^2), \qquad a(C) = -\frac{1}{2!}, \tag{3.351}$$

and the contribution is

$$-\frac{1}{2}(4 a_{1/2}(\beta)^6)^2. \tag{3.352}$$

Continuing in this manner one obtains an expansion of F in terms of the coefficients a_j:

$$F = -6 \ln c_0 - \sum a_{1/2}^k \, a_1^l \, a_{3/2}^m \cdots \tag{3.353}$$

For numerical purposes it may be reexpanded in powers of β or more conveniently in powers of

$$u = a_f(\beta). \tag{3.354}$$

Explicit expressions for this variable are

$$u = \frac{I_1(\beta)}{I_0(\beta)} = \frac{\beta}{2}\left(1 - \frac{\beta^2}{8} + \mathcal{O}(\beta^4)\right) \tag{3.355}$$

for gauge group U(1), and

$$u = \frac{I_2(\beta)}{I_1(\beta)} = \frac{\beta}{4}\left(1 - \frac{\beta^2}{24} + \mathcal{O}(\beta^4)\right) \tag{3.356}$$

for SU(2), where $I_n(\beta)$ are modified Bessel functions. For SU(3) it cannot be expressed in a simple way in terms of special functions, but for numerical purposes the following representation is often sufficient:

$$u = \frac{1}{3}\frac{c_3(\beta)}{c_1(\beta)} \tag{3.357}$$

$$c_3(\beta) = x + \frac{1}{2}x^2 + x^3 + \frac{5}{8}x^4 + \frac{13}{24}x^5 + \frac{77}{240}x^6 + \frac{139}{720}x^7 + \frac{19}{192}x^8$$

$$+ \frac{23}{480}x^9 + \frac{319}{15\,120}x^{10} + \frac{2\,629}{302\,400}x^{11} + \frac{16\,133}{4\,838\,400}x^{12}$$

$$+ \frac{17\,449}{14\,515\,200}x^{13} + \frac{35\,531}{87\,091\,200}x^{14} + \mathcal{O}(x^{15}) \tag{3.358}$$

$$c_1(\beta) = 1 + x^2 + \frac{1}{3}x^3 + \frac{1}{2}x^4 + \frac{1}{4}x^5 + \frac{13}{72}x^6 + \frac{11}{120}x^7 + \frac{139}{2\,880}x^8$$

$$+ \frac{19}{864}x^9 + \frac{23}{2\,400}x^{10} + \frac{29}{7\,560}x^{11} + \frac{2\,629}{1\,814\,400}x^{12}$$

$$+ \frac{1\,241}{2\,419\,200}x^{13} + \frac{17\,449}{101\,606\,400}x^{14} + \mathcal{O}(x^{15}), \tag{3.359}$$

where

$$x = \frac{\beta}{6}. \tag{3.360}$$

The strong-coupling expansion of the internal energy

$$E = \left\langle 1 - \frac{1}{\text{Tr}\,\mathbf{1}} \text{Tr}\, U_{\text{p}} \right\rangle \tag{3.361}$$

is obtained by differentiating F with respect to β:

$$E = \frac{1}{6} \frac{\partial}{\partial \beta} F(\beta). \tag{3.362}$$

We remark that

$$-\frac{\partial}{\partial \beta} \ln c_0 = 1 - u, \tag{3.363}$$

so that

$$E = 1 - u + \mathcal{O}(u^5). \tag{3.364}$$

3.4.5 Results from the strong-coupling expansion

In contrast to perturbation theory, which only yields asymptotic expansions, the strong-coupling expansion has a finite range of convergence [3.17, 3.9]. Within this range the strong-coupling expansion can be employed in a two-fold way to obtain information about lattice gauge theories. First of all it is possible to prove various properties of the theory by means of the strong-coupling expansion in a rigorous way [3.17, 3.9]. The most prominent examples are:

1. Existence of a mass gap,
   i.e. the plaquette–plaquette correlation function decays exponentially at strong couplings.

2. Static quark confinement,
   i.e. Wilson loops in the fundamental representation obey the area law at strong couplings.

The strong-coupling expansions of the mass gap and the string tension will be discussed in sections 3.5 and 3.6.

Secondly the strong-coupling expansion yields power series, which can be evaluated numerically in order to obtain estimates for the internal energy, mass gap, string tension, etc. These estimates are the more accurate the stronger the coupling is. They can be used e.g. to check Monte Carlo programs. On the other hand, it is possible to locate phase transitions with the help of these power series.

The strong-coupling expansion for the free energy has been calculated up to order β^{16} for various gauge groups in an arbitrary number of dimensions in [3.49, 3.50, 3.51, 3.52, 3.53, 3.54, 3.55]. For $d = 4$ dimensions the series have been extended by Wilson [3.56] up to order β^{22} for gauge groups Z_2, U(1) and SU(2). His results for the internal energy are

$$E = 1 - u - 4u^5 + 8u^7 - \frac{188}{3}u^9 + \frac{79\,784}{405}u^{11} - \frac{1\,639\,324}{1\,215}u^{13} + \frac{151\,535\,186}{25\,515}u^{15}$$

$$- \frac{485\,923\,556}{14\,175}u^{17} + \frac{201\,566\,719\,628}{1\,148\,175}u^{19} - 977\,457.232u^{21} \qquad (3.365)$$

for SU(2), and

$$E = 1 - u - 4u^5 + 6u^7 - \frac{182}{3}u^9 + \frac{3\,011}{24}u^{11} - \frac{842\,699}{720}u^{13} + \frac{7\,539\,329}{2\,160}u^{15}$$

$$- \frac{28\,232\,948\,341}{1\,088\,640}u^{17} + 94\,185.938\,57u^{19} - 629\,566.952\,8u^{21} \qquad (3.366)$$

for U(1). In $d = 3$ results are available up to order β^{20} for gauge groups U(1) and SU(2) [3.57].

The series expansions for the free energy of Z_n lattice gauge theories in four dimensions have been used together with the *duality mapping* to investigate the order of the phase transition [3.50]. For this purpose the power series are analysed with the help of Padé approximants. This method allows us to identify singularities of $F(\beta)$ in the complex β plane.

Analyses of the strong-coupling expansion of the internal energy for gauge groups U(1) and SU(2) have been presented in [3.58, 3.59, 3.60, 3.54]. The results on the singularity structure are in satisfactory agreement with those from Monte Carlo calculations. For more details about this topic, refer to the literature.

We shall return to the strong-coupling expansion in sections 3.5 and 3.6, where string tension and the mass gap are considered.

3.5 Static quark potential

3.5.1 *Wilson loop criterion*

In section 3.2.4 the static quark potential $V(R)$ has been introduced as a quantity related to the behaviour of large Wilson loops. This relation in turn is the basis for the Wilson loop criterion [3.4], which says that static quark confinement holds if Wilson loops obey the area law and the associated string tension is not zero. The Wilson loop criterion is very important since it allows us to distinguish different phases, with and without static quark confinement, by means of gauge invariant observables. Therefore the string tension has been used as an order parameter in many numerical and analytical investigations of lattice gauge theories.

In the following we prepare for deriving the relation between Wilson loops and the static quark potential. In order to make the discussion clear we temporarily return to the Minkowski space continuum. One can of course work entirely in the lattice formulation but in the continuum the arguments are formulated more easily. Ultraviolet divergencies are not important for the following considerations and we do not worry about them here.

The first thing to do is to introduce the notion of static quarks. It is related to the behaviour of states under gauge transformations and the proper framework for this is the Hamiltonian formalism for gauge fields (see e.g. [3.61]). In Minkowski space the Yang–Mills action is

$$S = -\frac{1}{4} \int d^4x \; F^a_{\mu\nu} F^{\mu\nu a} \,. \tag{3.367}$$

In the canonical formalism the momentum conjugate to $A^{\mu a}(x)$ is

$$\pi^a_\mu(x) = \frac{\delta S}{\delta(\partial_0 A^{\mu a}(x))} = -F^a_{0\mu}(x) \,. \tag{3.368}$$

There is no conjugate momentum for A_0 and the corresponding constraint $\pi_0 = 0$ is solved by choosing the temporal gauge

$$A_0 = 0 \,. \tag{3.369}$$

The only gauge transformations left which respect this gauge are the time-independent ones.

The momenta conjugate to the remaining components A^{ia} are

$$\pi^a_i = -F^a_{0i} = F^{0ia} \equiv -E^a_i \,, \tag{3.370}$$

where $E^a_i(x)$ is the non-Abelian electric field strength. The equations of motion can then be derived from the Hamiltonian

$$H = \frac{1}{2} \int d^3x \; \left(E^a_i E^a_i + B^a_i B^a_i \right) \,, \tag{3.371}$$

where the magnetic field strength is defined by

$$B_i = -\frac{1}{2}\epsilon_{ijk} F^{jk} \,. \tag{3.372}$$

The *Gauß law*

$$D^i E_i = 0 \tag{3.373}$$

does not belong to the equations of motion but has to be imposed as an initial condition. It is then valid at all times since

$$\frac{\partial}{\partial t} D^i E_i = 0 \tag{3.374}$$

follows from the equations of motion.

Now let us consider canonical quantization. In the coordinate representation the state vectors are represented by wave functionals

$$\Psi[A] \tag{3.375}$$

depending on the gauge field at time zero. The electric field strength becomes an operator which acts on these wave functionals as

$$E_i^a(x) = i\frac{\delta}{\delta A^{ia}(x)}. \tag{3.376}$$

Time independent gauge transformations $\Lambda(x)$ are represented by an operator $R(\Lambda)$ acting on wave functionals according to

$$(R(\Lambda)\Psi)[A_\mu] = \Psi[\Lambda^{-1}A_\mu\Lambda + \Lambda^{-1}\partial_\mu\Lambda]. \tag{3.377}$$

For infinitesimal gauge transformations

$$\Lambda = 1 + i\omega^a T_a \tag{3.378}$$

one finds

$$R(\Lambda) = 1 - i\omega^a\frac{1}{g}D^i E_i^a + \mathcal{O}(\omega^2), \tag{3.379}$$

which reveals the fact that $D^i E_i$ is the generator of local gauge transformations. Thus the Gauß law, which expresses the absence of external charges, is equivalent to the gauge invariance of the wave functional.

In order to introduce external charges we have to consider non-invariant wave functionals. Suppose wave functionals Ψ_α transform under local gauge transformations according to the fundamental representation of $\Lambda(x)$ for some fixed spatial point $\mathbf{x}$:

$$R(\Lambda)\Psi_\alpha = \Lambda_{\alpha\beta}(\mathbf{x})\Psi_\beta. \tag{3.380}$$

On such wave functionals the divergence of E acts as

$$D^i E_i^a = gT^a\delta^{(3)}(\mathbf{x}), \tag{3.381}$$

which means the presence of a static external charge at $\mathbf{x}$ in the fundamental representation of the gauge group. This is what we call a static quark. More generally states with several static charges in different representations can be characterized by means of the corresponding transformation law under gauge transformations.

On a Euclidean lattice the considerations above can be repeated in an appropriate way [3.62, 3.63]. As has been introduced in section 3.2.6, the wave functions $\Psi[\mathcal{U}]$ depend on the link variables $U(b)$ at time zero. The counterpart of the temporal gauge on the lattice is

$$U(b) = 1 \qquad \text{for } b \in B_{t+1,t}. \tag{3.382}$$

Let

$$U'(y, x) = \Lambda^{-1}(y) U(y, x) \Lambda(x) \tag{3.383}$$

be a time-independent gauge transformation. A state with a static quark at **x** and a static antiquark at **y** transforms as

$$\Psi_{\alpha\beta}[\mathcal{U}'] = \Lambda_{\alpha\gamma}(\mathbf{x})\Lambda^{-1}_{\delta\beta}(\mathbf{y})\Psi_{\gamma\delta}[\mathcal{U}]. \tag{3.384}$$

Owing to the gauge invariance of the Hamiltonian sectors with different distributions of external static charges decouple completely. Let us concentrate on the quark–antiquark sector defined above. Its states form a Hilbert space $\mathscr{H}_{\mathbf{xy}}$.

The aim is now to obtain the relation between the static potential and the Wilson loop [3.64]. Let $\Psi^{(n)}$ be a complete set of eigenvectors of the Hamiltonian H:

$$H\Psi^{(n)} = E_n \Psi^{(n)}, \tag{3.385}$$

with $\Psi^{(0)}$ being the ground state in $\mathscr{H}_{\mathbf{xy}}$. The energy E_0 of the ground state will depend on the location **x** and **y** of the static charges. If both points are lying on a common lattice axis the energy is a function of the distance R between **x** and **y** and we define the static quark potential through this energy:

$$V(R) \equiv E_0 = \min_{\mathscr{H}_{\mathbf{xy}}} H. \tag{3.386}$$

From an arbitrary state $\Psi \in \mathscr{H}_{\mathbf{xy}}$ the potential $V(R)$ can be obtained by means of

$$\langle\Psi|e^{-TH}|\Psi\rangle = \sum_n |\langle\Psi^{(n)}|\Psi\rangle|^2 e^{-TE_n}$$

$$\underset{T\to\infty}{\sim} |\langle\Psi^{(0)}|\Psi\rangle|^2 e^{-TV(R)}, \tag{3.387}$$

if Ψ has non-vanishing overlap with the ground state. We choose the following test-function for Ψ:

$$\Psi_{\alpha\beta}[\mathcal{U}] = U_{\alpha\beta}(\mathbf{x}, \mathbf{y})\Omega[\mathcal{U}], \tag{3.388}$$

where $\Omega[\mathcal{U}]$ is the gauge invariant vacuum wave function and $U(\mathbf{x}, \mathbf{y})$ is the parallel transporter for a straight path between **x** and **y**. The functional integral expression for the relevant matrix element is then

$$\langle\Psi|e^{-TH}|\Psi\rangle$$

$$= \frac{1}{Z} \int \prod_b \mathrm{d}U(b)\, \overline{U}_{\alpha\beta}(\mathbf{x} + T\hat{0}, \mathbf{y} + T\hat{0})\, U_{\alpha\beta}(\mathbf{x}, \mathbf{y})\, \exp(-S(U)). \tag{3.389}$$

Since the time-like link variables are unity in the present gauge, the product of parallel transporters above equals the parallel transporter $U(\mathscr{C}_{R,T})$ around a closed rectangular loop $\mathscr{C}_{R,T}$ of side-lengths R and T. Therefore

$$\langle \Psi | e^{-TH} | \Psi \rangle = \frac{1}{Z} \int \prod_b \mathrm{d}U(b) \; \mathrm{Tr} \, U(\mathscr{C}_{R,T}) \; \exp\left(-S(U)\right) = W(\mathscr{C}_{R,T}),$$

(3.390)

and, because both sides of the equation are gauge invariant, it holds in all gauges. From (3.387) the desired relation

$$V(R) \equiv - \lim_{T \to \infty} \frac{1}{T} \log W(\mathscr{C}_{R,T})$$

(3.391)

now follows.

Let us note here that Seiler [3.65] has proven that for large R the potential $V(R)$ cannot rise faster than linearly in R.

3.5.2 Strong-coupling expansion

In order to investigate the static quark potential at strong couplings one can apply the method of strong-coupling expansions to Wilson loops [3.48]. If the Boltzmann factor $\exp -S$ is expanded in terms of characters as in section 3.4.2 the Wilson loop is

$$W(\mathscr{C}) = \frac{c_0^{6\Omega}}{Z} \int \prod_b \mathrm{d}U(b) \; \mathrm{Tr} \, U(\mathscr{C}) \sum_{\mathscr{G}} \prod_{p \in |\mathscr{G}|} d_{r_p} a_{r_p} \chi_{r_p}(U_p)$$

$$\equiv \frac{c_0^{6\Omega}}{Z} Z(\mathscr{C}),$$

(3.392)

where

$$Z = c_0^{6\Omega} \sum_{\mathscr{G}} \Phi(\mathscr{G})$$

(3.393)

as before, and the numerator is a modified partition function

$$Z(\mathscr{C}) = \sum_{\mathscr{G}} \Phi(\mathscr{C}, \mathscr{G})$$

(3.394)

with

$$\Phi(\mathscr{C}, \mathscr{G}) = \int \prod_b \mathrm{d}U(b) \; \mathrm{Tr} \, U(\mathscr{C}) \prod_{p \in |\mathscr{G}|} d_{r_p} a_{r_p} \chi_{r_p}(U_p).$$

(3.395)

Owing to the presence of the additional factor $\mathrm{Tr} \, U(\mathscr{C})$ the selection rules for graphs contributing to this modified partition function $Z(\mathscr{C})$ differ from the ones discussed earlier in connection with the partition function Z:

1. for links not contained in $\mathscr{C}$ the selection rules are the same as in the case of Z,

2. if in a link $b \in \mathscr{C}$ several plaquettes of the graph meet, bearing representations $r_1, \ldots, r_n$, then the Kronecker product

$$r_1 \otimes \ldots \otimes r_n \otimes f \qquad (3.396)$$

must contain the trivial representation in its Clebsch–Gordan series. As a special case

3. the boundary of $\mathscr{G}$ may be contained in $\mathscr{C}$, such that the adjacent plaquettes carry $r = f$.

As a consequence the smallest graph $\mathscr{G}_0$ contributing to the numerator is a surface filling the curve $\mathscr{C}$ such that its boundary equals $\mathscr{C}$, and carrying representation $r = f$. In the case of a flat rectangular curve, which we are considering here, the number of plaquettes in $\mathscr{G}_0$ is equal to the area

$$A = TR. \qquad (3.397)$$

Successive integration of link variables as in (3.299) yields for SU(2)

$$\Phi(\mathscr{C}, \mathscr{G}_0) = \int dU \, \mathrm{Tr}\, U \, 2\, \mathrm{Tr}\, U \, (a_{1/2}(\beta))^A = 2u^A. \qquad (3.398)$$

The leading term in the strong-coupling expansion of the Wilson loop is thus

$$W(\mathscr{C}) = 2 \exp(-\alpha_0 A) + \cdots \qquad (3.399)$$

with

$$\alpha_0 = -\ln u = -\ln \frac{\beta}{4} + \ldots \qquad (3.400)$$

It obeys the area law with a string tension α_0.

Every other graph in the expansion of the numerator contains at least A plaquettes. Therefore its contribution also decays exponentially with the area A. It remains to show that the number of graphs does not grow too fast with the number of plaquettes. This has been done in [3.17]. They show that the strong-coupling expansion of $W(\mathscr{C})$ has a finite range of convergence and that within this range Wilson loops obey the area law.

From the expansion of Wilson loops an expansion for the string tension can be obtained. Consider for example the next-to-leading graph $\mathscr{G}_1$ depicted in fig. 3.6. It originates from the leading graph by shifting one of its plaquettes by one unit perpendicular to the surface and adding four new plaquettes to paste the gap. The contribution is

$$\Phi(\mathscr{C}, \mathscr{G}_1) = u^4 \Phi(\mathscr{C}, \mathscr{G}_0). \qquad (3.401)$$

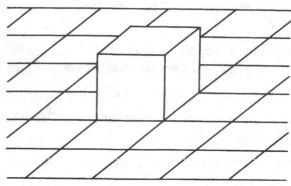

Fig. 3.6. The lowest non-trivial graph in the strong-coupling expansion of Wilson loops.

The number of such graphs is $4A$ in four dimensions. Up to fourth order in u we have

$$W(\mathscr{C}) = \Phi(\mathscr{C}, \mathscr{G}_0) \left\{ 1 + 4u^4 A + \cdots \right\}$$

$$= 2 \exp(-\alpha_0 A) \exp(4u^4 A + \cdots). \tag{3.402}$$

For the string tension this implies

$$\alpha = \alpha_0 - 4u^4 + \mathcal{O}(u^6). \tag{3.403}$$

In higher orders of the expansion, graphs appear which are accompanied by higher powers of A. It has been shown in [3.48] that the combinatorics of graphs is such that the contributions exponentiate in the form

$$W(\mathscr{C}) = C \exp[-\alpha A + \mu(R + T)]. \tag{3.404}$$

The perimeter term comes from graphs which are associated with the boundary $\mathscr{C}$. For the details of the proof we refer to the aforementioned article, but we would like to give a sketch of the essential ingredients.

The graphs contributing to $Z(\mathscr{C})$ decompose into connected components

$$\mathscr{G} = X_{\mathscr{C}} + X_1 + X_2 + \ldots \tag{3.405}$$

such that $X_{\mathscr{C}}$ is a polymer which is attached to the curve $\mathscr{C}$, and the other X_i are the usual polymers as they appear in the expansion of the partition function Z. Let Ξ be the flat rectangular surface whose boundary is $\mathscr{C}$. The polymer $X_{\mathscr{C}}$ can be considered as emerging from Ξ by modifying it locally at some places like in fig. 3.6. These local modifications are called decorations of Ξ. In general a graph $\mathscr{G}$ can be thought of as consisting of the plane Ξ with decorations Y_i and other usual polymers X_k not touching

Ξ. For its contribution one gets a product of the form

$$\Phi(\mathscr{C}, \mathscr{G}) = \Phi(\mathscr{C}, \mathscr{G}_0) \prod_i \Phi(Y_i) \prod_k \Phi(X_k). \tag{3.406}$$

The moment-cumulant formalism may now be applied to yield an expansion of $\ln Z(\mathscr{C})$ in terms of certain clusters, which are made from the Y_i and X_k. For the Wilson loop an expression

$$\ln \frac{W(\mathscr{C})}{\Phi(\mathscr{C}, \mathscr{G}_0)} = \sum_{C'} d'(C') \prod_i \Phi(Y_i) \prod_k \Phi(X_k) \tag{3.407}$$

results, where the clusters C' consists of connected sets of decorations Y_i and polymers X_k. In other words, one has an expansion in terms of local objects attached to the plane Ξ. Their bulk multiplicity is proportional to the area A. This leads to

$$\ln \frac{W(\mathscr{C})}{\Phi(\mathscr{C}, \mathscr{G}_0)} = -\Delta\alpha\, A + \mu(R + T) + \text{const.}, \tag{3.408}$$

and with

$$\alpha = \alpha_0 + \Delta\alpha = -\ln u + \Delta\alpha \tag{3.409}$$

the result (3.404) follows.

The correction $\Delta\alpha$ can be calculated as a power series in u or β. For gauge groups Z_2 and SU(2) in three or four dimensions, a list of all graphs contributing up to twelfth order is contained in [3.48]. The expansions for various groups and dimensions can be found in [3.66, 3.67, 3.68]. In four dimensions we have to fourteenth order (reintroducing the lattice spacing a)

$$-\alpha a^2 = \ln u + 4u^4 + 2u^6 + \frac{170}{3}u^8 + \frac{2\,125}{24}u^{10}$$

$$+ \frac{862\,619}{720}u^{12} + \frac{5\,754\,751}{2\,160}u^{14} \tag{3.410}$$

for U(1),

$$-\alpha a^2 = \ln u + 4u^4 + \frac{176}{3}u^8 + \frac{10\,936}{405}u^{10}$$

$$+ \frac{1\,532\,044}{1\,215}u^{12} + \frac{3\,596\,102}{5\,103}u^{14} \tag{3.411}$$

for SU(2), and

$$-\alpha a^2 = \ln u + 4u^4 + 12u^5 - 10u^6 - 36u^7 + \frac{391}{2}u^8 + \frac{1\,131}{10}u^9$$

$$+ \frac{2\,550\,837}{5\,120}u^{10} - \frac{5\,218\,287}{2\,048}u^{11} + \frac{285\,551\,579}{61\,440}u^{12} \tag{3.412}$$

for SU(3). Attempts have been made to estimate the ratio of the square root of the string tension to the lattice Λ-parameter

$$C_\alpha = \sqrt{\alpha}/\Lambda_{LAT} \tag{3.413}$$

from the strong-coupling expansion [3.67]. The idea was to match the strong-coupling expansion for αa^2 with the asymptotic scaling behaviour

$$\alpha a^2 = C_\alpha^2 \exp\left(-\frac{1}{\beta_0 g^2}\right) (\beta_0 g^2)^{-\beta_1/\beta_0^2}\{1 + \mathcal{O}(g^2)\} \tag{3.414}$$

in an intermediate range of couplings, where both curves become tangent to each other. This procedure was suggested by the fact that in numerical simulations [3.69] a rather abrupt crossover from strong coupling to weak coupling behaviour had been observed. The numerical result

$$C_\alpha \approx 270 \qquad \text{for SU(3)} \tag{3.415}$$

was in agreement with the Monte Carlo result of that time. It has been since recognized, however, that any extrapolation of the strong-coupling expansion for the string tension into the weak coupling region is prevented by the presence of the roughening transition (see below). Consequently the estimate above is afflicted by an unknown systematic error.

Let us close this section with a remark concerning the physics of the confinement mechanism. The surfaces contributing to the strong-coupling expansion of the Wilson loop may be considered as world-histories of a string connecting the static charges. This picture of a confining string between quark and antiquark is even more evident in the Hamiltonian strong-coupling expansion [3.70], where the dominant intermediate states can be viewed directly. In this way the strong-coupling expansion has, amongst other considerations, contributed to the intuitive picture of the confinement mechanism, where a tube of non-Abelian electric flux holds quark and antiquark together.

In Euclidean lattice gauge theory the width of the flux tube has been defined and investigated in [3.71]. It turns out that at rather strong couplings the tube is quite well localized and has a finite width, which is calculable in the strong-coupling expansion.

3.5.3   *Roughening transition*

The strong-coupling expansion of the string tension can be employed to evaluate the string tension numerically at large values of the bare coupling g and to compare with data from Monte Carlo calculations. An extrapolation to the weak coupling region, however, is hampered by a phenomenon characteristic for strings or random surfaces, respectively, namely the roughening transition [3.72]. In the context of lattice gauge theories it has been considered first in [3.73, 3.48, 3.74, 3.75, 3.71], see also [3.68, 3.76].

As has been shown in the previous section, in the strong-coupling expansion the Wilson loop expectation values are given by a sum over certain surfaces bounded by the Wilson loop. Sums of this type appear in statistical mechanics as partition functions for models of random surfaces or interfaces. In order to concentrate on the essential features, let us approximate the sum of interest by taking into account only those graphs which consist of a single surface bounded by the loop $\mathscr{C}$ and bearing the fundamental representation of the gauge group. The contribution of such a graph is proportional to u^A, where A is the number of plaquettes, i.e. the surface area. In the three-dimensional case one might consider a further approximation, in which all surfaces with overhangs are neglected. The graphs are then characterized by the height function $h(x_1, x_2)$, which specifies the distance of horizontal plaquettes from the x_1–x_2-plane, in which the rectangular loop $\mathscr{C}$ lies. Here x_1 and x_2 are coordinates of the midpoints of plaquettes. The area of such a surface is given by

$$A = \sum_{<x,y>} |h(x) - h(y)|. \tag{3.416}$$

If $h(x)$ is viewed as an integer valued field in two dimensions, the sum over such surfaces equals the partition function of a two-dimensional field theory with an action proportional to A, which is known as the *SOS (solid-on-solid) model*. This field theory has a phase transition which separates a massive phase, where the surfaces are essentially flat, from a massless phase, the so-called rough phase, where long-wavelength fluctuations of surfaces dominate [3.72]. This *roughening transition* is believed to be of infinite order, which means that the free energy has an essential (infinitely smooth) singularity there.

In terms of the gauge theory the massive phase corresponds to strong couplings, whereas the rough phase occurs below some finite critical value of g. Since the string tension is the analogue of the free energy per area of the SOS model, it would show a roughening singularity, but would not vanish in the rough phase.

In the three-dimensional case the roughening transition is known to be a very general phenomenon, which is not only present in the SOS model but in a variety of random surface models [3.72]. On the other hand, in the literature cited above, arguments have been presented that it also exists in four-dimensional gauge theories. A characteristic feature of the rough phase is that the suitably defined width of the confining string between static charges increases like $\ln R$ with the distance R between those charges.

The expected singularity in the string tension α as a function of the bare coupling provides a barrier for any analytical extrapolation of the strong-coupling series for α towards the weak-coupling region. The roughening transition, however, does not imply deconfinement.

3.5.4 *Numerical investigations*

The static potential and the string tension have been studied numerically by means of the Monte Carlo method (see chapter 7) in many investigations. There exists a vast literature on this subject and it is impossible to mention all relevant articles here. In this section we would just like to point out some specific aspects and refer to the reviews [3.77, 3.78] for more information.

In the Monte Carlo calculations one measures Wilson loops, in particular rectangular ones, numerically. A determination of the string tension would require first calculating the static potential $V(R)$ from $R \cdot T$ Wilson loops in the limit of large T and then extracting α from the large R asymptotics of the potential. Since such a procedure is computationally very demanding, Creutz [3.69] introduced a quantity which yields estimates for α from Wilson loops of moderate size. The *Creutz ratios* are defined by

$$\chi(I, J) \equiv -\ln\left(\frac{W(I, J)\, W(I-1, J-1)}{W(I, J-1)\, W(I-1, J)}\right), \qquad (3.417)$$

where $W(I, J) = W(\mathscr{C}_{I,J})$ is the expectation value of a rectangular Wilson loop of side lengths I and J. This definition ensures that constant factors and perimeter terms, as in (3.404), cancel out and that $\chi(I, J)$ would equal α identically if (3.404) would be true exactly. In reality this only holds for not too small I, J and an extrapolation procedure was also needed to obtain first numerical estimates.

Later work focussed directly on the potential, extracted from large Wilson loops [3.79]. The form of the potential was fitted by various *Ansätze*, representing modifications of a simple linear confinement plus Coulomb form:

$$V(R) = -\frac{A}{R} + \alpha R + \text{const.} \qquad (3.418)$$

The question of the rotational invariance of the potential has been studied too [3.80].

Latest estimates for the string tension for gauge group SU(3) [3.81] centre around

$$\sqrt{\alpha} \approx 100 \Lambda_{LAT}, \qquad (3.419)$$

but there are doubts concerning whether asymptotic scaling has been reached already.

3.6 Glueball spectrum

Classical gauge field theory does not contain any mass terms and is scale-invariant. Moreover, in perturbation theory the gluon propagator remains massless to all orders.

On the other hand, strongly coupled lattice gauge theory has a mass gap m, as the analysis of Osterwalder and Seiler [3.17] shows. This means that the lowest eigenstate of the Hamiltonian above the vacuum has a mass m. States with higher masses are to be expected too. If these masses scale near the continuum limit according to

$$am = C_m \exp\left(-\frac{1}{2\beta_0 g^2}\right) (\beta_0 g^2)^{-\beta_1/(2\beta_0^2)}\{1 + \mathcal{O}(g^2)\}, \qquad (3.420)$$

the corresponding states would survive the continuum limit and the quantized gauge theory would contain massive physical states. The existence of the corresponding particles, called *glueballs*, has been predicted in [3.82]. As the scaling formula reveals, the masses are of a purely non-perturbative nature (in the infinite-volume limit) and non-perturbative methods are required to calculate them.

The calculation of the spectrum of non-Abelian gauge theories is one of the central issues in this field. There are essentially three methods to tackle this problem:

1. Strong-coupling expansions,

2. Monte Carlo calculations, and

3. Finite-volume calculations.

The last one allows calculation of the spectrum in the continuum if the volume is small, and employs both perturbative and non-perturbative techniques [3.83]. In this section we shall consider the first two methods.

3.6.1 Glueball states

Physical states in the Hilbert space of lattice gauge theory are gauge invariant. They can be obtained by applying gauge invariant operators to the vacuum $|\Omega\rangle$. As particular candidates we may take Wilson loops, considered as multiplication operators on wave functions. Let us consider the simplest choice, which is the Wilson loop for an elementary plaquette lying in the $x_0 = 0$ timeslice. We denote such a plaquette by

$$\mathbf{p} = (\mathbf{x}; i, j), \qquad (3.421)$$

where $\mathbf{x} = (x_1, x_2, x_3)$ and $i, j \in \{1, 2, 3\}$. Applied to the vacuum the operator yields a state

$$\Psi_{ij}(\mathbf{x}) = \{\text{Tr}\, U(\mathbf{x}; i, j) - \langle \text{Tr}\, U(\mathbf{x}; i, j)\rangle\}\, |\Omega\rangle, \qquad (3.422)$$

where we have already subtracted the projection onto the vacuum itself.

The first thing to do now is to diagonalize the symmetries of the lattice theory in the space spanned by these states. Eigenstates of the momentum operators are formed as usual by means of Fourier transformation:

$$\Psi_{ij}(\mathbf{k}) = L^{-3/2} \sum_{\mathbf{x}} e^{i\mathbf{k}\mathbf{x}} \Psi_{ij}(\mathbf{x}). \qquad (3.423)$$

For the determination of masses it is sufficient to consider zero-momentum states, which we denote by

$$\Psi_{ij} = \Psi_{ij}(\mathbf{0}). \qquad (3.424)$$

In the continuum the spin of a state is characterized by the unitary irreducible representations of the group SU(2) in general, and by those of the rotation group SO(3) for bosonic states in particular. On a cubic lattice rotation symmetry is broken down to the symmetry group of a cube, the *cubic (or octahedral) group* **O** (see e.g. [3.84]). It is an exact symmetry group for the theory on the lattice and the eigenstates of the Hamiltonian have to be classified according to the unitary irreducible representations of **O**. The application of the representation theory of the cubic group to lattice gauge theory has been developed by Berg and Billoire [3.85].

The cubic group **O** has 24 elements, which uniquely correspond to the permutations of the four space diagonals of a cube. There are five inequivalent irreducible representations, which are denoted by A_1, A_2, E, T_1 and T_2, and which have dimensions 1, 1, 2, 3, and 3. Let D_J be the spin J representation of SO(3) for integer J. Because the cubic group is a subgroup of SO(3), its elements are also represented by D_J. Viewed as a representation of **O** this yields the *subduced representation* $D_J^{\mathbf{O}}$. It will in general no longer be irreducible and can be decomposed into the irreducible representations of **O**. Up to $J = 4$ one finds

$$D_0^{\mathbf{O}} = A_1, \qquad (3.425)$$
$$D_1^{\mathbf{O}} = T_1, \qquad (3.426)$$
$$D_2^{\mathbf{O}} = E \oplus T_2, \qquad (3.427)$$
$$D_3^{\mathbf{O}} = A_2 \oplus T_1 \oplus T_2, \qquad (3.428)$$
$$D_4^{\mathbf{O}} = A_1 \oplus E \oplus T_1 \oplus T_2. \qquad (3.429)$$

What does this mean for physical states? Consider for example a spin 2 particle in the continuum, described by a quintuplet of degenerate states. The lattice regularization will split this quintuplet into a doublet E and

Table 3.1. Irreducible representations of the cubic group **O**.

R	Dimension	Lowest spin J
A_1	1	0
A_2	1	3
E	2	2
T_1	3	1
T_2	3	2

a triplet T_2. Within the representations E and T_2, the masses will still be degenerate, but between these representations a splitting will occur. Thus we find different masses $m(E)$ and $m(T_2)$. As the continuum limit is approached by tuning the bare coupling g to zero, the mass ratio $m(E)/m(T_2)$ must converge to 1, in order that the full Euclidean symmetry is restored.

Complementary to this is the following aspect. If a lattice operator, which transforms according to some irreducible representation R of the cubic group, is applied to the vacuum, it will create a state, which is a superposition of various eigenstates of the Hamiltonian:

$$\Psi_R = \sum_\alpha c_\alpha^R \Psi_\alpha . \tag{3.430}$$

In the continuum limit any state Ψ_α belongs to some spin J multiplet. In this sense Ψ_R contains various spins J. Spin J can occur in the superposition only if R is contained in D_J^O. For example A_1 contains spin 0, spin 4 and higher spins. Most important is the lowest spin contained in R, since it usually belongs to the lowest mass and will therefore dominate correlation functions. The representations of the cubic group, their dimensions and the lowest spin content are summarized in table 3.1.

In addition to the transformations of the cubic group there are two further discrete symmetries. The first one is total space reflection. Its eigenvalues are the parity $P = \pm 1$. The cubic group combined with space reflection forms the 48-element group $\mathbf{O}_h = \mathbf{O} \times Z_2$, whose representations are labelled R^P.

Finally there is 'charge conjugation', which is equivalent to complex conjugation of Wilson loops. Its eigenvalues are called *C-parity* $C = \pm 1$. The states belonging to an irreducible representation of the lattice symmetry group are therefore labelled by R^{PC}.

Returning to the six particular states Ψ_{ij}, defined above, one finds that they carry three representations. The corresponding linear combinations are

$$A_1^{++} \quad : \quad \mathrm{Re}\,(\Psi_{12} + \Psi_{23} + \Psi_{31}) \tag{3.431}$$

$$E^{++} \quad : \quad \mathrm{Re}\,(\Psi_{13} - \Psi_{23}),\ \mathrm{Re}\,(\Psi_{13} + \Psi_{23} - 2\Psi_{12}) \tag{3.432}$$

$$T_1^{+-} \quad : \quad \mathrm{Im}\,\Psi_{12},\ \mathrm{Im}\,\Psi_{23},\ \mathrm{Im}\,\Psi_{31}. \tag{3.433}$$

They describe a scalar, a tensor, and an axial-vector glueball. For a gauge group with only real representations, as is the case for SU(2), the last one does not occur. For the physically interesting SU(3) it does.

The calculation of glueball masses is done via correlation functions. Define the timeslice observable at time t as

$$S_{ij}(t) = L^{-3/2} \sum_{\mathbf{x}} \mathrm{Tr}\, U((\mathbf{x}, t); i, j) \tag{3.434}$$

and let $S_\rho(t)$ be the linear combination belonging to some representation $\rho = R^{PC}$. Then

$$\langle S_\rho(t)\, S_\rho(0)\rangle_c = \langle \Psi_\rho | e^{-Ht} | \Psi_\rho \rangle = \sum_\alpha |c_\alpha^\rho|^2\, e^{-m_\alpha(\rho)t}. \tag{3.435}$$

For large t the lowest mass m_J, which belongs to some spin J glueball, dominates.

The group theoretic analysis of operators built from Wilson loops up to length 8 has been carried out in [3.85].

3.6.2 Strong-coupling expansion

In a pioneering work Kogut, Sinclair and Susskind [3.86] studied the glueball spectrum in the framework of Hamiltonian lattice gauge theory by means of the strong-coupling expansion. Using Padé approximants they extrapolated mass ratios to the continuum limit.

In this section we shall remain in the Euclidean framework. The results from the strong-coupling expansion are then directly comparable to Monte-Carlo calculations. In contrast to the case of the string tension, the strong-coupling expansions for glueball masses are not affected by the roughening transition and one may try to extrapolate them into the scaling region.

Strong-coupling expansions for glueball masses in Euclidean lattice gauge theories have been developed in [3.87]. In order to study the correlation function for two plaquettes p_1 and p_2, the partition function of a model where the coupling is β_i on plaquettes p_i, $i = 1, 2$, and β elsewhere, is considered. Then one has

$$\langle \mathrm{Tr}\, U_{p_1}\, \mathrm{Tr}\, U_{p_2}\rangle_c = (\mathrm{Tr}\,\mathbf{1})^2\, \frac{\partial^2}{\partial \beta_1 \partial \beta_2}\, \ln Z(\beta, \beta_1, \beta_2)|_{\beta_i = \beta}. \tag{3.436}$$

For $\ln Z(\beta, \beta_1, \beta_2)$ we have the cluster expansion from section 3.4.4. Every cluster contributing to the correlation function must contain both plaquettes p_1 and p_2.

Let us apply this to the correlation functions (3.435). The smallest relevant cluster is between two parallel plaquettes with the same coordinate $\mathbf{x}$, separated by t units in time. It is a single polymer X_0, namely a long tube connecting p_1 and p_2, containing $4t+2$ plaquettes in the fundamental representation of the gauge group. Its contribution to the correlation function is

$$Au^{4t} = Ae^{-m_0 t} \tag{3.437}$$

with

$$m_0 = -4 \ln u \tag{3.438}$$

and some constant A. Corrections to this leading term come from other clusters, which emerge from X_0 by adding local modifications. It can be shown in a way similar to that discussed in connection with the string tension that the corrections exponentiate in the form

$$\langle S_\rho(t) S_\rho(0) \rangle_c = Ae^{-m(\rho)t}, \tag{3.439}$$

$$m(\rho) = -4 \ln u + \sum_{k=1}^{\infty} m_k(\rho) u^k. \tag{3.440}$$

For gauge group SU(2) in four dimensions the expansions are up to eighth order

$$m(A_1^{++}) = -4 \ln u + 2u^2 - \frac{98}{3}u^4 - \frac{20\,984}{405}u^6 - \frac{151\,496}{243}u^8 \tag{3.441}$$

$$m(E^{++}) = -4 \ln u + 2u^2 - \frac{26}{3}u^4 + \frac{13\,036}{405}u^6 - \frac{28\,052}{243}u^8 \tag{3.442}$$

and for gauge group SU(3) one finds [3.87, 3.88]

$$m(A_1^{++}) = -4 \ln u - 3u + 9u^2 - \frac{27}{2}u^3 - 7u^4 - \frac{297}{2}u^5 + \frac{858\,827}{10\,240}u^6$$

$$+ \frac{47\,641\,149}{71\,680}u^7 - \frac{183\,140\,613}{40\,960}u^8 \tag{3.443}$$

$$m(E^{++}) = -4 \ln u - 3u + 9u^2 - \frac{27}{2}u^3 + 17u^4 - \frac{153}{2}u^5 + \frac{1\,104\,587}{10\,240}u^6$$

$$+ \frac{29\,577\,789}{71\,680}u^7 - \frac{90\,611\,973}{40\,960}u^8 \tag{3.444}$$

$$m(T_1^{+-}) = -4\ln u + 3u + \frac{9}{2}u^3 - \frac{99}{4}u^4 + \frac{33}{4}u^5 - \frac{36\,771}{1\,280}u^6$$

$$+ \frac{117\,897}{448}u^7 - \frac{1\,559}{2}u^8. \tag{3.445}$$

Results for other groups and other dimensions can be found in [3.87, 3.89]. The series displayed above have been obtained by explicit enumeration of all graphs up to eighth order. Decker [3.90] has developed a computer algorithm which undertakes this task, and has applied it to Z_2 lattice gauge theory in order to obtain more terms. Even more efficient appears to be the technique of Möbius inversion, which has allowed one to obtain the mass gap of three-dimensional Z_2 lattice gauge theory up to the 22nd order [3.91].

Results for glueball states with non-zero momentum are also available [3.92, 3.93]. They allow us to investigate the question of the restoration of Euclidean invariance.

The strong-coupling expansions for glueball masses can be compared with the results from numerical simulations in the strong-coupling regime. Moreover, assuming a rapid crossover from strong-coupling to scaling behaviour, one can also try to obtain estimates for the ratios $C_m = m/\Lambda_{LAT}$ or for ratios of glueball masses. Matching the strong-coupling series tangentially with the scaling function, as indicated in connection with the string tension series, one obtains

$$m(A_1^{++})/\Lambda_{LAT} = 130 \pm 60 \qquad \text{for SU(2)}. \tag{3.446}$$

For gauge group SU(3) the series behave more irregularly. Smit [3.94] has applied more sophisticated extrapolation methods in this case. Using the internal energy as an expansion variable and employing Padé and Padé-Borel extrapolation techniques, he observes that the resulting function shows indications of scaling over some range of couplings. His result for the scalar glueball is

$$m(A_1^{++})/\Lambda_{LAT} = 340 \pm 40 \qquad \text{for SU(3)}. \tag{3.447}$$

These numbers have to be regarded with reservations, however, since they would be influenced by any deviations from asymptotic scaling in the range of intermediate couplings, where the estimates are taken.

Glueball mass ratios, on the other hand, are affected by deviations from scaling only. Extrapolation techniques specific for mass ratios have been applied [3.95] and yield

$$\begin{aligned}
m(E^{++})/m(A_1^{++}) &\approx 1.25 & \text{for SU(2)} \\
m(E^{++})/m(A_1^{++}) &\approx 1 & \text{for SU(3)} \\
m(T_1^{+-})/m(A_1^{++}) &= 1.8 \pm 0.3 & \text{for SU(3)},
\end{aligned} \tag{3.448}$$

in agreement with the analysis of [3.86, 3.94].

As a final quantity related to glueballs let us mention the triple glueball coupling, which is defined in terms of the on-shell three-plaquette vertex function [3.96]. It determines the strength of the Yukawa potential

$$V(r) = -\frac{G^2}{4\pi} \frac{e^{-mr}}{r},$$

(3.449)

which effectively describes the interaction between non-relativistic glueballs. From low-order strong-coupling expansions its value has been estimated [3.96] for both SU(2) and SU(3) to be roughly

$$\frac{G^2}{4\pi} \approx 40,$$

(3.450)

which is surprisingly large. This number also governs the magnitude of finite-size effects on glueball masses calculated on finite lattices [3.97].

3.6.3 Monte Carlo results

In this section we would like to summarize some results related to glueballs, which have been obtained by means of the Monte Carlo method. First estimates on the mass gap in lattice gauge theories relied on simple plaquette–plaquette correlation functions [3.98] The analysis has then been refined using a variational technique, where the glueball states are obtained from optimized linear combinations of Wilson loops applied to the vacuum [3.99]. With the help of operators transforming according to various representations of the cubic group higher spin states have also been considered. Early results on the spectrum, based on these and other methods, are reviewed in [3.100, 3.101].

Concerning the ratio m/Λ_{LAT}, recent calculations give

$$m(A_1^{++})/\Lambda_{LAT} = 180 \pm 20$$

(3.451)

for SU(2) [3.102], and

$$m(A_1^{++})/\Lambda_{LAT} = 300 \pm 10$$

(3.452)

for SU(3) [3.103]. These numbers should be considered as preliminary, however, because it appears that asymptotic scaling has not yet been reached in the numerical calculations [3.102, 3.104].

The most reliable results are about mass ratios. Recent calculations have been compiled by van Baal and Kronfeld [3.105] and Schierholz [3.103]. It turns out that finite-size effects are not negligible if $mL \leq 9$ [3.105, 3.77]. In larger volumes one has for both SU(2) and SU(3)

$$m(E^{++})/m(A_1^{++}) = m(T_2^{++})/m(A_1^{++}) = 1.5 \pm 0.1.$$

(3.453)

Comparing masses to the string tension α, one gets

$$m(A_1^{++})/\sqrt{\alpha} = 3.8 \pm 0.2 \qquad \text{for SU(2)} \tag{3.454}$$

$$m(A_1^{++})/\sqrt{\alpha} = 3.25 \pm 0.2 \qquad \text{for SU(3)}. \tag{3.455}$$

For higher masses the results do not yet appear to be definite. Let us just quote [3.104]

$$m(T_1^{+-})/m(A_1^{++}) \approx 2 \tag{3.456}$$

for the spin 1 glueball in SU(3) gauge theory.

From finite-size effects the triple glueball coupling has also been estimated. The value [3.106]

$$\frac{G^2}{4\pi} = 60 \pm 25 \qquad \text{for SU(3)} \tag{3.457}$$

agrees with the strong-coupling calculation.

3.7 Phase structure of lattice gauge theory

Our main concern in this book is with the physically relevant gauge groups SU(2), SU(3) and U(1) in four dimensions. For SU(2) and SU(3) lattice gauge theory with Wilson's action in four dimensions there is much evidence for the absence of any bulk phase transition, and it is widely believed that at weak couplings a continuum limit with static quark confinement exists as a consequence of the analytic continuation of confinement at strong couplings (see section 3.4).

For other gauge groups or other space–time dimensions, however, various phase transitions show up, separating phases of different nature. In the following we would like to review briefly what is known about them. For the sake of lucidity we restrict ourselves to non-Abelian gauge groups SU(N) and Abelian gauge groups U(1) and Z_n. The latter are discrete subgroups of U(1), consisting of the elements

$$Z_n = \{e^{2\pi i k/n} \mid k = 0, \ldots, n - 1\}. \tag{3.458}$$

Unless stated otherwise, Wilson's action is assumed for the lattice gauge theory.

In order to distinguish different phases one has to consider *order parameters*. One of them is the Wilson loop in the fundamental representation of the gauge group, which has already been discussed at length. For large loops it cannot decay faster than according to the area law [3.65]. On the other hand, it always decays at least as fast as with the *perimeter law* [3.107]

$$W(\mathscr{C}_{R,T}) \underset{R,T \to \infty}{\sim} c\, e^{-\mu(R+T)}. \tag{3.459}$$

The asymptotic behaviour of Wilson loops can be employed to characterize different phases.

Another order parameter is the *'t Hooft loop* [3.108]. For its definition in the context of lattice gauge theory and further discussions see [3.13, 3.109, 3.110, 3.111]. In a certain sense the 't Hooft loop is dual to the Wilson loop and its asymptotic behaviour also serves as a characteristic for different phases.

Closely related to the 't Hooft loop is the *free energy of vortices* [3.112, 3.113, 3.109, 3.48, 3.114]. It measures the effect of changing the boundary conditions from periodic to twisted ones on the free energy in finite volumes.

Finally we mention the mass gap, which distinguishes massive from massless phases.

In general one encounters three types of phases in lattice gauge theory without matter fields. They are characterized in the following way:

Confinement phase: a massive phase in which the Wilson loop obeys the area law and the 't Hooft loop obeys the perimeter law,

Higgs phase: a massive phase in which the Wilson loop obeys the perimeter law and the 't Hooft loop obeys the area law. Alternatively it may be characterized through the free energy of vortices (see [3.114]),

Coulomb phase: a massless phase with perimeter laws for both Wilson and 't Hooft loops.

Phases different from the ones listed above may be possible, but they are the exception.

3.7.1 General results

Some general results are available, which have been established by rigorous methods. They are helpful in understanding the phase structure of lattice gauge theories.

1. Strong couplings:
   As has already been mentioned in section 3.4.5 there is a theorem due to Osterwalder and Seiler [3.17, 3.9] which says: for all compact gauge groups and all space–time dimensions there is a $\beta_1 > 0$ such that for $\beta < \beta_1$ the strong-coupling expansion converges and the lattice gauge theory is in the confinement phase.

2. Two dimensions:
   In two space–time dimensions gauge theory does not contain any dynamical degrees of freedom. The two-dimensional lattice gauge

theory is exactly solvable [3.115, 3.116, 3.117]. The Wilson loop obeys the area law and the mass gap is infinite.

3. Discrete groups:

   If the gauge group is discrete there is a finite gap in the action between its absolute minimum and the first value above it. For Z_n lattice gauge theory with Wilson's action it is

$$\Delta = 1 - \cos \frac{2\pi}{n}. \qquad (3.460)$$

   A simple energy–entropy argument [3.118] suggests that in more than two dimensions there is a phase transition at some coupling

$$\beta_c \approx \frac{c}{\Delta}. \qquad (3.461)$$

   In fact it can be shown rigorously that the model is in the Higgs phase at very large β. The proof consists of a Peierls argument for the Wilson loop and 't Hooft loop [3.119, 3.109, 3.9]. With respect to the free energy of vortices the proof is a special case of [3.114].

4. Gauge group U(1):

   4.1 Guth's theorem: The U(1) lattice gauge theory in four dimensions has a weak coupling phase, where it is in the Coulomb phase. Guth's original proof [3.120] showed the perimeter law for Wilson loops in the model with Villain's action. Other actions and 't Hooft loops are treated in [3.121, 3.9].

   4.2 Göpfert–Mack theorem: The U(1) lattice gauge theory (with Villain's action) in three dimensions has non-vanishing mass gap and string tension for all couplings. Mack and the late Göpfert [3.122] used iterated Mayer expansions to establish their result. Their proof stands out as a masterpiece of hard analysis. The numerical consequences of that work have been tested by means of the Monte Carlo method in [3.123].

3.7.2 Abelian gauge groups

From the results cited above and other analytical and numerical investigations a picture of the various phases of lattice gauge theories in different numbers of space–time dimensions has emerged, which we would like to summarize here. Let us begin with Abelian gauge groups Z_n and U(1).

In three dimensions the Z_n lattice gauge theory shows two different phases. At strong couplings it is in the confinement phase, whereas at weak couplings it is in the Higgs phase. The phase transition point β_c increases with n like

$$\beta_c \approx c(1 - \cos 2\pi/n)^{-1}. \qquad (3.462)$$

In the limit $n \to \infty$ it disappears and the U(1) lattice gauge theory is recovered, which is always in the confinement phase.

It goes without saying that for Z_n lattice gauge theory in four dimensions the confinement and Higgs phases are present too. For $n \geq 5$, however, a new phenomenon occurs. Between the two massive phases a Coulomb phase appears in a range $\beta_1 < \beta < \beta_2$ [3.124, 3.121]. Again β_2 goes to infinity with n, leaving the U(1) theory with a confinement and a Coulomb phase.

For a physical picture of the phases in the U(1) theory see [3.125, 3.126]. A continuum limit might be attainable in different ways. In any case it is guaranteed that in all dimensions $d \geq 3$ it is possible to take a limit at large β, which leads to the free quantized Maxwell field [3.127].

In five dimensions the Z_2 and U(1) theories appear to possess a confinement and a Higgs phase each.

Numerical work related to Z_n and U(1) can be found in [3.128, 3.129].

3.7.3 Non-Abelian gauge groups

The phase structure of non-Abelian lattice gauge theories has been discussed in [3.73] based on heuristic considerations. Together with available Monte Carlo results [3.128, 3.130] the following picture emerges.

In three and four dimensions the SU(N) lattice gauge theory is confining for all values of the coupling. For $N \geq 4$ in four dimensions, however, there are one or more phase transitions separating different confinement phases.

In five and more dimensions a Higgs phase is present at weak couplings, separated from the confinement phase by a first order phase transition.

4

Fermion fields

4.1 Fermionic variables

4.1.1 Creation and annihilation operators

In the canonical quantization formalism the creation and annihilation operators of fermions satisfy the anticommutation relations

$$\{a_i^+, a_j\} = \delta_{ij} , \qquad \{a_i, a_j\} = \{a_i^+, a_j^+\} = 0 . \qquad (4.1)$$

The indices $1 \leq i, j \leq N$ label the different states. The multi-fermion states are in the *Fock space*, which is built on the *vacuum state* satisfying

$$a_i|0\rangle = 0 \qquad (1 \leq i \leq N) . \qquad (4.2)$$

An orthonormal basis in the Fock space is defined by

$$a_i^+ a_j^+ a_k^+ \dots |0\rangle \qquad (i > j > k > \dots) . \qquad (4.3)$$

A matrix representation of the algebra of the fermionic creation and annihilation operators can be built up by N-tuple direct products of the Pauli matrices [4.1]. These satisfy, for $1 \leq i_1, i_2 \leq N$, the commutation relations

$$[\sigma_{(i_1)m_1}, \sigma_{(i_2)m_2}] = 2i\delta_{i_1 i_2}\epsilon_{m_1 m_2 m_3}\sigma_{(i_1)m_3} . \qquad (4.4)$$

In the 2^N dimensional space one can choose the basis where $\sigma_{(i)3}$, $1 \leq i \leq N$ is diagonal. Denoting the eigenvalues by $\xi = \pm 1$ we have

$$\sigma_{(i_j)3}|\xi_{i_1}, \xi_{i_2}, \dots, \xi_{i_N}\rangle = \xi_{i_j}|\xi_{i_1}, \xi_{i_2}, \dots, \xi_{i_N}\rangle . \qquad (4.5)$$

The vacuum state can be defined as $|0\rangle \equiv |-1, -1, \dots, -1\rangle$, and the matrix representation of the creation and annihilation operators is

$$a_i^+ = (-1)^{N-i}\sigma_{(N)3}\sigma_{(N-1)3} \cdots \sigma_{(i+1)3}\sigma_{(i)+} ,$$

$$a_i = (-1)^{N-i}\sigma_{(N)3}\sigma_{(N-1)3} \cdots \sigma_{(i+1)3}\sigma_{(i)-} . \qquad (4.6)$$

In most models of relativistic quantum field theory the fermionic bilinears play an important rôle. These are represented by

$$a_i^+ a_{i+k} = s(i, i+k)\sigma_{(i)+}\sigma_{(i+k)-} \, ,$$

$$s(i, i) = s(i, i+1) = 1 \, ,$$

$$s(i, i+k) = (-1)^{k+1}\sigma_{(i+k-1)3}\sigma_{(i+k-2)3}\cdots\sigma_{(i+1)3} \qquad (k \geq 2) \, . \qquad (4.7)$$

4.1.2 A simple example

In order to illustrate the type of Hamiltonians which appear in many relativistic quantum field theories let us consider the simple two-state fermion system described by the Hamiltonian

$$H_2 = m(a_1^+ a_1 + a_2^+ a_2) + K(a_1^+ a_2 + a_2^+ a_1) \, . \qquad (4.8)$$

Typically we want to determine the partition function $Z \equiv \mathrm{Tr}\, e^{-\beta H_2}$. The simplest way to obtain this is to introduce

$$a_1' \equiv \frac{1}{\sqrt{2}}(a_1 + a_2) \, , \qquad a_2' \equiv \frac{1}{\sqrt{2}}(-a_1 + a_2) \, . \qquad (4.9)$$

This brings H_2 to the diagonal form

$$H_2 = (m+K)a_1'^+ a_1' + (m-K)a_2'^+ a_2' \, . \qquad (4.10)$$

Inserting the complete set of states $|0\rangle, a_1'^+|0\rangle, a_2'^+|0\rangle, a_2'^+ a_1'^+|0\rangle$ in the trace, one immediately obtains

$$Z = \left(1 + e^{-\beta(m+K)}\right)\left(1 + e^{-\beta(m-K)}\right) \, . \qquad (4.11)$$

The same result can also be obtained in a different way, which resembles the path integral treatment of bosonic systems. Namely, we can divide the β-interval into T equal parts of length $\Delta\beta = \beta/T$ and write

$$Z = \mathrm{Tr}\, e^{-\beta H_2} = \lim_{T \to \infty} \mathrm{Tr}\,(1 - \Delta\beta H_2)^T$$

$$= \lim_{T \to \infty} \sum_{\xi^{(1)}\xi^{(2)}\dots\xi^{(T)}} \langle\xi^{(1)}|1 - \Delta\beta H_2|\xi^{(2)}\rangle$$

$$\cdot \langle\xi^{(2)}|1 - \Delta\beta H_2|\xi^{(3)}\rangle \cdots \langle\xi^{(T)}|1 - \Delta\beta H_2|\xi^{(1)}\rangle \, . \qquad (4.12)$$

Here in the last step complete sets of intermediate states $|\xi^{(i)}\rangle \equiv |\xi_1^{(i)}, \xi_2^{(i)}\rangle$ were inserted. The non-zero matrix elements occurring in (4.12) can be graphically represented as shown in fig. 4.1. Combining these matrix elements one obtains three types of contributions as shown by fig. 4.2. The values of the first two contributions are trivial. The sum of the third type of contributions can also be easily determined by counting the number of

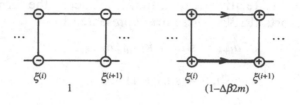

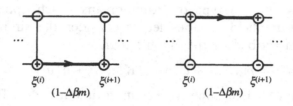

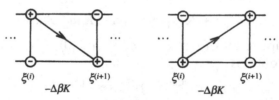

Fig. 4.1. Graphical representation of the non-zero matrix elements of $(1-\Delta\beta H_2)$. The upper horizontal line represents the first state, the lower one the second state. $\oplus$ stands for $\xi = +1$, $\ominus$ for $\xi = -1$. The line with an arrow shows the movement of the occupied state. The value of the matrix element is shown below the figures.

different possibilities for going from the upper line to the lower one and vice versa. In this way we obtain from (4.12) again the result in (4.11). The way of writing the partition function in (4.12) and the corresponding graphical representation of the contributions in fig. 4.2 is already similar to the path integral for bosons. The similarity can be further enhanced by the introduction of the concept of Grassmannian integrals [4.2].

4.1.3   Grassmann variables

For every fermion state $1 \leq i \leq N$ a pair of Grassmann variables (η_i, η_i^+) is introduced. These are totally anticommuting, that is instead of the

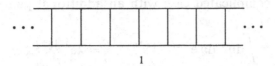

1

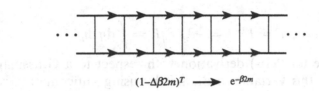

$(1-\Delta\beta2m)^T \longrightarrow e^{-\beta2m}$

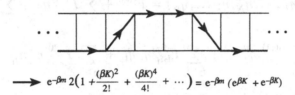

$$\longrightarrow e^{-\beta m} 2\left(1+\frac{(\beta K)^2}{2!}+\frac{(\beta K)^4}{4!}+\cdots\right)=e^{-\beta m}\left(e^{\beta K}+e^{-\beta K}\right)$$

Fig. 4.2. Three types of contributions to non-zero matrix elements in Z. The values of the contributions are given below the figures in the $T \to \infty$ limit.

algebra of creation and annihilation operators in (4.1) we have

$$\{\eta_i^+,\eta_j\}=\{\eta_i,\eta_j\}=\{\eta_i^+,\eta_j^+\}=0\,. \tag{4.13}$$

Since the square of every Grassmann variable is zero, functions of these variables are polynomials of finite degree.

Let us first consider, for simplicity, the case of a single Grassmann pair ($N=1$). In this case the most general function has the form

$$F(\eta^+,\eta)=F^{(00)}+F^{(01)}\eta+F^{(10)}\eta^++F^{(11)}\eta^+\eta\,. \tag{4.14}$$

Here the coefficients $F^{(\alpha\beta)}$ are assumed to be complex numbers, but later on we shall also consider more general cases. The integral of the function $F(\eta^+,\eta)$ is defined by

$$\int d\eta\,d\eta^+ F(\eta^+,\eta)\equiv-\int d\eta^+\,d\eta F(\eta^+,\eta)\equiv F^{(11)}\,. \tag{4.15}$$

A simple example of the application of this rule is

$$\int d\eta^+\,d\eta\,e^{-\lambda\eta^+\eta}=\int d\eta^+\,d\eta(1-\lambda\eta^+\eta)=\lambda\,. \tag{4.16}$$

A slightly more complicated case with an additional pair of Grassmann variables (ξ, ξ^+) is

$$\int \mathrm{d}\eta^+ \, \mathrm{d}\eta \, \mathrm{e}^{-\lambda\eta^+\eta+\xi^+\eta+\eta^+\xi} = \lambda \mathrm{e}^{\lambda^{-1}\xi^+\xi} \,. \tag{4.17}$$

Derivatives with respect to the Grassmann variables can also be defined:

$$\partial_{\eta^+} F(\eta^+, \eta) \equiv F^{(10)} + F^{(11)}\eta \,, \quad \partial_\eta F(\eta^+, \eta) \equiv F^{(01)} - F^{(11)}\eta^+ \,. \tag{4.18}$$

This implies

$$\partial_\eta \partial_{\eta^+} F = F^{(11)} = -\partial_{\eta^+}\partial_\eta F = \int \mathrm{d}\eta \, \mathrm{d}\eta^+ F \,. \tag{4.19}$$

The general rule for (left-) derivation with respect to a Grassmann variable is to bring this variable to the left by using anticommutativity and then omit it.

The generalization of (4.14) to a function of N pairs of Grassmann variables is

$$F[\eta^+, \eta] \equiv F(\eta_1^+, \ldots, \eta_N^+, \eta_1, \ldots, \eta_N) = F^{(00)} + \sum_{i_1} F_{i_1}^{(01)}\eta_{i_1} + \sum_{j_1} F_{j_1}^{(10)}\eta_{j_1}^+$$

$$+ \sum_{i_1, j_1} F_{(j_1 i_1)}^{(11)}\eta_{j_1}^+\eta_{i_1} + \sum_{i_2>i_1} F_{i_2 i_1}^{(02)}\eta_{i_2}\eta_{i_1} + \sum_{j_2>j_1} F_{j_2 j_1}^{(20)}\eta_{j_2}^+\eta_{j_1}^+$$

$$+ \sum_{j_1}\sum_{i_2>i_1} F_{(j_1 i_1)i_2}^{(12)}\eta_{i_2}\eta_{j_1}^+\eta_{i_1} + \sum_{j_2>j_1}\sum_{i_1} F_{(j_1 i_1)j_2}^{(21)}\eta_{j_2}^+\eta_{j_1}^+\eta_{i_1}$$

$$+ \sum_{j_2>j_1}\sum_{i_2>i_1} F_{(j_1 i_1)(j_2 i_2)}^{(22)}\eta_{j_2}^+\eta_{i_2}\eta_{j_1}^+\eta_{i_1} + \cdots$$

$$+ \sum_{j_N>\cdots>j_1}\sum_{i_N>\cdots>i_1} F_{(j_1 i_1)\cdots(j_N i_N)}^{(NN)}\eta_{j_N}^+\eta_{i_N}\cdots\eta_{j_1}^+\eta_{i_1} \,. \tag{4.20}$$

Here in the last sum only one independent non-zero combination of N Grassmann pairs appears, therefore one can define the following: $F^{(NN)} \equiv F_{(11)(22)\cdots(NN)}^{(NN)}$. The integral over N pairs of Grassmann variables is defined by

$$\int \mathrm{d}\eta_1 \, \mathrm{d}\eta_1^+ \, \mathrm{d}\eta_2 \, \mathrm{d}\eta_2^+ \cdots \mathrm{d}\eta_N \, \mathrm{d}\eta_N^+ \, F[\eta^+, \eta]$$

$$\equiv (-1)^N \int \mathrm{d}\eta_1^+ \, \mathrm{d}\eta_1 \, \mathrm{d}\eta_2^+ \, \mathrm{d}\eta_2 \cdots \mathrm{d}\eta_N^+ \, \mathrm{d}\eta_N \, F[\eta^+, \eta] \equiv F^{(NN)} \,. \tag{4.21}$$

An important example is the Gaussian integral appearing in many basic formulas of quantum field theory:

$$\int \mathrm{d}\eta_1^+ \, \mathrm{d}\eta_1 \cdots \mathrm{d}\eta_N^+ \, \mathrm{d}\eta_N \, \mathrm{e}^{-\sum_{i,j} \eta_j^+ A_{ji}\eta_i} = \det A$$

$$= \sum_{i_1 \cdots i_N} \epsilon^{12 \cdots N}_{i_1 i_2 \cdots i_N} A_{1 i_1} \cdots A_{N i_N} = \frac{1}{N!} \sum_{\substack{j_1 \cdots j_N \\ i_1 \cdots i_N}} \epsilon^{j_1 j_2 \cdots j_N}_{i_1 i_2 \cdots i_N} A_{j_1 i_1} \cdots A_{j_N i_N} \,, \qquad (4.22)$$

where in the explicit form of the determinant the following notation was introduced

$$\epsilon^{j_1 j_2 \cdots j_N}_{i_1 i_2 \cdots i_N} \equiv \left\{ \begin{array}{c} 1 \\ -1 \\ 0 \end{array} \right. \quad \text{if } (j_1 j_2 \cdots j_N) \text{ is}$$

$$\left\{ \begin{array}{c} \text{even} \\ \text{odd} \\ \text{no} \end{array} \right\} \quad \text{permutation of } (i_1 i_2 \cdots i_N) \,. \qquad (4.23)$$

The generalization of (4.17) to many Grassmann variables is

$$\int d\eta_1^+ \, d\eta_1 \cdots d\eta_N^+ \, d\eta_N \; e^{-\sum_{i,j} \eta_j^+ A_{ji} \eta_i + \sum_i \xi_i^+ \eta_i + \sum_i \eta_i^+ \xi_i}$$

$$= \det A \cdot e^{\sum_{i,j} \xi_j^+ A_{ji}^{-1} \xi_i} \,. \qquad (4.24)$$

Taking derivatives $\partial_{\xi_i} \partial_{\xi_j^+}$ of both sides n times at $\xi = \xi^+ = 0$ gives

$$\int d\eta_1^+ \, d\eta_1 \cdots d\eta_N^+ \, d\eta_N \; e^{-\sum_{i,j} \eta_j^+ A_{ji} \eta_i} \eta_{j_1} \eta_{i_1}^+ \cdots \eta_{j_n} \eta_{i_n}^+$$

$$= \det A \sum_{k_1 \cdots k_n} \epsilon^{k_1 k_2 \cdots k_n}_{j_1 j_2 \cdots j_n} A_{k_1 i_1}^{-1} \cdots A_{k_n i_n}^{-1} \,. \qquad (4.25)$$

The purpose of introducing Grassmann variables in quantum field theory is to write the traces over fermionic variables appearing in the partition function or, more generally in the generating functional for Green functions, in the form of an integral. In the Hamiltonian formulation the traces are defined in the Fock space of creation and annihilation operators. Therefore it is convenient to introduce 'Grassmannian coherent states' which are linear combinations of the elements of the Fock space with Grassmann variable coefficients. The Grassmann variables are assumed to commute with the creation and annihilation operators:

$$[\eta_i, a_j] = [\eta_i^+, a_j] = [\eta_i, a_j^+] = [\eta_i^+, a_j^+] = 0 \,. \qquad (4.26)$$

With this convention the Grassmannian coherent states are defined as

$$|\eta\rangle \equiv |\eta_1, \eta_2, \cdots, \eta_N\rangle \equiv e^{\sum_{i=1}^N a_i^+ \eta_i} |0\rangle$$

$$= |0\rangle + \sum_i a_i^+ |0\rangle \eta_i + \sum_{j>i} a_j^+ a_i^+ |0\rangle \eta_j \eta_i + \cdots \,. \qquad (4.27)$$

The corresponding bra-vector is, of course,

$$\langle \eta | = \langle 0 | + \sum_i \eta_i^+ \langle 0 | a_i + \sum_{j>i} \eta_i^+ \eta_j^+ \langle 0 | a_i a_j + \cdots . \tag{4.28}$$

These states satisfy $\langle 0 | \eta \rangle = 1$, and they are eigenvectors of the annihilation operators with Grassmann variables as eigenvalues:

$$a_i | \eta \rangle = \eta_i | \eta \rangle . \tag{4.29}$$

The decomposition of the unit operator in the Fock space in terms of the Grassmannian coherent states is

$$1 = \int d\eta_1^+ d\eta_1 \cdots d\eta_N^+ d\eta_N \, e^{-\sum_i \eta_i^+ \eta_i} | \eta_1, \cdots, \eta_N \rangle \langle \eta_1, \cdots, \eta_N | . \tag{4.30}$$

For two different (anticommuting) Grassmann-algebras $\{\eta, \eta^+\}$ and $\{\eta', \eta'^+\}$, the scalar product of the two corresponding coherent states is

$$\langle \eta | \eta' \rangle = 1 + \sum_i \eta_i^+ \eta_i' + \sum_{j>i} \eta_i^+ \eta_j^+ \eta_j' \eta_i' + \cdots = e^{\sum_i \eta_i^+ \eta_i'} . \tag{4.31}$$

From the definition one can calculate the matrix elements of some important types of operators, too. For instance, a bilinear operator gives

$$\langle \eta | \sum_{i,j} a_j^+ A_{ji} a_i | \eta' \rangle = \sum_{i,j} \eta_j^+ A_{ji} \eta_i' \, e^{\sum_k \eta_k^+ \eta_k'} . \tag{4.32}$$

The exponential of a bilinear operator has the matrix element

$$\langle \eta | e^{\sum_{i,j} a_j^+ A_{ji} a_i} | \eta' \rangle = \exp\left\{ \sum_{i,j} \eta_j^+ (e^A)_{ji} \eta_i' \right\} . \tag{4.33}$$

Using this relation and (4.22) one can show that the trace of such an operator is

$$\mathrm{Tr}\, e^{\sum_{i,j} a_j^+ A_{ji} a_i} = \int d\eta_1^+ d\eta_1 \cdots d\eta_N^+ d\eta_N \, e^{-\sum_i \eta_i^+ \eta_i}$$

$$\cdot \langle +\eta | e^{\sum_{i,j} a_j^+ A_{ji} a_i} | -\eta \rangle . \tag{4.34}$$

Here the negative sign in the Grassmannian coherent state is important. It will imply antiperiodicity in time for the Grassmann variables.

In order to illustrate the Grassmannian path integrals for fermions let us apply this formalism to the above simple model defined by the Hamiltonian (4.8). The partition function Z is given by

$$Z = \mathrm{Tr}\, e^{-\beta H_2} = \lim_{T \to \infty} \mathrm{Tr}\, (e^{-\Delta\beta H_2})^T$$

$$= \int d\eta_{1,T}^+ d\eta_{1,T} d\eta_{2,T}^+ d\eta_{2,T} \, e^{-\sum_{i=1}^2 \eta_{i,T}^+ \eta_{i,T}} \langle \eta_T | e^{-\beta H_2} | -\eta_T \rangle$$

$$= \lim_{T \to \infty} \int d\eta_{1,1}^+ \, d\eta_{1,1} \, d\eta_{2,1}^+ \, d\eta_{2,1} \cdots d\eta_{1,T}^+ \, d\eta_{1,T} \, d\eta_{2,T}^+ \, d\eta_{2,T}$$

$$e^{-\sum_{t=1}^T \sum_{i=1}^2 \eta_{i,t}^+ \eta_{i,t}} \langle \eta_T | e^{-\Delta\beta H_2} | \eta_{T-1} \rangle$$

$$\cdot \langle \eta_{T-1} | e^{-\Delta\beta H_2} | \eta_{T-2} \rangle \cdots \langle \eta_1 | e^{-\Delta\beta H_2} | -\eta_T \rangle . \tag{4.35}$$

Using (4.32), (4.33) we obtain

$$Z = \lim_{T \to \infty} \int d\eta_{1,1}^+ \, d\eta_{1,1} \, d\eta_{2,1}^+ \, d\eta_{2,1} \cdots d\eta_{1,T}^+ \, d\eta_{1,T} \, d\eta_{2,T}^+ \, d\eta_{2,T}$$

$$\cdot \exp \left\{ \sum_{t=1}^T \left[\eta_{1,t+1}^+ \eta_{1,t} + \eta_{2,t+1}^+ \eta_{2,t} - \eta_{1,t}^+ \eta_{1,t} - \eta_{2,t}^+ \eta_{2,t} \right. \right.$$

$$\left. \left. - \Delta\beta m (\eta_{1,t+1}^+ \eta_{1,t} + \eta_{2,t+1}^+ \eta_{2,t}) - \Delta\beta K (\eta_{1,t+1}^+ \eta_{2,t} + \eta_{2,t+1}^+ \eta_{1,t}) \right] \right\} . \tag{4.36}$$

The antiperiodicity is taken into account here by $\eta_{T+1} \equiv -\eta_1$. According to (4.22) this can be written as

$$Z = \lim_{T \to \infty} \det Q(\Delta\beta) , \tag{4.37}$$

where, with the short notations $\delta m \equiv \Delta\beta \, m$ and $\delta K \equiv \Delta\beta \, K$, the $2T \otimes 2T$ matrix Q is

$$Q(\Delta\beta) = \begin{pmatrix} 1 & 0 & 0 & \cdots & 1-\delta m & -\delta K \\ 0 & 1 & 0 & \cdots & -\delta K & 1-\delta m \\ -1+\delta m & \delta K & 1 & \cdots & 0 & 0 \\ \delta K & -1+\delta m & 0 & \cdots & 0 & 0 \\ 0 & 0 & -1+\delta m & \cdots & 0 & 0 \\ 0 & 0 & \delta K & \cdots & 0 & 0 \\ \cdots & \cdots & \cdots & \cdots & \cdots & \cdots \\ 0 & 0 & 0 & \cdots & 1 & 0 \\ 0 & 0 & 0 & \cdots & 0 & 1 \end{pmatrix} . \tag{4.38}$$

The determinant in (4.37) has a graphical 'polymer representation' which we consider next.

4.1.4 Polymer representation

Let us consider the determinant of a matrix Q which is given in the form

$$Q_{ji} \equiv M_i \delta_{ji} - K_{ji} \qquad (1 \le i, j \le N) . \tag{4.39}$$

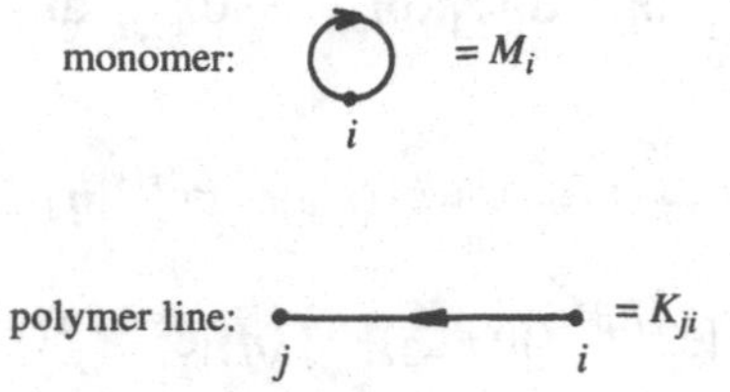

monomer: $\bigcirc$ $= M_i$
          i

polymer line: j •———←———• i $= K_{ji}$

Fig. 4.3. The building blocks of polymers: *'monomer'* at a single index point and a *'polymer line'* connecting two different index points.

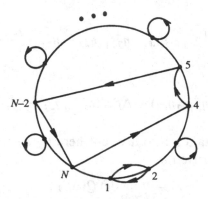

Fig. 4.4. An example of a polymer with index points arranged on a circle.

Here the diagonal elements of K are assumed to vanish: $K_{ii} = 0$. In terms of a Grassmann integral one can write

$$\det Q = \int d\eta_1^+ \, d\eta_1 \cdots d\eta_N^+ \, d\eta_N \; e^{-\sum_i M_i \eta_i^+ \eta_i + \sum_{i \neq j} \eta_j^+ K_{ji} \eta_i} \,. \qquad (4.40)$$

In order to represent the non-zero contributions to the determinant, the two building blocks shown in fig. 4.3 are needed. Using these one can draw *'polymers'* which consist of monomers and polymer lines in such a way that every index point ($1 \leq i \leq N$) has exactly one incoming and outgoing line. (A monomer is counted both as an incoming and an outgoing line.) The set of all polymers is denoted by C. The contribution of a polymer $z \in C$ is a product of the contributions of the monomers and polymer lines multiplied by a sign factor depending on the number of closed polymer loops:

$$R_z = (-1)^{\text{number of closed polymer loops}} \prod_i M_i \prod_{j,k} K_{kj} \,. \qquad (4.41)$$

For instance, the polymer in fig. 4.4 stands for

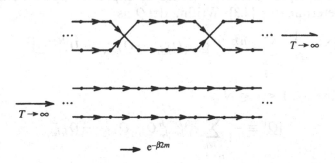

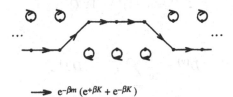

Fig. 4.5. A sample of polymer classes contributing to the polymer representation of det Q. In the limit $T \to \infty$ only the polymers without crossing survive.

$$R_{\text{fig. 4.4}} = (-1)^2 K_{12} K_{21} \cdot K_{45} K_{5,N-2} K_{N-2,N} K_{N4} \cdots \prod_{i \neq 1,2,4,5,N-2,N,\cdots} M_i \,.$$
$$(4.42)$$

As one can easily see from (4.40) by expanding the exponential function in power series and applying the rules of Grassmann integration, the determinant is equal to the sum of all polymer contributions:

$$\det Q = \sum_{z \in C} R_z \,. \tag{4.43}$$

The polymer representation of the determinant of the matrix Q in (4.38), which gives the partition function of the simple model defined by the Hamiltonian in (4.8), can be constructed according to these general rules. Let us now arrange the index points in two horizontal lines, similarly to fig. 4.2. The obtained types of polymers are shown in fig. 4.5, together with the $T \to \infty$ limit of the contributions. Note that the negative sign coming from antiperiodicity cancels the one due to the closed polymer line. Comparing fig. 4.2 and fig. 4.5 one can see the correspondence between the two $T \to \infty$ representations of the partition function: the first one directly in terms of the matrix elements of the Hamiltonian, the second one in terms of polymer contributions to the determinant, which is obtained from the Grassmannian path integral.

The polymer representation can also be expressed by some useful formulas for determinants [4.3]. Writing det Q as

$$\det Q = \frac{1}{N!} \sum_{\substack{j_1 \cdots j_N \\ i_1 \cdots i_N}} \epsilon^{j_1 j_2 \cdots j_N}_{i_1 i_2 \cdots i_N} Q_{j_1 i_1} Q_{j_2 i_2} \cdots Q_{j_N i_N} \equiv D^{(N)}[Q] , \qquad (4.44)$$

one can define for $1 \le v \le N$

$$D^{(v)}[Q] \equiv \frac{1}{v!} \sum_{\substack{j_1 \cdots j_v \\ i_1 \cdots i_v}} \epsilon^{j_1 j_2 \cdots j_v}_{i_1 i_2 \cdots i_v} Q_{j_1 i_1} Q_{j_2 i_2} \cdots Q_{j_v i_v} . \qquad (4.45)$$

Defining also

$$T^{(v)} \equiv (-1)^{v+1} \operatorname{Tr} Q^v , \qquad (4.46)$$

we have for $v \ge 1$

$$D^{(v)} = \frac{1}{v} \sum_{\mu=0}^{v-1} T^{(v-\mu)} D^{(\mu)} . \qquad (4.47)$$

After repeated application of this relation one obtains

$$D^{(v)}[Q] = \sum_{r=1}^{v} \frac{1}{r!} \sum_{\rho_1=1}^{v-r+1} \cdots \sum_{\rho_r=1}^{v-r+1} \delta_{v,\rho_1+\cdots+\rho_r} (-1)^{r+v}$$

$$\cdot \frac{\operatorname{Tr} Q^{\rho_1}}{\rho_1} \frac{\operatorname{Tr} Q^{\rho_2}}{\rho_2} \cdots \frac{\operatorname{Tr} Q^{\rho_r}}{\rho_r} . \qquad (4.48)$$

This may be applied to the expansion

$$\det [1 + \lambda R] = 1 + \lambda D^{(1)}[R] + \lambda^2 D^{(2)}[R] + \cdots + \lambda^N D^{(N)}[R] . \qquad (4.49)$$

Combining (4.48) and (4.49) one obtains an algebraic expression which corresponds to the polymer representation in (4.43).

4.2 Wilson fermions

The general procedure of defining a Euclidean path integral formulation of fermionic theories in terms of Grassmannian variables is illustrated by a very simple toy model in section 4.1. In the present section we make the first step towards the application of this method to quantum field theories by considering the free fermion field. It will, however, be clear that even in this very simple case the lattice formulation is far from being unique. In fact, there are infinitely many different possibilities. The two basic types of lattice formulations, applied in most cases, are the Wilson formulation [4.4] and the Kogut–Susskind staggered formulation [4.5]. The present section is devoted to the basic properties of free Wilson fermions.

4.2.1   Hamiltonian formulation

The Dirac Hamiltonian describing a single free fermion is

$$\mathcal{H}_{\text{Dirac}} = m\beta - i\alpha_k \frac{\partial}{\partial x_k} \, . \tag{4.50}$$

Here m is the mass, and the connection of the $4 \otimes 4$ matrices α_k ($k = 1, 2, 3$) and β to the Euclidean Dirac matrices is

$$\alpha_k \equiv i\gamma_4\gamma_k = \begin{pmatrix} 0 & \sigma_k \\ \sigma_k & 0 \end{pmatrix} , \qquad \beta \equiv \gamma_4 = \begin{pmatrix} 1 & 0 \\ 0 & -1 \end{pmatrix} . \tag{4.51}$$

The explicit representation is given here according to (8.8) and (8.9). The field operator describing the multi-fermion system will be denoted by $\chi_\alpha(\mathbf{x})$, where $\alpha = 1, 2, 3, 4$ is the Dirac index, and the three-component vector $\mathbf{x}$ denotes a space point. χ and χ^+ satisfy the anticommutation relations of fermionic creation and annihilation operators:

$$\{\chi_\alpha(\mathbf{x}), \chi_\beta(\mathbf{y})\} = \{\chi_\alpha^+(\mathbf{x}), \chi_\beta^+(\mathbf{y})\} = 0 \, ,$$

$$\{\chi_\alpha^+(\mathbf{x}), \chi_\beta(\mathbf{y})\} = \delta_{\alpha\beta}\delta^3(\mathbf{x} - \mathbf{y}) \, . \tag{4.52}$$

The Hamiltonian H of the free fermion field is obtained from the single particle Hamiltonian (4.50) as

$$H = \int d^3\mathbf{x} \, \chi_\beta^+(\mathbf{x}) \, \mathcal{H}_{\text{Dirac},\beta\alpha} \, \chi_\alpha(\mathbf{x}) \, . \tag{4.53}$$

Note the Einstein summation convention over repeated indices. With a symmetrized derivative this leads to the well-known form

$$H = \int d^3\mathbf{x} \, \chi_\gamma^+(\mathbf{x}) \left\{ m\beta + \frac{i}{2} \left(\overleftarrow{\frac{\partial}{\partial x_k}} - \overrightarrow{\frac{\partial}{\partial x_k}} \right) \alpha_k \right\}_{\gamma\delta} \chi_\delta(\mathbf{x}) \, . \tag{4.54}$$

In order to discretize the fermion field let us introduce a regular cubic lattice in space, with lattice spacing a and elementary cubes of volume a^3. On a finite L^3 lattice with total volume $V = (aL)^3$ the lattice points have integer indices $\mathbf{x} \equiv \{x_k; \ k = 1, 2, 3\}$ satisfying $0 \le x_k \le L - 1$. The continuous field will be replaced by discrete variables associated with the elementary cubes, and the volume integral becomes a finite sum:

$$\chi_\alpha(\mathbf{x}) \longrightarrow \chi_{\alpha\mathbf{x}} , \qquad \int d^3x \longrightarrow \sum_{\mathbf{x}} a^3 . \tag{4.55}$$

The discrete field variables are assumed to satisfy the anticommutation relations

$$\{\chi_{\alpha\mathbf{x}}, \chi_{\beta\mathbf{y}}\} = \{\chi_{\alpha\mathbf{x}}^+, \chi_{\beta\mathbf{y}}^+\} = 0 , \qquad \{\chi_{\alpha\mathbf{x}}^+, \chi_{\beta\mathbf{y}}\} = a^{-3}\delta_{\alpha\beta}\delta_{\mathbf{xy}} . \tag{4.56}$$

The continuous derivative can, of course, be discretized in infinitely many ways, but the simplest rule is

$$\frac{\partial \chi_\alpha(\mathbf{x})}{\partial x_k} \longrightarrow \frac{1}{a}\left(\chi_{\alpha\mathbf{x}+\hat{\mathbf{k}}} - \chi_{\alpha\mathbf{x}}\right) . \tag{4.57}$$

Here $\hat{\mathbf{k}}$ denotes a unit vector on the lattice in direction k. Applying these rules to the Hamiltonian (4.54) we obtain the naive discretized form

$$H = \sum_{\mathbf{x}} a^3 \left\{ \chi_{\gamma\mathbf{x}}^+ m\beta_{\gamma\delta}\chi_{\delta\mathbf{x}} + \frac{i}{2a}\sum_{k=1}^3 \left[\chi_{\gamma\mathbf{x}+\hat{\mathbf{k}}}^+ \alpha_{k,\gamma\delta}\chi_{\delta\mathbf{x}} - \chi_{\gamma\mathbf{x}}^+ \alpha_{k,\gamma\delta}\chi_{\delta\mathbf{x}+\hat{\mathbf{k}}} \right] \right\} . \tag{4.58}$$

The reason why this discretization is called 'naive' becomes clear later.

The free field Hamiltonian describes a collection of harmonic fermion oscillators in momentum space. In order to see this let us assume periodic boundary conditions in all three orthogonal directions of the L^3 cube (the question of periodic versus antiperiodic boundary conditions is discussed below, near (4.118)). In this case the allowed momentum components are discrete points in the *Brillouin zone*, namely

$$\mathbf{q} \equiv \left\{ q_k = \frac{2\pi}{L}v_k; \ k = 1, 2, 3 \right\} , \qquad 0 \le v_k \le L - 1 . \tag{4.59}$$

Owing to periodicity, instead of $0 \le q_k < 2\pi$, the Brillouin zone can also be equivalently defined in the interval $-\pi < q_k \le \pi$.

Let us define the Fourier-transformed field components by

$$\tilde{\chi}_{\alpha\mathbf{q}} \equiv \sum_{\mathbf{x}} a^3 e^{-iq_k x_k}\chi_{\alpha\mathbf{x}} , \qquad \chi_{\alpha\mathbf{x}} = \frac{1}{(aL)^3}\sum_{\mathbf{q}} e^{iq_k x_k}\tilde{\chi}_{\alpha\mathbf{q}} . \tag{4.60}$$

In terms of these the Hamiltonian (4.58) is

$$H = \frac{1}{(aL)^3}\sum_{\mathbf{q}} \tilde{\chi}_{\gamma\mathbf{q}}^+ \left\{ m\beta + \frac{1}{a}\alpha_k \sin q_k \right\}_{\gamma\delta} \tilde{\chi}_{\delta\mathbf{q}} . \tag{4.61}$$

This is similar to the diagonal form (4.10) of the simple two-state Hamiltonian H_2 defined in (4.8). The original form of H in (4.58) is of the same general type as the original form of H_2 in (4.8).

The matrix in curly brackets in (4.61) can be easily shown to have the eigenvalues (two of them positive, the other two negative):

$$E_{\mathbf{q}} = \pm \left(m^2 + \sum_{k=1}^{3} a^{-2} \sin^2 q_k \right)^{1/2} . \tag{4.62}$$

These are the possible energies of the free naive lattice fermion.

In the continuum limit, when the lattice spacing a tends to zero, (4.62) approaches the usual relativistic relation between energy and momentum. This is because the coordinate vector in physical units is $a\mathbf{x}$ and, correspondingly, the momentum vector is $\mathbf{p} \equiv a^{-1}\mathbf{q}$. For fixed momentum $\mathbf{p}$ and $a \rightarrow 0$ we have

$$E_{\mathbf{q}}^2 = m^2 + \mathbf{p}^2 + O(a^2) . \tag{4.63}$$

This is as expected, but still the 'naive' lattice Hamiltonian in (4.58) is not entirely satisfactory. Namely, due to $\sin(\pi + q) = -\sin q$ we have

$$E_{\mathbf{q}}^2 = E_{\mathbf{q}+\mathbf{q}_\pi}^2 , \tag{4.64}$$

where $\mathbf{q}_\pi$ is one of the eight vectors

$$\mathbf{q}_\pi = \{(0,0,0), \ (\pi,0,0), \ (0,\pi,0), \ (0,0,\pi),$$

$$(\pi,\pi,0), \ (\pi,0,\pi), \ (0,\pi,\pi), \ (\pi,\pi,\pi)\} . \tag{4.65}$$

These are the eight *corners* of the part of Brillouin zone satisfying $0 \le q_k \le \pi$ ($k = 1, 2, 3$). The consequence of (4.64) is that in the 'naive' lattice formulation there are eight fermion states per field component. This phenomenon is called *fermion doubling* on the lattice (although here it is, in fact, a 'fermion octupling'). The basic reason for fermion doubling is that the Dirac equation is of first order (for a more detailed discussion, see also section 4.4).

In the case of free fermions one could, perhaps, tolerate this unwanted proliferation of degrees of freedom, but in an interacting theory the extra fermions influence the physical content in a non-trivial way, because the additional states can be pair-produced through the interactions of the fermion field. For instance, even if the external particles in a process are the states at the zero $(0,0,0)$ corner of the Brillouin zone, the states at the other corners appear in virtual loops. In order to cure this disease of the 'naive' lattice fermion field, Wilson [4.4] introduced an additional second order term in the lattice Hamiltonian

$$H = \sum_{\mathbf{x}} a^3 \left\{ \chi_{\gamma\mathbf{x}}^+ \left(m + \frac{3r}{a} \right) \beta_{\gamma\delta} \chi_{\delta\mathbf{x}} \right.$$

$$+ \frac{i}{2a} \sum_{k=1}^{3} \left[\chi^{+}_{\gamma x + \hat{k}} \alpha_{k,\gamma\delta} \chi_{\delta x} - \chi^{+}_{\gamma x} \alpha_{k,\gamma\delta} \chi_{\delta x + \hat{k}} \right]$$

$$- \frac{r}{2a} \sum_{k=1}^{3} \left[\chi^{+}_{\gamma x + \hat{k}} \beta_{\gamma\delta} \chi_{\delta x} + \chi^{+}_{\gamma x} \beta_{\gamma\delta} \chi_{\delta x + \hat{k}} \right] \Bigg\} . \tag{4.66}$$

The *Wilson parameter* r is assumed to be in the interval $0 < r \le 1$. In momentum space, instead of (4.61), we now have

$$H = \frac{1}{(aL)^3} \sum_{\mathbf{q}} \tilde{\chi}^{+}_{\gamma \mathbf{q}} \left\{ m\beta + \frac{r}{a}\beta \sum_{k=1}^{3}(1 - \cos q_k) + \frac{1}{a}\alpha_k \sin q_k \right\}_{\gamma\delta} \tilde{\chi}_{\delta \mathbf{q}} , \tag{4.67}$$

and the corresponding energy eigenvalues are

$$E_{\mathbf{q}} = \pm \left\{ \left[m + \frac{r}{a} \sum_{k=1}^{3}(1 - \cos q_k) \right]^2 + \sum_{k=1}^{3} a^{-2} \sin^2 q_k \right\}^{1/2} . \tag{4.68}$$

In the continuum limit $a \to 0$ the mass m is replaced here by $m + 2ra^{-1}n_\pi$, where n_π is the number of momentum components equal to π. Therefore, the states with $n_\pi \ne 0$ become infinitely heavy. The only physical fermion state with finite energy is at the zero corner of the Brillouin zone.

4.2.2   *Euclidean formulation: Wilson action*

The general structure of the free fermion Hamiltonian H in (4.66) is the same as that of H_2 in (4.8). Therefore, the transformation of the partition function $Z = \text{Tr}\, e^{-\beta H}$ (with 'Euclidean' imaginary time β) to a Grassmannian path integral can be done in the same way as for H_2. The only difference is that the free fermion field has much more states, and the definition of the lowest energy *vacuum state* is more subtle, due to the interpretation of the negative energy states as antiparticle holes ('Dirac vacuum'). Here we follow [4.6], but for a different approach see also [4.7].

In order to define properly the particle and antiparticle states let us introduce the projection operators

$$P_{\pm} \equiv \frac{1}{2}(1 \pm \gamma_4) . \tag{4.69}$$

The fermion field $\bar{\chi}$ is defined, with explicit Dirac indices, or omitting Dirac indices in the usual row-matrix notation, as

$$\bar{\chi}_{\alpha x} \equiv \chi^{+}_{\beta x} \gamma_{4,\beta\alpha} \qquad \Longrightarrow \qquad \bar{\chi}_{x} \equiv \chi^{+}_{x} \gamma_4 . \tag{4.70}$$

According to (4.56) $\bar{\chi}$ satisfies the anticommutation relation, again both with and without explicit Dirac indices,

$$\{\chi_{\beta y}, \bar{\chi}_{\alpha x}\} = \gamma_{4,\beta\alpha} a^{-3} \delta_{yx} \qquad \Longrightarrow \qquad \{\chi_y, \bar{\chi}_x\} = \gamma_4 a^{-3} \delta_{yx} . \tag{4.71}$$

The creation and annihilation operators of fermions are then, respectively, $\overline{\chi}_{\mathbf{x}} P_+$ and $P_+ \chi_{\mathbf{x}}$. Similarly, the creation and annihilation operators of antifermions are, respectively, $P_- \chi_{\mathbf{x}}$ and $\overline{\chi}_{\mathbf{x}} P_-$. Therefore the *Dirac vacuum* $|0\rangle$ is defined by

$$P_+ \chi_{\mathbf{x}} |0\rangle = \overline{\chi}_{\mathbf{x}} P_- |0\rangle = 0 , \qquad (4.72)$$

and the only non-zero anticommutators are

$$\{P_+ \chi_{\mathbf{y}}, \overline{\chi}_{\mathbf{x}} P_+\} = P_+ a^{-3} \delta_{\mathbf{yx}} , \qquad \{P_- \chi_{\mathbf{y}}, \overline{\chi}_{\mathbf{x}} P_-\} = -P_- a^{-3} \delta_{\mathbf{yx}} . \qquad (4.73)$$

In the same way as in (4.13) let us introduce the totally anticommuting *Grassmann algebra* with elements $\psi_{\alpha \mathbf{x}} \Longrightarrow \psi_{\mathbf{x}}$ and $\overline{\psi}_{\alpha \mathbf{x}} \Longrightarrow \overline{\psi}_{\mathbf{x}}$:

$$\{\psi_{\mathbf{y}}, \psi_{\mathbf{x}}\} = \{\overline{\psi}_{\mathbf{y}}, \overline{\psi}_{\mathbf{x}}\} = \{\overline{\psi}_{\mathbf{y}}, \psi_{\mathbf{x}}\} = 0 . \qquad (4.74)$$

Following (4.26) these are commuting with the creation and annihilation operators: $[\psi_{...}, \chi_{...}] = 0$. The Grassmannian coherent states are now defined as

$$|\psi, \overline{\psi}\rangle \equiv \exp\left\{ a^3 \sum_{\mathbf{x}} [(\overline{\chi}_{\mathbf{x}} P_+ \psi_{\mathbf{x}}) + (\overline{\psi}_{\mathbf{x}} P_- \chi_{\mathbf{x}})] \right\} |0\rangle ,$$

$$\langle \psi, \overline{\psi}| \equiv \langle 0| \exp\left\{ a^3 \sum_{\mathbf{x}} [(\overline{\psi}_{\mathbf{x}} P_+ \chi_{\mathbf{x}}) + (\overline{\chi}_{\mathbf{x}} P_- \psi_{\mathbf{x}})] \right\} . \qquad (4.75)$$

The notation $(\overline{\chi} P \psi)$ means, as usual, a contraction of the row-matrix index of $\overline{\chi}$ and column-matrix index of ψ with the indices of the matrix P. These states satisfy $\langle 0| \psi, \overline{\psi}\rangle = 1$ and other relations analogous to (4.29)–(4.34). For instance, the decomposition of the unity in the Fock space in terms of these coherent states is

$$I = \int \prod_{\alpha, \mathbf{x}} \left\{ a^3 \, d\overline{\psi}_{\alpha \mathbf{x}} \, d\psi_{\alpha \mathbf{x}} \right\} \exp\left\{ -a^3 \sum_{\alpha, \mathbf{x}} \overline{\psi}_{\alpha \mathbf{x}} \psi_{\alpha \mathbf{x}} \right\} |\psi, \overline{\psi}\rangle \langle \psi, \overline{\psi}| . \qquad (4.76)$$

The representation of the *partition function*

$$Z \equiv \mathrm{Tr} \, e^{-\beta H} \qquad (4.77)$$

in terms of a Grassmannian path integral is obtained in the same way as in the simple toy model in (4.35)–(4.38). Without repeating every step of the derivation, only a few important elements will be included here. The imaginary time (or in case of a thermodynamic application the inverse temperature) β is divided into T small intervals of length $a_t = \beta/T$. In this way one obtains from the three-dimensional L^3 lattice a four-dimensional one with $L^3 \cdot T$ lattice points. The zero point energy in a field theory is usually subtracted from the Hamiltonian by introducing the *normal product*

$$: (\chi_{\mathbf{x}}^+ \beta \chi_{\mathbf{x}}) := (\overline{\chi}_{\mathbf{x}} P_+ \chi_{\mathbf{x}}) - \chi_{\alpha \mathbf{x}} (\overline{\chi}_{\mathbf{x}} P_-)_\alpha = (\overline{\chi}_{\mathbf{x}} P_+ \chi_{\mathbf{x}}) - (\chi_{\mathbf{x}}^T P_-^T \overline{\chi}_{\mathbf{x}}^T) . \qquad (4.78)$$

The normal ordered Hamiltonian satisfying $H|0\rangle = 0$ is

$$H = \sum_{\mathbf{x}} a^3 \left\{ : \left(\overline{\chi}_{\mathbf{x}} \left[m + \frac{3r}{a} \right] \chi_{\mathbf{x}} \right) : -\frac{1}{2a} \sum_{k=\pm 1}^{\pm 3} \left(\overline{\chi}_{\mathbf{x}+\hat{\mathbf{k}}} [r + \gamma_k] \chi_{\mathbf{x}} \right) \right\} . \quad (4.79)$$

Here (4.51) is used, the summation over positive and negative directions ($k = \pm 1, \pm 2, \pm 3$) is denoted by $\sum_{k=\pm 1}^{\pm 3}$, and the γ-matrices with negative indices are defined by $\gamma_{-k} \equiv -\gamma_k$. The Grassmann variables at the lattice point $x \equiv (\mathbf{x}, t)$ are $\psi_{\alpha \mathbf{x} t}$, $\overline{\psi}_{\alpha \mathbf{x} t}$. The corresponding Grassmann coherent state at time t can be denoted by $|\psi_t, \overline{\psi}_t\rangle$ for $0 \le t \le T - 1$. The antiperiodicity in time is implemented by $\psi_T \equiv -\psi_0$ and $\overline{\psi}_T \equiv -\overline{\psi}_0$. Then for the Hamiltonian (4.79) we have

$$\langle \psi_{t+1}, \overline{\psi}_{t+1} | (1 - a_t H) | \psi_t, \overline{\psi}_t \rangle = \langle \psi_{t+1}, \overline{\psi}_{t+1} | \psi_t, \overline{\psi}_t \rangle$$

$$\cdot \left\{ 1 - a_t \sum_{\mathbf{x}} a^3 \left[\left(m + \frac{3r}{a} \right) \left((\overline{\psi}_{\mathbf{x}t+1} P_+ \psi_{\mathbf{x}t}) + (\overline{\psi}_{\mathbf{x}t} P_- \psi_{\mathbf{x}t+1}) \right) \right. \right.$$

$$\left. \left. -\frac{1}{2a} \sum_{k=\pm 1}^{\pm 3} \left([\overline{\psi}_{\mathbf{x}+\hat{\mathbf{k}}t+1} P_+ + \overline{\psi}_{\mathbf{x}+\hat{\mathbf{k}}t} P_-][r + \gamma_k][P_+ \psi_{\mathbf{x}t} + P_- \psi_{\mathbf{x}t+1}] \right) \right] \right\} .$$

$$(4.80)$$

The coherent state matrix element can be written here as

$$\langle \psi_{t+1}, \overline{\psi}_{t+1} | \psi_t, \overline{\psi}_t \rangle =$$

$$\exp \left\{ a_t \sum_{\mathbf{x}} a^3 \frac{1}{2a_t} \left[(\overline{\psi}_{\mathbf{x}t+1} [1 + \gamma_4] \psi_{\mathbf{x}t}) + (\overline{\psi}_{\mathbf{x}t} [1 - \gamma_4] \psi_{\mathbf{x}t+1}) \right] \right\} . \quad (4.81)$$

Finally, in the limit $a_t \to 0$, the partition function is

$$Z = \lim_{a_t \to 0} \int \prod_{\alpha \mathbf{x} t} \left\{ a^3 \, \mathrm{d}\overline{\psi}_{\alpha \mathbf{x} t} \, \mathrm{d}\psi_{\alpha \mathbf{x} t} \right\} \exp \left\{ -\sum_{t=0}^{T-1} a_t \sum_{\mathbf{x}} a^3 \left[\frac{1}{a_t} (\overline{\psi}_{\mathbf{x}t} \psi_{\mathbf{x}t}) \right. \right.$$

$$+ \left(m + \frac{3r}{a} \right) \left((\overline{\psi}_{\mathbf{x}t+1} P_+ \psi_{\mathbf{x}t}) + (\overline{\psi}_{\mathbf{x}t} P_- \psi_{\mathbf{x}t+1}) \right)$$

$$-\frac{1}{2a_t} \left((\overline{\psi}_{\mathbf{x}t+1} [1 + \gamma_4] \psi_{\mathbf{x}t}) + (\overline{\psi}_{\mathbf{x}t-1} [1 - \gamma_4] \psi_{\mathbf{x}t}) \right)$$

$$\left. \left. -\frac{1}{2a} \sum_{k=\pm 1}^{\pm 3} \left([\overline{\psi}_{\mathbf{x}+\hat{\mathbf{k}}t+1} P_+ + \overline{\psi}_{\mathbf{x}+\hat{\mathbf{k}}t} P_-][r + \gamma_k][P_+ \psi_{\mathbf{x}t} + P_- \psi_{\mathbf{x}t+1}] \right) \right] \right\} .$$

$$(4.82)$$

In a short notation this can be written as

$$Z \equiv \int [d\overline{\psi}\, d\psi] e^{-S} \equiv \int [d\overline{\psi}\, d\psi] e^{-\overline{\psi}Q\psi} = \det Q \; . \qquad (4.83)$$

S is the *Euclidean lattice action* and Q is called the *fermion matrix*.

In quantum field theory we are always interested in quantities like e.g. the Green functions generated by $Z[\eta,\overline{\eta}]/Z$, where $Z[\eta,\overline{\eta}]$ is the partition function with external sources (see below). Therefore the source-independent normalization of Z is irrelevant. This allows simplification of the above form of the fermion action. Namely, one can multiply Q from the left and right by appropriate matrices of the form $(1 + a_t z_{L,R})$, with $z_{L,R}$ block-diagonal in time, in such a way that for $a_t \to 0$ the elements proportional to r and m are shifted to diagonal positions in time. Note that a physical energy corresponds in the limit $a_t \to 0$ to very long wavelengths of the order of a_t^{-1} in the time variation of lattice Green functions. Therefore, the t-dependence of the matrix elements of the fermion matrix Q matters only at large t-scales. This means that in the fermion action S for $a_t \to 0$ one can formally assume $\psi_{t+1} = \psi_t + O(a_t)$ and $\overline{\psi}_{t+1} = \overline{\psi}_t + O(a_t)$.

For a symmetric four-dimensional lattice with points $x \equiv (\mathbf{x}, t)$ and equal spacings in space and time directions $a_t = a$, at $r = 1$ and neglecting for the moment the boundaries, one obtains the *Wilson action*

$$S = \sum_x a^4 \left\{ \left(m + \frac{4}{a} \right) (\overline{\psi}_x \psi_x) - \frac{1}{2a} \sum_{\mu=\pm 1}^{\pm 4} (\overline{\psi}_{x+\hat{\mu}}[1 + \gamma_\mu]\psi_x) \right\} \; . \qquad (4.84)$$

It is remarkable that according to the above derivation the factor $(1 + \gamma_\mu)$, which is important for removing the lattice fermion doublers, is related to the projection operators in (4.69).

The form in (4.84) is symmetric with respect to the rotations and shifts of the four-dimensional hypercubical lattice. As we shall see below, in the continuum limit, when the lattice spacing a tends to zero, the physical Green functions become symmetric with respect to arbitrary continuous rotations and shifts in the four-dimensional Euclidean space. The same symmetries are kept if the above action is generalized to

$$S = \sum_x a^4 \left\{ \left(m + \frac{4r}{a} \right) (\overline{\psi}_x \psi_x) - \frac{1}{2a} \sum_{\mu=\pm 1}^{\pm 4} (\overline{\psi}_{x+\hat{\mu}}[r + \gamma_\mu]\psi_x) \right\} \; , \qquad (4.85)$$

for any $0 < r \leq 1$. This general form is related to the Hamiltonian in (4.66) also for $r \neq 1$, nevertheless, the above derivation of the lattice action from the corresponding Hamiltonian is only valid for $r = 1$. We shall see in section 4.2 below that, in fact, for $r \neq 1$ the lattice spectrum of the general Wilson action (4.85) contains additional time-doublers, in such a way that there are altogether 16 lattice fermion states per field components (instead of the 8 states present in the spectrum of the Hamiltonian (4.66)). Consequently, the positivity properties are also different for $r = 1$ and $r \neq 1$ (see below). Nevertheless, all these differences disappear in the continuum limit, and the Green functions tend to the Green functions of a free fermion field for every $0 < r \leq 1$. The case $r = 0$ (*naive lattice fermion action*) is special, because then the lattice fermion doublers also remain in the spectrum in the continuum limit.

In the above forms of the Wilson action the normalization of the fermion field was fixed originally by the anticommutation relations in the Hamiltonian formulation. In the Euclidean formulation one can use the freedom of the field normalizations and bring the lattice action to a particularly simple form by

$$a^{3/2}(am + 4r)^{1/2}\psi_x \Rightarrow \psi_x , \qquad a^{3/2}(am + 4r)^{1/2}\overline{\psi}_x \Rightarrow \overline{\psi}_x . \qquad (4.86)$$

Introducing the *hopping parameter K* as

$$K \equiv \frac{1}{2am + 8r} , \qquad (4.87)$$

and using the short notation $\sum_\mu \equiv \sum_{\mu=\pm 1}^{\pm 4}$ for the summation over the eight neighbours of a lattice point, one obtains

$$S = \sum_x \left\{ (\overline{\psi}_x \psi_x) - K \sum_\mu (\overline{\psi}_{x+\hat{\mu}}[r + \gamma_\mu]\psi_x) \right\} \equiv \sum_{xy} (\overline{\psi}_y Q_{yx} \psi_x) . \qquad (4.88)$$

This means that the matrix elements of the fermion matrix are

$$Q_{yx} = \delta_{yx} - K \sum_\mu \delta_{y,x+\hat{\mu}}(r + \gamma_\mu) . \qquad (4.89)$$

As noted before, this is for an infinite lattice without boundaries. A finite lattice with periodic or antiperiodic boundary conditions will be discussed after (4.112).

4.2.3 *Reflection positivity of the Wilson action*

The $r = 1$ Wilson action in (4.84) has been derived from the Hamiltonian in (4.66). It is, however, not necessary to start from a Hamiltonian formulation. One can immediately define a theory by the lattice action in Euclidean space. For instance, the generalization of (4.84) to $r \neq 1$ in (4.85) gives an equivalent continuum limit, therefore it is certainly admissible as a Euclidean description of the free fermion field. Of course, the physical interpretation of a relativistic quantum field theory always refers to Minkowski space, therefore one has to know under which conditions a Euclidean lattice action defines an acceptable quantum field theory, for instance, with an acceptable Hamiltonian. Among the axioms of quantum field theory formulated in Euclidean space [4.8] a fundamental rôle is played by *reflection positivity*, which is the main ingredient for establishing the existence of a positive semidefinite self-adjoint Hamiltonian. This axiom requires that there exists an antilinear mapping Θ which transforms an arbitrary function F of the fields at positive times into a function ΘF of fields at negative times such that

$$\langle (\Theta F) F \rangle \geq 0 . \tag{4.90}$$

If such a mapping exists, one can define a Hilbert space with positive norm and in this space a positive transfer matrix [4.9].

In principle, the reflection positivity condition has to be fulfilled only in the continuum limit, but usually it is better to start on the safe side, and impose it for finite lattice spacings. In this sense reflection positivity of the action (4.84) can also be questioned, because correction terms vanishing by some power of the lattice spacing were neglected in its derivation from the Hamiltonian. Similarly, reflection positivity of the more general action (4.85) with $r \neq 1$ can also be broken by terms vanishing in the continuum limit. Of course, the really interesting question is the reflection positivity of theories with interacting fermions. In the present section only the free fermion field is considered, but the basic steps of the proof of reflection positivity are also the same in the interacting case. Therefore extension to specific interactions will only be briefly mentioned in the appropriate chapters. For simplicity, we shall assume here that the time extension of the lattice is infinite.

There are two different ways of defining the time reflection of the lattice points: either one reflects with respect to the $t = 0$ timeslice ($t \rightarrow -t$) or with respect to the hyperplane between, say, the $t = 0$ and $t = 1$ timeslices ($t \rightarrow 1 - t$). The first can also be called 'reflection with respect to the $t = 0$ points' or in short 'site-reflection', the second 'reflection with respect to the $t = (0 \rightarrow 1)$ links' or 'link-reflection'.

Let us first consider the link-reflection. In this case the antilinear mapping Θ is defined on the fermion fields by

$$\Theta \psi_{\mathbf{x}t} = \overline{\psi}_{\mathbf{x}1-t} \gamma_4 , \qquad \Theta \overline{\psi}_{\mathbf{x}t} = \gamma_4 \psi_{\mathbf{x}1-t} . \tag{4.91}$$

The action of Θ on a general function of fermionic fields is defined by antilinearity and by the property that it reverses the order of Grassmannian variables. As a simple example, the mapping of a fermionic bilinear is

$$\Theta \left(\overline{\psi}_{\mathbf{x}t_x} \Gamma \psi_{\mathbf{y}t_y} \right) = \left(\overline{\psi}_{\mathbf{y}1-t_y} \gamma_4 \Gamma^+ \gamma_4 \psi_{\mathbf{x}1-t_x} \right) . \tag{4.92}$$

The first step in the proof of reflection positivity with respect to the mapping Θ in (4.91) is to decompose the $r \neq 1$ lattice action (4.88) into three components according to the time coordinates. Let us denote the field variables in the half-space with positive time $t \geq 1$ by $\psi^{(+)}$, $\overline{\psi}^{(+)}$, and in the other half-space $t \leq 0$ by $\psi^{(-)}$, $\overline{\psi}^{(-)}$. The action (4.88) can then be written as

$$S = S_+[\psi^{(+)}, \overline{\psi}^{(+)}] + S_-[\psi^{(-)}, \overline{\psi}^{(-)}] + S_c[\psi^{(+)}, \overline{\psi}^{(+)}, \psi^{(-)}, \overline{\psi}^{(-)}] . \quad (4.93)$$

For instance, the component S_c depending on the fields at the boundaries of both half-spaces is

$$S_c = -K \sum_{\mathbf{x}} \left[\left(\overline{\psi}_{\mathbf{x}0}^{(-)} [r - \gamma_4] \psi_{\mathbf{x}1}^{(+)} \right) + \left(\overline{\psi}_{\mathbf{x}1}^{(+)} [r + \gamma_4] \psi_{\mathbf{x}0}^{(-)} \right) \right] . \quad (4.94)$$

The pieces S_+ and S_- depending, respectively, only on the fields in the positive and negative half-space can be obtained from each other by Θ-mapping:

$$\Theta S_+[\psi^{(+)}, \overline{\psi}^{(+)}] = S_+^\dagger[\Theta \psi^{(+)}, \Theta \overline{\psi}^{(+)}] = S_-[\psi^{(-)}, \overline{\psi}^{(-)}] . \quad (4.95)$$

The notation $S_+^\dagger$ means that in S_+ one has to take the adjoint of complex matrices and reverse the order of Grassmannian variables (like in (4.92)). This property of the action is called *reflection symmetry*. It implies that the corresponding continuum Lagrangian density in Minkowski space is formally a Hermitean function of the field operators. Using (4.94) and (4.95) the expectation value of the general funtion F can be written as

$$\langle (\Theta F) F \rangle = Z^{-1} \int [\mathrm{d}\overline{\psi}^{(+)} \, \mathrm{d}\psi^{(+)}] \mathrm{e}^{-S_+[\psi^{(+)}, \overline{\psi}^{(+)}]} F[\psi^{(+)}, \overline{\psi}^{(+)}]$$

$$\cdot \int [\mathrm{d}(\Theta \overline{\psi}^{(+)}) \, \mathrm{d}(\Theta \psi^{(+)})] \mathrm{e}^{-S_+^\dagger[\Theta \psi^{(+)}, \Theta \overline{\psi}^{(+)}]} F^\dagger[\Theta \psi^{(+)}, \Theta \overline{\psi}^{(+)}]$$

$$\cdot \exp \left\{ K \sum_{\mathbf{x}} \left[\left(\psi_{\mathbf{x}1}^{(+)T} (1 - r\gamma_4)^T (\Theta \psi_{\mathbf{x}1}^{(+)})^T \right) + \left(\overline{\psi}_{\mathbf{x}1}^{(+)} (1 + r\gamma_4)(\Theta \overline{\psi}_{\mathbf{x}1}^{(+)}) \right) \right] \right\} .$$
$$(4.96)$$

The partition function Z is given by a similar integral to the one on the right hand side, only F is replaced in it by 1. Expanding the exponential and choosing a diagonal representation of γ_4 as in (8.9), one can easily see that the integral is a sum of positive terms for $K \geq 0$ and $|r| \leq 1$. Therefore the condition in (4.90) is satisfied. Note that negative values of r are not acceptable, because then in the continuum limit there will be states with energy $-\infty$ (see (4.140)).

For negative values of the hopping parameter, positivity would be valid if in (4.91) Θ would be replaced by $-\Theta$. Nevertheless K can be restricted to positive values anyway, because its sign is changed by the *staggered sign transformation*

$$\psi_x \to (-1)^{x_1 + x_2 + x_3 + x_4} \psi_x, \quad \overline{\psi}_x \to (-1)^{x_1 + x_2 + x_3 + x_4} \overline{\psi}_x . \quad (4.97)$$

($x_{1,2,3,4}$ are the lattice coordinates ot the point x.) Because of

$$(-1)^{x_1+x_2+x_3+x_4} = e^{\pm i\pi(x_1+x_2+x_3+x_4)} , \qquad (4.98)$$

after Fourier-transforming to momentum space, the transformation in (4.97) means a shift in all momentum components by $\pm\pi$. This is equivalent to going to the opposite corner of the Brillouin zone.

In case of the site-reflection the Θ-transformation of the fermion fields is

$$\Theta\psi_{xt} = \overline{\psi}_{x-t}\gamma_4 , \qquad \Theta\overline{\psi}_{xt} = \gamma_4\psi_{x-t} . \qquad (4.99)$$

In this case let us fix the Wilson parameter to $r = 1$ and decompose the action (4.84) in the form (4.88) into three terms according to

$$S = S_+[\psi^{(+)}, \overline{\psi}^{(+)}, \psi^{(0)}, \overline{\psi}^{(0)}] + S_-[\psi^{(-)}, \overline{\psi}^{(-)}, \psi^{(0)}, \overline{\psi}^{(0)}] + S_0[\psi^{(0)}, \overline{\psi}^{(0)}] .$$
$$(4.100)$$

Here the field variables with suffix $(+), (0), (-)$ refer to $t > 0$, $t = 0$, $t < 0$, respectively.

The component of the action depending only on the $t = 0$ fields $\psi_x \equiv \psi_{x0}$, $\overline{\psi}_x \equiv \overline{\psi}_{x0}$ is

$$S_0 = \sum_x \left\{ (\overline{\psi}_x\psi_x) - K \sum_{k=1}^3 \left[(\overline{\psi}_{x+\hat{k}}[1 + \gamma_k]\psi_x) + (\overline{\psi}_{x-\hat{k}}[1 - \gamma_k]\psi_x) \right] \right\} .$$
$$(4.101)$$

Let us write this in another form by using the variables [4.7]

$$\xi_x \equiv \frac{1}{2}(1 + \gamma_4)\psi_x , \qquad \xi_x^+ \equiv \overline{\psi}_x\frac{1}{2}(1 + \gamma_4) ,$$

$$\eta_x^T \equiv -\overline{\psi}_x\frac{1}{2}(1 - \gamma_4) , \qquad \eta_x^{+T} \equiv \frac{1}{2}(1 - \gamma_4)\psi_x . \qquad (4.102)$$

In the explicit representation of the Dirac matrices (8.8), (8.9) this leads to

$$S_0 = \sum_x \left\{ \xi_x^+\xi_x + \eta_x^+\eta_x - K \sum_{k=1}^3 \left[\xi_{x+\hat{k}}^+\xi_x + \xi_{x-\hat{k}}^+\xi_x + \eta_{x+\hat{k}}^+\eta_x + \eta_{x-\hat{k}}^+\eta_x \right] \right.$$

$$\left. +iK \sum_{k=1}^3 \left[\eta_{x+\hat{k}}^T\sigma_k\xi_x - \eta_{x-\hat{k}}^T\sigma_k\xi_x + \xi_{x+\hat{k}}^+\sigma_k\eta_x^{T+} - \xi_{x-\hat{k}}^+\sigma_k\eta_x^{T+} \right] \right\} . \qquad (4.103)$$

In a short matrix notation this can be written as

$$S_0 \equiv \xi^+B\xi + \eta^+B\eta + \eta^TC\xi + \xi^+C\eta^{T+} , \qquad (4.104)$$

where B and C are Hermitean matrices with both point and spinor indices.

Their matrix elements are given explicitly by

$$B_{\mathbf{yx}} = \delta_{\mathbf{yx}} - K \sum_{k=1}^{3} \left[\delta_{\mathbf{y,x+\hat{k}}} + \delta_{\mathbf{y,x-\hat{k}}} \right] ,$$

$$C_{\mathbf{yx}} = iK \sum_{k=1}^{3} \left[\delta_{\mathbf{y,x+\hat{k}}} \sigma_k - \delta_{\mathbf{y,x-\hat{k}}} \sigma_k \right] . \tag{4.105}$$

The essential property of the action (4.88) at $r = 1$ is that its positive time part S_+ depends only on the variables ξ and η, but not on ξ^+ and η^+. In contrast, S_- depends only on ξ^+ and η^+. The reflection symmetry relation between S_+ and S_- is

$$\Theta S_+[\psi^{(+)}, \overline{\psi}^{(+)}, \xi, \eta] = S_+^\dagger[\Theta\psi^{(+)}, \Theta\overline{\psi}^{(+)}, \Theta\xi, \Theta\eta]$$

$$= S_-[\psi^{(-)}, \overline{\psi}^{(-)}, \xi^+, \eta^+] . \tag{4.106}$$

Let us define the following Grassmannian integral of the function F of the variables at positive times:

$$\mathscr{I}_F[\xi, \eta] \equiv \int [\mathrm{d}\overline{\psi}^{(+)} \, \mathrm{d}\psi^{(+)}] e^{-S_+[\psi^{(+)}, \overline{\psi}^{(+)}, \xi, \eta]} F[\psi^{(+)}, \overline{\psi}^{(+)}] . \tag{4.107}$$

Using this we have

$$\langle (\Theta F) F \rangle = Z^{-1} \int [\mathrm{d}\xi^+ \, \mathrm{d}\xi \, \mathrm{d}\eta^+ \, \mathrm{d}\eta] e^{-S_0[\xi, \xi^+, \eta, \eta^+]} \mathscr{I}_F[\xi, \eta] \mathscr{I}_F^\dagger[\xi^+, \eta^+] . \tag{4.108}$$

The partition function Z is given by the corresponding integral with $F \equiv 1$.

For the positivity of the integral in (4.108) it is sufficient to show that the matrix B is positive, because then its square root $B^{1/2}$ is well defined and one can introduce

$$\xi' \equiv B^{1/2}\xi , \quad \xi^{+\prime} \equiv \xi^+ B^{1/2} , \quad \eta' \equiv B^{1/2}\eta , \quad \eta^{+\prime} \equiv \eta^+ B^{1/2} , \tag{4.109}$$

as new integration variables. In terms of these (4.108) can be transformed to the form

$$\langle (\Theta F) F \rangle = Z^{-1} \int [\mathrm{d}\xi^{+\prime} \, \mathrm{d}\xi' \, \mathrm{d}\eta^{+\prime} \, \mathrm{d}\eta'] e^{-\xi^{+\prime}\xi' - \eta^{+\prime}\eta'} \mathscr{I}_F[\xi', \eta'] \mathscr{I}_F^\dagger[\xi^{+\prime}, \eta^{+\prime}] , \tag{4.110}$$

with some new function $\mathscr{I}_F$. This integral is obviously positive, which proves the reflection positivity condition (4.90).

What remains is to prove the positivity of the $N \otimes N$ matrix B, where $N = 2V$ and V is the number of points at $t = 0$. As one can see in (4.105), the diagonal elements of B are all equal to 1. In addition, there are in every row (and in every column) exactly $6N$ non-zero non-diagonal elements with absolute value $|K|$. (Note that, according to the discussion in (4.97) one can always change the sign of the hopping parameter to $K \geq 0$.) From this immediately follows that the matrix B is positive for every

$$|K| < \frac{1}{6} . \tag{4.111}$$

As we can see, the proof of the site-reflection positivity is valid only for $r = 1$ and $|K| < 1/6$, whereas the link-reflection positivity could be proven in a much wider range, namely for every $|r| \leq 1$ and any K. However, one has to have in mind that the above proofs give only sufficient conditions but not necessary ones. Therefore the full range of positivity can, in principle, be wider.

If reflection positivity holds for either type of reflections, a bounded positive transfer matrix can be defined for time shifts with an even distance [4.9, 4.10]. Since the definition of the transfer matrices depends on Θ, they can be different for different Θ, and also the regions of bare parameter space where they are positive can, in principle, be different. At finite lattice spacings, for instance in numerical simulations, it is best to have reflection positivity of both kinds, because then for determination of the spectrum one can safely consider correlations at arbitrary time distances, and the positive transfer matrix is the usual one, which acts on wave functions defined at fixed times, e.g. $t = 0$ [4.7].

In the continuum limit the difference between the two time reflections in (4.91) and (4.99) disappears, therefore in order to guarantee the existence of a self-adjoint Hamiltonian with non-negative spectrum, it is sufficient to fulfill one of the two conditions. Let us note that, according to (4.125), in the continuum limit the hopping parameter tends to its critical value $K = 1/(8r)$. Therefore, as a matter of fact, the site-reflection positivity condition (4.111) holds for $r = 1$.

4.2.4 *Wilson fermion propagator and Green functions*

The propagator is an important Green function. It is a basic element of perturbation theory, and its poles in the energy show the particle content, therefore it is a central object to be determined in numerical simulations, too.

In order to derive the free Wilson fermion propagator, first one has to specify the boundary conditions. For instance, one can consider an infinite lattice without boundaries or a finite rectangular lattice with some conditions on its hyperplane boundaries. For fermions the most commonly used boundary conditions are periodic in the space directions and antiperiodic in time. The latter choice is due to the negative sign appearing in the expression of a trace as a Grassmann integral in (4.34).

(It is the usual minus sign associated with closed fermion loops.)

For the illustration of periodic and antiperiodic boundary conditions let us consider in detail the one-dimensional analogue of the Wilson fermion matrix (4.89) on a periodic or antiperiodic line (circle):

$$Q_{yx}^{(1)} \equiv \delta_{yx} - K\left[b_x^+ \delta_{y,x+1}(r+\gamma_1) + b_x^- \delta_{y,x-1}(r-\gamma_1)\right] . \tag{4.112}$$

For periodic boundary conditions the sign factors are equal to unity:

$$b_x^+ = b_x^- = 1 . \tag{4.113}$$

In the case of antiperiodicity on a circle of length L_1, if the coordinate x is chosen in the interval $0 \le x \le L_1 - 1$, and $x = L_1$ is the same as $x = 0$, then we have

$$b_x^+ \equiv \begin{cases} -1 & \text{for} \quad x = L_1 - 1 \\ +1 & \text{for} \quad x \ne L_1 - 1 \end{cases}, \qquad b_x^- \equiv \begin{cases} -1 & \text{for} \quad x = 0 \\ +1 & \text{for} \quad x \ne 0 \end{cases} . \tag{4.114}$$

Therefore the explicit matrix form of $Q^{(1)}$ is

$$Q_{yx}^{(1)} = \begin{pmatrix} 1 & -K(r-\gamma_1) & \cdots & 0 & \mp K(r+\gamma_1) \\ -K(r+\gamma_1) & 1 & \cdots & 0 & 0 \\ 0 & -K(r+\gamma_1) & \cdots & 0 & 0 \\ \cdots & \cdots & \cdots & \cdots & \cdots \\ 0 & 0 & \cdots & 1 & -K(r-\gamma_1) \\ \mp K(r-\gamma_1) & 0 & \cdots & -K(r+\gamma_1) & 1 \end{pmatrix} . \tag{4.115}$$

Here the upper (lower) sign refers to the periodic (antiperiodic) case.

The propagator $\Delta^{(1)}$ is the inverse of $Q^{(1)}$, therefore it satisfies

$$\sum_y \Delta_{zy}^{(1)} Q_{yx}^{(1)} = \delta_{zx} . \tag{4.116}$$

It is a translation invariant function, which has a Fourier representation

$$\Delta_{yx}^{(1)} = \Delta_{y-x}^{(1)} = \frac{1}{L_1} \sum_{k_1} e^{ik_1(y-x)} \tilde{\Delta}_{k_1}^{(1)} . \tag{4.117}$$

Substituting back into (4.116) in the periodic case one obtains the condition for the momentum k_μ (here for $\mu \equiv 1$):

$$k_\mu = \frac{2\pi}{L_\mu} \nu_\mu , \qquad \nu_\mu \in \{0, 1, 2, \dots, L_\mu - 1\} . \tag{4.118}$$

This shows that the propagator is a periodic function of its argument $(y - x)$ with period L_1: $\Delta_{y-x}^{(1)} = \Delta_{L_1+y-x}^{(1)}$. In the antiperiodic case the allowed discrete momenta are

$$k_\mu = \frac{2\pi}{L_\mu}\left(\nu_\mu + \frac{1}{2}\right) , \qquad \nu_\mu \in \{0, 1, 2, \dots, L_\mu - 1\} , \tag{4.119}$$

and the propagator is an antiperiodic function satisfying $\Delta^{(1)}_{y-x} = -\Delta^{(1)}_{L_1+y-x}$. In both periodic and antiperiodic cases (4.116) implies

$$\widetilde{\Delta}^{(1)}_{k_1}(1 - 2rK \cos k_1 + 2iK\gamma_1 \sin k_1) = 1 . \tag{4.120}$$

This shows that the momentum space propagator is the same in both periodic and antiperiodic cases, and it is a periodic function of k_1 with a period of 2π. As a consequence, the discrete values of the momenta are only determined up to an integer multiple of 2π. The values in (4.118) and (4.119) were chosen in the *Brillouin zone* between 0 and 2π. Another, more symmetric, choice of the Brillouin zone between $-\pi$ and $+\pi$ is, of course, also possible.

These considerations can easily be generalized to four dimensions. If, in the spatial directions, periodic boundary conditions are assumed, then for $\mu = 1, 2, 3$ we have (4.118). Antiperiodicity in time implies that the allowed values of the fourth component of the momentum satisfy (4.119) with $\mu = 4$. The propagator Δ_{yx} is the inverse of the fermion matrix (4.89), therefore we have

$$\sum_y \Delta_{zy} Q_{yx} = \delta_{zx} . \tag{4.121}$$

The Fourier representation is now

$$\Delta_{yx} = \Delta_{y-x} = \frac{1}{\Omega} \sum_k e^{ik \cdot (y-x)} \widetilde{\Delta}_k , \tag{4.122}$$

where $\Omega = L_1 L_2 L_3 L_4$ is the number of lattice points. The momentum space propagator satisfies

$$\widetilde{\Delta}_k \left(1 - 2rK \sum_{\mu=1}^{4} \cos k_\mu + 2iK \sum_{\mu=1}^{4} \gamma_\mu \sin k_\mu \right) = 1 . \tag{4.123}$$

The solution for $\widetilde{\Delta}_k$ can easily be derived from this. Using the notations in (8.40), and the relation (4.87), one obtains

$$\widetilde{\Delta}_k = \frac{1 - rK(8 - \hat{k}^2) - 2iK\gamma \cdot \overline{k}}{[1 - rK(8 - \hat{k}^2)]^2 + 4K^2\overline{k}^2} = (2K)^{-1} \frac{am + (r/2)\hat{k}^2 - i\gamma \cdot \overline{k}}{[am + (r/2)\hat{k}^2]^2 + \overline{k}^2} . \tag{4.124}$$

Apart from the factor $(2K)^{-1}$, which can be transformed away by an appropriate choice of the field normalization, this looks already quite similar to the Dirac fermion propagator in the continuum. In the limit of zero lattice spacing $a \to 0$ and finite mass m, according to (4.87), the hopping parameter K tends to its *critical value*

$$K_c = \frac{1}{8r} . \tag{4.125}$$

At the same time the physical momentum p_μ is related to the lattice momentum k_μ by $k_\mu = ap_\mu$, and p_μ is kept fixed, therefore $\bar{k}_\mu = \sin k_\mu \rightarrow ap_\mu$ and we have

$$\tilde{\Delta}_k \rightarrow (2aK)^{-1} \frac{m - i\gamma \cdot p}{m^2 + p^2} \,. \qquad (4.126)$$

Apart from the normalization factor this is the free fermion propagator.

In the derivation of (4.126) it was assumed that in the continuum limit the physical momentum is near the zero corner ($k_1 = k_2 = k_3 = k_4 = 0$) of the Brillouin zone. Of course, as we already know from the Hamiltonian formulation (see (4.64)–(4.65)), one has to look in the whole Brillouin zone. In particular, the reason of introducing the Wilson term was that in the naive lattice fomulation ($r = 0$) near the eight corners in (4.65) the energies were the same. The $r = 0$ Euclidean propagator is periodic in k_μ with a period of π in all four directions, therefore there are 16 equivalent corners of the Brillouin zone. The naive Euclidean fermion propagator describes the propagation of 16 fermions. In case of $r \neq 0$ for $k_4 \simeq 0$ and lattice three-momentum $\mathbf{k} = a\mathbf{p} + \mathbf{q}_\pi$ the mass is replaced by $m + 2ra^{-1}n_\pi$, where n_π is the number of momentum components equal to π in (4.65). Therefore the states with $n_\pi \neq 0$ become infinitely heavy if $a \rightarrow 0$, similarly to the Hamiltonian formulation.

The fate of the other eight 'time doublers' at $k_4 \simeq \pi$ (or equivalently at $k_4 \simeq -\pi$) is a bit more subtle, because finally what we have to investigate is the pole structure in the energy $E \equiv -ia^{-1}k_4$ (a physical state corresponds to a pole of the propagator in the energy variable). For simplicity, let us consider here only the case of zero three-momentum, because the discussion is similar also for $\mathbf{p} \neq \mathbf{0}$. The denominator of the propagator vanishes if the energy satisfies

$$\left[am - 2r \sinh^2 \left(\frac{aE}{2} \right) \right]^2 - \sinh^2 (aE) = 0 \,. \qquad (4.127)$$

The solutions of this equation in the general case $0 < r \leq 1$ will be discussed in detail below, after (4.138). Here let us consider only the simplest case, namely $r = 1$, when there are two roots:

$$E_{(\pm)} = \pm a^{-1} \log (1 + am) \,. \qquad (4.128)$$

In the continuum limit $a \rightarrow 0$ we have $E_{(\pm)} \rightarrow \pm m$. This is the same as the position of the energy poles in the continuum propagator, therefore there is only one fermion state. For $r = 1$ the time doubler is not propagating. This is in agreement with the above derivation of the $r = 1$ Wilson action by starting from the Hamiltonian formulation, where there are no time doublers either. We shall see after (4.138) that for $r \neq 1$ the situation is a bit different, namely there are two other poles corresponding to the time doubler. Nevertheless, in the continuum limit the energy of the time doubler becomes infinite, and only one fermion remains in the physical spectrum. This shows that the Wilson action (4.88) has the correct continuum limit for $0 < r \leq 1$.

The generating function of the *Green functions* of the free fermion field can be defined as $Z[\eta, \bar{\eta}]/Z[0,0]$, where $Z[\eta, \bar{\eta}]$ is the Grassmann integral

$$Z[\eta, \bar{\eta}] = \int [\mathrm{d}\bar{\psi}\, \mathrm{d}\psi] \exp\left\{ -\sum_{xy} (\bar{\psi}_y Q_{yx} \psi_x) + \sum_x [(\bar{\eta}_x \psi_x) - (\bar{\psi}_x \eta_x)] \right\}.$$

(4.129)

Here Q_{yx} is the fermion matrix in (4.89). Performing the integral according to (4.24) gives

$$Z[\eta, \bar{\eta}] = \det Q \cdot \exp\left\{ -\sum_{xy} (\bar{\eta}_y Q_{yx}^{-1} \eta_x) \right\},$$

(4.130)

therefore the partition function is $Z \equiv Z[0,0] = \det Q$. The two-point Green function is obtained as the second derivative

$$\langle \psi_y \bar{\psi}_x \rangle = \partial_{\bar{\eta}_y} \partial_{\eta_x} \left. \frac{Z[\eta, \bar{\eta}]}{Z[0,0]} \right|_{\eta=\bar{\eta}=0} = Q_{yx}^{-1} = \Delta_{yx}.$$

(4.131)

Here in the last step (4.121) was used. More generally, the Green function with n pairs of fermion variables is, according to (4.25),

$$\langle \psi_{y_1} \bar{\psi}_{x_1} \psi_{y_2} \bar{\psi}_{x_2} \cdots \psi_{y_n} \bar{\psi}_{x_n} \rangle = \partial_{\bar{\eta}_{y_1}} \partial_{\eta_{x_1}} \partial_{\bar{\eta}_{y_2}} \partial_{\eta_{x_2}} \cdots \partial_{\bar{\eta}_{y_n}} \partial_{\eta_{x_n}} \left. \frac{Z[\eta, \bar{\eta}]}{Z[0,0]} \right|_{\eta=\bar{\eta}=0}$$

$$= \sum_{z_1 \cdots z_n} \epsilon_{y_1 y_2 \cdots y_n}^{z_1 z_2 \cdots z_n} Q_{z_1 x_1}^{-1} Q_{z_2 x_2}^{-1} \cdots Q_{z_n x_n}^{-1},$$

(4.132)

where ϵ is defined in (4.23). The connected Green functions are defined as in section 2.1 by the derivatives of the generating function $W \equiv \log Z$. Since now we are dealing with a free field, except for

$$\langle \psi_y \bar{\psi}_x \rangle_c \equiv \langle \psi_y \bar{\psi}_x \rangle$$

(4.133)

all other connected Green functions with more pairs of fermion variables are zero.

An important quantity in fermionic theories is the expectation value of $(\bar{\psi}_x \psi_x)$, the so called *fermion condensate*. In the case of free Wilson fermions (4.131) and (4.124) imply that

$$\sum_\alpha \langle \psi_{\alpha x} \bar{\psi}_{\alpha x} \rangle = -\langle (\bar{\psi}_x \psi_x) \rangle = \mathrm{Tr}\, Q_{xx}^{-1} = \mathrm{Tr}\, \Delta_0 = \frac{1}{\Omega} \sum_k \mathrm{Tr}\, \tilde{\Delta}_k$$

$$= \frac{4}{2K\Omega} \sum_k \frac{am + (r/2)\hat{k}^2}{[am + (r/2)\hat{k}^2]^2 + \bar{k}^2}.$$

(4.134)

4.2.5   Timeslices of Wilson fermions

In numerical simulations the properties of the mass spectrum are usually determined by the investigation of the *timeslices* of the two-point function, which are obtained by summing over the spatial extension of the lattice for fixed times. The real interest is, of course, in the mass spectrum of interacting theories, but it is useful to demonstrate how the method works in the case of the free fermion field.

Up to now the energy spectrum of a free Wilson lattice fermion was only determined for $r = 1$ in (4.128). Here we shall consider the general case $0 < r \leq 1$. The timeslice correlation function belonging to the fermion two-point function with zero spatial momentum is defined on a lattice with $V \equiv L_1 L_2 L_3$ points in a timeslice and $\Omega \equiv VT$ points in total as

$$C(t) \equiv \frac{1}{V} \sum_{\mathbf{x}} \langle \psi_{\mathbf{x},t} \overline{\psi}_{0,0} \rangle = \frac{1}{\Omega V} \sum_{x_0, t_0} \sum_{\mathbf{x}} \langle \psi_{\mathbf{x}_0 + \mathbf{x}, t_0 + t} \overline{\psi}_{\mathbf{x}_0, t_0} \rangle$$

$$= \frac{1}{V} \sum_{\mathbf{x}} \Delta_{\mathbf{x},t} = \frac{1}{\Omega} \sum_{k_4} e^{itk_4} \tilde{\Delta}_{0,k_4} . \tag{4.135}$$

Timeslices with non-zero spatial momentum can be defined similarly, but they will not be considered here, because the determination of the energy spectrum can be done in the same way as for zero momentum. Using the explicit form of the momentum space propagator in (4.124) and the relations in (8.40) one obtains

$$C(t) = \frac{1}{2K\Omega} \sum_{k_4} e^{itk_4} \frac{am + 2r \sin^2(k_4/2) - i\gamma_4 \sin k_4}{[am + 2r \sin^2(k_4/2)]^2 + \sin^2 k_4} . \tag{4.136}$$

For evaluation of this formula it is advantageous to consider first the limit of infinite time extension $T \equiv L_4 \to \infty$ when, according to (4.118) or (4.119), the density of discrete momentum points in the Brillouin zone becomes infinite, and therefore one can replace the momentum sum by an integral:

$$\frac{1}{T} \sum_{k_4} \longrightarrow \int_0^{2\pi} \frac{dk_4}{2\pi} \longrightarrow \int_{-\pi}^{\pi} \frac{dk_4}{2\pi} . \tag{4.137}$$

The integrand in (4.136) has poles in the energy variable $aE \equiv -ik_4$ if the denominator vanishes. The condition for this is given by (4.127). The case $r = 1$ has already been considered separately in (4.128). For $r \neq \pm 1$ one can easily show that the roots of (4.127) are at

$$aE = \pm aE_1 \qquad \text{and} \qquad aE = i\pi \pm aE_2 , \tag{4.138}$$

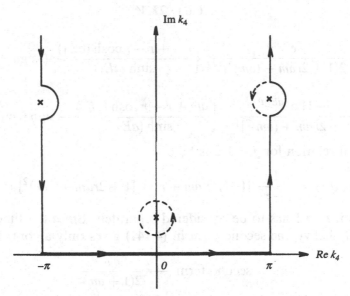

Fig. 4.6. The deformation of the integration contour in the complex k_4 plane for $t > 0$. The crosses show the position of the poles. First the vertical pieces are added which cancel each other due to periodicity, and then the contour is deformed to enclose the poles. For $t < 0$ the procedure is similar, but the poles are at $\text{Im } k_4 < 0$.

where aE_1 and aE_2 are given by

$$aE_1 = \log\left\{\frac{1}{1+r}\left[[1 + 2ram + (am)^2]^{1/2} + am + r\right]\right\},$$

$$aE_2 = \log\left\{\frac{1}{1-r}\left[[1 + 2ram + (am)^2]^{1/2} + am + r\right]\right\}. \qquad (4.139)$$

In the limit $r = 1$ (4.128) is reproduced by the root $\pm aE_1$. At the same time $aE_2 \to \infty$, that is the time doubler becomes infinitely heavy for finite lattice spacing. In the continuum limit $a \to 0$, for $r \neq \pm 1$, the energies tend to

$$E_1 \to m, \qquad E_2 \to m + a^{-1}\log\frac{1+r}{1-r}. \qquad (4.140)$$

This shows that for $0 < r < 1$ the time doubler becomes infinitely heavy in the continuum limit.

In order to perform the integral over k_4 one can use the Cauchy theorem and deform the integration contour according to fig. 4.6 (see also [4.11]). The result is

$$C(t) \cdot 2KV =$$

$$\frac{e^{-aE_1|t|}}{2[1 + 2ram + (am)^2]^{1/2}} \left\{ \frac{am + r - r\cosh(aE_1)}{\sinh(aE_1)} + \text{sign}(t)\gamma_4 \right\}$$

$$+ \frac{(-1)^t e^{-aE_2|t|}}{2[1 + 2ram + (am)^2]^{1/2}} \left\{ \frac{am + r + r\cosh(aE_2)}{\sinh(aE_2)} - \text{sign}(t)\gamma_4 \right\} . \quad (4.141)$$

A useful relation for $j = 1, 2$ is here

$$\cosh(aE_j) = \frac{1}{1 - r^2} \left[(-1)^j r(am + r) + [1 + 2ram + (am)^2]^{1/2} \right] . \quad (4.142)$$

The limit $r \to 1$ has to be considered separately. Since the time doubler is infinitely heavy, the second term in (4.141) gives only a contribution

$$\text{second term} \longrightarrow \frac{\delta_{t0}}{2(1 + am)} . \quad (4.143)$$

Both terms in (4.141) have a characteristic exponential behaviour in the time distance t with an exponent proportional to the real part of the poles in energy. In the Dirac index space there is always a constant term plus a term proportional to γ_4. One can also see that for $r \neq 1$ the contribution of the time doubler in the second term has an oscillating sign, whereas the physical particle in the first term gives always a positive contribution. This is in accordance with the fact that in the $r \neq 1$ case only the link-reflection positivity could be proven (see after (4.91)). For $r = 1$ the time doubler is absent, there is no oscillating contribution, both link-reflection and site-reflection positivity could be proven (for small enough values of am), and the positive transfer matrix acts on wave functions defined at fixed times [4.7].

The result for the timeslice correlations in (4.141) holds in the case of an infinite time extension $T = \infty$. For finite T one can use

$$\sum_{m=-\infty}^{+\infty} e^{imTk_4} = \frac{1}{T} \sum_{v_4=0}^{T-1} 2\pi\delta\left(k_4 - \frac{2\pi}{T}v_4\right) . \quad (4.144)$$

Therefore, defining

$$b \equiv \left\{ \begin{array}{ll} +1 & \text{for periodicity in time} \\ -1 & \text{for antiperiodicity in time} \end{array} \right. , \quad (4.145)$$

we have for a periodic function $F(k_4)$ with period 2π

$$\frac{1}{T} \sum_{k_4} e^{itk_4} F(k_4) = \sum_{m=-\infty}^{+\infty} b^m \frac{1}{2\pi} \int_{-\pi}^{\pi} dk_4 \, e^{ik_4(t+mT)} F(k_4) . \quad (4.146)$$

Having in mind that

$$C(t + T) = \begin{cases} +C(t) & \text{for periodicity in time} \\ -C(t) & \text{for antiperiodicity in time} \end{cases}, \qquad (4.147)$$

for $0 \le t \le T$ one obtains the relations

$$\sum_{m=-\infty}^{+\infty} b^m e^{-\mathscr{E}|t+mT|} = \frac{e^{-\mathscr{E}t} + b e^{-\mathscr{E}(T-t)}}{1 - b e^{-\mathscr{E}T}},$$

$$\sum_{m=-\infty}^{+\infty} b^m \operatorname{sign}(t+mT) e^{-\mathscr{E}|t+mT|} = \frac{e^{-\mathscr{E}t} - b e^{-\mathscr{E}(T-t)}}{1 - b e^{-\mathscr{E}T}} - \delta_{t0} + b \delta_{tT} . \quad (4.148)$$

These have to be inserted in (4.141) instead of $e^{-\mathscr{E}|t|}$ and $\operatorname{sign}(t) e^{-\mathscr{E}|t|}$, respectively, with $\mathscr{E} = aE_{1,2}$. In contrast to the case of scalar fields, where the periodic time dependence for finite T is always given by a cosh function, one can see that the dependence on t is described here sometimes by a cosh and sometimes by a sinh function.

In an interacting fermionic theory one can try to fit the large t-dependence of the two-point function timeslices by the above form. Since every contribution is decreasing exponentially by the corresponding mass, at large time distances $t \gg 1$ the states with large masses can be neglected. If the lowest mass state is separated from the rest by a sufficiently large mass gap, then at large t one can fit by a single state. A successful fit with a correct Dirac structure and an exponential decrease $e^{-m|t|}$ is interpreted as the contribution of a fermionic particle of mass m.

4.3 Kogut–Susskind staggered fermions

We have seen in section 4.2 that the naive discretization of the Dirac equation leads to a replication of fermionic states. This is called 'lattice fermion doubling'. The naively discretized free Hamiltonian describes in the continuum limit eight degenerate fermion species (see (4.65)). In the Euclidean formulation there is an additional duplication by 'time doublers', therefore the naive free fermion lattice action is equivalent in the continuum limit to 16 free fermion fields. The doublers appear in the fermion propagator at different non-zero corners of the Brillouin zone. The real problem is, of course, not the replication of the free field. The interesting question is, whether the fermion doubling can be avoided in the continuum limit of interacting theories, where the fermion doublers can, in principle, be pair-produced. In section 4.2 the 15 fermion doublers were decoupled by a large mass in the continuum limit ('Wilson lattice fermions'). An opposite way of dealing with this problem is to interpret the extra fermions as new physical 'flavours'. This idea was introduced by Susskind and his collaborators [4.5]. It turned out that the replication can be reduced from 16 to 4 by assigning only a single fermion field component to every lattice site, and the remaining four fermion species can be interpreted in the continuum limit as physical flavours.

4.3.1   From naive to staggered fermions

In order to illustrate the idea of reduction of the number of fermion doublers let us first consider the Dirac equation in (1+1), that is one space and one time, dimensions. The Dirac Hamiltonian operator corresponding to (4.50) is in this case

$$\mathscr{H}_{\text{Dirac}} = m\beta - i\alpha_1 \frac{\partial}{\partial x_1} . \tag{4.149}$$

In (1+1) dimensions the representation of the two Dirac matrices $\gamma_{1,2}$ and of β and α_1 can be chosen as

$$\gamma_1 = \sigma_2 , \quad \gamma_2 = \sigma_3 , \quad \beta \equiv \gamma_2 = \begin{pmatrix} 1 & 0 \\ 0 & -1 \end{pmatrix} , \quad \alpha_1 \equiv i\gamma_2\gamma_1 = \begin{pmatrix} 0 & 1 \\ 1 & 0 \end{pmatrix} . \tag{4.150}$$

Let us denote the time derivative by a dot. The components of the lattice Dirac equation in the above representation can be written as

$$i\dot{\chi}_{1,x} = m\chi_{1,x} - \frac{i}{2a}\left(\chi_{2,x+1} - \chi_{2,x-1}\right) ,$$

$$i\dot{\chi}_{2,x} = -m\chi_{2,x} - \frac{i}{2a}\left(\chi_{1,x+1} - \chi_{1,x-1}\right) . \tag{4.151}$$

Here the x-derivative is discretized by the difference of the two neighbouring points (a is, as usual, the lattice spacing). Other discretizations would also be possible, but this choice is particularly interesting, because in this case the lattice variables form two independent sets: the first component of χ on odd (even) sites is connected only to the second component on even (odd) sites. Therefore, one can try to reduce the fermion degrees of freedom on the lattice by considering only one of these two *staggered* structures. For instance, let us define a single component lattice fermion field by

$$\zeta_x \equiv \begin{cases} \chi_{1,x} & \text{if} \quad x = \text{odd} \\ \chi_{2,x} & \text{if} \quad x = \text{even} \end{cases} . \tag{4.152}$$

This seems to describe only right-moving waves because, for instance, in the massless ($m = 0$) case in the continuum limit we have $\dot{\zeta}_x = -\partial_x\zeta_x$. However, defining another field by going to the other corner of the Brillouin zone as

$$\zeta'_x \equiv (-1)^x\zeta_x = e^{\pm i\pi x}\zeta_x , \tag{4.153}$$

we get $\dot{\zeta}'_x = +\partial_x \zeta'_x$, that is left-moving waves. This shows that in (1+1) dimensions the single component *staggered lattice fermion* in the continuum limit describes a single Dirac fermion species. It is remarkable that this is in some sense a non-local description: the two spinor components of the fermion sit on different sites. One can imagine that the spatial lattice is divided into blocks of two points, and each block goes to a single point in the continuum limit. The same trick can be applied also in (3+1) dimensions (see, for instance, the paper of Susskind in [4.5]). Putting only one out of the four spinor components on a site, the number of degrees of freedom is divided by four. Therefore the eight fermions at the eight corners of the Brillouin zone (see (4.65)) are reduced to two flavours.

One could proceed by looking for an appropriate Euclidean lattice action which corresponds to the Kogut–Susskind staggered fermion Hamiltonian. However, the requirement of the four-dimensional discrete Euclidean rotation symmetry has to be implemented and it leads to the appearance of time doublers. Therefore it is simpler to start from the Euclidean invariant naive lattice fermion action and to try to reduce the degrees of freedom there [4.12]–[4.17]. The naive fermion action is given by (4.85) with $r = 0$. Denoting the naive lattice fermion field here by Ψ_x, $\overline{\Psi}_x$ we have

$$S = \sum_x a^4 \left\{ m(\overline{\Psi}_x \Psi_x) + \frac{1}{2a} \sum_{\mu=1}^{4} \left[(\overline{\Psi}_x \gamma_\mu \Psi_{x+\hat{\mu}}) - (\overline{\Psi}_{x+\hat{\mu}} \gamma_\mu \Psi_x) \right] \right\} . \quad (4.154)$$

Here γ_μ is the usual Hermitean and unitary Euclidean Dirac matrix satisfying the *Clifford algebra* anticommutation relations in (8.6). The following discussion is in four dimensions ($d = 4$), but everything can be repeated also in any even d dimensions, where the number of Dirac matrices is d and the Clifford algebra is represented by $C \otimes C$ matrices with $C = 2^{d/2}$.

The reduction of degrees of freedom to a single fermion field component can be achieved by a *spin diagonalization*, which consists in performing a local change of fermionic variables

$$\Psi_x = A_x \psi_x , \qquad\qquad \overline{\Psi}_x = \overline{\psi}_x A_x^+ . \quad (4.155)$$

Here A_x is a $4 \otimes 4$ unitary matrix diagonalizing all the γ-matrices in the action (4.154), in the sense that

$$A_x^+ \gamma_\mu A_{x+\hat{\mu}} \equiv \Delta_\mu(x) \in U(1)^{\otimes 4} . \quad (4.156)$$

$\Delta_\mu(x)$ is a diagonal unitary matrix belonging to $U(1)^{\otimes 4}$. There exist different solutions for A_x, for instance, [4.14]

$$A_x = A_x^{(0)} \equiv \gamma_1^{x_1} \gamma_2^{x_2} \gamma_3^{x_3} \gamma_4^{x_4} , \quad (4.157)$$

where (x_1, x_2, x_3, x_4) are the components of the lattice site four-vector x. In this case $\Delta_\mu(x)$ is a multiple of the unity $\mathbf{1}$, namely

$$\Delta_\mu(x) = \Delta_\mu^{(0)}(x) \equiv \alpha_{x\mu}\mathbf{1} ,$$

$$\alpha_{x\mu} = (-1)^{x_1 + \cdots + x_{\mu-1}} \quad (\mu = 1, 2, 3, 4) . \qquad (4.158)$$

In terms of the new fields $\psi_x, \overline{\psi}_x$ the naive lattice action (4.154) is

$$S = \sum_x a^4 \left\{ m(\overline{\psi}_x \psi_x) + \frac{1}{2a} \sum_{\mu=1}^{4} \left[(\overline{\psi}_x \Delta_\mu(x) \psi_{x+\hat{\mu}}) - (\overline{\psi}_{x+\hat{\mu}} \Delta_\mu^+(x) \psi_x) \right] \right\} . \qquad (4.159)$$

Since this is diagonal in the Dirac indices, the different components are decoupled and one can keep only a single field component per site.

There are many possible choices of unitary matrices A_x which realize spin diagonalization [4.5, 4.12, 4.14, 4.18, 4.19, 4.20], therefore a natural question is: to what extent are different choices equivalent? An important condition pointed out in [4.20] is that on every plaquette (that is for $\mu \neq v$) the product of $\Delta_\mu(x)$ has to satisfy

$$\Delta_\mu(x)\Delta_v(x + \hat{\mu})\Delta_\mu^+(x + \hat{v})\Delta_v^+(x) = -\mathbf{1} . \qquad (4.160)$$

This is due to $\gamma_\mu \gamma_v \gamma_\mu \gamma_v = -1$ for $\mu \neq v$. (Note that on links in negative directions one can define $\Delta_{-\mu}(x + \hat{\mu}) \equiv \Delta_\mu^+(x)$.) If (4.160) is true, then the field $V_\mu(x)$ defined on the links by

$$V_\mu(x) \equiv \Delta_\mu(x)\Delta_\mu^{(0)+}(x) \qquad (4.161)$$

satisfies

$$V_\mu(x)V_v(x + \hat{\mu})V_\mu^+(x + \hat{v})V_v^+(x) = \mathbf{1} . \qquad (4.162)$$

This implies that the same holds for an arbitrary closed loop consisting of consecutive links, and therefore $V_\mu(x)$ is a 'pure gauge' field in $U(1)^{\otimes 4}$, which can be written as

$$V_\mu(x) = S_x^+ S_{x+\hat{\mu}} . \qquad (4.163)$$

Here S_x is a field in $U(1)^{\otimes 4}$ defined on lattice points. This implies that every $\Delta_\mu(x)$ satisfying (4.160) is equivalent to $\Delta_\mu^{(0)}(x)$ in the sense that

$$\Delta_\mu(x) = S_x^+ \Delta_\mu^{(0)}(x)S_{x+\hat{\mu}} , \qquad A_x = A_x^{(0)}S_x . \qquad (4.164)$$

Conversely, an arbitrary diagonal field $S_x \in U(1)^{\otimes 4}$ on the lattice points can be used to define according to (4.164) a $\Delta_\mu(x)$, which is suitable for the spin diagonalization of the naive lattice fermion action.

After the spin diagonalization in (4.159) one can reduce the number of fermion field components per site to one. One can also keep the original four components or, in fact, also introduce an arbitrary number of components f describing $4f$ flavours of Dirac fermions (as we shall see in the next subsection). In the massless case $m = 0$, the obtained lattice fermion action has an exact global $U_o(f) \otimes U_e(f)$ chiral symmetry, which is defined by the transformations

$$
\left.\begin{array}{l}
\psi'_x = U_{o(e)}\psi_x \\[2mm]
\overline{\psi}'_x = \overline{\psi}_x U^+_{e(o)}
\end{array}\right\} \quad \text{for } x \text{ odd (even) .} \tag{4.165}
$$

In the massive case ($m \neq 0$) the symmetry group is reduced to the diagonal subgroup defined by $U_o = U_e$.

For a single component ($f = 1$), the exact chiral symmetry is $U(1)_o \otimes U(1)_e$ and there are four flavours of Dirac fields. Choosing $\Delta_\mu(x)$ as in (4.158) and keeping the field normalization general, we have the free *staggered fermion action*

$$
S = \sum_x \left\{ M(\overline{\psi}_x \psi_x) + K \sum_{\mu=1}^{4} \alpha_{x\mu} \left[(\overline{\psi}_x \psi_{x+\hat{\mu}}) - (\overline{\psi}_{x+\hat{\mu}} \psi_x) \right] \right\} . \tag{4.166}
$$

The ratio of the two coefficients gives the mass in lattice units:

$$
am = \frac{M}{2K} . \tag{4.167}
$$

Otherwise M and K can be changed by redefining the field normalization. For instance, (4.88) with $r = 0$ corresponds to the choice $M = 1$. In lattice perturbation theory the usual normalization is fixed by $K = \frac{1}{2}$, $M = am$.

4.3.2 Flavours of staggered fermions

In (4.166) the naive lattice fermion action is reduced to a single component field, and the Dirac matrices are represented by the x-dependent sign factor $\alpha_{x\mu}$. The reduction of the number of degrees of freedom by a factor of four suggests that, instead of the 16 naive lattice fermion species, the staggered fermion action describes four 'flavours' of fermions. The discussion of the staggered fermion in (1+1) dimensions in the previous subsection suggests that the spinor and flavour components can be associated with the 16 corners of 2^4 hypercubes on the lattice. Since the reason for the multiplication of the fermion states is the appearance of poles at different corners of the Brillouin zone in momentum space, one can also try to identify the flavours in momentum space. Both these ways were followed in the literature: see, respectively, [4.15, 4.17, 4.21] and [4.12, 4.22, 4.23].

In order to find the spinor-flavour interpretation of the field components in 2^4 hypercubes, let us assume that the number of points in a hypercubic lattice is even in every direction: $L_\mu = 2L'_\mu$ ($\mu = 1, 2, 3, 4$). The coordinates of the lattice points are assumed, as usual, to lie in the interval $0 \leq x_\mu \leq$

$L_\mu - 1$. Let us introduce the coordinates of the hypercubic blocks y_μ by

$$x_\mu = 2y_\mu + \eta_\mu , \qquad \eta_\mu = 0, 1 , \tag{4.168}$$

where $0 \leq y_\mu \leq L'_\mu - 1$. The sum over lattice points can be written as a sum over hypercubes and points within hypercubes, that is

$$\sum_x = \sum_y{}' \sum_\eta . \tag{4.169}$$

Here the prime on the y-sum is a shorthand notation for the different summation limits. The sign factor $\alpha_{x\mu}$ in (4.158) can obviously be considered also as a function of the η-variables within the hypercubes: $\alpha_{\eta\mu} \equiv \alpha_{x\mu}$.

The new fields with definite flavour $q_y^{\alpha a}$ and $\overline{q}_y^{a\alpha}$ depend on the block coordinate y and on the *Dirac spinor* ($1 \leq \alpha \leq 4$) and *flavour* ($1 \leq a \leq 4$) indices:

$$q_y^{\alpha a} \equiv \frac{1}{8} \sum_\eta \Gamma_{\eta;\alpha a} \psi_{2y+\eta} , \qquad \overline{q}_y^{a\alpha} \equiv \frac{1}{8} \sum_\eta \overline{\psi}_{2y+\eta} \Gamma^+_{\eta;a\alpha} . \tag{4.170}$$

The matrices Γ_η are defined by

$$\Gamma_\eta \equiv \gamma_1^{\eta_1} \gamma_2^{\eta_2} \gamma_3^{\eta_3} \gamma_4^{\eta_4} . \tag{4.171}$$

These satisfy the relations

$$\frac{1}{4} \mathrm{Tr} \left(\Gamma^+_\eta \Gamma_{\eta'} \right) = \delta_{\eta\eta'} , \qquad \frac{1}{4} \sum_\eta \Gamma^+_{\eta;b\beta} \Gamma_{\eta;\alpha a} = \delta_{ba} \delta_{\beta\alpha} , \tag{4.172}$$

therefore the inverse expressions are

$$\psi_{2y+\eta} = 2\,\mathrm{Tr} \left(\Gamma^+_\eta q_y \right) , \qquad \overline{\psi}_{2y+\eta} = 2\,\mathrm{Tr} \left(\overline{q}_y \Gamma_\eta \right) . \tag{4.173}$$

Using these relations one can write the lattice action (4.166) in terms of the new fields. For instance, the mass term is

$$\sum_x (\overline{\psi}_x \psi_x) = 16 \sum_y{}' (\overline{q}_y \mathbf{1} \otimes \mathbf{1} q_y) . \tag{4.174}$$

The notation $\gamma \otimes t$ stands for the direct product of the matrix γ acting on the Dirac indices, and the matrix t acting on the flavour indices. In order to obtain the kinetic term one uses relations like

$$\gamma_\mu \Gamma_\eta = \delta_{0\eta_\mu} \alpha_{\eta\mu} \Gamma_{\eta+\hat{\mu}} + \delta_{1\eta_\mu} \alpha_{\eta\mu} \Gamma_{\eta-\hat{\mu}} ,$$

$$\gamma_5 \Gamma_\eta \gamma_5 = (-1)^{\eta_1 + \eta_2 + \eta_3 + \eta_4} \Gamma_\eta . \tag{4.175}$$

Defining two different lattice derivatives on the block lattice by

$$\Delta_\mu f_y \equiv \frac{1}{4} \left(f_{y+\hat{\mu}} - f_{y-\hat{\mu}} \right) \longrightarrow a\partial_\mu f_y ,$$

$$\delta_\mu f_y \equiv \frac{1}{4}\left(f_{y+\hat{\mu}} + f_{y-\hat{\mu}} - 2f_y\right) \longrightarrow a^2 \partial_\mu^2 f_y \,, \tag{4.176}$$

the lattice action (4.166) can be written in terms of the new fields as

$$S = 16\sum_y{}' \left\{ M\left(\bar{q}_y \mathbf{1} \otimes \mathbf{1} q_y\right) + 2K\sum_{\mu=1}^4 \left(\bar{q}_y [\gamma_\mu \otimes \mathbf{1}\Delta_\mu - \gamma_5 \otimes t_5 t_\mu \delta_\mu] q_y\right) \right\} \,. \tag{4.177}$$

Here the flavour matrices t_μ are defined by

$$t_\mu \equiv \gamma_\mu^T = t_\mu^+ \qquad (\mu = 1, 2, \ldots, 5) \,. \tag{4.178}$$

In the continuum limit the first part of the kinetic term, which is diagonal in flavour, tends to the usual kinetic term of four Dirac fermion flavours. The second part is a *lattice artifact* of order $O(a)$, which vanishes in the continuum limit. (See, for instance, the propagator in momentum space given below.) For finite lattice spacings this term breaks the $U(4) \otimes U(4)$ continuous chiral symmetry of four massless flavours to $U(1)_o \otimes U(1)_e$ defined in (4.165). As can be seen from (4.173) and the second equation in (4.175), on the flavour basis the operators P_e (P_o) projecting on the even (odd) sites of the original local basis are

$$P_e = \frac{1}{2}(\mathbf{1} \otimes \mathbf{1} + \gamma_5 \otimes t_5) \,, \qquad P_o = \frac{1}{2}(\mathbf{1} \otimes \mathbf{1} - \gamma_5 \otimes t_5) \,. \tag{4.179}$$

The $U(1)_o \otimes U(1)_e$ transformation on the flavour basis is, according to (4.165),

$$q_y' = (U_o P_o + U_e P_e) q_y \,, \qquad \bar{q}_y' = \bar{q}_y (P_o U_e^+ + P_e U_o^+) \,. \tag{4.180}$$

It is easy to see that for $M = 0$ this is a symmetry of the action S in the form (4.177).

In order to derive the staggered fermion propagator in the flavour basis, let us write the action as

$$S = \sum_{y_1, y_2}{}' \left(\bar{q}_{y_2} Q_{y_2 y_1}^q q_{y_1}\right) \,, \tag{4.181}$$

where the fermion matrix Q^q is

$$Q_{y_2 y_1}^q = 16 M \delta_{y_2 y_1} \mathbf{1} \otimes \mathbf{1} + 8K \sum_{\mu=1}^4 \left[\left(\delta_{y_2+\hat{\mu}, y_1} - \delta_{y_2-\hat{\mu}, y_1}\right) \gamma_\mu \otimes \mathbf{1}\right.$$

$$\left. - \left(\delta_{y_2+\hat{\mu}, y_1} + \delta_{y_2-\hat{\mu}, y_1} - 2\delta_{y_2 y_1}\right) \gamma_5 \otimes t_5 t_\mu\right] \,. \tag{4.182}$$

One can proceed similarly to (4.122) for Wilson fermions. The propagator is the inverse Δ^q of the fermion matrix Q^q, given by

$$\Delta_{y_2 y_1}^q = \Delta_{y_2 - y_1}^q = \frac{1}{\Omega'}\sum_k{}' e^{ik\cdot(y_2 - y_1)} \tilde{\Delta}_k^q \,. \tag{4.183}$$

The Fourier transformation to momentum space is done here on the block-lattice, therefore $\Omega' = L'_1 L'_2 L'_3 L'_4$ is the number of 2^4 blocks, and the momentum sum runs over the Brillouin zone defined in (4.118) or (4.119) with L_μ replaced by L'_μ. One can easily see that the propagator in momentum space is

$$\tilde{\Delta}^q_k = \left\{ 16M\mathbf{1} \otimes \mathbf{1} + 16K \sum_{\mu=1}^4 \left[\mathrm{i}\sin k_\mu \gamma_\mu \otimes \mathbf{1} + (1 - \cos k_\mu)\gamma_5 \otimes t_5 t_\mu \right] \right\}^{-1}$$

$$= \frac{M\mathbf{1} \otimes \mathbf{1} - K \sum_{\mu=1}^4 \left(\mathrm{i}\bar{k}_\mu \gamma_\mu \otimes \mathbf{1} + \tfrac{1}{2}\hat{k}^2 \gamma_5 \otimes t_5 t_\mu \right)}{16\left(M^2 + K^2 \hat{k}^2 \right)} . \tag{4.184}$$

Here in the second line the notations in (8.40) were used.

For the physical interpretation of the momentum space propagator it is better to introduce the mass by (4.167) and write it as

$$\tilde{\Delta}^q_k = \frac{am\mathbf{1} \otimes \mathbf{1} - \sum_{\mu=1}^4 \left[\tfrac{1}{2}\mathrm{i}\bar{k}_\mu \gamma_\mu \otimes \mathbf{1} + \tfrac{1}{4}\hat{k}^2 \gamma_5 \otimes t_5 t_\mu \right]}{32K\left[(am)^2 + \tfrac{1}{4}\hat{k}^2 \right]} . \tag{4.185}$$

This form shows that the momentum components in units of the original ('fine') lattice are $p_\mu \equiv k_\mu/2$, $(\mu = 1, 2, 3, 4)$. The momentum component k_μ in lattice units satisfies $0 \leq k_\mu \leq 2\pi$ or $-\pi \leq k_\mu \leq \pi$. Therefore the momentum in physical units p_μ/a is in the interval

$$-\frac{\pi}{2a} \leq \frac{p_\mu}{a} \leq \frac{\pi}{2a} . \tag{4.186}$$

In the continuum limit $a \to 0$, $p_\mu = O(a)$ the propagator has a pole only near $p_\mu = 0$. Near $|p_\mu| = \pi/2$, $|k_\mu| = \pi$ there are no poles at all. In this limit the term proportional to $\gamma_5 \otimes t_5 t_\mu$ can be neglected, because compared to the other terms in the nominator it has an extra factor a. Hence, apart from a normalization factor, the propagator tends to the continuum Dirac propagator of four diagonal fermion flavours. This shows that in the continuum limit the single component staggered fermions do describe four independent flavours of Dirac fermions. According to (4.170) the Dirac spinor components of the four flavours reside on 2^4 hypercubic blocks of the lattice. The consequence is that the cut-off on the physical momentum components in (4.186) is lower by a factor of two than it would be on the same lattice, for instance, for scalar particles or for Wilson fermions.

An alternative approach to defining the four flavours of staggered fermions is to start directly in momentum space [4.12, 4.22, 4.23]. Let us write the staggered fermion action (4.166) in terms of the momentum dependent fields

$$\widetilde{\psi}_p = \sum_x e^{-ix \cdot p} \psi_x \,, \qquad \widetilde{\overline{\psi}}_p = \sum_x e^{ix \cdot p} \overline{\psi}_x \,. \tag{4.187}$$

Choosing the field normalization such that $K = \frac{1}{2}$, $M = am$, we have

$$S = \frac{1}{\Omega} \sum_p \left\{ am(\widetilde{\overline{\psi}}_p \widetilde{\psi}_p) + \sum_{\mu=1}^4 i \sin p_\mu (\widetilde{\overline{\psi}}_{p+\pi_{(\mu)}} \widetilde{\psi}_p) \right\} \,. \tag{4.188}$$

The x-dependent sign factor $\alpha_{x\mu}$ representing the γ-matrices is included here by the momentum four-vector $\pi_{(\mu)}$ with components

$$\pi_{(\mu),\nu} = \begin{cases} \pi & \text{if} \quad \nu < \mu \\ 0 & \text{if} \quad \nu \geq \mu \end{cases} \,. \tag{4.189}$$

According to (4.158) we have

$$\alpha_{x\mu} = (-1)^{x_1 + \cdots x_{\mu-1}} = e^{i\pi_{(\mu)} \cdot x} \,. \tag{4.190}$$

The possible values of the momentum components in lattice units p_μ ($\mu = 1, 2, 3, 4$) are given by (4.118) or (4.119) with the original lattice size L_μ. Owing to periodicity, one can always choose $-\pi \leq p_\mu \leq \pi$. In p_μ the poles of the propagator occur not only near $p_\mu = 0$, but also at the other corners of the Brillouin zone with $p_\mu = 0, \pi$. Therefore let us map 16 subregions in p into the full region $-\pi \leq k_\mu \leq \pi$ by

$$p_\mu = \frac{k_\mu}{2} + p_{H,\mu} \pmod{2\pi} \,. \tag{4.191}$$

Here the momentum p_H with index H belongs to the ordered set of different four-vector indices $H \equiv \{\mu_1, \ldots, \mu_h\}$, $(\mu_1 < \mu_2 < \cdots < \mu_h)$. For $0 \leq h \leq 4$ there are altogether 16 different such sets. The components of p_H are defined by

$$p_{H,\mu} = \begin{cases} \pi & \text{if} \quad \mu \in H \\ 0 & \text{otherwise} \end{cases} \,. \tag{4.192}$$

The summation over the momenta can be written as

$$\sum_p = \sum_k{}' \sum_H \,. \tag{4.193}$$

The prime on the k-sum takes correctly into account that the number of allowed values for k is 16 times smaller than for p.

Let us define in momentum space the field components with index H as

$$\widetilde{\psi}_{Hk} \equiv \frac{1}{4}\widetilde{\psi}_{k/2+p_H} , \qquad \widetilde{\overline{\psi}}_{Hk} \equiv \frac{1}{4}\widetilde{\overline{\psi}}_{k/2+p_H} . \qquad (4.194)$$

In terms of these, the momentum space action (4.188) can be written as

$$S = \frac{1}{\Omega'}\sum_k{}' \sum_{H,H'} \widetilde{\overline{\psi}}_{Hk} \left\{ am\delta_{HH'} + \sum_{\mu=1}^4 i\Gamma_{\mu,HH'} \sin\frac{k_\mu}{2} \right\} \widetilde{\psi}_{H'k} . \qquad (4.195)$$

The four $16 \otimes 16$ matrices Γ_μ ($\mu - 1, 2, 3, 4$) can be transformed by a unitary transformation to the form $\gamma_\mu \otimes \mathbf{1}$ [4.22, 4.23], where the indices of the first direct product factor are, by definition, the Dirac indices ($1 \le \alpha \le 4$) and those of the second factor the flavour indices ($1 \le a \le 4$). On the new basis the action has the form

$$S = \frac{1}{\Omega'}\sum_k{}' \sum_{ab\alpha\beta} \widetilde{\overline{\psi}}_k^{a\alpha} \left\{ am\mathbf{1} \otimes \mathbf{1} + \sum_{\mu=1}^4 i \sin\frac{k_\mu}{2}\gamma_\mu \otimes \mathbf{1} \right\}_{a\alpha,\beta b} \widetilde{\psi}_k^{\beta b} . \qquad (4.196)$$

In the continuum limit, when $k \to 0$, this obviously tends to the free action of four independent Dirac fermions.

In order to compare the two ways of defining the four flavours of staggered fermions, let us also rewrite the action (4.177) in momentum space. Defining the blocked fields in momentum space by

$$\widetilde{q}_k \equiv 4\sum_y{}' e^{-iy\cdot k} q_y , \qquad \widetilde{\overline{q}}_k \equiv 4\sum_y{}' e^{iy\cdot k}\overline{q}_y , \qquad (4.197)$$

we get instead of (4.196)

$$S = \frac{1}{\Omega'}\sum_k{}' \sum_{ab\alpha\beta} \widetilde{\overline{q}}_k^{a\alpha} \Big\{ am\mathbf{1} \otimes \mathbf{1}$$

$$+ \sum_{\mu=1}^4 \left[\frac{i}{2} \sin k_\mu \gamma_\mu \otimes \mathbf{1} + \frac{1}{2}(1 - \cos k_\mu)\gamma_5 \otimes t_5 t_\mu \right] \Big\}_{a\alpha,\beta b} \widetilde{q}_k^{\beta b} . \qquad (4.198)$$

The action in (4.196) has for $m = 0$ the full exact $U(4) \otimes U(4)$ chiral symmetry of four massless flavours, also for finite lattice spacings. The action (4.198), which was originally defined by 2^4 blocking in x-space, has only the $U(1)_o \otimes U(1)_e$ continuous symmetry, because the term proportional to $\gamma_5 \otimes t_5 t_\mu$ breaks $U(4) \otimes U(4)$. However, the inverse propagator in (4.198) is continuous at the edges $|k_\mu| = \pi$ of the Brillouin zone, whereas the one in (4.196) is discontinuous. The discontinuity is possible, because the definition of flavours in momentum space does not correspond to a local definition in coordinate space. (In this respect see also section 4.4.) In fact, the unitary transformation connecting $\widetilde{\psi}_k$ to $\widetilde{q}_k$ is momentum dependent

[4.15, 4.17] and hence non-local in x. Since in the continuum limit the chiral symmetry breaking term in (4.198) vanishes, for zero lattice spacing both flavour bases become equivalent, and the unitary transformation between $\widetilde{\psi}_k$ and $\widetilde{q}_k$ tends to the identity.

4.3.3   Connection to the Dirac–Kähler equation

The introduction of the flavour basis for staggered fermions in the previous subsection seems somewhat *ad hoc*. The construction is based on the block lattice of 2^4 hypercubes, but other decompositions of the fine lattice may come to ones mind, which could lead to a different number of flavours, not just four. The form of the transformation matrix Γ_η in (4.171) may, however, suggest that there is some deep mathematical meaning behind it.

A geometric interpretation of the staggered fermion field on the lattice can be found [4.24, 4.25] on the basis of a differential geometric generalization of the Dirac equation due to Kähler [4.26]. In order to introduce this approach let us briefly repeat the formulation of the continuum Dirac–Kähler equation (for more details see [4.24, 4.27, 4.28]). The Dirac–Kähler equation is a linear field equation formulated in terms of differential forms. The inhomogeneous complex *differential form* Φ is a linear combination of antisymmetric *wedge products* $\mathrm{d}x_\mu \wedge \mathrm{d}x_\nu = -\mathrm{d}x_\nu \wedge \mathrm{d}x_\mu$ of the coordinate differentials:

$$\Phi = \sum_{h=0}^{4} \sum_{\mu_1,\cdots,\mu_h=1}^{4} \frac{1}{h!} \Phi_{\mu_1,\cdots,\mu_h}(x)\, \mathrm{d}x_{\mu_1} \wedge \ldots \wedge \mathrm{d}x_{\mu_h} \equiv \sum_H \Phi(x)_H\, \mathrm{d}x_H \ . \quad (4.199)$$

Here the index H runs, in the same way as in the previous subsection, over the ordered sets of different four-vector indices $H \equiv \{\mu_1,\ldots,\mu_h\}$, $(\mu_1 < \mu_2 < \cdots < \mu_h)$. The *main automorphism* $\mathscr{A}$ of the differential forms is defined by

$$\mathscr{A}\, \mathrm{d}x_H = (-1)^h\, \mathrm{d}x_H \ , \qquad \mathscr{A}(\Phi_1 \wedge \Phi_2) = (\mathscr{A}\Phi_1) \wedge (\mathscr{A}\Phi_2) \ . \quad (4.200)$$

The *contraction operator* $e_\mu\neg$ satisfies

$$e_\mu\neg 1 = 0 \ , \quad e_\mu\neg\, \mathrm{d}x_\nu = \delta_{\mu\nu} \ ,$$

$$e_\mu\neg(\Phi_1 \wedge \Phi_2) = (e_\mu\neg\Phi_1) \wedge \Phi_2 + (\mathscr{A}\Phi_1) \wedge (e_\mu\neg\Phi_2) \ . \quad (4.201)$$

An important notion is the bilinear associative *Clifford product* of differential forms defined by

$$\mathrm{d}x_\mu \vee \mathrm{d}x_\nu = \mathrm{d}x_\mu \wedge \mathrm{d}x_\nu + \delta_{\mu\nu} \ , \quad (4.202)$$

or by

$$\mathrm{d}x_\mu \vee \Phi = \mathrm{d}x_\mu \wedge \Phi + e_\mu\neg\Phi \ . \quad (4.203)$$

The 'exterior differentiation operator' d and 'codifferentiation operator' δ can be constructed as

$$\mathrm{d}\Phi = \mathrm{d}x_\mu \wedge \partial_\mu \Phi \,, \qquad \delta\Phi = -e_\mu \neg \partial_\mu \Phi \,. \tag{4.204}$$

These are nilpotent: $\mathrm{d}^2 = \delta^2 = 0$, and because $(\mathrm{d} - \delta)^2 = -(\mathrm{d}\delta + \delta\mathrm{d}) = \Box$, the operator $(\mathrm{d} - \delta)$ is a square root of the Laplace operator, in the same way as the Dirac operator. According to (4.203) we have

$$(\mathrm{d} - \delta)\Phi = \mathrm{d}x_\mu \vee \partial_\mu \Phi \,, \tag{4.205}$$

and the *Dirac–Kähler equation* with mass m is

$$(\mathrm{d} - \delta + m)\Phi = 0 \,. \tag{4.206}$$

As we shall see, the Dirac–Kähler equation is equivalent to four independent simultaneous Dirac equations with the same mass m. This is due to the equivalence of the definition of the Clifford product in (4.202) with the anticommutation relations defining the algebra of Dirac matrices γ_μ (see in (8.6)). In order to derive the explicit connection let us define

$$\gamma_H \equiv \gamma_{\mu_1} \cdots \gamma_{\mu_h} \,, \tag{4.207}$$

and introduce a new basis of differential forms $Z_a^{(b)}$ $(a, b = 1, 2, 3, 4)$. In a $4 \otimes 4$ matrix notation this is

$$\left(Z_a^{(b)} \right) \equiv Z \equiv \sum_H (\gamma_H)^T (-1)^{h(h-1)/2} \, \mathrm{d}x_H \,. \tag{4.208}$$

On this basis the left regular representation of the Clifford algebra of differential forms is decomposed into four copies of irreducible representations of the γ-algebra. This can be seen because

$$\mathrm{d}x_\mu \vee Z = \gamma_\mu^T Z \,, \qquad Z \vee \mathrm{d}x_\mu = Z\gamma_\mu^T \,. \tag{4.209}$$

Decomposing the general differential form Φ on this basis as

$$\Phi = \sum_H \Phi(x)_H \, \mathrm{d}x_H = \sum_{a,b} \Psi(x)_a^{(b)} Z_a^{(b)} \,, \tag{4.210}$$

the relation between the old and new components is

$$\Phi(x)_H = \mathrm{Tr}\left(\Psi(x)\gamma_H^+ \right) \,, \qquad \Psi(x)_a^{(b)} = \frac{1}{4} \sum_H \Phi(x)_H (\gamma_H)_{ab} \,. \tag{4.211}$$

It can be easily seen that the Dirac–Kähler equation (4.206) is transformed on this new basis into

$$\left(\gamma_\mu \partial_\mu + m \right) \Psi(x)^{(b)} = 0 \,. \tag{4.212}$$

Since this holds for $b = 1, 2, 3, 4$, we have the above announced equivalence to four simultaneous degenerate Dirac equations.

The advantage of considering the Dirac–Kähler equation is that there is a natural way to find a lattice approximation to a field theory which is formulated in terms of differential forms. The correspondence between continuum and lattice equations is provided by the known mapping of differential forms on to linear functions on the space of some lattice elements [4.29]. In this context the lattice is considered as a *cell complex*. The 'cells' are h-dimensional elements of the lattice: $h = 0, 1, 2, 3, 4$ corresponds, respectively, to a point, link, plaquette, cube and hypercube. An h-cell $C_{x,H}^{(h)}$ can be indexed by $(x, \mu_1, \dots, \mu_h) \equiv (x, H)$, where x is a point index and, as usual, H is an ordered set of different four-vector indices $H \equiv \{\mu_1, \dots, \mu_h\}$ ($\mu_1 < \mu_2 < \cdots < \mu_h$). An '$h$-chain' is a formal linear combination of h-cells:

$$C^{(h)} \equiv \sum_{x,H} \alpha_{x,H} C_{x,H}^{(h)} . \tag{4.213}$$

A *chain* C is any h-chain. An 'h-cochain' is a linear functional defined on h-chains: $\Phi^{(h)}(C^{(h)})$. A differential form is represented on the lattice by a *cochain* $\Phi(C)$. The 'boundary of a cell' $\triangle C_{x,H}^{(h)}$ and 'coboundary of a cell' $\nabla C_{x,H}^{(h)}$ are defined by their intuitive geometric meaning [4.24]. These are linear operators acting on chains. The corresponding operators acting on the dual space are the 'dual boundary' $\check{\triangle}$ and 'dual coboundary' $\check{\nabla}$ defined, respectively, by

$$(\check{\triangle}\Phi)(C) \equiv \Phi(\triangle C) , \qquad (\check{\nabla}\Phi)(C) \equiv \Phi(\nabla C) . \tag{4.214}$$

The dual boundary operator corresponds to the exterior differentiation d of differential forms, the dual coboundary to the codifferentiation δ, therefore the Dirac–Kähler equation (4.206) on the lattice is

$$(\check{\triangle} - \check{\nabla} + m)\Phi = 0 . \tag{4.215}$$

The cochain Φ in this equation can be represented by a linear combination of the basic cochains $\mathrm{d}x_{x,H}$ defined by

$$\mathrm{d}x_{x,H}[x', H'] \equiv \delta_{xx'}\delta_{HH'} . \tag{4.216}$$

Therefore we have

$$\Phi = \sum_{x,H} \Phi_{x,H}\, \mathrm{d}x_{x,H} . \tag{4.217}$$

The expansion coefficients $\Phi_{x,H}$ are the components of the staggered fermion field in the Dirac–Kähler representation, which can either be associated with the cells (x, H) of the lattice or, in an obvious way, with the points of a 2^4 hypercube situated at x. The appearance of the index H in three different contexts is remarkable: either it denotes the corners of the Brillouin zone in momentum space as in (4.191), or the types of cells of the lattice, or the points of a 2^4 hypercubic block.

4.4　　Nielsen–Ninomiya theorem and mirror fermions

A remarkable feature of lattice regularization is the appearance of several fermion species per fermion field in the lattice action. In the case of the naive discretization of the Dirac fermion action in four dimensions 16 fermion species appear. By an appropriate modification of the action one can achieve that 15 'fermion doublers' become infinitely heavy in the continuum limit (see section 4.2). Another way of reducing the number of lattice fermions is to put only a single field component on every lattice site (see section 4.3). Up to now only free fermions were considered, therefore the question arises, whether the removal of the extra fermion doubler species is also successful after introducing some interactions. In some cases, namely in the electroweak sector of the standard model, the left- and right-handed chiral fermion components have to be treated differently. For instance, a massless neutrino in the minimal standard model does not have a right-handed component at all. The appearance of the lattice fermion doublers is obviously a potential difficulty for the lattice regularization of the chiral electroweak sector (see chapter 6).

The Wilson and staggered lattice fermions are, of course, only two simple examples for the discretization of the fermion field. Therefore one has to consider the properties of lattice fermion propagators, especially the possibility of removing the doublers, more generally.

4.4.1　　*Doublers in lattice fermion propagators*

In the case of the naive fermion action (4.154) the existence of the doublers is due to the *spectrum doubling symmetry* [4.30, 4.31, 4.12]. This symmetry is defined by the transformations

$$\Psi'_x = e^{-ix\cdot\pi_H} M_H \Psi_x \,, \qquad \overline{\Psi}'_x = \overline{\Psi}_x M_H^+ e^{ix\cdot\pi_H} \,, \qquad (4.218)$$

where $\pi_H \equiv p_H$ is defined in (4.192) and the $4 \otimes 4$ matrices M_H for $H = \{\mu_1, \ldots, \mu_h\}$ are given by

$$M_H \equiv M_{\mu_1} M_{\mu_2} \cdots M_{\mu_h} \,, \qquad M_\mu \equiv i\gamma_5\gamma_\mu = M_\mu^+ = M_\mu^{-1} \,. \qquad (4.219)$$

Using the relation

$$M_H^+ \gamma_\mu M_H = e^{\pm i\pi_{H,\mu}} \gamma_\mu \,, \qquad (4.220)$$

one can easily see that the transformation in (4.218) is a symmetry of the naive lattice action (4.154). In momentum space (with the Fourier transformation defined according to (8.38)) the transformation (4.218) exchanges the corners of the Brillouin zone:

$$\widetilde{\Psi}'_p = M_H \widetilde{\Psi}_{p+\pi_H} \,, \qquad \widetilde{\overline{\Psi}}'_p = \widetilde{\overline{\Psi}}_{p+\pi_H} M_H^+ \,. \qquad (4.221)$$

The proliferation of fermion species in the case of naive lattice fermions gives rise to a representation of the spectrum doubling symmetry group. This representation is, however, not irreducible. Its reduction by the 'spin diagonalization' led in section 4.3 to the staggered fermions (see also [4.12]). The rôle of the Wilson term in the Wilson action is to break the spectrum doubling symmetry.

In the case of more general lattice fermion actions the spectrum doubling symmetry is, in general, also broken. The question is, what can be said about the doubling under some reasonable general assumptions. In order to answer this question let us first start with different sets of assumptions about the lattice fermion action, in order to see their consequences. Let us first consider a general lattice fermion action in Euclidean momentum space:

$$S = \frac{1}{\Omega} \sum_p \widetilde{\overline{\Psi}}_p F(p) \widetilde{\Psi}_p .$$ (4.222)

The $4 \otimes 4$ matrix $F(p)$ is the inverse fermion propagator (for examples see sections 4.2 and 4.3). Its zeros correspond to the poles of the propagator, therefore specify the fermionic particle content of the theory. About $F(p)$ let us assume that the following properties hold:

Reflection positivity: in momentum space it follows from (4.99) that

$$\Theta \widetilde{\Psi}_{\mathbf{p}p_4} = \widetilde{\overline{\Psi}}_{\mathbf{p}-p_4} \gamma_4 , \qquad \Theta \widetilde{\overline{\Psi}}_{\mathbf{p}p_4} = \gamma_4 \widetilde{\Psi}_{\mathbf{p}-p_4} .$$ (4.223)

According to (4.92) this implies

$$F(p) = \gamma_4 F^+(\mathbf{p}, -p_4) \gamma_4 .$$ (4.224)

Invariance under the cubic group: the cubic symmetry of the action on the hypercubic lattice implies that (4.224) holds for $4 \to \mu = 1, 2, 3, 4$:

$$F(p) = \gamma_\mu F^+(R_{(\mu)}p) \gamma_\mu ,$$ (4.225)

with $(R_{(\mu)}p)_v = p_v(1 - 2\delta_{\mu v})$.

Chiral invariance: in the massless case the transformation

$$\widetilde{\Psi}'_p = e^{-i\alpha_5\gamma_5} \widetilde{\Psi}_p , \qquad \widetilde{\overline{\Psi}}'_p = \widetilde{\overline{\Psi}}_p e^{-i\alpha_5\gamma_5} ,$$ (4.226)

with real α_5, leaves the action invariant. This is equivalent to

$$F(p) = -\gamma_5 F(p) \gamma_5 .$$ (4.227)

Applying (4.225) subsequently for $\mu = 1, 2, 3, 4$, and comparing with (4.227) we get

$$F(p) = -F(-p) .$$ (4.228)

(Note that this assumption is not satisfied by the Wilson fermion action, which explicitly breaks chiral symmetry. Later on we shall also consider more general sets of assumptions.)

Locality: the interaction in coordinate space vanishes sufficiently fast for large distances, implying the continuity of the inverse propagator in momentum space $F(p)$.

Since for periodic or antiperiodic boundary conditions $F(p)$ is a periodic function with periods $p_\mu \to p_\mu + 2\pi$, $\mu = 1, 2, 3, 4$, the property (4.228) implies that $F(p)$ has to vanish for every $p_\mu = 0, \pm\pi$. Therefore, under the above assumptions there are always at least 16 poles in the fermion propagator at the corners of the Brillouin zone. In general, the poles of the propagator imply the existence of physical particles which, in the absence of some selection rules due to symmetries, can be pair-produced by the interactions. The same conclusions can even be achieved under weaker assumptions, for instance in the case of some non-local interactions, too (see [4.32]). Note that this derivation does not apply to staggered fermions, because $F(p)$ has been assumed to be a $4 \otimes 4$ matrix.

An interesting question, particularly in the context of the lattice formulation of the electroweak standard model, is the *chirality* of the lattice fermion doublers. For simplicity, let us also assume invariance with respect to space-reflection (*P*-parity) [A1]:

$$P\widetilde{\Psi}_p = \sigma_P \gamma_4 \widetilde{\Psi}_{-\mathbf{p}p_4} , \qquad \widetilde{\overline{\Psi}}_p P = \sigma_P^* \widetilde{\overline{\Psi}}_{-\mathbf{p}p_4} \gamma_4 , \qquad (4.229)$$

with $\sigma_P \sigma_P^* = 1$. Combined with the above assumptions this implies that the inverse propagator $F(p)$ can be written as

$$F(p) = \sum_{\mu=1}^{4} \mathrm{i}f_\mu(p)\gamma_\mu = \sum_{\mu=1}^{4} \mathrm{i}f_\mu(p)\gamma_\mu(P_L + P_R) , \qquad (4.230)$$

with real $f_\mu(p)$. According to (8.12), P_L and P_R project, respectively, on left-handed and right-handed chirality. As an example, the free naive propagator in (4.154) corresponds to $f_\mu(p) = \sin p_\mu$.

From the above locality assumption, it follows that $f_\mu(p)$ defines a continuous vector field on the four-dimensional hypertorus $T_4 = S_1 \otimes S_1 \otimes S_1 \otimes S_1$. A particle corresponds to a zero of this vector field: $f_1(\overline{p}) = f_2(\overline{p}) = f_3(\overline{p}) = f_4(\overline{p}) = 0$. Since, due to the assumed chiral invariance, the chiral parts of $F(p)$ in (4.230) are independent (the propagator does not change chirality), one can consider a fermion field with only one, say L-handed, chirality. In the vicinity of the zero the inverse propagator behaves as

$$F(p) = \sum_{\mu,\nu=1}^{4} \mathrm{i}\gamma_\mu J_{\mu\nu}(\overline{p})(p_\nu - \overline{p}_\nu) + O(p - \overline{p})^2 , \qquad (4.231)$$

where J is the derivative matrix: $J_{\mu\nu} \equiv \partial f_\mu / \partial p_\nu$. In the continuum limit $a \to 0$, the factor multiplying $\mathrm{i}\gamma_\mu$ is the physical momentum in lattice units: ak_μ. This means that one has to perform a transformation from the old momenta p_μ to the physical momenta k_μ. This transformation preserves chirality for $\det J > 0$, but it changes chirality for $\det J < 0$. The simplest example of a zero with $\det J > 0$ in the left-handed inverse propagator is

$$i\gamma_\mu(p_\mu - \bar{p}_\mu)P_L \; . \qquad (4.232)$$

In this case, between p_μ and k_μ no coordinate rotation is needed at all, and the corresponding pole of the propagator obviously describes a left-handed chiral fermion. A simple example of a zero with $\det J < 0$ is

$$i\gamma_\mu(-1)^{\delta_{\mu 1}}(p_\mu - \bar{p}_\mu)P_L \; . \qquad (4.233)$$

The sign factor can be transformed away by the spectrum doubling transformation M_1 in (4.219), but due to

$$M_1^+ i\gamma_\mu(-1)^{\delta_{\mu 1}}P_L M_1 = i\gamma_\mu P_R \; , \qquad (4.234)$$

the chirality becomes opposite. This zero of the inverse propagator describes a right-handed chiral fermion. It is clear from this that the spectrum doubling transformations M_H for $h = odd$ change the chirality, the ones with $h = even$ preserve it.

We see that the number of left-handed and right-handed fermions is determined by the number of zeros with *index +1* (det > 0), and *index -1* (det < 0). The Poincaré–Hopf theorem implies that the sum of indices of the zeros of a continuous vector field on a compact manifold is equal to the Euler-number of the manifold. For the hypertorus T_4 the Euler-number is zero, therefore [4.33] under the above assumptions *there is always an equal number of left- and right-handed particles in the lattice fermion propagator.*

The number 16 derived above for the doublers implies eight left-handed and eight right-handed fermions, but this is due to the assumed cubic symmetry. Keeping the form (4.230) for the inverse propagator, but releasing the requirement of cubic symmetry, one can achieve also a single pair of left-handed and right-handed fermions (i. e. a single Dirac field). A simple example is given in [4.34]:

$$f_{1,2,3} = \sin p_{1,2,3} \; , \quad f_4 = \sin p_4 + \lambda \sum_{n=1}^{3} \sin^2 \frac{p_n}{2} \; , \qquad (4.235)$$

with positive λ. Of course, the loss of cubic invariance is a disadvantage. For the smaller number of propagating fermions one has to pay by a larger number of bare parameters to tune in the lattice action.

The above simple proof for the existence of an equal number of left-handed and right-handed fermions in lattice fermion propagators can be generalized, that is the assumptions can be less stringent. In particular, the chirality and/or space reflection symmetry can be omitted from the set of assumptions. The latter is important in the electroweak sector where parity is not conserved. A very general proof of the existence of an equal number of left-handed and right-handed fermions in lattice fermion propagators was given by Nielsen and Ninomiya [4.35] in the Hamiltonian formulation with Weyl fermions (no parity symmetry assumed). The assumptions about the Hamiltonian

$$H = \sum_{\mathbf{x},\mathbf{y}} \chi_{\mathbf{x}}^+ H(\mathbf{x} - \mathbf{y})\chi_{\mathbf{y}} \qquad (4.236)$$

are:

- *Translation invariance*: H is a function of $\mathbf{x} - \mathbf{y}$ only;

- *Locality*: $H(\mathbf{x} - \mathbf{y})$ vanishes fast enough for large $|\mathbf{x} - \mathbf{y}|$ in such a way that its Fourier transform has continuous derivatives;

- *Hermiticity* of H.

It is further assumed that there is a set of exactly conserved charges Q in the theory, defining the quantum numbers of fermions, such that the charges are

- locally defined (given by a sum of a density);

- have discrete eigenvalues ('quantum numbers');

- are bilinear in the fermion fields.

In this case the *Nielsen–Ninomiya theorem* tells us that in the fermion propagator *there are an equal number of left-handed and right-handed particles for every set of quantum numbers.*

The proofs given in [4.35] are based on homotopy theory (algebraic topology) or differential topology. Another proof in [4.36], emphasizing the connection to the conservation of chiral charge, uses differential geometry.

As an example of the implications of the Nielsen–Ninomiya theorem let us consider the *hypercharge* quantum number Y in the electroweak theory, which occurs in the definition of the electric charge Q_{em} in connection with the third component of the left-handed isospin T_{L3}:

$$Q_{em} = T_{L3} + \frac{1}{2}Y \ . \qquad (4.237)$$

For instance, in the lepton doublet consisting of the neutrino v and electron e, we have

$$Y(v_L) = Y(e_L) = -1 \ ; \qquad Y(e_R) = -2 \ . \qquad (4.238)$$

According to the Nielsen–Ninomiya theorem, on the lattice there are always fermions with opposite chirality, that is there exists a doublet (N, E) with

$$Y(N_R) = Y(E_R) = -1 \ ; \qquad Y(E_L) = -2 \ . \qquad (4.239)$$

Since the particles (E, N) have the same quantum numbers as the leptons (v, e), but they have opposite chirality, one can call them *mirror leptons*. The content of the Nielsen–Ninomiya theorem is that on the lattice *the fermion spectrum consists of fermion–mirror-fermion pairs*.

The Nielsen–Ninomiya theorem holds under rather general assumptions, but one has to keep in mind that it refers to fermion propagation on the lattice. It does not say anything neither about the behaviour of masses in the continuum limit nor about couplings. Therefore, one way to escape its physical implications is to *decouple* the superfluous fermion states from the set of physical particles. This can be achieved, for instance in QCD, by giving the fermion doublers a mass proportional to the cut-off. In chiral models of the electroweak sector such a decoupling is usually very difficult to arrange without symmetries. Nevertheless, sometimes it can be achieved in the continuum limit, due to the *Golterman–Petcher shift symmetry* of some lattice fermion actions [4.37]. This will be discussed in more detail in section 6.2.

The phenomenon of fermion doubling on the lattice has far-reaching consequences: it is crucial in the description of the axial anomaly (see next subsection), it plays an important rôle in reproducing the spontaneously broken chiral symmetry in QCD (see section 5.3), and it is a serious obstacle for defining the minimal electroweak Standard Model on the lattice (see section 6.2).

4.4.2   Doublers and the axial anomaly

We have seen in the previous subsection that, under rather mild assumptions about the lattice action, there are always an equal number of left- and right-handed fermion poles in the lattice fermion propagator. The *axial charge* Q_5 is defined in the continuum field theory by the axial vector current

$$J_5(x)_\mu \equiv \overline{\psi}(x)\gamma_\mu\gamma_5\psi(x) \,, \tag{4.240}$$

as the integral of the density:

$$Q_5 \equiv \int d^3x \, J_5(\mathbf{x}, t)_4$$

$$= \int d^3x \, \overline{\psi}(\mathbf{x}, t)\gamma_4\gamma_5\psi(\mathbf{x}, t) = \int d^3x \, \psi^+(\mathbf{x}, t)\gamma_5\psi(\mathbf{x}, t) \,. \tag{4.241}$$

The vector current and (vector) charge Q is defined similarly:

$$J(x)_\mu \equiv \overline{\psi}(x)\gamma_\mu\psi(x) \,,$$

$$Q \equiv \int d^3x \, J(\mathbf{x}, t)_4$$

$$= \int \mathrm{d}^3 x \, \overline{\psi}(\mathbf{x}, t) \gamma_4 \psi(\mathbf{x}, t) = \int \mathrm{d}^3 x \, \psi^+(\mathbf{x}, t) \psi(\mathbf{x}, t) . \qquad (4.242)$$

According to (4.241) Q_5 is the number of right-handed fermions minus the number of left-handed fermions minus the number of right-handed antifermions plus the number of left-handed antifermions (see (8.11)). For massless fermions Q_5 is conserved in the classical field theory, not only in the free case, but also in interacting theories like electrodynamics or σ-models. The classical chiral symmetry is, however, broken in the quantum field theory by the *chiral anomaly* [4.38, 4.39, 4.40]. For instance, in massless QED the divergence of the axial vector current is given by

$$\partial_\mu J_5(x)_\mu = \frac{g^2}{16\pi^2} \epsilon_{\mu\nu\rho\sigma} F(x)_{\mu\nu} F(x)_{\rho\sigma} , \qquad (4.243)$$

where $F(x)_{\mu\nu} = \partial_\mu A(x)_\nu - \partial_\nu A(x)_\mu$ is the field strength tensor. There is no way to avoid the physical consequences of the axial anomaly. One can also define a conserved axial vector current, but it is not gauge invariant, and the physical consequences are unchanged [A3]. In perturbation theory the axial anomaly appears as a consequence of the linear divergence in the triangle graph with an internal fermion loop, and either three axial vector current insertions or one axial vector and two vector current insertions. More generally, one can say that in the effective action the 1-loop quantum corrections to a chirally symmetric classical action are not chiral symmetric.

It is an interesting question, how the axial anomaly appears in the lattice regularization. In general, the fermion doublers do contribute to the anomaly. Since the doublers appear with opposite chiralities, one could superficially conclude that on the lattice the anomaly is zero. In order to decide this question one first has to define the vector and axial vector currents on the lattice. For definiteness, let us consider here a Wilson lattice fermion in an external ('classical') electromagnetic field

$$U_{x\mu} = \exp\{ig A_\mu(x)\} , \qquad (4.244)$$

defined on the links $(x, x+\hat{\mu})$. With real $A_\mu(x)$ the gauge field variable $U_{x\mu}$ is an element of a compact U(1) group (g is the gauge coupling constant). According to (4.85) the lattice action is

$$S = \sum_x a^4 \left\{ \left(m + \frac{4r}{a} \right) (\overline{\psi}_x \psi_x) - \frac{1}{2a} \sum_{\mu=\pm 1}^{\pm 4} (\overline{\psi}_{x+\hat{\mu}}[r + \gamma_\mu] U_{x\mu} \psi_x) \right\} . \qquad (4.245)$$

Here the external gauge field on negative direction links is defined by

$$U_{x,-\mu} \equiv U^*_{x-\hat{\mu},\mu} . \qquad (4.246)$$

The above action is locally gauge invariant with respect to the transformations

$$\psi'_x = e^{-i\alpha_x}\psi_x \,, \quad \overline{\psi}'_x = \overline{\psi}_x e^{i\alpha_x} \,, \quad U'_{x\mu} = e^{-i\alpha_{x+\hat{\mu}}}U_{x\mu}e^{i\alpha_x} \,. \tag{4.247}$$

The global (i. e. rigid) U(1) symmetry acting on the fermion fields is

$$\psi'_x = e^{-i\alpha}\psi_x \,, \quad \overline{\psi}'_x = \overline{\psi}_x e^{i\alpha} \,. \tag{4.248}$$

In the continuum field theory from this global U(1) symmetry one constructs the corresponding Noether current, which is conserved according to Noether's theorem. In the corresponding Euclidean quantum field theory the global symmetry implies Ward–Takahashi identities, which express the consequences of the symmetry in terms of the Green functions containing the composite current operators [A3,A5].

In order to derive the Ward–Takahashi identities in the lattice theory defined by the action S in (4.245), let us write S as

$$S = \sum_x a^4 \left\{ m(\overline{\psi}_x\psi_x) - \frac{1}{2a}\sum_{\mu=1}^{4} \left[(\overline{\psi}_{x+\hat{\mu}}[r + \gamma_\mu]U_{x\mu}\psi_x) \right.\right.$$

$$\left.\left. +(\overline{\psi}_x[r - \gamma_\mu]U^+_{x\mu}\psi_{x+\hat{\mu}}) - 2r(\overline{\psi}_x\psi_x) \right] \right\} \,. \tag{4.249}$$

Let us consider the partition function

$$Z \equiv \int [\mathrm{d}\overline{\psi}\,\mathrm{d}\psi]e^{-S[\psi,\overline{\psi},U]} \,. \tag{4.250}$$

Performing an x-dependent infinitesimal transformation on the fermion fields according to (4.247), we have

$$\psi_x = (1 + i\alpha_x)\psi'_x \,, \quad \overline{\psi}_x = \overline{\psi}'_x(1 - i\alpha_x) \,. \tag{4.251}$$

In the case of a general linear change of integration variables $\psi' = A\psi$, $\overline{\psi}' = \overline{\psi}\,\overline{A}$ the rules of Grassmann integration in (4.21) imply

$$[\mathrm{d}\overline{\psi}\,\mathrm{d}\psi] = \det\overline{A}\det A[\mathrm{d}\overline{\psi}'\,\mathrm{d}\psi'] \,. \tag{4.252}$$

In the present case, up to first order in α_x, the product of determinants is equal to 1, therefore introducing the new integration variables in (4.250) gives

$$Z \equiv \int [\mathrm{d}\overline{\psi}'\,\mathrm{d}\psi']e^{-S[\psi',\overline{\psi}',U]} \left\{ 1 - i\sum_x \alpha_x \left[\frac{S\overleftarrow{\partial}}{\partial\psi'_x}\psi'_x - \overline{\psi}'_x\frac{\overrightarrow{\partial}S}{\partial\overline{\psi}'_x} \right] \right\} + O(\alpha_x^2) \,. \tag{4.253}$$

Omitting the primes on the integration variables and subtracting (4.250), due to the arbitrariness of α_x, we get

$$\left\langle \Delta_\mu^b J_{x\mu} \right\rangle = 0 \,. \tag{4.254}$$

Here a summation over $\mu = 1, 2, 3, 4$ is understood, the backward lattice derivative in (8.42) is used, and the *vector current* $J_{x\mu}$ is defined as

$$J_{x\mu} \equiv \frac{1}{2} \left\{ (\overline{\psi}_{x+\hat{\mu}}[r + \gamma_\mu] U_{x\mu} \psi_x) - (\overline{\psi}_x [r - \gamma_\mu] U_{x\mu}^+ \psi_{x+\hat{\mu}}) \right\} \,. \tag{4.255}$$

The *lattice Ward–Takahashi identity* for the vector current in (4.254) is the consequence of the global symmetry (4.248) in the quantized theory. In the continuum limit it implies the exact conservation of the vector current operator and of the electric charge. Note that in (4.250) one could also consider an integrand containing an additional arbitrary function $F[\psi, \overline{\psi}, U]$, which occurs in the expectation value $\langle F \rangle$. By the same procedure one could then derive a similar relation as (4.254), also in this more general case.

Let us now turn to the global chiral symmetry

$$\psi_x' = e^{-i\alpha_5\gamma_5} \psi_x \,, \qquad \overline{\psi}_x' = \overline{\psi}_x e^{-i\alpha_5\gamma_5} \,, \tag{4.256}$$

which is an exact symmetry in the massless continuum theory. The term proportional to r in the Wilson lattice fermion action, however, breaks chiral symmetry explicitly. Similarly to the case of the vector current in (4.251), let us introduce new fermionic variables in the functional integral by an x-dependent infinitesimal chiral transformation

$$\psi_x = (1 + i\alpha_{5x}\gamma_5)\psi_x' \,, \qquad \overline{\psi}_x = \overline{\psi}_x'(1 + i\alpha_{5x}\gamma_5) \,. \tag{4.257}$$

Instead of (4.253) we now have

$$Z \equiv \int [d\overline{\psi}' \, d\psi'] e^{-S[\psi',\overline{\psi}',U]} \left\{ 1 - i \sum_x \alpha_{5x} \left[\frac{S \overleftarrow{\partial}}{\partial \psi_x'} \gamma_5 \psi_x' + \overline{\psi}_x' \gamma_5 \frac{\overrightarrow{\partial} S}{\partial \overline{\psi}_x'} \right] \right\} \,. \tag{4.258}$$

The *axial vector current* $J_{x\mu}^5$ can be defined as

$$J_{x\mu}^5 \equiv \frac{1}{2} \left\{ (\overline{\psi}_{x+\hat{\mu}} \gamma_\mu \gamma_5 U_{x\mu} \psi_x) + (\overline{\psi}_x \gamma_\mu \gamma_5 U_{x\mu}^+ \psi_{x+\hat{\mu}}) \right\} \,. \tag{4.259}$$

The *anomalous Ward–Takahashi identity* for the axial vector current on the lattice follows from (4.258) as

$$\left\langle \Delta_\mu^b J_{x\mu}^5 - 2am(\overline{\psi}_x \gamma_5 \psi_x) - r \sum_{\mu=1}^{4} \left\{ 2(\overline{\psi}_x \gamma_5 \psi_x) - \frac{1}{2} \left[(\overline{\psi}_{x+\hat{\mu}} \gamma_5 U_{x\mu} \psi_x) \right. \right. \right.$$

$$+(\overline{\psi}_x\gamma_5 U^+_{x\mu}\psi_{x+\hat\mu}) + (\overline{\psi}_x\gamma_5 U_{x-\hat\mu,\mu}\psi_{x-\hat\mu}) + (\overline{\psi}_{x-\hat\mu}\gamma_5 U^+_{x-\hat\mu,\mu}\psi_x)\Big]\Big\}\Big\rangle = 0 \ .$$

$$(4.260)$$

Note that a term proportional to r, which is analogous to a term included in the vector current (4.255), is not included here in the definition of $J^5_{x\mu}$. In the case of the vector current the motivation was to have an exact conservation already for finite lattice spacings (not only in the continuum limit). Since the axial vector current is not conserved anyway, it is better to choose its definition by simplicity.

In order to investigate the continuum limit of the anomalous Ward–Takahashi identity, let us write it in a form where the left hand side tends to the expectation value of the divergence of the axial vector current:

$$\left\langle a^{-1}\Delta^b_\mu J^5_{x\mu}\right\rangle = 2m\left\langle\overline{\psi}_x\gamma_5\psi_x\right\rangle + \left\langle X_x\right\rangle \ . \qquad (4.261)$$

Here we use the notation

$$X_x \equiv \frac{r}{a}\sum_{\mu=1}^{4}\left\{2(\overline{\psi}_x\gamma_5\psi_x) - \frac{1}{2}\Big[(\overline{\psi}_{x+\hat\mu}\gamma_5 U_{x\mu}\psi_x)\right.$$

$$+(\overline{\psi}_x\gamma_5 U^+_{x\mu}\psi_{x+\hat\mu}) + (\overline{\psi}_x\gamma_5 U_{x-\hat\mu,\mu}\psi_{x-\hat\mu}) + (\overline{\psi}_{x-\hat\mu}\gamma_5 U^+_{x-\hat\mu,\mu}\psi_x)\Big]\Big\} \ . \quad (4.262)$$

The form of the axial vector Ward–Takahashi identity in (4.261) also makes clear that the Wilson term in the action, proportional to r, breaks the chiral symmetry even for massless fermions ($m = 0$). Using a formula analogous to (4.131) one can express the right hand side of (4.261) in terms of the matrix elements of the full fermion propagator in the background gauge field. In the continuum limit $a \to 0$ one can show [4.31, 4.41, 4.42], that if the mass and the typical momentum q carried by the external gauge field satisfy

$$\frac{r}{a} \gg m, |q| \ , \qquad (4.263)$$

then for $r > 0$ the limit of the triangle graph contributions to the last term in (4.261), which arise from the fermion doubler regions, reproduces the correct axial anomaly in (4.243):

$$\lim_{a\to 0} X_x = \frac{g^2}{16\pi^2}\epsilon_{\mu\nu\rho\sigma}F(x)_{\mu\nu}F(x)_{\rho\sigma} \ . \qquad (4.264)$$

The same happens also in other theories with Abelian and non-Abelian external and quantized gauge fields, and with Wilson or staggered fermions. For instance, for the non-Abelian anomaly see [4.43, 4.44].

A physical understanding of the appearance of the axial anomaly in lattice regularization can be achieved in the Hamiltonian approach [4.45].

In a continuum regularization the dynamical origin of the axial anomaly can be explained by the point split axial current definition. The creation of a pseudoscalar fermion–antifermion bound state by the straightforward process of increasing the energy of a fermion in the Dirac sea above the sea level is forbidden, because the chiralities of the fermion and antifermion in the pseudoscalar state are opposite (and the axial charges Q_5 are the same). This means that the axial charge would change during this process. However, since in field theory the Dirac sea has no bottom (there is an infinity of energy levels), the creation of the pseudoscalar state can proceed by collectively shifting up in the Dirac sea the energies of one chirality and at the same time shifting down the energies of the other chirality. The well-defined point-split axial vector current is, however, insensitive to the energy levels near minus infinity, therefore it is insensitive to the compensation of the axial charge at the bottom of the Dirac sea. That is why it is not conserved. On the lattice with Wilson fermions the Wilson term suppresses pair production at the bottom of the Dirac sea, the contributions from $p = \pm\pi/a$ cannot cancel the contributions from the physical region at $p = 0$. The contribution of the physical region in the continuum limit becomes the same as in continuum regularizations.

4.5   QED on the lattice

Quantum Electrodynamics (QED) in the framework of renormalized perturbation theory is very successful in describing the electromagnetic interactions of electrons, muons etc. to a high precision [A2]. It is a part of the standard electroweak model, which describes all phenomena of both electromagnetic and weak interactions in the presently known energy range up to about 100 GeV. Therefore, the non-perturbative lattice study of QED is not motivated by some unexplained phenomena or by the lack of theoretical tools for the extraction of numerical predictions from the basic equations of the theory. The motivation at present is to try to improve the theoretical understanding of the general mathematical properties of this type of quantum field theories.

An apparent mathematical inconsistency in QED is the existence of the so-called 'Landau pole' in the perturbative behaviour of the renormalized coupling constant as a function of the cut-off [4.46]. The Callan–Symanzik β-function is defined [A3] as

$$\beta(\alpha) = -\Lambda \left(\frac{\partial \alpha}{\partial \Lambda} \right)_{e,m_R} . \tag{4.265}$$

Here α is the renormalized fine structure constant, Λ is the cut-off, and the derivative is taken at fixed bare coupling e and renormalized mass m_R.

The lowest orders in perturbation theory (up to four loops) in QED with one fermion species in the $\overline{\text{MS}}$ scheme are given by [4.47]

$$\beta(\alpha) \equiv \alpha \sum_{\nu=1}^{\infty} \beta_\nu \alpha^\nu \;,$$

$$\beta_1 = \frac{2}{3\pi} \;, \qquad \beta_2 = \frac{1}{2\pi^2} \;, \qquad \beta_3 = -\frac{31}{144\pi^3} \;,$$

$$\beta_4 = -\left(\frac{5570}{243} + \frac{832}{9}\zeta(3)\right)\frac{1}{128\pi^4} \;. \tag{4.266}$$

The dependence of α on the cut-off is obtained from the differential equation

$$\frac{d\alpha}{d\log\Lambda} = -\beta(\alpha) \;. \tag{4.267}$$

Let us consider here, for simplicity, only the one-loop approximation to the β-function (proportional to β_1). In this case the solution with $\alpha_0 \equiv e^2/(4\pi)$ is

$$\alpha\left(\frac{\Lambda}{m_R}\right) = \frac{\alpha_0}{1 + \alpha_0\beta_1 \log\left(\Lambda/m_R\right)} \;. \tag{4.268}$$

Therefore, in this approximation, the infinite cut-off limit of α is zero for any bare coupling α_0. The continuum theory is consistent only for vanishing renormalized coupling. In other words, according to the 1-loop β-function, the continuum limit of QED is trivial, similarly to the case of scalar ϕ^4 theory (see chapter 2). The two-loop contribution proportional to β_2 does not change this conclusion qualitatively.

Instead of the fine structure constant α on the lattice one usually considers the renormalized coupling squared e_R^2 defined by

$$e_R^2 \equiv 4\pi\alpha \;. \tag{4.269}$$

The β-function also determines the change of e_R^2 as a function of the renormalization scale μ ('running coupling constant', see section 1.7). The differential equation for $e_R^2(\mu)$ is, with $\beta_{e^2} \equiv 4\pi\beta$,

$$\frac{de_R^2(\mu)}{d\log\mu} = \beta_{e^2}(e_R^2(\mu)) \;. \tag{4.270}$$

In the 1-loop approximation we have

$$e_R^2(\mu) = \frac{e_R^2(\mu_0)}{1 - e_R^2(\mu_0)(\beta_1/4\pi)\log\left(\mu/\mu_0\right)} \;. \tag{4.271}$$

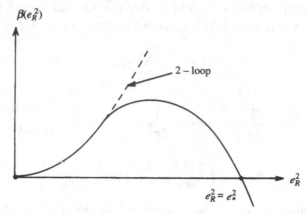

Fig. 4.7. The behaviour of the Callan–Symanzik β-function in QED if there is a non-trivial ultraviolet fixed point at $e_R^2 = e_*^2$. The dashed line qualitatively shows the 2-loop β-function.

This shows that, in this approximation, $e_R^2(\mu)$ becomes infinite at the scale

$$\mu_{\text{Landau}} = \mu_0 \exp\left\{\frac{4\pi}{\beta_1 e_R^2(\mu_0)}\right\} , \qquad (4.272)$$

if it is equal to $e_R^2(\mu_0)$ at the scale μ_0. The position of the *Landau pole* on the right hand side of (4.271) is changed by the 2-loop contribution to

$$\mu_{\text{Landau}} = \mu_0 \left[\frac{\beta_2}{4\pi\beta_1}e_R^2(\mu_0)\right]^{\beta_2/\beta_1^2} \exp\left\{\frac{4\pi}{\beta_1 e_R^2(\mu_0)}\right\} \left(1 + O[e_R^2(\mu_0)]\right) .$$
$$(4.273)$$

The higher order corrections are contained here in the last parentheses. Substituting here the physical value of the renormalized coupling $(e_R^2(\mu_0) = 4\pi/137)$ one obtains a very very high scale. Therefore, in QED the mathematically inconsistent energy range is very far away from any reasonable scale.

The appearance of the Landau pole shows a mathematical inconsistency of renormalized perturbation theory in QED. This inconsistency can be resolved if the full β-function qualitatively behaves as shown in fig. 4.7. In this case the zero of the β-function at $e_R^2 = e_*^2$ is an *ultraviolet stable fixed point* (UVFP). This means that the solution of the differential equation (4.270) for the running coupling constant $e_R^2(\mu)$ always tends to e_*^2 for $\mu \to \infty$. The zero of the Callan–Symanzik β-function in (4.267) implies that it is possible to tune the bare coupling α_0 near $\alpha_{0*} \equiv e_*^2/(4\pi)$ in such a way that for infinite cut-off α has an arbitrary finite limit. Therefore, if a UVFP at e_*^2 exists, the continuum limit is non-trivial. Note that the 3-loop β-function looks actually like fig. 4.7, due to the opposite sign of the

3-loop term in (4.266). Nevertheless, the 3-loop zero is at such large value of e_R^2, that the perturbative approximation is not reliable there. Besides, the 3-loop contribution does also depend on the regularization scheme, unlike the existence or non-existence of a zero.

The zero of the β-function may be associated with a phase transition of QED in the space of bare parameters (m, e). A hint for the existence of a critical (second order phase transition) point was obtained by Miranski from the solution of a truncated Schwinger–Dyson equation in the ladder approximation [4.48]. At this critical point $e^2 = e_c^2 = e_*^2$, which is at strong coupling, the chiral symmetry of the massless $(m = m_R = 0)$ theory is spontaneously broken, in such a way that the vacuum expectation value $\langle (\overline{\psi}_x \psi_x) \rangle$ becomes non-zero for $m = 0$, $e^2 \geq e_c^2$. In the ladder approximation the critical coupling corresponds to $\alpha = \pi/3$. In the phase with spontaneously broken chiral symmetry the Goldstone theorem implies the existence of a massless pseudoscalar state. This massless state is the strong-coupling limit of the lowest singlet parapositronium state of weakly coupled QED. For small bare masses, near the spontaneously broken phase, there is a light pseudo-Goldstone state with almost zero mass. This means that almost the whole rest mass of the constituent fermion–antifermion ('electron–positron') pair is compensated by the binding energy.

Of course, the physical value of the fine structure constant is far too small for providing such a strong binding (the physical parapositronium is a weakly bound state). Nevertheless, it is not exluded that as a function of some external parameters, like temperature or external electromagnetic field (or perhaps as a function of the number of fermion species), the critical point of chiral phase transition can become physically relevant.

4.5.1 Lattice actions

The fermionic part of the lattice action for QED can be constructed either with Wilson fermions (see section 4.2) or with staggered fermions (see section 4.3). Let us start with the Wilson formulation. According to chapter 3 the gauge field variable $U_{x\mu}$ is put on the links of the lattice connecting two neighbouring sites: $(x, x + \hat{\mu})$. $U_{x\mu}$ is an element of a compact U(1) group and can be written as

$$U_{x\mu} = \exp\left\{ieA_\mu(x)\right\}. \tag{4.274}$$

Here $A_\mu(x)$ is a real number corresponding to the continuum vector potential $(A_\mu(x) \equiv aA(x)_\mu)$, and e is the bare (electromagnetic) gauge coupling. In the continuum $U_{x\mu}$ corresponds to the 'path dependent phase factor' between the points x and $x + \hat{\mu}$. On links with negative direction

$U_{x\mu}$ is defined by

$$U_{x,-\mu} \equiv U^*_{x-\hat{\mu},\mu} \,. \tag{4.275}$$

Similarly to (4.245), but with a general fermion field normalization, the fermionic part of the QED lattice action with Wilson fermions is

$$S_{WI} = \sum_x \left\{ M(\overline{\psi}_x \psi_x) - K \sum_{\mu=\pm 1}^{\pm 4} (\overline{\psi}_{x+\hat{\mu}}[r + \gamma_\mu]U_{x\mu}\psi_x) \right\} \,. \tag{4.276}$$

In the free case ($e = 0$) the fermion mass in lattice units (am) is given by

$$am + 4r = \frac{M}{2K} \,. \tag{4.277}$$

After the introduction of the electromagnetic interaction this relation is changed ('renormalized'). In general, am will be a function of e^2, r and M/K. Typical normalization conventions are: $K = \frac{1}{2}$ in lattice perturbation theory, and $M = 1$ in numerical simulations. In the continuum limit the action (4.276) is expected to describe a single Dirac fermion. By adding N_f such actions (or considering a fermion field with N_f components) one can describe QED with N_f fermion *flavours*.

The fermionic part of the QED lattice action can also be constructed with staggered fermions. In this case, with a single staggered fermion field, $N_f = 4$ flavours are described. The global symmetry of the free massive action is the diagonal subgroup ($U_o = U_e$) of $U(1)_o \otimes U(1)_e$ in (4.165). This global symmetry is gauged in the fermionic action, with a general field normalization corresponding to (4.166), as

$$S_{ST} = \sum_x \left\{ M(\overline{\psi}_x \psi_x) + K \sum_{\mu=1}^{4} \alpha_{x\mu} \left[(\overline{\psi}_x U^*_{x\mu} \psi_{x+\hat{\mu}}) - (\overline{\psi}_{x+\hat{\mu}} U_{x\mu} \psi_x) \right] \right\} \,. \tag{4.278}$$

Here the fermion mass in the non-interacting case is given by (4.167), and the sign factor $\alpha_{x\mu}$ is defined by (4.158). Note that in the staggered case the gauge interaction is included on the 'local' basis. In the Dirac–Kähler formulation, discussed at the end of section 4.3, a more natural choice would be to define interactions on the 'flavour' basis (see [4.24]). However, similarly to QCD, if the gauge interaction is defined on the fields with definite flavour, then a mass term proportional to $\sum_\mu \mathbf{1} \otimes t_\mu$ is induced by the interaction [4.49]. Therefore, in this case such a term has to be included in the lattice action for consistency, and has to be appropriately tuned in the continuum limit.

Turning now to the gauge field part of the lattice action, there are again several possibilities. One can consider a *compact formulation* based on compact link variables $U_{x\mu} \in U(1)$. In this case the Wilson action for

Abelian gauge fields is, according to chapter 3 (see also [4.50]):

$$S_W = \frac{1}{2e^2} \sum_x \sum_{\mu,\nu=1}^{4} \left[1 - U_{xv}^* U_{x+\hat{\nu},\mu}^* U_{x+\hat{\mu},\nu} U_{x\mu} \right]$$

$$= \frac{1}{2e^2} \sum_x \sum_{\mu,\nu=1}^{4} \left[1 - \cos\left(eF_{\mu\nu}(x)\right) \right] = \beta \sum_x \sum_{1\leq\mu<\nu\leq4} \left[1 - \cos\left(eF_{\mu\nu}(x)\right) \right] .$$

$$(4.279)$$

Here $\beta \equiv 1/e^2$, and the lattice field strength tensor is defined with (8.42) as

$$F_{\mu\nu}(x) \equiv \Delta_\mu^f A_\nu(x) - \Delta_\nu^f A_\mu(x) = A_\nu(x+\hat{\mu}) - A_\nu(x) - A_\mu(x+\hat{\nu}) + A_\mu(x)$$

$$= -F_{\nu\mu}(x) . \qquad (4.280)$$

Note that the summation in the last form of the action in (4.279) goes over all positively oriented plaquettes of the lattice.

In the case of Abelian gauge fields it is also possible to define a *non-compact formulation* of the lattice action [4.51], which is based on the field strength tensor $F_{\mu\nu}(x)$:

$$S_{NC} = \frac{1}{4} \sum_x \sum_{\mu,\nu=1}^{4} F_{\mu\nu}(x) F_{\mu\nu}(x) . \qquad (4.281)$$

This is a direct discretization of the Euclidean action of the Abelian gauge field in continuum. Here the real link variables $A_\mu(x)$ have a range $(-\infty, +\infty)$. Therefore, the integration over gauge variables in the path integral is

$$\int [dA] \equiv \prod_x \prod_{\mu=1}^{4} \int_{-\infty}^{+\infty} dA_\mu(x) . \qquad (4.282)$$

In the compact formulation with the action S_W the integration over $A_\mu(x)$ goes, of course, only over the interval $0 \leq eA_\mu(x) \leq 2\pi$.

The whole lattice action is a combination of some of the fermionic parts with some of the gauge parts:

$$S = \{S_{WI}, S_{ST}\} + \{S_W, S_{NC}\} . \qquad (4.283)$$

Both fermionic and gauge parts of the lattice action are invariant under the *local gauge transformation* with real $\alpha_x \equiv \alpha(x)$:

$$\psi_x' = e^{-ie\alpha_x} \psi_x , \quad \overline{\psi}_x' = \overline{\psi}_x e^{ie\alpha_x} , \quad U_{x\mu}' = e^{-ie\alpha_{x+\hat{\mu}}} U_{x\mu} e^{ie\alpha_x} . \qquad (4.284)$$

Here the real link variables are transformed according to

$$A_\mu(x)' = A_\mu(x) - \Delta_\mu^f \alpha(x) , \qquad (4.285)$$

and $F_{\mu\nu}(x)' = F_{\mu\nu}(x)$ is invariant. The integrand in the path integral is constant along the orbits of gauge transformations, hence in the non-compact case, when the integration is performed over an infinite range (4.282), the integral is not convergent. Therefore a *gauge fixing term* S_{GF} has to be added to the lattice action $S_{NC} \to S_{NC} + S_{GF}$, where one can take, for instance,

$$S_{GF} = \frac{\lambda}{2} \sum_x \sum_{\mu,\nu=1}^{4} \left(\Delta_{\mu}^b A_{\mu}(x)\right)\left(\Delta_{\nu}^b A_{\nu}(x)\right) , \qquad (4.286)$$

with a positive *gauge fixing parameter* λ. Another possibility is to fix the gauge degrees of freedom completely, which means that in the path integral $e^{-S_{GF}}$ is replaced by a delta-function enforcing some gauge fixing condition.

The lattice realization of the *Lorentz gauge* condition is

$$\Delta_{\mu}^b A_{\mu}(x) = \sum_{\mu=1}^{4} (A_{\mu}(x) - A_{\mu}(x - \hat{\mu})) = 0 . \qquad (4.287)$$

This can also be easily implemented on the lattice numerically, because an appropriate gauge transformation function $\alpha(x)$ is [4.52]

$$\alpha(x) = \sum_y D(x - y) \sum_{\mu=1}^{4} \Delta_{\mu}^b A_{\mu}(y) . \qquad (4.288)$$

Here $D(x)$ is the lattice Green function

$$D(x) \equiv -\frac{1}{4\Omega} \sum_{k \neq 0} \frac{e^{ik \cdot x}}{\sum_{\mu=1}^{4} \sin^2 (k_{\mu}/2)} . \qquad (4.289)$$

Owing to the omission of zero momentum in the sum, on a finite lattice $D(x)$ satisfies $\Delta D(x) = \delta_{x0} + O(1/\Omega)$.

One has to keep in mind that the condition (4.287) does not yet fix the gauge completely. This is because a gauge transformation with $\alpha(x)$ obeying (4.287) can always be replaced by $\alpha(x) + \omega(x)$, if $\omega(x)$ is a harmonic function which satisfies the lattice Laplace equation $\Delta_{\mu}^b \Delta_{\mu}^f \omega(x) = \Delta\omega(x) = 0$. On a finite lattice with extensions L_{μ} ($\mu = 1, 2, 3, 4$) such an $\omega(x)$ is given, for instance, by

$$\omega(x) = \omega_0 + \sum_{\mu=1}^{4} x_{\mu} \frac{2\pi n_{\mu}}{eL_{\mu}} , \qquad (4.290)$$

with constant ω_0 and integer n_{μ}. A way to achieve a complete gauge fixing is to transform first to the unique temporal gauge on a maximal tree [4.6] and then, by using $\alpha(x)$ in (4.288), to a gauge satisfying (4.287). This unique gauge can be called the *TL gauge* [4.52].

4.5.2 Analytic results

Given some of the above lattice actions for QED, one can define lattice perturbation theory similarly to the non-Abelian case in chapters 3 and 5. This turns out to be a valid regularization scheme for QED in the sense that renormalizability can be proven, and the renormalized S-matrix is the same, to all orders of perturbation theory, as in a continuum regularization scheme, like for instance dimensional regularization. This was first proven in the Wilson fermion formulation by Sharatchandra [4.53].

In the bare parameter space (e^2, am) bare perturbation theory, which is an expansion in powers of e^2, gives a good approximation near $e^2 = 0$, provided that the lattice spacing a is not too small, in the sense that $\log (am_R)^{-1}$ is not too large (m_R is the renormalized fermion mass in physical units). For $\log (am_R)^{-1} \gg 1$ one has to rearrange the series in powers of the renormalized coupling squared e_R^2. For small e_R^2 this renormalized perturbation series gives a good approximation for appropriately defined renormalized Green functions, if the considered physical energy range is not too broad. The point $(e^2 = 0, am = 0)$ in the bare parameter space is the *Gaussian Fixed Point* (GFP). At this point we have a free field theory with massless fields, that is also $e_R^2 = 0$. Near the GFP the β-function, which for fixed bare coupling describes the dependence of the renormalized coupling on $\log (am_R)^{-1}$, is dominated by the universal 1- and 2-loop contributions. In case of N_f fermion flavours these are given by the coefficients [4.47]

$$\beta_1 = \frac{2N_f}{3\pi}, \qquad \beta_2 = \frac{N_f}{2\pi^2}. \qquad (4.291)$$

Since the β-function vanishes at the GFP, and β_1 is positive, the GFP is an *infrared stable fixed point* (IRFP), similarly to scalar ϕ^4 models (see chapter 2). Therefore, near the GFP the lines of constant physics (with $e_R^2 = $ const.) qualitatively behave as shown in the left hand corner of fig. 4.8.

In the case of the compact gauge action (S_W), another simple limit is the *strong-coupling limit* $\beta = 1/e^2 = 0$. In this case, instead of the link variables $A_\mu(x)$, it is better to use $\varphi_{x\mu} \equiv eA_\mu(x)$ ($eA_\mu(x)$ is reasonable only for small e). As one can see in (4.279), for $\beta = 0$ the gauge part of the action vanishes: $S_W = 0$, hence the gauge variables on the links are independent from each other. Therefore, the integral over $\varphi_{x\mu}$ can be performed. Let us consider here, for simplicity, the case of staggered fermions with fermion action S_{ST} in (4.278) (Wilson fermions with S_{WI} in (4.276) can be treated similarly). The integral over the link variables can be performed in the general case with f flavours of staggered fermions

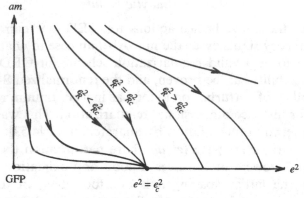

Fig. 4.8. A possible scenario for the behaviour of the lines of constant renormalized coupling in lattice QED. The arrows point in the direction of decreasing lattice spacing. The point e_c^2 on the $m = 0$ axis is the critical point for the chiral phase transition. The origin $(m = 0, e^2 = 0)$ is the Gaussian fixed point (GFP).

[4.54]. For $f = 1$ it is

$$\frac{1}{2\pi} \int_0^{2\pi} \mathrm{d}\varphi_{x\mu} \exp\left\{-K\alpha_{x\mu}[(\overline{\psi}_x e^{-\mathrm{i}\varphi_{x\mu}}\psi_{x+\hat{\mu}}) - (\overline{\psi}_{x+\hat{\mu}}e^{\mathrm{i}\varphi_{x\mu}}\psi_x)]\right\}$$

$$= 1 - K^2(\overline{\psi}_x\psi_{x+\hat{\mu}})(\overline{\psi}_{x+\hat{\mu}}\psi_x) = 1 + K^2(\overline{\psi}_x\psi_x)(\overline{\psi}_{x+\hat{\mu}}\psi_{x+\hat{\mu}}) . \qquad (4.292)$$

It is clear from this that all observables in the strong-coupling limit can be expressed as functions of the composite field

$$\sigma_x \equiv \overline{\psi}_x\psi_x . \qquad (4.293)$$

The four-fermion product appearing in (4.292) is chiral symmetric with respect to the $U(1)_o \otimes U(1)_e$ transformation in (4.165). In the massless case $M = am = 0$ this chiral symmetry of the effective action at $\beta = 0$ is, however, spontaneously broken by a 'chiral long range order' of σ_x in the vacuum state, as was proven in [4.55]. On the other hand, at $am = 0$ and $e^2 = 0$ the vacuum expectation value of $\overline{\psi}\psi$ is zero, as one sees from the formula valid for free staggered fermions:

$$\langle\psi_x\overline{\psi}_x\rangle = \frac{1}{\Omega'}\sum_k{}' \frac{am}{(am)^2 + \sum_{\mu=1}^4 \sin^2(k_\mu/2)} . \qquad (4.294)$$

This can be obtained, similarly to (4.134), by using (4.184). Since in the massless limit $\langle\overline{\psi}\psi\rangle$ remains zero in every finite order of perturbation theory, it is plausible that between $(e^2 = 0, am = 0)$ and $(e^2 = \infty, am = 0)$ there is a critical point at $e^2 = e_c^2$, where the chiral symmetry gets spontaneously broken.

Besides the proof of spontaneous chiral symmetry breaking, another important exact result in the strong-coupling limit concerns *charge renormalization*. It is well-known that the bare charge e in QED is larger than the observable, renormalized charge defined through the low energy Thomson-scattering cross section. The vacuum polarization partially screens the bare charge seen at short distances. Let us define the charge renormalization constant Z_3 by

$$Z_3 \equiv \frac{e_R^2}{e^2} . \tag{4.295}$$

In perturbation theory with Pauli–Villars cut-off Λ we have [A3]:

$$Z_3 = 1 - \frac{e^2}{6\pi^2} \log \frac{\Lambda}{m_R} + O(e^4) . \tag{4.296}$$

This is, indeed, smaller than 1, showing charge screening. For large Λ, however, the series in e^2 is divergent. Still one can show [A2] that

$$0 \le Z_3 \le 1 . \tag{4.297}$$

The proof in continuum QED is, however, somewhat formal. An exact proof of the 'charge renormalization inequality' (4.297) in lattice QED with non-compact gauge action S_{NC} (4.281) was given by Lüscher [4.56].

For definiteness let us consider here the Wilson fermion formulation. The proof of (4.297) is based on the *vector Ward–Takahashi identity*. Including now also an arbitrary observable $O[\psi\overline\psi A]$ one can show, similarly to (4.254), that

$$\left\langle O \Delta_\mu^b J_{x\mu} \right\rangle = \left\langle \overline\psi_x \frac{\overrightarrow{\partial} O}{\partial \overline\psi_x} - \frac{O \overleftarrow{\partial}}{\partial \psi_x} \psi_x \right\rangle . \tag{4.298}$$

Here the vector current with a general field normalization is

$$J_{x\mu} \equiv K \left\{ (\overline\psi_{x+\hat\mu}[r + \gamma_\mu] U_{x\mu}\psi_x) - (\overline\psi_x[r - \gamma_\mu] U_{x\mu}^* \psi_{x+\hat\mu}) \right\} . \tag{4.299}$$

Another ingredient of the proof is the *photon field equation of motion*, which is obtained by taking the derivative of the integrand in the path integral with respect to $A_\mu(x)$. The integral is zero by partial integration, and one obtains

$$\left\langle O(\Delta_\rho^b F_{\rho\mu}(x) + \lambda \Delta_\mu^f \Delta_\rho^b A_\rho(x)) \right\rangle = -\mathrm{i}e \left\langle J_{x\mu} O \right\rangle - \left\langle \frac{\partial O}{\partial A_\mu(x)} \right\rangle . \tag{4.300}$$

The inequality in (4.297) follows from (4.298), (4.300) and from reflection positivity [4.56].

The charge renormalization inequality implies that in more than four dimensions lattice QED has a trivial, non-interacting continuum limit. If in four dimensions there is a fixed point at the chiral phase transition

$e^2 = e_c^2$, then in the continuum limit the renormalized coupling is bounded by

$$e_R^2 \leq e_c^2 .$$ (4.301)

In fig. 4.7, which shows the non-trivial UVFP of the Callan–Symanzik β-function, this means that the component $e_R^2 > e_*^2 = e_c^2$ is absent, due to the singularity at e_c^2.

4.5.3   Non-perturbative studies

The first numerical studies in lattice QED concentrated on the question of the chiral phase transition. In the staggered fermion formulation (4.278) with the compact gauge action (4.279) this turned out to be of first order, which is not suitable for the definition of a continuum limit [4.57]. However, the action $S_{ST} + S_{NC}$ has a second order chiral phase transition [4.51]. For smaller number of fermion flavours, especially for $N_f = 2$, some hints of a non-trivial continuum limit were also seen in [4.51]. This is also supported by numerical results in the *quenched approximation*, when the fermion determinant in the bosonic path integral is omitted. These observations seem to substantiate the non-trivial behaviour suggested by the truncated ladder approximation [4.48]. Nevertheless, one has to keep in mind that in the numerical simulations $N_f = 2$ is achieved by an approximation, taking the square root of the $N_f = 4$ fermion determinant. In the case $N_f = 4$ the unquenched numerical results [4.58, 4.59] are consistent with a second order chiral phase transition, but the critical exponents are within errors equal to the mean field ones. Therefore, according to [4.58, 4.59], a trivial continuum limit is probable.

An interesting question is, whether the chiral phase transition in strong-coupling QED can become physically relevant in some unusual physical situations. In principle it is possible that as a function of some external parameters, such as temperature, external fields, the critical coupling comes down to the known small value of the fine structure constant $\alpha = 1/137$. In the case of a strong external Coulomb field, occurring near high-Z nuclei, this does not seem to happen, because in this case the critical point is shifted to even stronger couplings (the chiral symmetry breaking condensate is suppressed by a Coulomb field) [4.60]. There is a better chance of this occurring for strong magnetic fields, which may potentially bring the critical coupling down to $\alpha = 1/137$ [4.61].

A detailed study of the question of the continuum limit can be performed by investigating the *lines of constant physics* (LCP) in the space of bare parameters. A possible scenario for the qualitative behaviour of the lines with constant renormalized coupling is shown in fig. 4.8. Near the Gaussian fixed point (GFP) the behaviour corresponds to an IRFP, which means that the LCPs do not end at the GFP, as in asymptotically free theories, but behave similarly to the LCPs in scalar ϕ^4 models (see chapter 2). For renormalized couplings satisfying $e_R^2 \le e_{R_c}^2 = e_*^2$ the LCPs end at the critical point of chiral phase transition $e^2 = e_c^2$. For non-zero $e_{R_c}^2$ this would mean a non-trivial continuum limit with $e_R^2 \le e_{R_c}^2$. For $e_R^2 > e_{R_c}^2$ the LCPs in fig. 4.8 end at non-zero lattice fermion mass $am_R > 0$ on the line $(e^2 > e_c^2, am = 0)$, therefore in this case the cut-off cannot be removed.

Owing to the charge renormalization inequality (4.297) the maximal renormalized coupling in the continuum limit has to satisfy $e_{R_c}^2 \le e_c^2$. Since numerical data with four flavours ($N_f = 4$) give $\beta_c = e_c^{-2} \simeq 0.195$ [4.59], this means for the fine structure constant $\alpha \le e_c^2/(4\pi) \simeq 0.41$. At such a small value one would expect that the β-function is still well approximated by the first two terms in the perturbative series. This would mean that, at least for $N_f = 4$, the β-function cannot have a non-trivial zero, and therefore the continuum limit of lattice QED is trivial.

The LCPs were also investigated numerically both in an analytic approximation [4.62] and in Monte Carlo simulations [4.59]. For the lines of constant renormalized coupling fig. 4.8 is supported with $e_{R_c}^2 \simeq 0$. However, it turned out that all the lines with constant ratio of the pseudoscalar boson mass to the fermion mass seem to go to the critical point $(e_c^2, am = 0)$. This means that for larger couplings they do not coincide with the lines of constant renormalized couplings. In other words, the mass ratios in the critical region are not functions of the renormalized coupling. This is quite a surprise, because in weak coupling QED the mass of the pseudoscalar positronium bound states is always assumed to be given by the renormalized electron mass and by the fine structure coupling constant. Lattice QED shows that this is, in fact, only an approximation valid for weak couplings. For much stronger couplings than the physical value of α, the pseudoscalar parapositronium state becomes light, and its coupling to the fermion is an essential independent parameter of the dynamics.

This suggests that strong-coupling QED has to be considered in a wider space of bare couplings, which also includes Yukawa coupling of the pseudoscalar bound state to the fermion. A simple special case of the inclusion of Yukawa coupling is to consider four-fermion interaction [4.63, 4.58, 4.62, 4.64]. (This is obtained at zero scalar hopping parameter after integrating out the scalar fields, see chapter 6.) In the literature the chirally invariant four-fermion interaction of the Nambu–Jona-Lasinio model [4.65] is usually considered:

$$\frac{G}{2}\left[(\overline{\psi}_x \psi_x)(\overline{\psi}_x \psi_x) - (\overline{\psi}_x \gamma_5 \psi_x)(\overline{\psi}_x \gamma_5 \psi_x)\right] . \tag{4.302}$$

In the numerical solution of the truncated Schwinger–Dyson equations, which include the effect of vacuum polarization [4.62], the LCPs do not seem to go to a critical point even in this enlarged bare parameter space. They tend to $G\Lambda^2 \to \infty$ for finite fermion mass in lattice units. This means that the fermion always decouples in the continuum limit, leading to a trivial bosonic theory.

These non-perturbative studies of renormalization suggest that the continuum limit of lattice QED is trivial. Nevertheless, further investigations are necessary. In particular, since the issue of the continuum limit leads to an enlarged theory containing also Yukawa couplings, the question can only be decided in a broader context, which requires also understanding of the Yukawa couplings on the lattice.

5
Quantum chromodynamics

Quantum Chromodynamics (QCD) is believed by most physicists to be the correct theory of the strong nuclear force. Besides the qualitatively correct description of the hadron spectrum by bound states of light (*u*-, *d*- and *s*-) and heavy (*c*- and *b*-) quarks, the main source of confidence is the successful application of quark–gluon perturbation theory to calculate the cross-sections of many short distance (large transverse momentum) processes (see, for instance, [A4]).

The outstanding property of QCD, which is the basis of the success of perturbation theory, is *asymptotic freedom*. According to it the effective quark–gluon and gluon–gluon couplings become small at short distances. This unique property is a consequence of the self-coupling of gluons, the quanta of the non-Abelian SU(3) colour gauge field. The counterpart of asymptotic freedom at high energies is *infrared slavery* at low energies, which is due to the increase of the effective coupling at 'long' distances (of the order 1 *fermi*). The consequence is the *confinement* of quarks and gluons inside colour singlet hadrons. The basic property of confinement and the approximate SU(3) ⊗ SU(3) flavour symmetry of light quarks qualitatively describes many features of the hadron spectrum and of other low energy strong interaction phenomena. But due to the strong couplings, a precise quantitative description of these phenomena is difficult. This spoils to some extent even the predictions of perturbative QCD, because the confinement acts also on the initial and final states of every short distance 'hard' process. Therefore the transformation of the perturbative predictions at the parton (i. e. quark and gluon) level to the hadronic incoming and outgoing states cannot be done without some knowledge of 'soft' confinement phenomena.

The description of the long distance strong colour force requires non-perturbative methods. The main motivation for the introduction of lattice gauge theory by Wilson [5.1, 5.2] was to formulate QCD non-pertur-

231

batively, in order to explain confinement, and to allow for a numerical determination of the hadron spectrum and of other low energy hadron properties. A successful calculation of low energy hadronic parameters in lattice QCD can, in principle, also yield convincing evidence for QCD to be a valid description of strong interactions.

5.1   Lattice action and continuum limit

5.1.1   *Lattice actions*

The Euclidean lattice action of QCD depends on the $SU(3)$ colour gauge field $U_{x\mu} \in SU(3)$ on lattice links, and on the Grassmannian quark fields ψ_{qx}, $\overline{\psi}_{qx}$ on lattice sites. Here, as usual, x is a lattice point and $\mu = \pm 1, \pm 2, \pm 3, \pm 4$ are the directions of its eight neighbouring points $(x + \hat{\mu})$ on the four-dimensional hypercubic lattice. On opposite direction links we define

$$U_{x,-\mu} \equiv U^{+}_{x-\hat{\mu},\mu} \,, \tag{5.1}$$

and the pure gauge field lattice action is, according to chapter 3, a sum over plaquettes

$$S_g[U] = \beta \sum_{\mathrm{p}} \left(1 - \frac{1}{3} \operatorname{Re} \operatorname{Tr} U_{\mathrm{p}} \right) \,. \tag{5.2}$$

Here $\beta \equiv 6g^{-2}$ is proportional to the inverse bare gauge coupling squared, and the plaquette variable is

$$U_{\mathrm{p}} \equiv U^{+}_{x\nu} U^{+}_{x+\hat{\nu},\mu} U_{x+\hat{\mu},\nu} U_{x\mu} \,. \tag{5.3}$$

The quark fields $\psi_{qx} \equiv \psi_{qx\alpha c}$ and $\overline{\psi}_{qx} \equiv \overline{\psi}_{qx\alpha c}$ have, besides the lattice point index x, also the flavour index $q = u, d, c, s, t, b$, the Dirac spinor index $\alpha = 1, 2, 3, 4$ and $SU(3)$ colour index $c = 1, 2, 3$. The whole Wilson lattice action for QCD is

$$S[U, \psi, \overline{\psi}] = S_g[U] + S_q[U, \psi, \overline{\psi}] \,, \tag{5.4}$$

where the quark field dependent part is

$$S_q[U, \psi, \overline{\psi}] = \sum_{q,x} \left\{ (\overline{\psi}_{qx}\psi_{qx}) - K_q \sum_{\mu=\pm 1}^{\pm 4} (\overline{\psi}_{qx+\hat{\mu}}[r_q + \gamma_\mu] U_{x\mu}\psi_{qx}) \right\} \,. \tag{5.5}$$

This is the $SU(3)$ colour gauged counterpart of the free Wilson fermion action studied in section 4.2 (see (4.88)). The spinor and colour indices are not written out explicitly here, as in most cases in this chapter. The *hopping parameter K_q* and *Wilson fermion parameter r_q* can depend on the flavour index q. In fact, since the quark masses in nature are different and, according to section 4.2, the bare quark masses are determined by K_q and r_q, one usually needs different hopping parameters for different flavours. In most cases the Wilson parameters are chosen to be $r_q \equiv 1$. (This can be done, because r_q is assumed to be an irrelevant parameter in the continuum limit, as long as it is non-zero.) Therefore, for N_f quark flavours the $(N_f + 1)$ independent bare parameters of the QCD Wilson lattice action are β and $K_1, K_2, \ldots, K_{N_f}$.

Reflection positivity of the Wilson action (5.4) can be proven similarly as in the case of pure gauge theory (section 3.2) and free Wilson fermions (section 4.2).

In the limit of N_f equal quark masses (approximately true for $q = u, d, s$), the action (5.4) has an exact global $U(N_f)$ flavour symmetry defined by $U \in U(N_f)$

$$\psi'_{qx} = U^{-1}_{qq'} \psi_{q'x} , \qquad \overline{\psi}'_{qx} = \overline{\psi}_{q'x} U_{q'q} . \tag{5.6}$$

The spontaneously broken chiral symmetry for zero quark mass will be discussed in section 5.3. It can only be realized in the continuum limit, because at finite lattice spacing the Wilson term proportional to r_q explicitly breaks the conservation of the axial vector currents (see at the end of section 4.4).

Another possibility for the quark field dependent part of the QCD lattice action is to take staggered fermions (see section 4.3). The staggered fermion field $\psi_x, \overline{\psi}_x$ can describe four quark flavours on the 2^4 hypercubic blocks of the lattice. Since there is only a single component per lattice point, the 16 components in a 2^4 block describe the $4 \cdot 4 = 16$ Dirac spinor components of four flavours. The three colour states are represented by three different staggered fields, therefore the staggered quark fields are $\psi_x \equiv \psi_{xc}$ and $\overline{\psi}_x \equiv \overline{\psi}_{xc}$ $(c = 1, 2, 3)$. The quark part of the QCD lattice action with staggered quarks is then

$$S_q[U, \psi, \overline{\psi}] = \sum_x \left\{ am(\overline{\psi}_x \psi_x) \right.$$

$$\left. + \frac{1}{2} \sum_{\mu=1}^{4} \alpha_{x\mu} \left[(\overline{\psi}_x U^+_{x\mu} \psi_{x+\hat{\mu}}) - (\overline{\psi}_{x+\hat{\mu}} U_{x\mu} \psi_x) \right] \right\} . \tag{5.7}$$

This is the SU(3) colour gauge invariant version of the free staggered fermion action defined in section 4.3 (see (4.166), (4.158)). The quark field normalization is fixed here by $K = \frac{1}{2}$ therefore, according to (4.167), am is the bare quark mass in lattice units. Since the mass term $am(\overline{\psi}_x \psi_x)$ is flavour independent, it is expected that in the continuum limit the action (5.7) describes four degenerate quark flavours with an exact U(4) global flavour symmetry for general am, and a spontaneously broken $U(4) \otimes U(4)$ chiral flavour symmetry in the massless case $am = 0$. Nevertheless, according to section 4.3, at non-zero lattice spacing the exact symmetry of the staggered kinetic term is only $U(1)_o \otimes U(1)_e$. The rest of the chiral $U(4) \otimes U(4)$ flavour symmetry group is broken by the lattice regularization. Therefore the restoration of the flavour symmetry in the continuum limit is a non-trivial dynamical question. Note that the flavour independence of the mass term in the action (5.7) can be lifted by adding non-local, gauge field dependent mass terms to it [5.3].

The flavour content of the staggered fermion field in the action (5.7) is not displayed explicitly. The transformation to the *flavour basis* can be performed according to (4.170). An equivalent flavour formulation of the staggered fermions can also be directly given in the framework of the Dirac–Kähler equation (see at the end of section 4.3). In the Dirac–Kähler framework it seems more natural to define the colour gauge interaction on the flavour basis. The fields with definite quark flavour $q_y^{\alpha a}, \bar{q}_y^{a\alpha}$ are introduced in (4.170)–(4.173). In terms of them the quark part of the QCD lattice action is

$$S_q[U, q, \bar{q}] = 16 \sum_y{}' \left\{ am \left(\bar{q}_y \mathbf{1} \otimes \mathbf{1} q_y \right) \right.$$

$$\left. + \sum_{\mu=1}^4 \left(\bar{q}_y [\gamma_\mu \otimes \mathbf{1} \Delta_\mu(U) - \gamma_5 \otimes t_5 t_\mu \delta_\mu(U)] q_y \right) \right\} . \tag{5.8}$$

Here the normalization is fixed again by $K = \frac{1}{2}$, and the summation $\sum_y'$ is performed over the 2^4 blocks labelled by y (see (4.168) and (4.169)). The covariant lattice derivatives $\Delta_\mu(U)$ and $\delta_\mu(U)$ are defined, following (4.176), by

$$\Delta_\mu(U) f_y \equiv \frac{1}{4} \left(U_{y\mu}^+ f_{y+\hat{\mu}} - U_{y-\hat{\mu},\mu} f_{y-\hat{\mu}} \right) ,$$

$$\delta_\mu(U) f_y \equiv \frac{1}{4} \left(U_{y\mu}^+ f_{y+\hat{\mu}} + U_{y-\hat{\mu},\mu} f_{y-\hat{\mu}} - 2 f_y \right) . \tag{5.9}$$

As shown by the notation, the gauge field $U_{y\mu}$ is associated now with the links of the block lattice with lattice spacing twice as large as the lattice spacing of the original 'fine' lattice. Therefore also in the pure gauge part $S_g[U]$ of the action, $\sum_p$ means a summation over the plaquettes of the block lattice (with points y).

Although in the case of the free staggered fermion field the 'local' formulation in terms of $\psi_x, \bar{\psi}_x$ and the 'block' or 'Dirac–Kähler' formulation in terms of $q_y, \bar{q}_y$ are equivalent, after introducing the gauge interaction in (5.7), and respectively, (5.8), the two become inequivalent. From the point of view of bare mass parameters the Dirac–Kähler formulation is similar to the Wilson fermion formulation (5.5). For instance, the flavour dependent mass terms defined on the 2^4 blocks are independent of the gauge field [5.4], unlike in the local case. In addition, in the Dirac–Kähler case the $U(1)_o \otimes U(1)_e$ symmetry does not prevent the gauge interaction from producing non-zero mass counterterms [5.4], therefore the bare mass parameters have to be tuned in the continuum limit, similarly to the Wilson fermion case (see fig. 5.7a). This is different from the local staggered formulation, where $U(1)_o \otimes U(1)_e$ forbids the renormalization of the mass term, hence a non-trivial tuning of am is not necessary (fig. 5.7b). Finally, the main advantage of the local staggered formulation is the presence of an exact zero mass Goldstone boson on the lattice at $am = 0$. This is a consequence of the spontaneous breaking of $U(1)_o \otimes U(1)_e$. In the Dirac–Kähler formulation there is no massless Goldstone boson [5.5]. Therefore, in most applications of staggered lattice QCD, the local formulation (5.7) is used.

5.1.2   Quenched approximation

In lattice QCD the expectation values of different physical quantities are given in terms of path integrals over the gauge ($U_{x\mu}$) and quark ($\psi_x, \overline{\psi}_x$) field variables:

$$\langle F \rangle = \frac{\int [dU \, d\overline{\psi} \, d\psi] e^{-S_g - S_q} F[U, \psi, \overline{\psi}]}{\int [dU \, d\overline{\psi} \, d\psi] e^{-S_g - S_q}} . \tag{5.10}$$

Here $F[U, \psi, \overline{\psi}]$ is an arbitrary function of the field variables, for instance, with n pairs of quark fields

$$F[U, \psi, \overline{\psi}] = \psi_{y_1} \overline{\psi}_{x_1} \psi_{y_2} \overline{\psi}_{x_2} \cdots \psi_{y_n} \overline{\psi}_{x_n} A[U] . \tag{5.11}$$

The integral over the fermionic Grassmann variables can be performed according to (4.25). In particular, we have

$$\int [d\overline{\psi} \, d\psi] e^{-S_q} = \det Q[U] , \tag{5.12}$$

where $Q[U]$ is the *quark matrix* in the quark part of the action:

$$S_q[U, \psi, \overline{\psi}] \equiv \sum_{xy} (\overline{\psi}_y Q[U]_{yx} \psi_x) . \tag{5.13}$$

For definiteness, let us consider now Wilson fermions with a single quark flavour and $r = 1$, when (5.5) implies

$$Q[U]_{yx} = \delta_{yx} - K \sum_\mu \delta_{y, x+\hat{\mu}} (1 + \gamma_\mu) U_{x\mu} . \tag{5.14}$$

This satisfies the relation

$$Q_{yx} = \gamma_5 Q_{xy}^\dagger \gamma_5 , \qquad (5.15)$$

where $Q_{xy}^\dagger$ denotes the adjoint with respect to the Dirac spinor and colour indices. Therefore the *quark determinant* is real, because

$$\det Q = \det Q^+ = (\det Q)^* . \qquad (5.16)$$

Here Q^+ means a complete adjoint, that is with respect to the point indices, too. If we introduce the *effective gauge action* $S_{eff}[U]$ by

$$S_{eff}[U] \equiv S_g[U] - \log \det Q[U] = S_g[U] - \text{Tr} \log Q[U] \qquad (5.17)$$

then, after performing the Grassmannian integrals in the nominator and denominator, (5.10) and (5.11) give

$$\langle \psi_{y_1} \overline{\psi}_{x_1} \psi_{y_2} \overline{\psi}_{x_2} \cdots \psi_{y_n} \overline{\psi}_{x_n} A[U] \rangle =$$

$$= Z^{-1} \int [dU] e^{-S_{eff}[U]} A[U] \sum_{z_1 \cdots z_n} \epsilon_{y_1 y_2 \cdots y_n}^{z_1 z_2 \cdots z_n} Q[U]_{z_1 x_1}^{-1} Q[U]_{z_2 x_2}^{-1} \cdots Q[U]_{z_n x_n}^{-1} .$$

$$(5.18)$$

Here the partition function Z is defined as

$$Z \equiv \int [dU] e^{-S_{eff}[U]} = \int [dU] e^{-S_g[U]} \det Q[U] . \qquad (5.19)$$

In perturbation theory the expansion of the quark determinant in powers of the gauge coupling generates the closed fermion loops. This can be seen if the quark matrix in (5.14) is written as

$$Q[U]_{yx} \equiv Q_{yx}^{(0)} - V[U]_{yx} . \qquad (5.20)$$

Here the first term is the free Wilson fermion matrix in (4.89):

$$Q_{yx}^{(0)} = \delta_{yx} - K \sum_{\mu} \delta_{y,x+\hat{\mu}}(1 + \gamma_\mu) , \qquad (5.21)$$

and the interaction term is

$$V[U]_{yx} = K \sum_{\mu} \delta_{y,x+\hat{\mu}}(1 + \gamma_\mu)(U_{x\mu} - 1) . \qquad (5.22)$$

The decomposition (5.20) gives for the quark determinant

$$\det Q[U] = \det Q^{(0)} \cdot \det \{Q^{(0)-1} Q[U]\}$$

$$= \det Q^{(0)} \cdot \det \{1 - Q^{(0)-1} V[U]\} . \qquad (5.23)$$

Using this relation, and denoting the quark propagator as usual by $Q^{(0)-1} \equiv \Delta$ (see (4.121)–(4.124) in section 4.2), the partition function Z in

Fig. 5.1. The effective gauge interaction vertices including quark loops in the effective gauge action S_{eff}. The lines are quark propagators and the blobs stand for $(U_{x\mu} - 1)$.

(5.19) can be written as

$$Z = \det Q^{(0)} \int [dU] \exp \left\{ -S_g[U] - \sum_{j=1}^{\infty} \frac{1}{j} \text{Tr} \left(\Delta V[U] \right)^j \right\} . \tag{5.24}$$

If one also applies the same transformation to the Boltzmann factor $\exp(-S_{eff})$ in (5.18), then in the expectation values the gauge field independent factor $\det Q^{(0)}$ cancels. The terms of the sum over j can be considered as effective interaction vertices for the gauge field. The jth effective gauge interaction vertex is the trace over a product of j quark propagators times j factors $V[U]$. Since the trace is a sum over all quark indices (including lattice points), the jth term represents a quark loop with j quark propagators (a graphical representation is given in fig. 5.1). We shall see in (5.50)–(5.52), how $(U_{x\mu} - 1)$ in $V[U]$ can be expanded in powers of the gauge coupling g (in fact, the expansion starts with a term linear in g). Therefore the perturbative expansion of the quark determinant does, indeed, generate the closed quark loops.

There are some phenomenological facts in low energy hadron physics, like the OZI rule [5.6] or the approximate linearity of Regge trajectories [5.7], which suggest that closed quark loops have only a small effect. This is the basis of the so-called *quenched approximation* [5.8, 5.9], which amounts to replacing the quark determinant by a gauge field independent constant:

$$\det Q[U] \Longrightarrow \text{const.} \tag{5.25}$$

From the point of view of numerical simulations this is an enormous simplification, because the gauge field updating can be done by the pure gauge action $S_g[U]$. The only numerical task left is to evaluate the full quark propagators $Q[U]^{-1}$ in (5.18) on the gauge configurations. (For a discussion of the difficulties of the fermionic simulations see section 7.4.)

According to (5.22) the interaction part in the action $V[U]$ is proportional to the hopping parameter K, therefore the quenched approximation (5.25) is equivalent to take the zero hopping parameter limit in the quark determinant. Since in the $K = 0$ limit the quark matrix is $Q[U]_{yx} = \delta_{yx}$, the full quark propagator $Q[U]^{-1}$ is local. This corresponds to infinitely heavy static quarks in the quark determinant. Some corrections to the $K = 0$ limit can be calculated in the hopping parameter expansion (see next subsection). The correction terms can also be expressed in general by expectation values of fluctuations in the pure gauge theory [5.10]. Expanding the Boltzmann factor $\exp(-S_{eff})$ in powers of $\mathrm{Tr} \log Q \equiv T$, for instance, in the case of the expectation value of a purely gluonic quantity $A[U]$ we have

$$\langle A \rangle = \langle A \rangle_0 + \langle (A - \langle A \rangle_0)(T - \langle T \rangle_0) \rangle_0$$

$$+ \frac{1}{2} \left\langle (A - \langle A \rangle_0)(T - \langle T \rangle_0)^2 \right\rangle_0 + \cdots . \qquad (5.26)$$

Here the expectation values $\langle \cdots \rangle_0$ are defined by the pure gauge action Boltzmann factor $\exp(-S_g)$. As one can see, the first two corrections are given by the pure gauge theory expectation values of the fluctuations of the quantity $A[U]$ times the powers of the fluctuations of the logarithm of the quark determinant. In higher orders the precise numerical evaluation of the corrections becomes more and more difficult.

5.1.3   Hopping parameter expansion

A useful approximation to lattice QCD with Wilson fermions is the expansion in powers of the hopping parameter. Let us now consider, for definiteness, N_f degenerate flavours with common hopping parameter K and Wilson parameter $r > 0$. In the quark part of the lattice action (5.5) the flavour index q will not explicitly be displayed here. The effective gauge field action (5.17) is in this case

$$S_{eff}[U] = S_g[U] - \mathrm{Tr} \log (1 - KM[U]) \equiv S_g[U] + S_{eff}^q[U] . \qquad (5.27)$$

Here the quark matrix in (5.14) was written as

$$Q[U] = 1 - KM[U] , \qquad (5.28)$$

with the *hopping matrix*

$$M[U]_{yx} = \sum_\mu \delta_{y,x+\hat{\mu}}(r + \gamma_\mu)U_{x\mu} . \qquad (5.29)$$

The hopping parameter expansion of the quark part of the effective action in (5.27) is

$$S_{eff}^q[U] = -\mathrm{Tr} \log (1 - KM[U]) = \sum_{l=1}^{\infty} \frac{K^l}{l} \mathrm{Tr} M[U]^l . \qquad (5.30)$$

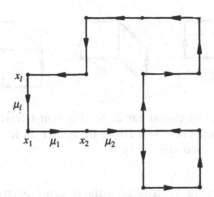

Fig. 5.2. A closed loop contribution to the hopping parameter expansion of the effective action.

Using the explicit form of the hopping matrix in (5.29) we get

$$\text{Tr}\, M[U]^l = N_f \sum_{x_1\mu_1...x_l\mu_l} \delta_{x_1,x_l+\hat{\mu}_l}\delta_{x_l,x_{l-1}+\hat{\mu}_{l-1}} \cdots \delta_{x_2,x_1+\hat{\mu}_1}$$

$$\cdot \text{Tr}_c \left[U_{x_l\mu_l} \cdots U_{x_2\mu_2} U_{x_1\mu_1} \right] \text{Tr}_s \left[(r+\gamma_{\mu_l}) \cdots (r+\gamma_{\mu_2})(r+\gamma_{\mu_1}) \right] \ . \qquad (5.31)$$

Here $\text{Tr}_c [...]$ denotes a trace over the colour and $\text{Tr}_s [...]$ over the Dirac spinor indices. The trivial trace over the flavours gives an overall factor N_f. Due to the δ-functions in space, this can be represented by a sum over closed loops on the lattice (see fig. 5.2). Obviously, on a hypercubic lattice, only even powers of $M[U]$ contribute: $l = 2, 4, \ldots$. The lowest order non-zero contribution is

$$\frac{K^2}{2} \text{Tr}\, M[U]^2 = \frac{K^2}{2} N_f \sum_{x\mu} \text{Tr}_c \left[U_{x+\hat{\mu},-\mu} U_{x\mu} \right] \text{Tr}_s \left[(r-\gamma_\mu)(r+\gamma_\mu) \right]$$

$$= 48K^2 N_f \Omega(r^2-1) \ . \qquad (5.32)$$

Here we used (5.1) and $\gamma_{-\mu} = -\gamma_\mu$. The result is an uninteresting constant, which cancels in expectation values. The K^4 contribution is, using the notation (8.32),

$$\frac{K^4}{4} \text{Tr}\, M[U]^4 = 360K^4 N_f \Omega(r^2-1)^2 + \frac{K^4}{4}$$

$$\cdot N_f \sum_{x\mu} \sum_{v\neq\pm\mu} \text{Tr}_c \left[U_{xv}^+ U_{x+\hat{v},\mu}^+ U_{x+\hat{\mu},v} U_{x\mu} \right] \text{Tr}_s \left[(r-\gamma_v)(r-\gamma_\mu)(r+\gamma_v)(r+\gamma_\mu) \right]$$

$$= 360K^4 N_f \Omega(r^2-1)^2 + 8K^4 N_f(r^4-2r^2-1) \sum_p \text{Re}\,\text{Tr}\, U_p \ . \qquad (5.33)$$

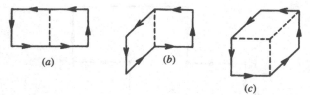

Fig. 5.3.   The three types of closed curves for the non-trivial K^6 contributions in the effective gauge action. The $r = 1$ Dirac trace factors for type (a), (b) and (c) are, respectively, -32, -16 and -16.

Apart from uninteresting constants, this is equivalent to a shift in the value of the bare parameter β of the pure gauge action, because the relevant contribution to the quark part of the effective gauge field action is

$$S^q_{eff}[U] = 24K^4N_f(1 + 2r^2 - r^4)\sum_P \left(1 - \frac{1}{3}\operatorname{Re}\operatorname{Tr}U_P\right) + O(K^6) \ . \quad (5.34)$$

Comparing to (5.2), the shift in β turns out to be

$$\Delta\beta = 24K^4N_f(1 + 2r^2 - r^4) \ . \quad (5.35)$$

The higher order contributions to the hopping parameter expansion of the effective action can also be given as a sum over closed curves of the colour trace of the products of link variables along that curve times the corresponding Dirac trace. (For the K^6 order see fig. 5.3.) The higher orders, however, are difficult to evaluate, because the number of closed curves grows exponentially with the length [5.11].

Besides the effective gauge action the other ingredients in the expectation value (5.18) are the matrix elements of the inverse quark matrix ('full quark propagator') $Q[U]^{-1}_{yx}$. The hopping parameter expansion for this is simply

$$Q[U]^{-1}_{yx} = (1 - KM[U])^{-1}_{yx} = \sum_{l=0}^{\infty} K^l M[U]^l_{yx} \ . \quad (5.36)$$

The matrix elements of the lth power of the hopping matrix $M[U]^l_{yx}$ are given, similarly to (5.31), by sums over curves of length l connecting the points x and y.

The representation of the hopping parameter expansion of the full quark propagator in terms of a sum over curves can be used to calculate numerically the hadron spectrum in the quenched approximation up to some relatively high orders (say, K^{10}) [5.12]. In order to determine numerically the hopping parameter expansion coefficients, one has to compute the expectation values of some closed gauge field loops on the gauge field configurations generated by the pure gauge action $S_g[U]$. The extension to high orders is, however, prevented by the fast increase of the number of different loops.

A method to overcome this difficulty was proposed in [5.13]. In order to obtain the required matrix elements of $M[U]^l$ on a given gauge field configuration, an iterative numerical method is applied. Writing out this time also the Dirac spinor $(\alpha, \beta, \ldots)$ and colour $(c, d, \ldots)$ indices, and using the explicit form of the hopping matrix in (5.29), the iteration equation is

$$M[U]^l_{y\beta d,x\alpha c} = \sum_{z\gamma e} M[U]_{y\beta d,z\gamma e} M[U]^{l-1}_{z\gamma e,x\alpha c}$$

$$= \sum_{\mu\gamma e} (r - \gamma_\mu)_{\beta\gamma} U^+_{y\mu,de} M[U]^{l-1}_{y+\hat{\mu}\gamma e,x\alpha c} \,. \tag{5.37}$$

This shows how the matrix element of the lth power $M[U]^l_{yx}$ can be calculated from the matrix elements of the $(l-1)$th power at the eight neighbouring points, namely $M[U]^{l-1}_{y+\hat{\mu},x}$. By this iterative method very high order expansions are possible. The only limitation is the accumulation of round-off errors, and eventually storage space, if the expansion coefficients of a very large number of different quantities are collected. In practice, at $\beta = 5.7$ on a 16^4 lattice, good results for the meson and baryon spectrum can be obtained in K^{32} order [5.14, 5.15]. (For a review see also [5.16].)

5.1.4 Strong gauge coupling limit

A useful qualitative insight in non-perturbative lattice QCD can be obtained in the infinitely strong bare gauge coupling limit $\beta \equiv 6g^{-2} = 0$ ($g = \infty$). The most important qualitative features of the hadron spectrum, namely confinement and spontaneous breakdown of the chiral flavour symmetry, are manifest at infinite gauge coupling. In fact, one of the main original motivations of formulating QCD on the lattice was to explain confinement [5.1]. The spontaneous chiral symmetry breakdown and the emergence of the Nambu–Goldstone bosons explaining the lightness of the physical pion can also be qualitatively understood in the staggered fermion formulation at $\beta = 0$ [5.17, 5.18, 5.19].

Technically speaking, the important simplification at $\beta = 0$ is that the gauge variables on different links become independent, and hence can be integrated out explicitly. The leftover pure fermion system is, however, still non-trivial. In the case of Wilson fermions the $\beta = 0$ limit can be combined with the partial resummation of the hopping parameter expansion [5.20]. This leads to a 'random walk' representation of meson and baryon propagators, and shows that at $\beta = 0$ the lightest hadrons are the pseudoscalar mesons. This is because the 'critical' hopping parameter for zero pseudoscalar meson masses (with $r = 1$) is $K_{cr}(\pi) = \frac{1}{4}$. The same for the vector mesons is $K_{cr}(\rho) = 1/(2\sqrt{3})$, and for the baryons $K_{cr}(B) = \frac{1}{2}$. In the case of staggered fermions at $\beta = 0$ one can use a systematic expansion in powers of $1/d$, the inverse of the number of space–time dimensions [5.19].

Besides reproducing the main qualitative features of the hadron spectrum, the strong-coupling limit of lattice QCD also gives a way to understand and control the numerical simulations at $\beta \neq 0$. In order to have a better description it is also possible to calculate the first non-trivial (order β) corrections in a strong gauge coupling expansion [5.21, 5.22].

The integration over the gauge link variables in the $\beta = 0$ limit requires the evaluation of the following type of group integrals:

$$H_G(\varphi, \overline{\varphi}, \chi, \overline{\chi}) \equiv \int_G dU \exp\{(\overline{\chi} U \varphi) - (\overline{\varphi} U^+ \chi)\} . \tag{5.38}$$

Here dU is the invariant Haar measure on the gauge group G, and $\varphi, \overline{\varphi}, \chi$ and $\overline{\chi}$ are Grassmann variables. In order to be more general than required for QCD, we now consider the gauge groups $G = U(N)$ or $G = SU(N)$ [5.23]. Therefore the Grassmann variables have N components. Let us first consider $G = U(N)$. In this case the integration over the centre of the group in (5.38) tells that in the expansion of the exponential only terms with equal powers of U and U^+ contribute. Therefore the result can be written in the form

$$H_{U(N)} = \sum_{k=0}^{N} \alpha_k \left[(\overline{\varphi}\varphi)(\overline{\chi}\chi)\right]^k . \tag{5.39}$$

The summation stops at N because of the nilpotency of the Grassmann variables. The coefficients α_k can be determined if one multiplies (5.38) and (5.39) by a factor $\exp(\overline{\chi}\chi)$ and integrates over $\chi, \overline{\chi}$ according to the rules in (4.21)–(4.25):

$$\int d\chi \, d\overline{\chi} \int_G dU \exp\{(\overline{\chi}\chi) + (\overline{\chi} U \varphi) - (\overline{\varphi} U^+ \chi)\} = \exp(\overline{\varphi}\varphi)$$

$$= \int d\chi \, d\overline{\chi} \int_G dU \exp(\overline{\chi}\chi) \sum_{k=0}^{N} \alpha_k \left[(\overline{\varphi}\varphi)(\overline{\chi}\chi)\right]^k = \sum_{k=0}^{N} \alpha_k \frac{N!}{(N-k)!}(\overline{\varphi}\varphi)^k . \tag{5.40}$$

This implies that

$$\alpha_k = \frac{(N-k)!}{N!k!} . \tag{5.41}$$

For $G = SU(N)$ the only difference is that the sum over the centre only requires the equality of the number of U and U^+ factors modulo N. This allows an additional pair of terms, which can be easily determined to give

$$H_{SU(N)} = \sum_{k=0}^{N} \frac{(N-k)!}{N!k!} [(\overline{\varphi}\varphi)(\overline{\chi}\chi)]^k + \frac{1}{N!} \left[(\overline{\chi}\varphi)^N + (-1)^N (\overline{\varphi}\chi)^N \right] . \quad (5.42)$$

In order to see how to apply this to the $\beta = 0$ integration over the gauge fields, let us consider, for simplicity, staggered quark fields. (The treatment of Wilson quarks is similar, only somewhat more involved due to the extra Dirac spinor indices.) The $\beta = 0$ partition function belonging to the action (5.7) is

$$Z \equiv \int [dU \, d\overline{\psi} \, d\psi] e^{-S_q[U,\psi,\overline{\psi}]} = \int [d\overline{\psi} \, d\psi] e^{-am \sum_x (\overline{\psi}_x \psi_x)}$$

$$\cdot \prod_x \prod_{\mu=1}^{4} \int dU_{x\mu} \exp \left\{ \frac{\alpha_{x\mu}}{2} (\overline{\psi}_{x+\hat{\mu}} U_{x\mu} \psi_x) - \frac{\alpha_{x\mu}}{2} (\overline{\psi}_x U_{x\mu}^+ \psi_{x+\hat{\mu}}) \right\} . \quad (5.43)$$

By using (5.42) Z can be expressed as a function of the composite 'meson' (M_x) and 'baryon' ($B_x, \overline{B}_x$) fields, which are defined with summation convention over the colour indices as

$$M_x \equiv (\overline{\psi}_x \psi_x) = \overline{\psi}_{xc} \psi_{xc} ,$$

$$B_x \equiv \psi_{x1} \psi_{x2} \psi_{x3} = \frac{1}{3!} \epsilon_{cde} \psi_{xc} \psi_{xd} \psi_{xe} ,$$

$$\overline{B}_x \equiv \overline{\psi}_{x1} \overline{\psi}_{x2} \overline{\psi}_{x3} = \frac{1}{3!} \epsilon_{cde} \overline{\psi}_{xc} \overline{\psi}_{xd} \overline{\psi}_{xe} . \quad (5.44)$$

The result is

$$Z = \int [d\overline{\psi} \, d\psi] e^{-am \sum_x M_x} \prod_x \prod_{\mu=1}^{4} \left\{ 1 + \frac{1}{3 \cdot 4} M_{x+\hat{\mu}} M_x \right.$$

$$+ \frac{1}{12 \cdot 16} (M_{x+\hat{\mu}} M_x)^2 + \frac{1}{36 \cdot 64} (M_{x+\hat{\mu}} M_x)^3$$

$$\left. - \frac{\alpha_{x\mu}}{8} (\overline{B}_{x+\hat{\mu}} B_x - \overline{B}_x B_{x+\hat{\mu}}) \right\} . \quad (5.45)$$

This shows that lattice QCD at infinite bare gauge coupling is equivalent to a system of colour-neutral mesons and baryons. The fermionic path integral in (5.45) can be represented by a summation over monomers, dimers and baryonic loops [5.23, 5.24].

5.1.5   Lattice perturbation theory

As a consequence of the property of *asymptotic freedom* in QCD it is possible to use perturbation theory to calculate the cross-sections of many short distance or light cone processes. In order to perform the perturbative calculations a regularization scheme is needed, which allows for the removal of the infinities of perturbation theory by renormalization. In principle the lattice formulation of QCD is a valid perturbative regularization scheme [5.25], but in practice other schemes, like dimensional regularization, are much simpler to use [A3,A4]. The reason to develop lattice perturbation theory for QCD is primarily not the calculation of differential cross-sections of quarks and gluons at large transverse momenta. The necessity of perturbative calculations in lattice QCD arises in the continuum limit. A consequence of asymptotic freedom is that also the bare coupling becomes weak for small lattice spacings. This allows for a control of numerical simulation results by comparing them in the vicinity of the continuum limit to lattice perturbation theory. For instance, the magnitude of lattice artifacts in lattice perturbation theory gives useful hints for the expected size of scale breaking effects in the continuum limit. The *improved actions*, for minimizing lattice artifacts, are also determined by perturbative calculations on the lattice [5.26, 5.27]. Another aspect is to determine the relation of the lattice regularization scheme to other regularization schemes, like for instance the popular $\overline{\text{MS}}$ scheme in dimensional regularization (see section 3.3).

In the present section SU(N) lattice gauge theory with Wilson fermions (quarks) in the N-dimensional fundamental representation will be considered. The Feynman rules in the pure gauge sector are derived in chapter 3. The SU(N) gauge field on the links is written as

$$U_{x\mu} \equiv \exp\left\{ ig \sum_{r=1}^{N^2-1} \frac{1}{2}\lambda_r A_\mu^r(x) \right\} , \qquad (5.46)$$

with a real *gluon field* $A_\mu^r(x)$, the bare gauge coupling g and $\frac{1}{2}\lambda_r$ ($r = 1, 2, \ldots, N^2 - 1$) as the generators of the Lie algebra in the fundamental representation. (For properties of the λ_r Gell-Mann matrices see (8.14)–(8.21).) The momentum space propagator of the gluon field with gauge fixing parameter α is

$$\tilde{\Delta}^{(\alpha)}(k)_{sv,r\mu} = \frac{\delta_{sr}}{\hat{k}^2}\left[\delta_{v\mu} - (1-\alpha)\frac{\hat{k}_v\hat{k}_\mu}{\hat{k}^2} \right] , \qquad (5.47)$$

where the notations in (8.40) are used. The Wilson fermion propagator is derived in (4.121)–(4.124). For perturbation theory, however, the normalization of the fermion field introduced in (4.86), which is useful in numerical simulations, is not convenient. An appropriate choice, instead of (5.5), is defined in the quark part of the action (see (4.85)) by

$$S_q[U, \psi, \overline{\psi}] = \sum_x \left\{ (am + 4r)(\overline{\psi}_x \psi_x) - \frac{1}{2} \sum_{\mu=\pm 1}^{\pm 4} (\overline{\psi}_{x+\hat{\mu}}[r + \gamma_\mu] U_{x\mu} \psi_x) \right\}.$$

(5.48)

This corresponds to the momentum space fermion propagator

$$\tilde{\Delta}_p = \frac{am + \frac{1}{2}r\hat{p}^2 - i\gamma \cdot \overline{p}}{(am + \frac{1}{2}r\hat{p}^2)^2 + \overline{p}^2} = \left(am + \frac{r}{2}\hat{p}^2 + i\gamma \cdot \overline{p} \right)^{-1}$$

(5.49)

(compare to (4.124)).

The lowest orders in g in (5.46) are, with a summation convention on repeated SU(N) indices $r, s, t, \ldots$,

$$U_{x\mu} = 1 + ig\frac{1}{2}\lambda_r A^r_\mu(x)$$

$$- \frac{g^2}{4}\left(\frac{1}{N}A^r_\mu(x)A^r_\mu(x) + d_{rst}\frac{1}{2}\lambda_t A^r_\mu(x)A^s_\mu(x) \right) + O(g^3).$$

(5.50)

Therefore the interaction part of the fermion action in (5.13), (5.20), (5.22) is

$$\mathscr{V} \equiv \sum_{xy}(\overline{\psi}_y V[U]\psi_x) = \frac{1}{2}\sum_x \sum_{\mu=1}^4 \left\{ \left(\overline{\psi}_{x+\hat{\mu}}[r + \gamma_\mu] \left[ig\frac{1}{2}\lambda_r A^r_\mu(x) \right. \right. \right.$$

$$\left. \left. - \frac{g^2}{4}\left(\frac{1}{N}A^r_\mu(x)A^r_\mu(x) + d_{rst}\frac{1}{2}\lambda_t A^r_\mu(x)A^s_\mu(x) \right) + O(g^3) \right] \psi_x \right)$$

$$+ \left(\overline{\psi}_{x-\hat{\mu}}[r - \gamma_\mu] \left[-ig\frac{1}{2}\lambda_r A^r_\mu(x - \hat{\mu}) - \frac{g^2}{4}\left(\frac{1}{N}A^r_\mu(x - \hat{\mu})A^r_\mu(x - \hat{\mu}) \right. \right. \right.$$

$$\left. \left. \left. + d_{rst}\frac{1}{2}\lambda_t A^r_\mu(x - \hat{\mu})A^s_\mu(x - \hat{\mu}) \right) + O(g^3) \right] \psi_x \right) \right\}.$$

(5.51)

In terms of the Fourier-transformed fields in (8.38)–(8.41), with $\overline{p} = p + k$ respectively $\overline{p} = p + k_1 + k_2$, this can be written as

$$\mathscr{V} = ig\int_p \int_k \sum_{\mu=1}^4 \left(\tilde{\overline{\psi}}_{\overline{p}}\frac{1}{2}\lambda_r \tilde{A}^r_\mu(k) \left[\gamma_\mu \cos\frac{(p + \overline{p})_\mu}{2} - ir\sin\frac{(p + \overline{p})_\mu}{2} \right] \tilde{\psi}_p \right)$$

$$- \frac{g^2}{4}\int_p \int_{k_1} \int_{k_2} \sum_{\mu=1}^4 \left(\tilde{\overline{\psi}}_{\overline{p}} \left[\frac{1}{N}\tilde{A}^r_\mu(k_1)\tilde{A}^r_\mu(k_2) + d_{rst}\frac{1}{2}\lambda_t \tilde{A}^r_\mu(k_1)\tilde{A}^s_\mu(k_2) \right] \right.$$

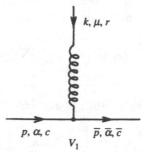

Fig. 5.4. The quark–quark–gluon vertex in lattice perturbation theory. Quarks are denoted by full lines, gluons by curly lines.

$$\cdot \left[r \cos \frac{(p + \bar{p})_\mu}{2} - i\gamma_\mu \sin \frac{(p + \bar{p})_\mu}{2} \right] \tilde{\psi}_p \right) + O(g^3) \,. \tag{5.52}$$

Therefore, including the usual $n!$ factor in a vertex with n identical legs, the vertices corresponding to figs. 5.4 and 5.5 are [5.28]:

$$V_1 = ig \left(\tfrac{1}{2}\lambda_r \right)_{\bar{c}c} \left[\gamma_\mu \cos \frac{(p + \bar{p})_\mu}{2} - ir \sin \frac{(p + \bar{p})_\mu}{2} \right]_{\bar{\alpha}\alpha} \,,$$

$$V_2 = -\frac{g^2}{2} \delta_{\mu_1\mu_2} \left(\frac{1}{N} \delta_{r_1 r_2} + d_{r_1 r_2 t} \tfrac{1}{2}\lambda_t \right)_{\bar{c}c}$$

$$\cdot \left[r \cos \frac{(p + \bar{p})_\mu}{2} - i\gamma_\mu \sin \frac{(p + \bar{p})_\mu}{2} \right]_{\bar{\alpha}\alpha} \,. \tag{5.53}$$

As one can see, besides the quark–quark–gluon (qqg) vertex in fig. 5.4, there is also the qqgg vertex in fig. 5.5. In fact, continuing the expansion of $(U_{x\mu} - 1)$ in (5.50) to higher orders, one gets quark–quark–n-gluon vertices for any n. Also the qqg vertex contains an additional component besides the usual term proportional to γ_μ. All the additional vertices are, however, irrelevant from the point of view of renormalization, because they contain higher powers of the lattice spacing than a^4. This can be seen by noting that the momenta $(p, k, \ldots)$ are proportional to a, the gluon legs $(A_\mu^r(x))$ to a and the quark legs (ψ_x or $\overline{\psi}_x$) to $a^{3/2}$.

As an example, let us consider now the 1-loop contributions to the quark self-energy [5.20, 5.29, 5.30]. The two contributing graphs are shown in fig. 5.6. Using (5.47), (5.49) and (5.53) the inverse propagator $\tilde{\Gamma}_p$ and self-energy Σ_p as functions of the quark four-momentum p are given up to this order by

$$\tilde{\Gamma}_p \equiv \tilde{\Delta}_p^{-1} - \Sigma_p = am + \frac{r}{2}\hat{p}^2 + i\gamma \cdot \bar{p} + g^2 \frac{(N^2 - 1)}{2N}$$

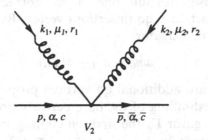

Fig. 5.5. The quark–quark–gluon–gluon vertex in lattice perturbation theory.

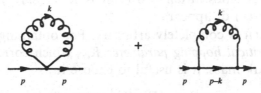

Fig. 5.6. One-loop quark self-energy graphs in lattice perturbation theory.

$$\cdot \int_k \frac{1}{\hat{k}^2} \left\{ \frac{1}{2} \sum_{\mu=1}^{4} \left[1 - (1-\alpha)\frac{\hat{k}_\mu^2}{\hat{k}^2} \right] (r \cos p_\mu - i\gamma_\mu \bar{p}_\mu) \right.$$

$$+ \left[\left(am + \frac{r}{2}\widehat{(p-k)}^2 \right)^2 + \overline{(p-k)}^2 \right]^{-1} \sum_{\mu,\nu=1}^{4} \left[\delta_{\mu\nu} - (1-\alpha)\frac{\hat{k}_\mu \hat{k}_\nu}{\hat{k}^2} \right]$$

$$\cdot \left[\gamma_\nu \cos\left(p - \frac{k}{2}\right)_\nu - ir \sin\left(p - \frac{k}{2}\right)_\nu \right] \left[am + \frac{r}{2}\widehat{(p-k)}^2 - i\gamma \cdot \overline{(p-k)} \right]$$

$$\left. \cdot \left[\gamma_\mu \cos\left(p - \frac{k}{2}\right)_\mu - ir \sin\left(p - \frac{k}{2}\right)_\mu \right] \right\} + O(g^4) . \qquad (5.54)$$

A condensed form of the self-energy for $\alpha = 1$ can be found in [5.31].

For fixed r and α the self-energy is a function of the momentum p and bare parameters g^2 and am: $\Sigma_p \equiv \Sigma_p(g^2, am)$. The bare quark mass in lattice units am is connected to the hopping parameter by $K = (8r + 2am)^{-1}$ (see (4.87)). The dependence of Σ_p on am appears in the quark propagators. There one can use the freedom of splitting the fermion action into a free quadratic part plus the interaction part. Namely, one can add a mass counterterm $\delta(am)(\overline{\psi}_x \psi_x)$ to the quadratic part defining the propagator, which is assumed to be a power series in the bare coupling:

$$\delta(am) \equiv d_1 g^2 + d_2 g^4 + \cdots . \qquad (5.55)$$

In order to compensate for this, one has to subtract the same term again in the interaction part, as a qq (insertion) vertex. By this nothing changes, but in every quark propagator

$$am_0 \equiv am + \delta(am) \tag{5.56}$$

appears, and there are additional qq vertices proportional to $\delta(am)$. This is equivalent to a reshuffling of the bare perturbation series. For instance, in the inverse propagator $\tilde{\Gamma}_p$ the first am coming from $\tilde{\Delta}_p^{-1}$ remains as it is in (5.54), but in the 1-loop contribution $-\Sigma_p(g^2, am_0)$ the original bare mass am is everywhere replaced by am_0. Therefore if $\delta(am)$ is positive, then the spurious singularities at $am = 0$ (that is $K = (8r)^{-1}$) in the original self-energy $\Sigma_p(g^2, am)$ disappear.

Up to now $\delta(am)$ is completely arbitrary. For obtaining a perturbative estimate of the *critical hopping parameter* K_{cr}, which corresponds to zero renormalized quark mass, it is useful to choose it as

$$\delta(am) = -\Sigma_{p=0}(g^2, am_0 = 0)$$

$$= g^2 \frac{(N^2-1)r}{4N} \int_k \frac{1}{\hat{k}^2} \left[3 + \frac{\hat{k}^2(4 - \frac{1}{4}\hat{k}^2) - \bar{k}^2}{\bar{k}^2 + \frac{1}{4}r^2(\hat{k}^2)^2} \right] + O(g^4) . \tag{5.57}$$

In this case the zero momentum inverse propagator $\tilde{\Gamma}_{p=0}$ vanishes at

$$am = -\delta(am) = \Sigma_{p=0}(g^2, am_0 = 0) . \tag{5.58}$$

Since $\delta(am)$ is a function of g^2 alone, this means that $\tilde{\Gamma}_{p=0}$ vanishes along a line in the (g^2, am) (or (g^2, K)) plane. This line is the critical line, where the renormalized quark mass, and due to the Goldstone theorem, also the pion mass, vanishes. The argument for this is based on the definition of the renormalized quark mass m_R at zero four-momentum as $m_R \equiv a^{-1}Z_2\tilde{\Gamma}_{p=0}$, with Z_2 as the quark wave function renormalization factor. Note, however, that such a definition is problematic in QCD (unlike in the symmetric phase of the ϕ^4 theory in chapter 2), because there are serious infrared singularities reflecting quark confinement in perturbation theory. Our assumption is that $\tilde{\Gamma}_{p=0} = 0$ is the condition for zero Goldstone-pion mass. This can also be formulated in the $a \to 0$ continuum limit as the vanishing of the linearly divergent contribution to the renormalized quark mass m_R. An important consequence of (5.57) is that this definition of the critical line in the bare parameter space is gauge invariant: the α-dependence present in (5.54) at $am \neq 0$ and/or $p \neq 0$ disappears at $am = p = 0$.

In terms of the hopping parameter the critical line is given up to $O(g^2)$ by

$$K_{cr} = \left[8r + 2\Sigma_{p=0}(g^2, am_0 = 0) \right]^{-1}$$

$$= \frac{1}{8r} \left\{ 1 + g^2 \frac{(N^2-1)}{16N} \int_k \frac{1}{\bar{k}^2} \left[3 + \frac{\hat{k}^2(4 - \frac{1}{4}\hat{k}^2) - \bar{k}^2}{\bar{k}^2 + \frac{1}{4}r^2(\hat{k}^2)^2} \right] + O(g^4) \right\} . \quad (5.59)$$

The quantity m_0 defined in (5.56) can be considered as the real *bare quark mass* (instead of the original bare mass m). In terms of the hopping parameter m_0 is obtained from (5.58) and (5.59) as

$$m_0 = \frac{1}{2a} \left(\frac{1}{K} - \frac{1}{K_{cr}} \right) . \quad (5.60)$$

The lattice sums (or integrals in the limit of an infinitely large lattice) can be evaluated numerically [5.32]:

$$\mathscr{I}_1 \equiv \int_k \frac{1}{\bar{k}^2} \equiv \frac{1}{\Omega} \sum_k \frac{1}{\bar{k}^2} = 0.1549\ldots ,$$

$$\mathscr{I}_2(r) \equiv \int_k \frac{\hat{k}^2(4 - \frac{1}{4}\hat{k}^2) - \bar{k}^2}{\bar{k}^2[\bar{k}^2 + \frac{r^2}{4}(\hat{k}^2)^2]} \xrightarrow{r=1} 0.1865\ldots . \quad (5.61)$$

For $N = 3$ colours and $r = 1$ this gives

$$K_{cr} = \frac{1}{8}(1 + g^2 \cdot 0.1085 + \cdots) = \frac{1}{8}(1 + 0.651\beta^{-1} + \cdots) . \quad (5.62)$$

In the region of numerical simulation data ($g \simeq 1$, $\beta \simeq 6$) (5.62) does not give a good approximation [5.33]. This can be due to higher orders or to non-perturbative contributions. In fact, since we are here calculating a quantity multiplied by a, which is divergent linearly with a^{-1} in the continuum limit, the appearance of finite non-perturbative contributions, behaving like $\exp(-\mathrm{const.}/g^2)$, is possible. As we shall see in the next subsection, a has exactly the behaviour $\exp(-\mathrm{const.}/g^2)$ for $g^2 \to 0$. Therefore the calculation of quantities with a power divergence is in general not reliable in lattice perturbation theory. In spite of this, sometimes reasonably good results can be obtained if, instead of the bare parameter g^2, the series is expressed in powers of some physical coupling, for instance, the renormalized coupling defined by the static potential energy [5.33].

5.1.6 Continuum limit

In the present section the continuum limit of SU(N) lattice QCD with N_f degenerate quark flavours will be considered. In this case, besides the gauge coupling, there is only a single quark mass parameter. (The generalization to several mass parameters for non-degenerate quark flavours

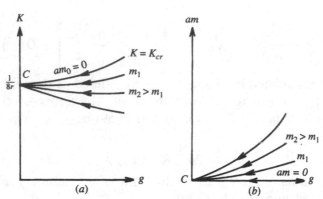

Fig. 5.7. The lines of constant renormalized quark mass for Wilson quarks in the (g, K)-plane (a), and for staggered quarks in the (g, am)-plane (b). The arrows point in the direction of decreasing lattice spacing. The continuum limit is reached at point C.

is straightforward.) From the point of view of asymptotic freedom the continuum limit is similar to pure non-Abelian gauge theory (see chapter 3). The presence of a mass parameter besides the coupling is also characteristic of the ϕ^4 theory discussed in chapter 2. As a consequence of asymptotic freedom, the continuum limit is at vanishing bare gauge coupling ($g = 0$, $\beta = \infty$). The bare quark mass parameter in lattice units, namely am_0 given by (5.60) for Wilson quarks, or $am = 0$ in the action (5.7) for staggered quarks, also tends to zero, because of $a \to 0$. Therefore the lines of constant renormalized quark mass behave qualitatively as shown by fig. 5.7.

In order to describe the continuum limit in more detail let us first restrict ourselves to the special case of zero quark mass. For the definition of a physical coupling let us consider the static energy of an infinitely heavy external quark–antiquark source pair ('static potential'), which can be calculated from the expectation values of Wilson loops (see chapter 3). In the continuum limit this can be written as

$$V(r) = -\frac{g_{POT}^2(r)}{4\pi r}\frac{(N^2 - 1)}{2N} + V_0 \, . \tag{5.63}$$

In this formula V_0 is a constant self-energy and $g_{POT}(r)$ is the *running coupling constant* at the physical distance r. The value of V_0 can be fixed, for instance, at large distances by requiring that the energy of the heavy quark–antiquark system is equal either to zero or twice the energy of the lowest bound state of a light dynamical quark in the field of the infinitely heavy quark. An alternative way to get rid of V_0 is to consider

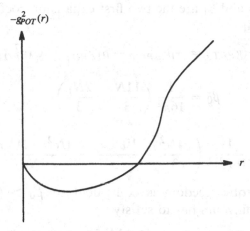

Fig. 5.8. The qualitative behaviour of the running coupling constant squared $g_{POT}^2(r)$ as a function of the distance r. The spike at $r = 0$ is the consequence of asymptotic freedom. The nearly quadratic part at intermediate distances corresponds to the string-like behaviour. The linear behaviour at large r is due to the screening of the external colour charges by virtual quark–antiquark pairs. The slope of this linear part depends on the choice of V_0.

the force, which is the derivative of $V(r)$ with respect to r. Although the Coulomb-like form with a constant g_{POT}^2 is a good approximation only at small distances, we consider (5.63) to be valid at arbitrary r by definition. The qualitative behaviour of $g_{POT}^2(r)$ is shown by fig. 5.8.

g_{POT} is a possible definition of a renormalized coupling of QCD, depending on the distance scale r. Correspondingly it satisfies the renormalization group equation

$$\frac{\mathrm{d}g_{POT}}{\mathrm{d}\log r} = r\frac{\mathrm{d}g_{POT}}{\mathrm{d}r} = -\beta_{POT}(g_{POT}) , \qquad (5.64)$$

with a particular Callan–Symanzik β-function $\beta_{POT}(g_{POT})$ which belongs to the running coupling constant $g_{POT}(r)$. The solution of this equation is analogous to the corresponding one considered in section 3.3, and can be expressed as

$$r\Lambda_{POT} \equiv (\beta_0 g_{POT}^2)^{-\beta_1/(2\beta_0^2)} e^{-1/(2\beta_0 g_{POT}^2)}$$

$$\cdot \exp\left\{-\int_0^{g_{POT}} \mathrm{d}g \left[\frac{1}{\beta_{POT}(g)} + \frac{1}{\beta_0 g^3} - \frac{\beta_1}{\beta_0^2 g}\right]\right\}$$

$$= (\beta_0 g_{POT}^2)^{-\beta_1/(2\beta_0^2)} e^{-1/(2\beta_0 g_{POT}^2)}(1 + O(g_{POT}^2)) . \qquad (5.65)$$

Here the lambda parameter Λ_{POT} appears as an integration constant.

The coefficients β_0 and β_1 are the two first expansion coefficients of β_{POT}, namely [A3,A5,A8]:

$$\beta_{POT}(g_{POT}) = -\beta_0 g_{POT}^3 - \beta_1 g_{POT}^5 - O(g_{POT}^7) ,$$

$$\beta_0 = \frac{1}{16\pi^2} \left(\frac{11N}{3} - \frac{2N_f}{3} \right) ,$$

$$\beta_1 = \frac{1}{(16\pi^2)^2} \left(\frac{34N^2}{3} - \frac{10NN_f}{3} - \frac{(N^2-1)N_f}{N} \right) . \tag{5.66}$$

Note that asymptotic freedom is equivalent to $\beta_0 > 0$, therefore the number of quark flavours has to satisfy

$$N_f < \frac{11N}{2} . \tag{5.67}$$

Since QCD at zero quark mass has only a single free parameter, any physical quantity P is a function of $g_{POT}(r)$. The arbitrariness of the reference scale (here distance) r is expressed by the *renormalization group equation (RGE)*

$$\left\{ -r\frac{\partial}{\partial r} + \beta_{POT}(g_{POT})\frac{\partial}{\partial g_{POT}} \right\} P(r, g_{POT}; x) = 0 . \tag{5.68}$$

The general physical quantity $P(r, g_{POT}; x)$ is considered here as a function of the scale r, the renormalized coupling $g_{POT}(r)$, and possibly some kinematical variables summarized by x. A general solution of the RGE (5.68) is a function of the standard solution $\Lambda_{POT} = \Lambda_{POT}(r, g_{POT})$, which can be obtained from (5.65) as a function of r and g_{POT}. The coupling is dimensionless, therefore the lambda parameter Λ_{POT} has a dimension of a mass. As a consequence, for instance, any hadron mass m is given by $m = c_m \Lambda_{POT}$, where c_m is a dimensionless number. A hadronic scattering amplitude $A(s, t)$, with a dimension of mass squared, is $A(s, t) = \Lambda_{POT}^2 f_A(s/\Lambda_{POT}^2, t/\Lambda_{POT}^2)$, where f_A is a function of its dimensionless arguments. Similar expressions hold for any other physical quantity.

Of course, it is not necessary to choose for the running coupling just $g_{POT}(r)$. There are also other physical quantities which are proportional to some other running coupling $g(\mu)$ at mass scale μ. Examples of such quantities are, for instance, different scattering amplitudes at momenta proportional to μ. (If instead of the mass scale or momentum scale μ a distance scale is used, as in $g_{POT}(r)$, then the corresponding mass scale is defined to be $\mu \equiv r^{-1}$.) The renormalization group equation is in general

$$\left\{ \mu\frac{\partial}{\partial \mu} + \beta(g)\frac{\partial}{\partial g} \right\} P(\mu, g; x) = 0 . \tag{5.69}$$

According to section 1.7 the connection between two possible choices, g_I with the Callan–Symanzik β-function $\beta_I(g_I)$ and g_{II} with $\beta_{II}(g_{II})$, is given by

$$g_{II} = g_I z(g_I), \qquad \mu_{II} = \mu_I. \tag{5.70}$$

The function $z(g_I)$ is assumed to have the expansion

$$z(g_I) = 1 + a_1 g_I^2 + a_2 g_I^4 + \cdots. \tag{5.71}$$

This implies that the two β-functions are related by

$$\beta_{II}(g_{II}) = \beta_I(g_I)\frac{dg_{II}}{dg_I} = \left[\beta_I(g_I)\left(z(g_I) + g_I \frac{dz}{dg_I}\right)\right]_{g_I = g_I(g_{II})}. \tag{5.72}$$

As a consequence, the first two expansion coefficients β_0 and β_1 are universal:

$$\beta_{0,II} = \beta_{0,I} \equiv \beta_0, \qquad \beta_{1,II} = \beta_{1,I} \equiv \beta_1. \tag{5.73}$$

The relation between the two Λ-parameters is the same as in pure gauge theory (see section 3.3):

$$\Lambda_{II}/\Lambda_I = \exp\left\{\frac{1}{2\beta_0}\left(\frac{1}{g_I^2} - \frac{1}{g_{II}^2}\right) + O(g_I^2)\right\} = \exp\left\{\frac{a_1}{\beta_0}\right\}. \tag{5.74}$$

The first equality here can be easily proven from the explicit form of $\Lambda_{I,II}$ in (5.65). The second one follows from the fact that the derivative of Λ_{II}/Λ_I with respect to g_I is zero. The consequence of this formula is that the ratio of the Λ-parameters can be determined by a 1-loop perturbative calculation [5.34].

The connection of the Λ-parameter Λ_{POT} to the popular $\Lambda_{\overline{MS}}$ can be read off from the perturbative expansion of $g_{POT}^2(r)$ in powers of $g_{\overline{MS}}^2(\mu)$ [5.35]:

$$g_{POT}^2(r) = g_{\overline{MS}}^2(\mu)\left\{1 + \frac{g_{\overline{MS}}^2(\mu)}{16\pi^2}\left[\frac{22N}{3}\left(\log(\mu r) + \gamma\right) + \frac{31N}{9}\right.\right.$$

$$\left.\left. - \frac{4N_f}{3}\left(\log(\mu r) + \gamma\right) - \frac{10N_f}{9}\right] + O(g_{\overline{MS}}(\mu)^4)\right\}. \tag{5.75}$$

Here $\gamma = 0.577215665\ldots$ is Euler's constant. The physical quantities as functions of $g_{\overline{MS}}(\mu)$ also satisfy the RGE (5.69). The corresponding β-function can be denoted by $\beta_{\overline{MS}}(g_{\overline{MS}})$, and for the ratio of Λ-parameters one can use (5.74) to obtain

$$\Lambda_{POT}/\Lambda_{\overline{MS}} = \exp\left\{\gamma + \frac{31N - 10N_f}{66N - 12N_f}\right\}. \tag{5.76}$$

This gives, for instance, for QCD with three massless quarks ($N = N_f = 3$) $\log(\Lambda_{POT}/\Lambda_{\overline{MS}}) = 0.966105\ldots$.

After this preparation let us now return to the discussion of the continuum limit of lattice QCD with massless quarks. One imagines that along the critical line for zero quark mass (fig. 5.7) the lattice spacing a is a function of the bare gauge coupling g. Inversely, as a function of the lattice spacing the bare coupling is $g(a)$. The existence of the continuum limit requires that for $a \to 0$ the physical quantities $P(a, g)$ become independent of the lattice spacing. (This circumstance is often referred to as 'scaling'.) Therefore they satisfy the RGE

$$\left\{ -a\frac{\partial}{\partial a} + \beta_{LAT}(g)\frac{\partial}{\partial g} \right\} P(a, g) = O(a) . \tag{5.77}$$

On the right hand side of this equation some 'scale breaking lattice artifacts' appear, which vanish for $a \to 0$ as some power of the lattice spacing a. These lattice artifacts in general also depend on the particular physical quantity P. $\beta_{LAT}(g)$ is the lattice β-function, which defines the corresponding Λ-parameter $\Lambda_{LAT} = \Lambda_{LAT}(a, g)$ in analogy to (5.65) by

$$a\Lambda_{LAT} \equiv (\beta_0 g^2)^{-\beta_1/(2\beta_0^2)} e^{-1/(2\beta_0 g^2)}$$

$$\cdot \exp\left\{ -\int_0^g dh \left[\frac{1}{\beta_{LAT}(h)} + \frac{1}{\beta_0 h^3} - \frac{\beta_1}{\beta_0^2 h} \right] \right\}$$

$$= (\beta_0 g^2)^{-\beta_1/(2\beta_0^2)} e^{-1/(2\beta_0 g^2)} (1 + O(g^2)) . \tag{5.78}$$

Apart from the first two universal coefficients β_0 and β_1, the β-function $\beta_{LAT}(g)$ and Λ-parameter Λ_{LAT} depend on the choice of the lattice action.

In order to obtain the ratio of Λ_{LAT} to, say, Λ_{POT} one has to calculate the static potential in bare perturbation theory up to 1-loop order:

$$g_{POT}^2(r) = g^2(a)\left\{ 1 + g^2(a)2\beta_0 \left(\log\left(\frac{r}{a}\right) + A_1 \right) + O(g(a)^4) \right\} . \tag{5.79}$$

According to (5.74) this corresponds to

$$\Lambda_{POT}/\Lambda_{LAT} = e^{A_1} . \tag{5.80}$$

Similarly, one can also determine the ratio of the lattice Λ-parameter to $\Lambda_{\overline{MS}}$. In QCD with massless Wilson quarks and $r = 1$ one obtains [5.36]:

$$\Lambda_{LAT}/\Lambda_{\overline{MS}} = \exp\left\{ \frac{1}{\beta_0} \left(\frac{1}{16N} - N \cdot 0.084978\ldots + \frac{N_f}{2} \cdot 0.006887\ldots \right) \right\} . \tag{5.81}$$

In fact, it is easier to calculate this by the background field method [5.37], than to determine A_1 in (5.79) by lattice perturbation theory from the lattice Coulomb potential between static quarks. The value in (5.81), for instance, for $N = N_f = 3$ is $\log(\Lambda_{LAT}/\Lambda_{\overline{MS}}) = -3.926263\ldots$. Combining (5.76) and (5.81) one can indirectly determine also $\Lambda_{POT}/\Lambda_{LAT}$. The obtained value is consistent with the results of an approximate numerical calculation done directly on the Wilson loops [5.38].

It is worth emphasizing that the 1-loop static potential in lattice perturbation theory given by (5.63) and (5.79) not only determines the relation of the Λ-parameters Λ_{LAT} and Λ_{POT} by the value of the constant A_1, but above all shows how the bare coupling $g(a)$ depends on the lattice spacing a. This dependence is given by the relation (5.78), where Λ_{LAT} is a constant, namely the only dimensionful free parameter of lattice QCD with massless quarks. The function $g(a)$ is thus determined by lattice perturbation theory from the short distance behaviour of the static potential. It is an important assumption that the same function $g(a)$ is valid also for the whole potential at all distances, for instance, also for the region characterized by the string tension. In addition, the same has to hold not only for the potential, but also for an arbitrary physical quantity. This *scaling hypothesis* has to be true up to order $O(a)$, at least if a and $g(a)$ are small enough.

Up to now the special case of zero quark mass was considered. The generalization to massive quarks implies the increase of the number of free parameters. In the present section only the case of a single quark mass parameter is explicitly considered. In this case a single suitably chosen mass ratio can fix the renormalized (physical) quark mass. This means that, for instance, in fig. 5.7 the lines of constant renormalized quark mass can be characterized by the ratio of, say, the lowest pseudoscalar boson mass to the lowest baryon mass. In reality the quark masses are different from each other. Considering (u, d, c, s, t, b) quarks, there are six free renormalized quark mass parameters, which can be fixed, for instance, by specifying the mass ratios of p/π, n/π, K/π, D/π, B/π and T/π (π = pion, p = proton, n = neutron, K = kaon, D, B, T = some bosons containing c-, b- and t-quarks, respectively).

The quark mass also appears in the RGE corresponding to (5.69). In a mass-independent renormalization scheme, where the Callan–Symanzik β- and γ-functions are defined in the massless theory, one can still use a homogeneous RGE [A5]:

$$\left\{\mu\frac{\partial}{\partial\mu} + \beta(g)\frac{\partial}{\partial g} - \gamma(g)m\frac{\partial}{\partial m}\right\} P(\mu, g, m; x) = 0 . \tag{5.82}$$

Here μ is the mass scale, g the physical gauge coupling and m the physical quark mass. P is a general physical quantity depending on (μ, g, m) and the kinematical variables summarized by x. The β-function $\beta(g)$ has the same small g expansion as in (5.66). The γ-function is similarly given by the series

$$\gamma(g) = \gamma_0 g^2 + \gamma_1 g^4 + O(g^6) \,,$$

$$\gamma_0 = \frac{1}{16\pi^2} \frac{3(N^2 - 1)}{N} \,. \tag{5.83}$$

The RGE (5.82) has two independent standard solutions: the Λ-parameter $\Lambda(\mu, g)$ and the *renormalization group invariant (RGI) quark mass* $M(m, g)$:

$$\Lambda \equiv \mu(\beta_0 g^2)^{-\beta_1/(2\beta_0^2)} e^{-1/(2\beta_0 g^2)} \exp\left\{ -\int_0^g dh \left[\frac{1}{\beta(h)} + \frac{1}{\beta_0 h^3} - \frac{\beta_1}{\beta_0^2 h} \right] \right\} \,,$$

$$M = m(2\beta_0 g^2)^{-\gamma_0/(2\beta_0)} \exp\left\{ \int_0^g dh \left[\frac{\gamma(h)}{\beta(h)} + \frac{\gamma_0}{\beta_0 h} \right] \right\} \,. \tag{5.84}$$

Λ and M are also called standard *renormalization group invariants*. Every physical quantity in massive QCD is a function of Λ and M.

In order to study the scheme dependence of the RGE, let us introduce the dimensionless measures for the quark mass m and RGI quark mass M, respectively, as

$$l \equiv \frac{m}{\mu} \,, \qquad \lambda \equiv \frac{M}{\Lambda} \,. \tag{5.85}$$

Let us write the RGE (5.82) in a more general and more convenient form, omitting the kinematical variables x, as

$$\left\{ \mu \frac{\partial}{\partial \mu} + \beta^{(g)}(g, \lambda) \frac{\partial}{\partial g} + \beta^{(l)}(g, \lambda) l \frac{\partial}{\partial l} \right\} P(\mu, g, l) = 0 \,. \tag{5.86}$$

The dependence of the renormalization scheme on λ means that these are now, in general, mass-dependent schemes. $\beta^{(g)}$ and $\beta^{(l)}$ can be considered either as functions of (g, λ) or of (g, l). Comparing (5.82) and (5.86) one can establish the relation to the original, mass-independent Callan–Symanzik functions:

$$\beta^{(g)}(g, \lambda) = \beta(g) \,, \qquad \beta^{(l)}(g, \lambda) = -1 - \gamma(g) \,. \tag{5.87}$$

Let us now imagine that there are two schemes for the definition of the gauge coupling and quark mass, which are denoted by the indices I and II. The reference mass scales are assumed to be identical: $\mu_I = \mu_{II} \equiv \mu$ and, similarly to (5.70) and (5.71), the relations between (g_I, l_I) and (g_{II}, l_{II}) have the form

$$g_{II} = g_I z^{(g)}(g_I, \lambda) = g_I \left(1 + a_1^{(g)}(\lambda) g_I^2 + \cdots \right) \,,$$

$$l_{II} = l_I z^{(l)}(g_I, \lambda) = l_I \left(1 + a_1^{(l)}(\lambda) g_I^2 + \cdots \right) \,. \tag{5.88}$$

Since the RGE (5.86) holds in both renormalization schemes, the relations between the Callan–Symanzik functions are

$$\beta_{II}^{(g)}(g_{II}, \lambda) = \beta_I^{(g)} \frac{dg_{II}}{dg_I} = \left[\beta_I^{(g)}(g_I, \lambda) \left(z^{(g)} + g_I \frac{\partial z^{(g)}}{\partial g_I} \right) \right]_{g_I = g_I(g_{II}, \lambda)},$$

$$\beta_{II}^{(l)}(g_{II}, \lambda) = \left[\beta_I^{(l)}(g_I, \lambda) + \beta_I^{(g)}(g_I, \lambda) \frac{\partial \log z^{(l)}}{\partial g_I} \right]_{g_I = g_I(g_{II}, \lambda)}. \tag{5.89}$$

This implies that the lowest expansion coefficients β_0, β_1 and γ_0 are scheme-independent. The relation between the standard RGIs is now

$$\Lambda_{II}/\Lambda_I = \exp \left\{ \frac{a_1^{(g)}(\lambda)}{\beta_0} \right\}, \qquad M_{II} = M_I. \tag{5.90}$$

For the ratio of the Λ-parameters this corresponds to (5.74) in the single variable case. In addition, as one can see, the RGI quark masses turn out to be scheme-independent.

In the continuum limit of massive lattice QCD the physical quantities become independent of the lattice spacing a. This is expressed by a RGE similar to (5.77), namely

$$\left\{ -a \frac{\partial}{\partial a} + \beta_{LAT}^{(g)}(g, \lambda) \frac{\partial}{\partial g} + \beta_{LAT}^{(l_0)}(g, \lambda) l_0 \frac{\partial}{\partial l_0} \right\} P(a, g, l_0) = O(a). \tag{5.91}$$

In this equation g is the bare gauge coupling, and $l_0 \equiv am_0$ the bare quark mass in lattice units. The lattice spacing a can be considered to be a function of (g, λ) or (g, l_0). The lattice β-functions $\beta_{LAT}^{(g)}(g, \lambda)$ and $\beta_{LAT}^{(l_0)}(g, \lambda)$ depend, in general, on the dimensionless RGI quark mass parameter λ defined in (5.85). The lowest expansion coefficients of $\beta_{LAT}^{(g)}$ and $\beta_{LAT}^{(l_0)}$, namely β_0, β_1 and γ_0, are the universal ones given by (5.87), (5.66) and (5.83), but the higher ones do, in general, also depend on the choice of the lattice action and bare quark mass definition. For a more detailed discussion of the quark mass parameters of some QCD lattice actions see also [5.30, 5.39].

5.2 Hadron spectrum

The first task in numerical simulations of lattice QCD is to determine the lowest hadron masses. Since most of these masses are experimentally known to a high precision, the comparison to the numerical results is an important check of non-perturbative QCD. The first step in the numerical determination of other hadronic parameters, like matrix elements or thermodynamic properties, is also the measurement of the hadron masses.

5.2.1   *Hadronic two-point functions*

The hadron masses can be extracted from hadronic two-point functions in a similar way as in scalar theories (chapter 2) and pure gauge theories (chapter 3). The vacuum expectation value of the time-ordered product of two hadron operators, after Wick-rotation to Euclidean times y_4 and x_4, is given by

$$\langle 0|T\{B(y_4)A(x_4)\}|0\rangle = \begin{cases} \langle 0|B(y_4)A(x_4)|0\rangle & \text{if} \quad y_4 \geq x_4 \\ \pm\langle 0|A(x_4)B(y_4)|0\rangle & \text{if} \quad x_4 > y_4 \end{cases}$$

$$= \langle B(y_4)A(x_4)\rangle$$

$$\equiv \lim_{T\to\infty} \left(\text{Tr}\, e^{-TH}\right)^{-1} \cdot \begin{cases} \text{Tr}\left[B(0)e^{-(y_4-x_4)H}A(0)e^{-(T-y_4+x_4)H}\right] \\ \pm\text{Tr}\left[A(0)e^{-(x_4-y_4)H}B(0)e^{-(T-x_4+y_4)H}\right] \end{cases} . \quad (5.92)$$

Here H is the Hamiltonian and the sign $+$ $(-)$ stands for bosonic (fermionic) operators. In the Euclidean time direction the lattice extension is T and periodic (or antiperiodic) boundary conditions are assumed.

Let us first consider in more detail the bosonic case. Inserting in the traces complete sets of intermediate states, for large T and $|y_4 - x_4|$ only the states with lowest energy contribute, and we have

$$\langle B(y_4)A(x_4)\rangle \quad \xrightarrow{T,|y_4-x_4|\to\infty}$$

$$\cdot \begin{cases} c_{BA}e^{-(y_4-x_4)E_{BA}} + c_{AB}e^{-(T-y_4+x_4)E_{AB}} \\ c_{AB}e^{-(x_4-y_4)E_{AB}} + c_{BA}e^{-(T-x_4+y_4)E_{BA}} \end{cases} , \quad (5.93)$$

where E_{BA} and E_{AB} are, respectively, the energies of the lowest contributing states $|n_{BA}\rangle$ and $|n_{AB}\rangle$ in lattice units, if the vacuum energy is set equal to zero by definition. The constant factors c_{BA} and c_{AB} are given by the matrix elements between the vacuum and the lowest states:

$$c_{BA} = \langle 0|B(0)|n_{BA}\rangle\langle n_{BA}|A(0)|0\rangle \, ,$$

$$c_{AB} = \langle 0|A(0)|n_{AB}\rangle\langle n_{AB}|B(0)|0\rangle \, . \quad (5.94)$$

In the special case of $B(y_4) = A(y_4)^+$ these are real and equal to each other: $c \equiv c_{BA} = c_{AB}$, and the corresponding energies are also equal: $E \equiv E_{BA} = E_{AB}$, therefore

$$\langle A(y_4)^+A(x_4)\rangle \quad \xrightarrow{T,|y_4-x_4|\to\infty} \quad c\left[e^{-|y_4-x_4|E} + e^{-(T-|y_4-x_4|)E}\right] \, . \quad (5.95)$$

Besides this dominant contribution with the lowest energy there are, of course, also subdominant ones with the next lowest energies in the sector with the quantum numbers specified by the operator $A(x_4)$. On a finite lattice the energy spectrum is discrete, therefore at large T and $|y_4 - x_4|$ the subdominant contributions are exponentially suppressed.

In most applications the local operators are summed over the space-like coordinates. These *timeslices* (S_t, $t = 1, 2, \ldots, T$) are projecting out the intermediate states with zero spatial momentum. Therefore, if there is a single-particle state in the given channel, then the lowest energy is equal to its mass m_A, and the asymptotic behaviour for large time separations is

$$\langle S_{t_2}^+ S_{t_1} \rangle \xrightarrow{T, |t_2 - t_1| \to \infty} c_0 \left[e^{-|t_2 - t_1| m_A} + e^{-(T - |t_2 - t_1|) m_A} \right]$$

$$= 2c_0 e^{-T m_A / 2} \cosh \left[m_A \left(\frac{T}{2} - |t_2 - t_1| \right) \right] . \tag{5.96}$$

The mass m_A can be determined by fitting the expectation value of the timeslices at large time distances by such an asymptotic form.

The asymptotic formula (5.96) can also be understood as a consequence of the Källen–Lehmann representation of two-point functions [A3]. In Euclidean coordinates in infinite space we have, for simplicity, for a spinless real bosonic field $\varphi(x)$,

$$\langle 0 | T \{ \varphi(y) \varphi(x) \} | 0 \rangle = \int_{m_0^2}^{\infty} \mathrm{d}m^2 \, \rho(m^2) \Delta_E(y - x; m^2) . \tag{5.97}$$

Here $\rho(m^2)$ is a positive spectral weight function and

$$\Delta_E(x; m^2) \equiv \int \frac{\mathrm{d}^4 k}{(2\pi)^4} \frac{e^{ik \cdot x}}{m^2 + k^2} \tag{5.98}$$

is the Euclidean scalar propagator with mass m. Stable single-particle states with mass m_n contribute to $\rho(m^2)$ by a term proportional to a δ-function: $Z_n \delta(m^2 - m_n^2)$, whereas multiparticle intermediate states give a continuum contribution. (The factor Z_n multiplying the δ-function gives the field normalization, as defined in (8.50).) By an integration over three-space one obtains

$$\int \mathrm{d}^3 \mathbf{x} \, \langle 0 | T \{ \varphi(\mathbf{y}, y_4) \varphi(\mathbf{x}, x_4) \} | 0 \rangle = \int_{m_0}^{\infty} \mathrm{d}m \, \rho(m^2) e^{-m|y_4 - x_4|} . \tag{5.99}$$

For large Euclidean time separations the lowest mass m_0 dominates, and we have

$$m_0 = - \lim_{|y_4 - x_4| \to \infty} \frac{1}{|y_4 - x_4|} \log \int \mathrm{d}^3 \mathbf{x} \, \langle 0 | T \{ \varphi(\mathbf{y}, y_4) \varphi(\mathbf{x}, x_4) \} | 0 \rangle . \tag{5.100}$$

This corresponds to the $T \to \infty$ limit of (5.96).

In the reduction formulas (8.58)–(8.64), if the time-ordered products in them are continued to imaginary times, the particle poles are reproduced by contributions which are exponentially decreasing in Euclidean time separation. (Note that after the continuation $x_0 = -ix_4$ the Fourier transform becomes a Laplace transform.)

Another possibility in (5.99) is to perform a Fourier transformation to momentum space:

$$\int dx_4\, e^{-ip_4x_4} \int d^3x\, \langle 0|T\{\varphi(\mathbf{y}, y_4)\varphi(\mathbf{x}, x_4)\}|0\rangle = \int_{m_0^2}^{\infty} dm^2\, \frac{\rho(m^2)}{m^2 + p_4^2} . \quad (5.101)$$

This shows the particle poles in the (real) energy variable $E \equiv -ip_4$. Both (5.100) and (5.101) can be used to extract the lowest masses from the expectation value of timeslices. Of course, one has to keep in mind that these are valid on infinite lattices in the continuum limit, therefore on a finite lattice there are corrections arising from lattice artifacts and finite volume effects.

In the case of fermionic timeslice operators for baryons $(B_{t_2}, \overline{B}_{t_1})$ the asymptotics at large time separations is again dominated by the contribution of the lowest mass state in the given channel. In the continuum limit one can take the $am \to 0$ limit of the timeslices of free Wilson fermions derived in section 4.2 (see (4.141)), namely

$$\langle B_{t_2}\overline{B}_{t_1}\rangle \propto \frac{1}{2}[1 + \text{sign}\,(t_2 - t_1)\gamma_4]e^{-|t_2-t_1|m_B} . \quad (5.102)$$

This is valid for $T = \infty$. m_B is the lowest mass in lattice units, which belongs to the quantum numbers specified by B_t. The factor $(1 \pm \gamma_4)/2$ is the Dirac projection matrix arising from the spin sum at zero spatial momentum. In general, the simplest operators used for the description of the baryons are coupled to both parities, therefore the asymptotic form contains the lowest contributions of both parities (with masses m_{B+} and m_{B-}, respectively). Since the sign of γ_4 is reversed for negative parity states, we get, with some constant factors c_{B+} and c_{B-},

$$\langle B_{t_2}\overline{B}_{t_1}\rangle \quad \xrightarrow{|t_2-t_1|\to\infty} \quad c_{B+}[1 + \text{sign}\,(t_2 - t_1)\gamma_4]e^{-|t_2-t_1|m_{B+}}$$

$$+ c_{B-}[1 - \text{sign}\,(t_2 - t_1)\gamma_4]e^{-|t_2-t_1|m_{B-}} . \quad (5.103)$$

According to (4.148), for finite time extension T with periodic $(b = +1)$ or antiperiodic $(b = -1)$ boundary conditions, this becomes for $0 \le t_2 - t_1 \le T$

$$\langle B_{t_2}\overline{B}_{t_1}\rangle \quad \xrightarrow{T,(t_2-t_1)\to\infty}$$

$$(1 + \gamma_4)\left[c_{B+}e^{-(t_2-t_1)m_{B+}} + c_{B-}be^{-(T-t_2+t_1)m_{B-}}\right]$$

$$+ (1 - \gamma_4) \left[c_{B+} b e^{-(T - t_2 + t_1) m_{B+}} + c_{B-} e^{-(t_2 - t_1) m_{B-}} \right] . \qquad (5.104)$$

In order to project out the channels with different quantum numbers the appropriate hadronic operators have to be constructed from the quark and gluon fields. The choice of the composite operators is to a large extent arbitrary. In fact, for given values of the coupling constants one has to find the optimal operator, which has a strong enough coupling to the hadron in question and, at the same time, can be evaluated without too great difficulties. The simplest local operators will be considered in the present subsection. The optimization of the space-like extension will be dealt with in the next subsection.

Let us now restrict ourselves to Wilson quarks and to the ground state mesons and baryons (in the sense of SU(6)) containing u, d and s quarks. (The staggered quarks are substantially more complicated, due to the non-conservation of flavours, and will not be considered here.) The spin dependence of the operators is dictated in this case by the relativistic generalization of SU(6) symmetry (for a review and references, see [5.40]). The $J^{PC} = 0^{-+}$ pseudoscalar mesons are described by bilinear quark–antiquark composite operators like, for instance,

$$\mathcal{M}_x^{\pi^+} \equiv \bar{d}_{x\alpha c} \gamma_{5,\alpha\beta} u_{x\beta c} , \qquad \mathcal{M}_x^{K^+} \equiv \bar{s}_{x\alpha c} \gamma_{5,\alpha\beta} u_{x\beta c} . \qquad (5.105)$$

Here $u_{x\alpha c}$, $d_{x\alpha c}$ and $s_{x\alpha c}$ stand for the $q = u$, $q = d$ and $q = s$ flavour components of the quark field $\psi_{qx\alpha c}$, respectively. (As usual, $\alpha, \beta, \ldots$ are the Dirac spinor, and $c, d, \ldots$ the SU(3) colour indices, both with summation convention assumed.) The corresponding 1^{--} vector meson fields are ($k = 1, 2, 3$):

$$\mathcal{M}_{xk}^{\rho^+} \equiv \bar{d}_{x\alpha c} \gamma_{k,\alpha\beta} u_{x\beta c} , \qquad \mathcal{M}_{xk}^{K^{*+}} \equiv \bar{s}_{x\alpha c} \gamma_{k,\alpha\beta} u_{x\beta c} . \qquad (5.106)$$

For the baryons the trilinear (three-quark) composite operators can be chosen in different ways (see, for example, [5.41]). Non-relativistic operators can be constructed in the non-relativistic representation of the Dirac matrices given by (8.8) and (8.9) [5.42]. Relativistic baryon operators involve the *charge conjugation* Dirac matrix C, which satisfies

$$C \gamma_\mu C^{-1} = -\gamma_\mu^T ,$$

$$- C = C^T = C^{-1} = C^+ . \qquad (5.107)$$

In the non-relativistic Dirac matrix representation a possible choice is

$$C \equiv \gamma_4 \gamma_2 = -\gamma_1 \gamma_3 \gamma_5 = \begin{pmatrix} 0 & -i\sigma_2 \\ -i\sigma_2 & 0 \end{pmatrix} = \begin{pmatrix} 0 & 0 & 0 & -1 \\ 0 & 0 & 1 & 0 \\ 0 & -1 & 0 & 0 \\ 1 & 0 & 0 & 0 \end{pmatrix} .$$

$$(5.108)$$

In terms of this and the totally antisymmetric SU(3) unit tensor ϵ_{cde} the relativistic spin-$\frac{1}{2}$ baryon octet operators are, for instance,

$$\mathcal{B}^{p}_{x\alpha} \equiv \epsilon_{cde}(C\gamma_5)_{\beta\gamma}u_{x\alpha c}(u_{x\beta d}d_{x\gamma e} - d_{x\beta d}u_{x\gamma e}) \,,$$

$$\mathcal{B}^{\Sigma^+}_{x\alpha} \equiv \epsilon_{cde}(C\gamma_5)_{\beta\gamma}u_{x\alpha c}(u_{x\beta d}s_{x\gamma e} - s_{x\beta d}u_{x\gamma e}) \,,$$

$$\mathcal{B}^{\Lambda}_{x\alpha} \equiv \epsilon_{cde}(C\gamma_5)_{\beta\gamma}[u_{x\alpha c}(d_{x\beta d}s_{x\gamma e} - s_{x\beta d}d_{x\gamma e}) + d_{x\alpha c}(s_{x\beta d}u_{x\gamma e} - u_{x\beta d}s_{x\gamma e})$$

$$-2s_{x\alpha c}(u_{x\beta d}d_{x\gamma e} - d_{x\beta d}u_{x\gamma e})] \,,$$

$$\mathcal{B}^{\Xi^0}_{x\alpha} \equiv \epsilon_{cde}(C\gamma_5)_{\beta\gamma}s_{x\alpha c}(s_{x\beta d}u_{x\gamma e} - u_{x\beta d}s_{x\gamma e}) \,. \tag{5.109}$$

Following the usual rules in Minkowski space, one can also define the overlined operators, for instance,

$$\overline{\mathcal{B}}^{p}_{y\delta} \equiv \epsilon_{fgh}(C\gamma_5)_{\epsilon\varphi}\overline{u}_{y\delta f}(\overline{d}_{yeg}\overline{u}_{y\varphi h} - \overline{u}_{yeg}\overline{d}_{y\varphi h}) \,. \tag{5.110}$$

For the spin-$\frac{3}{2}$ decouplet one can consider

$$\mathcal{B}^{\Delta^{++}}_{xk\alpha} \equiv \epsilon_{cde}(C\gamma_k)_{\beta\gamma}u_{x\alpha c}u_{x\beta d}u_{x\gamma e} \,, \qquad \mathcal{B}^{\Omega^-}_{xk\alpha} \equiv \epsilon_{cde}(C\gamma_k)_{\beta\gamma}s_{x\alpha c}s_{x\beta d}s_{x\gamma e} \,,$$

$$\mathcal{B}^{\Sigma^{*+}}_{xk\alpha} \equiv \epsilon_{cde}(C\gamma_k)_{\beta\gamma}(u_{x\alpha c}u_{x\beta d}s_{x\gamma e} + u_{x\alpha c}s_{x\beta d}u_{x\gamma e} + s_{x\alpha c}u_{x\beta d}u_{x\gamma e}) \,,$$

$$\mathcal{B}^{\Xi^{*0}}_{xk\alpha} \equiv \epsilon_{cde}(C\gamma_k)_{\beta\gamma}(s_{x\alpha c}s_{x\beta d}u_{x\gamma e} + s_{x\alpha c}u_{x\beta d}s_{x\gamma e} + u_{x\alpha c}s_{x\beta d}s_{x\gamma e}) \,. \tag{5.111}$$

The expectation value of two-point functions of composite hadron operators can be determined on the basis of (5.18). This leads to products of the quark propagator $Q^{-1}[U;K]$ evaluated in some gauge configuration $[U]$ at some hopping parameter K. For the flavours u, d and s one has to take the hopping parameter values K_u, K_d and K_s, respectively. (The small difference between u and d quark masses is, however, usually neglected: $K_u = K_d$.) Writing out indices explicitly, let us introduce

$$U_{x\alpha c,y\beta d} \equiv Q^{-1}[U, K = K_u]_{x\alpha c,y\beta d} \,, \qquad D_{x\alpha c,y\beta d} \equiv Q^{-1}[U, K = K_d]_{x\alpha c,y\beta d} \,,$$

$$S_{x\alpha c,y\beta d} \equiv Q^{-1}[U, K = K_s]_{x\alpha c,y\beta d} \,. \tag{5.112}$$

Then, for instance, for $\langle \mathcal{M}^{\pi^+}_x \mathcal{M}^{\pi^-}_y \rangle$ and $\langle \mathcal{M}^{\rho^+}_{xk} \mathcal{M}^{\rho^-}_{yl} \rangle$ one has to calculate, respectively,

$$\langle -\text{Tr}_{sc}\{\gamma_5 D_{yx}\gamma_5 U_{xy}\}\rangle_{S_{eff}} \,, \qquad \langle -\text{Tr}_{sc}\{\gamma_l D_{yx}\gamma_k U_{xy}\}\rangle_{S_{eff}} \,, \tag{5.113}$$

where Tr_{sc} stands for a trace over spin and colour indices, and $\langle \ldots \rangle_{S_{eff}}$ denotes an expectation value with respect to the effective gauge action S_{eff}, as shown by (5.17) and (5.18). Similar formulas apply to all flavour non-diagonal mesons. For mesons like $\eta, \eta', \omega, \phi, \ldots$, however, some combination of flavour diagonal composite operators, such as

$$\mathscr{M}_x^{(\bar{u}u)} \equiv \bar{u}_{x\alpha c} \Gamma_{\alpha\beta} u_{x\beta c} , \tag{5.114}$$

is needed (with some Dirac matrix Γ). For the expectation value $\langle \mathscr{M}_x^{(\bar{u}u)} \mathscr{M}_y^{(\bar{u}u)} \rangle$ the general expression (5.18) involves the combination

$$\langle \mathrm{Tr}_{sc}\{\Gamma U_{xx}\} \mathrm{Tr}_{sc}\{\Gamma U_{yy}\} - \mathrm{Tr}_{sc}\{\Gamma U_{yx}\Gamma U_{xy}\} \rangle_{S_{eff}} . \tag{5.115}$$

In order to obtain baryon masses the necessary combinations of quark propagators have to be determined. For the proton (and similarly for Ξ^0 and Σ^+) these are

$$\epsilon_{cde}\epsilon_{fgh}(C\gamma_5)_{\beta\gamma}(C\gamma_5)_{\epsilon\varphi}$$

$$\cdot (U_{x\alpha c,y\delta f} U_{x\beta d,y\epsilon g} D_{x\gamma e,y\varphi h} + U_{x\alpha c,y\epsilon f} U_{x\beta d,y\delta g} D_{x\gamma e,y\varphi h}) . \tag{5.116}$$

For the Λ-baryon they are

$$\epsilon_{cde}\epsilon_{fgh}(C\gamma_5)_{\beta\gamma}(C\gamma_5)_{\epsilon\varphi}[U_{x\alpha c,y\delta f} D_{x\beta d,y\epsilon g} S_{x\gamma e,y\varphi h}$$

$$+ D_{x\alpha c,y\delta f} U_{x\beta d,y\epsilon g} S_{x\gamma e,y\varphi h} + 4 S_{x\alpha c,y\delta f} U_{x\beta d,y\epsilon g} D_{x\gamma e,y\varphi h}$$

$$- U_{x\alpha c,y\epsilon f} D_{x\beta d,y\delta g} S_{x\gamma e,y\varphi h} - D_{x\alpha c,y\epsilon f} U_{x\beta d,y\delta g} S_{x\gamma e,y\varphi h}$$

$$- 2 U_{x\alpha c,y\epsilon f} D_{x\beta d,y\varphi g} S_{x\gamma e,y\delta h} - 2 D_{x\alpha c,y\epsilon f} U_{x\beta d,y\varphi g} S_{x\gamma e,y\delta h}$$

$$- 2 S_{x\alpha c,y\epsilon f} D_{x\beta d,y\varphi g} U_{x\gamma e,y\delta h} - 2 S_{x\alpha c,y\epsilon f} U_{x\beta d,y\varphi g} D_{x\gamma e,y\delta h}] . \tag{5.117}$$

For the Δ^{++} baryon (and similarly for Ω^-) the propagator combinations are

$$\epsilon_{cde}\epsilon_{fgh}(C\gamma_k)_{\beta\gamma}(C\gamma_k)_{\epsilon\varphi}$$

$$\cdot (U_{x\alpha c,y\delta f} U_{x\beta d,y\epsilon g} U_{x\gamma e,y\varphi h} + 2 U_{x\alpha c,y\epsilon f} U_{x\beta d,y\delta g} U_{x\gamma e,y\varphi h}) . \tag{5.118}$$

For the Σ^{*+} baryon (and similarly for Ξ^{*0}) they are

$$\epsilon_{cde}\epsilon_{fgh}(C\gamma_k)_{\beta\gamma}(C\gamma_k)_{\epsilon\varphi}$$

$$\cdot (U_{x\alpha c,y\delta f} U_{x\beta d,y\epsilon g} S_{x\gamma e,y\varphi h} + 2 U_{x\alpha c,y\epsilon f} U_{x\beta d,y\delta g} S_{x\gamma e,y\varphi h}) . \tag{5.119}$$

The calculation of the above combinations of quark propagator matrix elements is facilitated by the relation

$$Q_{yx}^{-1} = \gamma_5 Q_{xy}^{-1\dagger} \gamma_5 , \tag{5.120}$$

which follows from (5.15). This implies that, for instance, the charged pion two-point function in (5.113) can be written as

$$\langle -\mathrm{Tr}_{sc}\{D_{xy}^{\dagger} U_{xy}\} \rangle_{S_{eff}} . \tag{5.121}$$

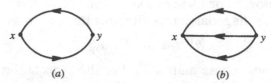

Fig. 5.9. The quark line graphs for the two-point function of local, flavour non-singlet meson (*a*) and ground state baryon (*b*) operators.

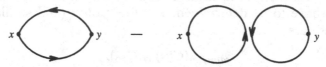

Fig. 5.10. The quark line graphs for the two-point function of local, flavour singlet meson operators.

Besides being simple to compute, this form can also be used to derive some rigorous inequalities for hadron masses [5.43].

The quark propagator configurations can also be graphically represented by drawing oriented lines between the lattice points connected by the propagator matrix elements. The resulting graphs are reminescent of the coordinate space Feynman graphs in perturbative QCD, but only the quark lines directly connected to the external sources are visible. The gluon lines attached to the quark lines and the virtual quark loops are not shown, because they are generated by the evaluation of the expectation value of the quark propagators according to (5.18). Such graphs can be called *quark line graphs*. For the simple flavour singlet mesons in (5.113) the relevant quark line graph is shown in fig. 5.9(a). For the baryons in (5.116)–(5.119) the corresponding graph is in fig. 5.9(b). In the case of the flavour diagonal mesons in (5.115) there are two graphs, as shown by fig. 5.10. The numerical evaluation of the second graph in this figure is rather demanding. This is because one cannot reduce it to the calculation of the quark propagator from a single initial point by use of the relation (5.120).

The final aim of the numerical hadron mass calculations in lattice QCD is to extrapolate the results to the continuum limit in infinitely large volumes. The difficulty is, of course, to satisfy the requirement of physically large volumes at physically small lattice spacings, because the number of lattice sites becomes very large. Besides the large spatial extension one has to choose a sufficiently large lattice also in the time direction, in order to be able to extract the lowest masses [5.44]. Things get particularly difficult if large mass ratios are involved like, for instance, in the case of Δ-baryon and pion with $m_\Delta/m_\pi \simeq 9$.

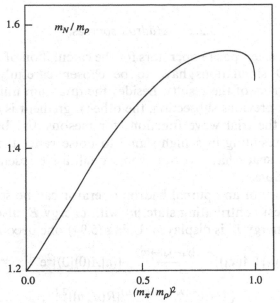

Fig. 5.11. The qualitative dependence of the mass ratio m_N/m_ρ on $(m_\pi/m_\rho)^2$ according to simple potential models.

Another kind of problem of a more conceptual type is that many hadrons are not stable, but decay rather fast to several other hadrons. Therefore some of them appear as rather broad resonances. For suggestions to deal with this problem see [5.45, 5.46, 5.47].

Both this problem and the problem of large mass ratios is less severe, if the u- and d-quark masses are larger than the physical value. Namely, in this case the pions are relatively heavier, their masses are closer to the ρ-meson mass. In fact, for quark masses nearly as large as the s-quark mass the lightest 0^{-+} pseudoscalar meson gets heavier than half of the ρ-meson mass, and therefore the ρ-meson becomes stable. Heavier quark masses are easier also from the technical point of view, because the necessary quark matrix inversions are faster (see section 7.4).

Therefore it is almost inevitable that studies start at relatively heavy quark masses, and then one subsequently tries to push the u- and d-quark masses to smaller and smaller values. It is interesting to observe the quark mass dependence of different hadron mass ratios. The best way to represent the results is to concentrate on the interrelations of different mass ratios. For instance, one can plot m_N/m_ρ versus m_π/m_ρ (called an 'Edinburgh plot' [5.48]), or m_N/m_ρ versus $(m_\pi/m_\rho)^2$ ('APE plot' [5.49]). The numerically measured points can be compared with the predictions of simple, approximate potential models (see fig. 5.11). For a recent review of the results of hadron spectrum calculations see [5.50].

5.2.2   Hadron sources

The quark–gluon composite operators for the calculation of hadron masses in lattice QCD simulations have to be chosen carefully, in order to minimize the errors of the results. Besides the quantum number structure discussed in the previous subsection, the other ingredient is the coordinate dependence of the trial wave functions for mesons and baryons. For a strong overlap resulting in a high signal to noise ratio, the quark–gluon distributions in space have to bear some qualitative resemblance to the true wave functions.

The importance of an optimal hadron operator can be seen if in (5.95), besides the lowest contributing state $|n\rangle$ with energy E, also some higher state $|n'\rangle$ with energy E' is displayed. Using (5.94) one becomes for $T = \infty$

$$\langle A(y_4)^+ A(x_4)\rangle \xrightarrow{|y_4-x_4|\to\infty} |\langle n|A(0)|0\rangle|^2 e^{-|y_4-x_4|E}$$

$$\cdot \left\{1 + e^{-|y_4-x_4|(E'-E)}|R(n',n)|^2\right\} , \qquad (5.122)$$

where $R(n',n) \equiv \langle n'|A(0)|0\rangle/\langle n|A(0)|0\rangle$. This is dominated by the leading term if $|R(n',n)|^2 \exp\{-|y_4-x_4|(E'-E)\}$ is much smaller than 1. Therefore if $|R(n',n)|^2$ is not much less than 1, a large time separation $|y_4 - x_4| \gg 1$ is necessary, and there the signal to noise ratio is lower than at $|y_4 - x_4| = O(1)$.

Let us now specialize, for definiteness, to the case of flavour non-singlet mesons. (The flavour singlet case and the baryon operators can be treated similarly.) As a generalization of (5.105), (5.106) one can consider the composite quark–antiquark timeslice operator (at time t)

$$\mathcal{M}_{\Gamma t}^{(\bar{q}_2 q_1)} \equiv \sum_{\mathbf{x}_1 \mathbf{x}_2} F_t(\mathbf{x}_2,\mathbf{x}_1)\overline{\psi}_{q_2 \mathbf{x}_2 t a_2}\Gamma_{a_2 a_1}\psi_{q_1 \mathbf{x}_1 t a_1} . \qquad (5.123)$$

Here the flavours q_2 and q_1 are assumed to be different from each other, and the spinor and colour indices are abbreviated as $a \equiv \alpha c$. If such indices occur in pairs, a summation convention is followed. The spin–colour structure is specified by the matrix Γ, and the spatial wave function $F_t(\mathbf{x}_2,\mathbf{x}_1)$, called the *smearing function*, is usually assumed to be rotation invariant. According to the general formula (5.18) we have

$$-\left\langle \mathcal{M}_{\Gamma t}^{(\bar{q}_2 q_1)} \mathcal{M}_{\Gamma' t'}^{(\bar{q}_1 q_2)}\right\rangle = \left\langle \sum_{\mathbf{x}_1 \mathbf{x}_2} F_t(\mathbf{x}_2,\mathbf{x}_1) \sum_{\mathbf{x}_1' \mathbf{x}_2'} F_{t'}(\mathbf{x}_2',\mathbf{x}_1') \right.$$

$$\left. \cdot \Gamma'_{a_2' a_1'} Q^{(2)-1}_{\mathbf{x}_1' t' a_1', \mathbf{x}_2 t a_2} \Gamma_{a_2 a_1} Q^{(1)-1}_{\mathbf{x}_1 t a_1, \mathbf{x}_2' t' a_2'}\right\rangle_{S_{eff}} , \qquad (5.124)$$

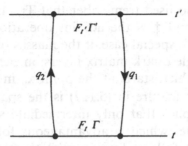

Fig. 5.12. The quark line graph for a general non-local flavour non-singlet ($q_2 \neq q_1$) meson two-point function. The spatial coordinates on the timeslices t and t' are, respectively, $\mathbf{x}_{1,2}$ and $\mathbf{x}'_{1,2}$.

with $Q^{(i)-1}$ ($i = 1, 2$) as the quark propagator for quark flavour q_i. The quark line graph representing this is depicted in fig. 5.12.

The numerical evaluation of (5.124) requires the knowledge of the quark propagator matrix elements between two timeslices t and t'. In the usual iterative algorithms discussed in section 7.4 one solves the equation $Qp = v$ in (7.152) for a given initial vector v, in order to obtain $p = Q^{-1}v$. From this one can easily obtain the matrix element $(w, p) = (w, Q^{-1}v)$ for the given v and any vector w. As a consequence, for the calculation of the matrix element combination in (5.124), for general F_t, $F_{t'}$, Γ and Γ', one has to perform the quark matrix inversion $24V$ times (where V is the number of points in a timeslice). For instance, for fixed $\mathbf{x}_1$ and a_1 one takes the vector $u_{\mathbf{x}a} \equiv \delta_{\mathbf{x}\mathbf{x}_1}\delta_{aa_1}$ and solves $Q^{(2)}p = F_t\Gamma u$ for p, and then $Q^{(1)}p' = F_{t'}\Gamma'p$ for p', and finally one performs the sum of the components $p'_{\mathbf{x}_1 a_1}$ in timeslice t over $\mathbf{x}_1$ and a_1.

The number $24V$ may be quite large, therefore one might wish to reduce the number of necessary inversions by some special choice of the smearing functions F_t and $F_{t'}$. A possible way to do this is to assume that at least one of them, for instance F_t, is factorized in the space coordinates:

$$F_t(\mathbf{x}_2, \mathbf{x}_1) \equiv f_t^{(2)}(\mathbf{x}_2)f_t^{(1)}(\mathbf{x}_1) . \qquad (5.125)$$

It is easy to see that in this case, for the calculation of (5.124) for arbitrary Γ and Γ', only 24 quark matrix inversions are necessary (two per spin and colour indices).

A simple factorized choice, which corresponds to local operators, is

$$F_t(\mathbf{x}_2, \mathbf{x}_1) = \delta_{\mathbf{x}_2\mathbf{x}_0}\delta_{\mathbf{x}_1\mathbf{x}_0} , \qquad F_{t'}(\mathbf{x}'_2, \mathbf{x}'_1) = \delta_{\mathbf{x}'_2\mathbf{x}'_1} . \qquad (5.126)$$

In (5.124) this leads to

$$-\langle \mathcal{M}_{\Gamma t}^{(\bar{q}_2 q_1)} \mathcal{M}_{\Gamma' t'}^{(\bar{q}_1 q_2)} \rangle = \left\langle \sum_{\mathbf{x}'} \mathrm{Tr}_{sc} \left[\gamma_5 \Gamma' Q_{\mathbf{x}'t', \mathbf{x}_0 t}^{(2)-1} \Gamma \gamma_5 Q_{\mathbf{x}'t', \mathbf{x}_0 t}^{(1)-1\dagger} \right] \right\rangle_{S_{eff}} , \qquad (5.127)$$

where (5.120) was also used (remember that Tr_{sc} means a trace over spin and colour indices, and † is the adjoint operation in spin and colour). This shows that in this special case, if the masses of the flavours q_1 and q_2 are equal, only a single quark matrix inversion per spin and colour index values is necessary, which starts at the point $\mathbf{x}_0$ in timeslice t.

The other pleasant feature in (5.127) is the spatial sum over $\mathbf{x}'$ in the timeslice t'. This implies that only intermediate states with zero spatial momentum contribute, which is advantageous for the extraction of the mass. In addition, it follows from spatial translation invariance in the timeslice t, that for $\gamma_4 \Gamma' \gamma_4 = -\Gamma^\dagger$ the operator at t' is the negative adjoint of the operator at t. Therefore all the exponential contributions on the right hand side of (5.127) have positive coefficients (like in (5.95), (5.96) or (5.122)). This is a very useful property, because then the mass estimates obtained from the logarithm of timeslice ratios decrease monotonically with increasing time separation $|t' - t|$.

Another simple choice [5.49] for the smearing functions is the totally factorized form

$$F_t(\mathbf{x}_2, \mathbf{x}_1) = f_t(\mathbf{x}_2) f_t^*(\mathbf{x}_1) , \qquad F_{t'}(\mathbf{x}'_2, \mathbf{x}'_1) = f_{t'}^*(\mathbf{x}'_2) f_{t'}(\mathbf{x}'_1) . \qquad (5.128)$$

These can be called 'shell model wave functions' [5.51]. In this case (5.124), together with (5.120), gives

$$-\langle \mathcal{M}_{\Gamma t}^{(\bar{q}_2 q_1)} \mathcal{M}_{\Gamma' t'}^{(\bar{q}_1 q_2)} \rangle = \left\langle \sum_{\mathbf{x}_1 \mathbf{x}_2 \mathbf{x}'_1 \mathbf{x}'_2} \mathrm{Tr}_{sc} \left[\gamma_5 \Gamma' f_{t'}(\mathbf{x}'_1) Q_{\mathbf{x}'_1 t', \mathbf{x}_2 t}^{(2)-1} f_t(\mathbf{x}_2) \right. \right.$$

$$\left. \left. \cdot \Gamma \gamma_5 \left(f_{t'}(\mathbf{x}'_2) Q_{\mathbf{x}'_2 t', \mathbf{x}_1 t}^{(1)-1} f_t(\mathbf{x}_1) \right)^\dagger \right] \right\rangle_{S_{eff}} . \qquad (5.129)$$

This implies that for equal quark masses ($Q^{(1)} = Q^{(2)}$) one quark matrix inversion per spin and colour index values is sufficient. For $\gamma_4 \Gamma' \gamma_4 = -\Gamma^\dagger$ and time-independent smearing ($f_t = f_{t'}$), the coefficients of the exponential contributions are positive.

A natural choice for factorized smearing functions is a Gaussian [5.51]. The dispersion in it can be optimized for large overlap, or better, for a small statistical error of the mass. Note that for non-local gauge field independent smearing functions one cannot integrate over the gauge field without fixing the gauge degrees of freedom. A possible choice for a smooth physical gauge is the Coulomb gauge [5.49].

Another important class of smearing functions, which is motivated by physical intuition, has the form

$$F_t(\mathbf{x}_2, \mathbf{x}_1) = \phi_t(\mathbf{x}_2 - \mathbf{x}_1) \delta_{\mathbf{x}_1 \mathbf{x}_0} , \qquad F_{t'}(\mathbf{x}'_2, \mathbf{x}'_1) = \phi_{t'}(\mathbf{x}'_2 - \mathbf{x}'_1) . \qquad (5.130)$$

Since ϕ_t and $\phi_{t'}$ only depend on the coordinate differences, they can really be interpreted as bound state wave functions. In addition, in timeslice t one of the constituents is fixed at position x_0, therefore in (5.124) two quark matrix inversions per spin and colour indices are sufficient. In timeslice t' there is a summation over the centre of mass position of the bound state, therefore only intermediate states with zero total spatial momenta contribute. Another advantage of these 'bound state smearing functions' is that they can easily be made gauge invariant, therefore a gauge fixing is not necessary.

A simple and practical way to define $\phi_t(x)$ is to apply the following iterative scheme [5.52]: $\phi_t(x) = \phi_t^{(N)}(x)$ and for $n = 0, 1, \ldots, N - 1$

$$\phi_t^{(n+1)}(x) = \phi_t^{(n)}(x) + \kappa \sum_{k=1}^{3} \left\{ U_{xtk}^+ \phi_t^{(n)}(x + \hat{k}) + U_{x-\hat{k}tk} \phi_t^{(n)}(x - \hat{k}) \right\} . \quad (5.131)$$

Gauge invariance is guaranteed by the inclusion of the spatial gauge link variables U_{xtk} ($k = 1, 2, 3$). The same iteration applies also to $\phi_{t'}(x')$. In the first step in timeslice t one starts at site $x = x_0$ by $\phi_t^{(0)}(x) = \delta_{xx_0}$. In the timeslice t' the initial vector is the solution p of the first quark matrix inversion $Q^{(2)} p = \phi_t^{(N)}$.

These smearing functions have two parameters: the 'smearing parameter' κ and the number of iterations N. The x-distribution in $\phi_t(x)$ is on average rotation invariant and approximately Gaussian. The dispersion has to be optimized to reach a minimal statistical error of the obtained hadron mass. An improvement can be achieved by replacing the gauge link variables everywhere in the timeslices t and t' (but not in the timeslices in between) by the averaged gauge links defined in (7.67).

A related possibility is to choose the smearing function $\phi_t(x)$ to be the three-dimensional scalar propagator defined over the gauge field links U_{xtk} [5.53]. The relation to (5.131) is quite close because, assuming convergence at the scalar hopping parameter value κ, the scalar propagator can be calculated by an iteration analogous to (5.36) or (5.37) (see also the Jacobi iteration in (7.153)), namely for $n = 0, 1, \ldots$

$$\phi_t^{(n+1)}(x) = \phi_t^{(0)}(x) + \kappa \sum_{k=1}^{3} \left\{ U_{xtk}^+ \phi_t^{(n)}(x + \hat{k}) + U_{x-\hat{k}tk} \phi_t^{(n)}(x - \hat{k}) \right\} . \quad (5.132)$$

The corresponding matrix element of the scalar propagator with hopping parameter κ is given by $\phi_t(x) = \lim_{n \to \infty} \phi_t^{(n)}(x)$. This yields a bound state wave function decaying exponentially at large space-like separations. Of course, instead of the Jacobi iteration, the scalar propagator can also be calculated by any other matrix inversion method.

5.2.3 Heavy quark systems

A possibility to deal with heavy quark–antiquark bound states on the lattice is to assume that such systems can be described by a non-relativistic potential model. The static quark–antiquark potential can be numerically calculated from the expectation values of Wilson loops (see chapter 3). Although non-relativistic potential models of charmonium and bottomonium systems are quite successful, the potential picture relies on approximations like the absence of dynamical gluons and the instantaneous interaction mediated by a static gluon field. These approximations cannot be systematically improved. The retardation effects in the interaction, for instance, cannot be properly handled. A direct simulation of heavy quarkonia is in principle possible, but in practice it would be very difficult, because in the continuum limit the mass of the heavy quark in lattice units (am_Q) has to go to zero (m_Q denotes the mass of the heavy quark $Q = c, b, t$). But even if, for instance, the pion mass in lattice units is as small as, say, $am_\pi = 0.1$, the charm quark mass is $am_c \simeq 1.0$ and the bottom quark mass $am_b \simeq 3.0$. That is, the mass of the heavy quarks is still large in lattice units. (If one would choose am_π much smaller, then, in order to avoid large finite size effects due to pions and other light quark states, huge lattices were needed.)

A different approach of simulating heavy quark systems on the lattice [5.54, 5.55] retains the dynamics of the light quarks and gluons exactly, and treats the heavy quarks by a systematic $1/m_Q$ expansion. The zeroth-order approximation is the *static* or *non-relativistic limit*, and the corrections proportional to the powers of $1/m_Q$ can be systematically included. In this way one takes advantage of the fact that the energy scale of the large quark mass plays only a minor rôle in the dynamics of the non-relativistic system.

The conceptually clearest way to formulate the static or non-relativistic approximation is in terms of an effective field theory. This amounts to transform the relativistic quark theory into an equivalent non-relativistic theory by integrating out all quantum fluctuations at momentum scales of order m_Q or larger. This is in the spirit of the renormalization group strategy, and has been used successfully in high precision analyses of QED bound states [5.56]. The static effective field theory is expected to reproduce the results of the full theory to all orders of the QCD coupling α_s, and to any fixed order in $1/m_Q$ (for a proof at zeroth order in $1/m_Q$ see [5.57]).

The essential difference between static and non-relativistic theories is that in the latter the non-relativistic kinetic energy of heavy quarks is incorporated at lowest order, whereas in the former it is taken into account only as a $1/m_Q$ correction. The non-relativistic limit is appropriate for the description of $Q\bar{Q}$ meson states ('heavy quarkonia'). The static approximation can be the starting point of simulations of heavy–light ($Q\bar{q}$ or $\bar{Q}q$) states, because in these mesons the momentum scale is set by the dynamics of the light quark. (A QED analogue is the hydrogen atom, where the internal momentum is of the order of $\alpha m_{\text{electron}}$, and largely independent of m_{proton}.) As a consequence, in heavy–light mesons the internal kinetic energy of the heavy quark is negligible relative to that of the light quark. Starting by the static approximation, the systematic expansion in inverse powers of the heavy quark mass incorporates two terms of order $1/m_Q$. These are the dimension-five operators describing the non-relativistic kinetic energy and the chromomagnetic moment interaction of the heavy quarks. The coefficients of these operators in the Lagrangian of the static effective field theory must be fixed by matching amplitudes in the effective theory to their counterparts in the full theory [5.58].

The Euclidean lattice formulation of the gluonic and light quark part of the static and non-relativistic effective action is identical to the usual QCD lattice action (5.4)–(5.5). The part depending on the Grassmannian heavy quark fields ψ_{Qx} and ψ_{Qx}^+ and heavy antiquark fields $\psi_{\bar{Q}x}$ and $\psi_{\bar{Q}x}^+$ can be written as [5.54, 5.55, 5.59]

$$S_{Q+\bar{Q}} = S_{Q+\bar{Q}}^{stat} + S_{Q+\bar{Q}}^{kin} + S_{Q+\bar{Q}}^{magn} + \cdots . \tag{5.133}$$

Here $S_{Q+\bar{Q}}^{stat}$ is the static limit of the heavy quark action, $S_{Q+\bar{Q}}^{kin}$ is the kinetic part of order $1/m_Q$, and $S_{Q+\bar{Q}}^{magn}$ is the chromomagnetic moment part, which is also of order $1/m_Q$, and the dots stand for higher order corrections in $1/m_Q$. The static part is

$$S_{Q+\bar{Q}}^{stat} = \sum_x \left\{ (\psi_{Qx}^+ \psi_{Qx}) - (\psi_{Qx+\hat{4}}^+ U_{x4} \psi_{Qx}) \right.$$

$$\left. -(\psi_{\bar{Q}x}^+ \psi_{\bar{Q}x}) + (\psi_{\bar{Q}x-\hat{4}}^+ U_{x-\hat{4},4}^+ \psi_{\bar{Q}x}) \right\} . \tag{5.134}$$

Note that, as usual, ψ_Q corresponds to the annihilation operator of quarks, ψ_Q^+ to the creation operator of quarks, but $\psi_{\bar{Q}}$ to the creation operator of antiquarks and $\psi_{\bar{Q}}^+$ to the annihilation operator of antiquarks. The kinetic part of heavy quarks can be written as

$$S_{Q+\bar{Q}}^{kin} = \frac{1}{2am_Q} \sum_x \left\{ \sum_{j=1}^{3} \left[(\psi_{Qx+\hat{j}}^+ U_{xj} \psi_{Qx}) + (\psi_{Qx-\hat{j}}^+ U_{x-\hat{j},j}^+ \psi_{Qx}) \right. \right.$$

$$\left. \left. -2(\psi_{Qx}^+ \psi_{Qx}) \right] \right\} - \frac{1}{2am_Q} \sum_x \left\{ \psi_{\bar{Q}}^+, \psi_Q \Longrightarrow \psi_{\bar{Q}}^+, \psi_{\bar{Q}} \right\} . \tag{5.135}$$

The discretization of the chromomagnetic moment part can be chosen differently. One of the possibilities is the following:

$$
S_{Q+\bar{Q}}^{magn} = \frac{i\kappa_Q}{2am_Q} \sum_x \left\{ \frac{1}{8} \sum_{j,k,l=1}^{3} \psi_{Qx}^+ \sigma_j \epsilon_{jkl} \left[U_{xl}^+ U_{x+\hat{l},k}^+ U_{x+\hat{k},l} U_{xk} \right. \right.
$$

$$
+ U_{xk}^+ U_{x-\hat{l}+\hat{k},l} U_{x-\hat{l},k}^+ U_{x-\hat{l},l}^+ + U_{x-\hat{k},k} U_{x-\hat{k},l}^+ U_{x-\hat{k}+\hat{l},k}^+ U_{x,l}
$$

$$
\left. \left. + U_{x-\hat{l},l} U_{x-\hat{k}-\hat{l},k} U_{x-\hat{k}-\hat{l},l}^+ U_{x-\hat{k},k}^+ \right] \psi_{Qx} \right\}
$$

$$
- \frac{i\kappa_Q}{2am_Q} \sum_x \left\{ \psi_{\bar{Q}}^+, \psi_Q \implies \psi_{\bar{Q}}^+, \psi_{\bar{Q}} \right\} . \tag{5.136}
$$

This has the advantage that the quark fields are at the same point. The four terms in squared brackets are the product of gauge link variables along four plaquettes touching the point x in the plane (k, l) ('clover terms').

The bare parameters in the action (5.133) are the bare heavy quark mass in lattice units am_Q and the bare chromomagnetic moment κ_Q. The coefficient of the static component is put equal to unity by the choice of heavy quark field normalization. In the vicinity of the continuum limit the bare parameters of the lattice action can be approximately related to the parameters of the continuum effective field theory action by a perturbative calculation. A similar perturbative matching can also be applied to the renormalization of composite current operators.

From the point of view of numerical simulations, the heavy quark action $S_{Q+\bar{Q}}$ is an enormous simplification relative to the full QCD action. Since the heavy quark propagator is of first order in time derivatives, it can be obtained as a solution of an initial value problem in a single pass through the lattice. The determinant of the heavy quark matrix has only little effect on the gauge dynamics. In a first approximation it can be completely neglected, and for better accuracy it can be taken into account as a small perturbation by using equations like (5.26). A further simplification is that the problem of fermion doubling does not arise in the large quark mass limit in $S_{Q+\bar{Q}}$.

The heavy quark effective theory on the lattice is a potentially very useful tool for the calculation of heavy quark bound state spectra and current matrix elements of hadronic states containing heavy quarks.

5.3 Broken chiral symmetry on the lattice

An important feature of low energy strong interactions is the broken SU(2) ⊗ SU(2) (or, to a lower accuracy, the SU(3) ⊗ SU(3)) chiral symmetry. Different elements of this symmetry, like the approximate SU(3) flavour symmetry of the hadron spectrum, or PCAC for pions, played a very important rôle in the historical development of the basic concepts underlying the theory of strong interactions.

In the framework of QCD the chiral symmetry is explained by the smallness of the u-, d- and s-quark masses. On a typical strong interaction scale represented, for instance, by $\Lambda_{\overline{MS}} \simeq 200$ MeV the up- and down-quark masses are rather small ($m_u, m_d \simeq 10$ MeV). This means that the real world closely resembles a hypothetical world with $m_u = m_d = 0$. In such a world the equality $m_u = m_d$ implies an exact global U(2)$_V$ = U(1)$_V \otimes$ SU(2)$_V$ flavour symmetry, where U(1)$_V$ expresses the conservation of quark (fermion) number, and SU(2)$_V$ is the isospin symmetry [A4]. At zero quark mass this symmetry is extended to U(2)$_V \otimes$ U(2)$_A$ = U(1)$_V \otimes$ U(1)$_A \otimes$ SU(2)$_V \otimes$ SU(2)$_A$, which is generated by the vector and axial vector currents. The additional U(1)$_A$ symmetry is explicitly broken by quantum effects (by the 'axial anomaly'), as will be discussed later on in this section.

The SU(2)$_A$ axial isospin symmetry is spontaneously broken by the vacuum expectation value of the scalar quark densities. The consequence of this is the appearance of three massless pseudoscalar bosons in the zero quark mass ($m_u = m_d = 0$) limit. This is due to the Goldstone theorem [5.60, 5.61],[A3], which says that for every broken generator of a spontaneously broken global symmetry there is a corresponding zero-mass scalar (or pseudoscalar) state in the particle spectrum.

In a somewhat worse approximation the strange quark mass ($m_s \simeq 150$ MeV) can also be neglected. In this case the chiral flavour symmetry is extended from SU(2)$_V \otimes$ SU(2)$_A$ to SU(3)$_V \otimes$ SU(3)$_A$. In the continuum theory this symmetry is generated by the eight vector ($V(x)_{S\mu}$) and eight axial vector ($A(x)_{S\mu}$) currents ($S = 1, 2, \ldots, 8$; $\mu = 0, 1, 2, 3$):

$$V(x)_{S\mu} \equiv \overline{\psi}(x)\gamma_\mu \frac{\lambda_S}{2}\psi(x) \,, \quad A(x)_{S\mu} \equiv \overline{\psi}(x)\gamma_\mu\gamma_5 \frac{\lambda_S}{2}\psi(x) \,. \tag{5.137}$$

Here $\psi(x) \equiv \{\psi(x)_{q\alpha c}; q = u, d, s; \alpha = 1, 2, 3, 4; c = 1, 2, 3\}$ is the continuum quark field with flavour (q), Dirac (α) and colour (c) indices, and λ_S is the Gell-Mann matrix of SU(3) acting on flavour indices. (In order to distinguish it from the Gell-Mann matrices $\lambda_{r,s,t,\ldots}$ in colour space, we denote the flavour octet indices by capital letters.) The corresponding charge operators are

$$Q_{VS}(x_0) \equiv \int d^3\mathbf{x}\, V(x_0, \mathbf{x})_{S0} \, , \quad Q_{AS}(x_0) \equiv \int d^3\mathbf{x}\, A(x_0, \mathbf{x})_{S0} \, . \qquad (5.138)$$

These are the 16 generators of the symmetry group $SU(3)_V \otimes SU(3)_A$.

A basic assumption in the theory of strong interactions is the *current algebra hypothesis* about the equal time commutation relations of currents [5.62, 5.63]:

$$[V(x_0, \mathbf{x})_{S0}, V(x_0, \mathbf{y})_{T,\mu=0}] = \mathrm{i} f_{STU} \delta^3(\mathbf{x} - \mathbf{y}) V(x)_{U,\mu=0} \, ,$$

$$[V(x_0, \mathbf{x})_{S0}, A(x_0, \mathbf{y})_{T,\mu=0}] = \mathrm{i} f_{STU} \delta^3(\mathbf{x} - \mathbf{y}) A(x)_{U,\mu=0} \, ,$$

$$[A(x_0, \mathbf{x})_{S0}, A(x_0, \mathbf{y})_{T,\mu=0}] = \mathrm{i} f_{STU} \delta^3(\mathbf{x} - \mathbf{y}) V(x)_{U,\mu=0} \, . \qquad (5.139)$$

The extension of these relations to $\mu \neq 0$ is also possible, but then some additional local 'Schwinger terms' appear on the right hand side [5.63]. The integrated version of (5.139) with respect to $\int d^3\mathbf{x}$ expresses the transformation properties of the current operators with respect to $SU(3)_V \otimes SU(3)_A$.

In the case of exactly zero mass quarks ($m_u = m_d = m_s = 0$) the pseudoscalar Goldstone bosons are exactly massless, the charges in (5.138) are time (x_0) independent, and the divergence of the vector and axial vector currents vanishes: $\partial/\partial x_\mu V(x)_{S\mu} = \partial/\partial x_\mu A(x)_{S\mu} = 0$. In the real world, however, the u-, d- and s-quarks have small masses, which break the $SU(3)_V \otimes SU(3)_A$ symmetry explicitly. Owing to this symmetry breaking the pseudoscalar Goldstone bosons acquire a (small) non-zero mass. They are no longer real Goldstone bosons, only 'quasi-Goldstone bosons'. Another consequence of the explicit chiral symmetry breaking by the quark masses is that the charges $Q_{VS}(x_0)$ and $Q_{AS}(x_0)$ do depend on the time coordinate x_0, and the current divergences are not exactly zero. The approximate conservation of the axial vector current is expressed by the *PCAC hypothesis* ('partial conservation of the axial vector currents') [5.64], which states that the divergence of the axial vector current is an interpolating field of the corresponding pseudoscalar meson, with a smooth off-mass-shell extrapolation near zero four-momentum. For instance, for the π^+ field operator $\varphi_{\pi^+}(x)$ we have in the exact $SU(2)_V$ ($m_u = m_d$) limit

$$\frac{\partial}{\partial x_\mu} \{A(x)_{1\mu} - \mathrm{i} A(x)_{2\mu}\} = f_\pi m_\pi^2 \varphi_{\pi^+}(x) \, , \qquad (5.140)$$

where m_π is the pion mass, and $f_\pi \simeq 132$ MeV is the pion decay constant. A consequence of (5.140) is that for a π^+ state $|\pi^+(p)\rangle$ with four-momentum p we have

$$\langle 0|A(x)_{1\mu} - \mathrm{i} A(x)_{2\mu}|\pi^+(p)\rangle = \langle 0| \{A(x)_{1\mu} - \mathrm{i} A(x)_{2\mu}\} a_{\pi^+}^+(\mathbf{p})|0\rangle$$

$$= \mathrm{i} f_\pi p_\mu \mathrm{e}^{-\mathrm{i} x * p} \, . \tag{5.141}$$

(For the normalization of the single-particle states see (8.45), (8.50).)

Using the current algebra (5.139) and PCAC relation (5.140) one can derive many sum rules and soft pion theorems [5.63],[A3], which can also be obtained from an effective chiral Lagrangian with non-linear realization of the chiral symmetry in terms of the quasi-Goldstone boson fields [5.65].

In order to reproduce the results of broken chiral symmetry in lattice QCD, one has to introduce on the lattice vector and axial vector currents, and consider the continuum limit of equal time commutation relations (5.139), as well as PCAC relations like (5.140). This is best done by exploiting the Ward–Takahashi identities for the expectation values of currents. In principle, this can be done both in the Wilson fermion and staggered fermion formulation of lattice QCD. Here we shall concentrate on the Wilson formulation, which has the advantage of being exactly flavour symmetric for equal mass quarks. The advantage of staggered fermions is that for zero quark mass a $U(1)_A$ subgroup of the axial flavour symmetry is not broken explicitly by the regularization (see section 4.3, and in particular (4.165)). But in this case most elements of the vector-like flavour symmetry group are explicitly broken, and the number of quark flavours is restricted to multiples of four. As a consequence, the realization of the $SU(3)_V \otimes SU(3)_A$ chiral symmetry as a whole is simpler for Wilson quarks.

5.3.1 Ward–Takahashi identities

Let us first recall some particular type of *Ward–Takahashi (W–T) identities* in the continuum [A3,A5]. Let $J(x)_\mu$ be some vector or axial vector current operator and $\phi(x)$ any local bosonic operator. Then from the definition of the time-ordered product,

$$T\{J(x)_\mu \phi(y)\} \equiv \Theta(x_0 - y_0)J(x)_\mu \phi(y) + \Theta(y_0 - x_0)\phi(y)J(x)_\mu \, , \tag{5.142}$$

it follows that

$$\frac{\partial}{\partial x_\mu} T\{J(x)_\mu \phi(y)\} = T\left\{\frac{\partial}{\partial x_\mu}J(x)_\mu \phi(y)\right\} + \delta(x_0 - y_0)[J(x)_0, \phi(y)] \, . \tag{5.143}$$

Similarly, for a pair $(\phi(x), \chi(x))$ of either bosonic or fermionic operators

$$T\{J(x)_\mu \phi(y)\chi(z)\} \equiv \Theta(x_0 - y_0)\Theta(y_0 - z_0)J(x)_\mu \phi(y)\chi(z)$$

$$+\Theta(y_0 - x_0)\Theta(x_0 - z_0)\phi(y)J(x)_\mu\chi(z) + \Theta(y_0 - z_0)\Theta(z_0 - x_0)\phi(y)\chi(z)J(x)_\mu$$

$$\pm\Theta(x_0 - z_0)\Theta(z_0 - y_0)J(x)_\mu\chi(z)\phi(y) \pm \Theta(z_0 - x_0)\Theta(x_0 - y_0)\chi(z)J(x)_\mu\phi(y)$$

$$\pm \Theta(z_0 - y_0)\Theta(y_0 - x_0)\chi(z)\phi(y)J(x)_\mu \, , \tag{5.144}$$

where the plus (minus) signs refer to a bosonic (fermionic) pair, we have

$$\frac{\partial}{\partial x_\mu} T\{J(x)_\mu \phi(y)\chi(z)\} = T\left\{\frac{\partial}{\partial x_\mu} J(x)_\mu \phi(y)\chi(z)\right\}$$

$$+ T\{\delta(x_0 - y_0)[J(x)_0, \phi(y)]\chi(z)\} + T\{\delta(x_0 - z_0)\phi(y)[J(x)_0, \chi(z)]\} . \quad (5.145)$$

Taking the vacuum expectation values of (5.143) and (5.145) one obtains the corresponding W–T identities. Similar identities also hold for the time-ordered products of currents with more operators. Since on the right hand side the equal time commutators of the current $J(x)_0$ appear and, as is well known, the vacuum expectation value of time-ordered products is given by the corresponding expectation value in the Euclidean path integral formulation, by considering the corresponding W–T identities on the lattice one can formulate the consequences of the current algebra hypothesis.

The derivation of the lattice W–T identities for flavour currents is based on a change of the Grassmannian integration variables, very similarly to the derivation of (4.254) and (4.260) in section 4.4. One considers the expectation value of some arbitrary function $\mathcal{O}[\psi, \overline{\psi}, U]$ of the field variables

$$\langle \mathcal{O} \rangle = Z^{-1} \int [\mathrm{d}U \, \mathrm{d}\overline{\psi} \, \mathrm{d}\psi] \mathcal{O}[\psi, \overline{\psi}, U] e^{-S_g[U] - S_q[U, \psi, \overline{\psi}]} , \quad (5.146)$$

where the QCD lattice action is given by (5.2) and (5.48). The new integration variables ψ'_x and $\overline{\psi}'_x$ are introduced by an x-dependent infinitesimal chiral flavour transformation

$$\psi_x = \left(1 + i\sum_S \frac{\lambda_S}{2}\alpha_{VSx} + i\sum_S \gamma_5 \frac{\lambda_S}{2}\alpha_{ASx}\right)\psi'_x ,$$

$$\overline{\psi}_x = \overline{\psi}'_x \left(1 - i\sum_S \frac{\lambda_S}{2}\alpha_{VSx} + i\sum_S \gamma_5 \frac{\lambda_S}{2}\alpha_{ASx}\right) . \quad (5.147)$$

The invariance of (5.146) with respect to this transformation gives

$$\left\langle \mathcal{O}\left[\left(\frac{S_q \overleftarrow{\partial}}{\partial \psi_x}\frac{\lambda_S}{2}\psi_x\right) - \left(\overline{\psi}_x \frac{\lambda_S}{2}\frac{\overrightarrow{\partial} S_q}{\partial \overline{\psi}_x}\right)\right]\right\rangle$$

$$= \left\langle \left(\frac{\mathcal{O} \overleftarrow{\partial}}{\partial \psi_x}\frac{\lambda_S}{2}\psi_x\right) - \left(\overline{\psi}_x \frac{\lambda_S}{2}\frac{\overrightarrow{\partial} \mathcal{O}}{\partial \overline{\psi}_x}\right)\right\rangle ,$$

$$\left\langle \mathcal{O}\left[\left(\frac{S_q \overleftarrow{\partial}}{\partial \psi_x}\gamma_5\frac{\lambda_S}{2}\psi_x\right) + \left(\overline{\psi}_x \gamma_5\frac{\lambda_S}{2}\frac{\overrightarrow{\partial} S_q}{\partial \overline{\psi}_x}\right)\right]\right\rangle$$

$$= \left\langle \left(\frac{\mathcal{O} \overleftarrow{\partial}}{\partial \psi_x} \gamma_5 \frac{\lambda_S}{2} \psi_x \right) + \left(\overline{\psi}_x \gamma_5 \frac{\lambda_S}{2} \frac{\overrightarrow{\partial} \mathcal{O}}{\partial \overline{\psi}_x} \right) \right\rangle . \tag{5.148}$$

Introducing the *conserved flavour vector current* as

$$V_{Sx\mu}^{con} \equiv \frac{1}{2} \left\{ \left(\overline{\psi}_{x+\hat{\mu}} [r + \gamma_\mu] U_{x\mu} \frac{\lambda_S}{2} \psi_x \right) - \left(\overline{\psi}_x [r - \gamma_\mu] U_{x\mu}^+ \frac{\lambda_S}{2} \psi_{x+\hat{\mu}} \right) \right\} , \tag{5.149}$$

and taking into account that the bare quark mass parameter am is a diagonal matrix in flavour space, one can write the first equation in (5.148) as [5.29]

$$\left\langle \mathcal{O} \Delta_\mu^b V_{Sx\mu}^{con} \right\rangle + \left\langle \left(\frac{\mathcal{O} \overleftarrow{\partial}}{\partial \psi_x} \frac{\lambda_S}{2} \psi_x \right) - \left(\overline{\psi}_x \frac{\lambda_S}{2} \frac{\overrightarrow{\partial} \mathcal{O}}{\partial \overline{\psi}_x} \right) \right\rangle$$

$$= \left\langle \mathcal{O}(\overline{\psi}_x \left[am, \frac{\lambda_S}{2} \right] \psi_x) \right\rangle . \tag{5.150}$$

Here Δ_μ^b denotes the backward lattice derivative (8.42), and in the first term the summation convention over $\mu = 1, 2, 3, 4$ is applied. For $\mathcal{O} = 1$ and degenerate flavours (when $[am, \lambda_S] = 0$), in the continuum limit (when $\Delta_\mu^b \to a\partial/\partial x_\mu$ and $V_{Sx\mu}^{con} \to a^3 V(x)_{S\mu}$), this gives us that $\langle 0|\partial/\partial x_\mu V(x)_{S\mu}|0\rangle = 0$, in accordance with the conservation of the vector current. This shows how the symmetries in quantum field theories are reflected by the W–T identities: the existence of a symmetry is expressed by the fact that the derivatives of the action in (5.148) can be written as the divergence of a conserved current.

In the case of the flavour axial vector currents the W–T identities corresponding to (5.150) contain explicit symmetry breaking terms, which are proportional to the Wilson parameter r in the action (5.5). Therefore on the lattice it is not possible to define an exactly conserved axial vector current. The conservation of the axial vector current can only be restored in the continuum limit (if at all). The conventional definition is either to use the local axial vector current

$$A_{Sx\mu}^{loc} \equiv \left(\overline{\psi}_x \gamma_\mu \gamma_5 \frac{\lambda_S}{2} \psi_x \right) , \tag{5.151}$$

or, introducing the non-local combinations appearing in (5.148),

$$A_{Sx\mu} \equiv \frac{1}{2} \left\{ \left(\overline{\psi}_{x+\hat{\mu}} \gamma_\mu \gamma_5 U_{x\mu} \frac{\lambda_S}{2} \psi_x \right) + \left(\overline{\psi}_x \gamma_\mu \gamma_5 U_{x\mu}^+ \frac{\lambda_S}{2} \psi_{x+\hat{\mu}} \right) \right\} . \tag{5.152}$$

The corresponding vector currents

$$V_{Sx\mu}^{loc} \equiv \left(\overline{\psi}_x \gamma_\mu \frac{\lambda_S}{2} \psi_x \right) ,$$

$$V_{Sx\mu} \equiv \frac{1}{2} \left\{ \left(\overline{\psi}_{x+\hat{\mu}} \gamma_\mu U_{x\mu} \frac{\lambda_S}{2} \psi_x \right) + \left(\overline{\psi}_x \gamma_\mu U_{x\mu}^+ \frac{\lambda_S}{2} \psi_{x+\hat{\mu}} \right) \right\} \tag{5.153}$$

are also useful. Other combinations appearing in (5.148) are

$$X_{VSx} \equiv \frac{r}{2} \sum_{\mu=1}^{4} \left\{ \left(\overline{\psi}_x U_{x\mu}^+ \frac{\lambda_S}{2} \psi_{x+\hat{\mu}} \right) + \left(\overline{\psi}_x U_{x-\hat{\mu},\mu} \frac{\lambda_S}{2} \psi_{x-\hat{\mu}} \right) \right.$$

$$\left. - \left(\overline{\psi}_{x+\hat{\mu}} U_{x\mu} \frac{\lambda_S}{2} \psi_x \right) - \left(\overline{\psi}_{x-\hat{\mu}} U_{x-\hat{\mu},\mu}^+ \frac{\lambda_S}{2} \psi_x \right) \right\} ,$$

$$X_{ASx} \equiv \frac{r}{2} \sum_{\mu=1}^{4} \left\{ 4 \left(\overline{\psi}_x \gamma_5 \frac{\lambda_S}{2} \psi_x \right) \right.$$

$$- \left(\overline{\psi}_x \gamma_5 U_{x\mu}^+ \frac{\lambda_S}{2} \psi_{x+\hat{\mu}} \right) - \left(\overline{\psi}_x \gamma_5 U_{x-\hat{\mu},\mu} \frac{\lambda_S}{2} \psi_{x-\hat{\mu}} \right)$$

$$\left. - \left(\overline{\psi}_{x+\hat{\mu}} \gamma_5 U_{x\mu} \frac{\lambda_S}{2} \psi_x \right) - \left(\overline{\psi}_{x-\hat{\mu}} \gamma_5 U_{x-\hat{\mu},\mu}^+ \frac{\lambda_S}{2} \psi_x \right) \right\} . \tag{5.154}$$

Using these notations, (5.148) can be written as [5.29]

$$\left\langle \mathcal{O} \Delta_\mu^b V_{Sx\mu} \right\rangle + \left\langle \left(\frac{\mathcal{O} \overleftarrow{\partial}}{\partial \psi_x} \frac{\lambda_S}{2} \psi_x \right) - \left(\overline{\psi}_x \frac{\lambda_S}{2} \frac{\overrightarrow{\partial} \mathcal{O}}{\partial \overline{\psi}_x} \right) \right\rangle$$

$$= \left\langle \mathcal{O} \left(\overline{\psi}_x \left[am, \frac{\lambda_S}{2} \right] \psi_x \right) + \mathcal{O} X_{VSx} \right\rangle ,$$

$$\left\langle \mathcal{O} \Delta_\mu^b A_{Sx\mu} \right\rangle + \left\langle \left(\frac{\mathcal{O} \overleftarrow{\partial}}{\partial \psi_x} \gamma_5 \frac{\lambda_S}{2} \psi_x \right) + \left(\overline{\psi}_x \gamma_5 \frac{\lambda_S}{2} \frac{\overrightarrow{\partial} \mathcal{O}}{\partial \overline{\psi}_x} \right) \right\rangle$$

$$= \left\langle \mathcal{O} \left(\overline{\psi}_x \gamma_5 \left\{ am, \frac{\lambda_S}{2} \right\} \psi_x \right) + \mathcal{O} X_{ASx} \right\rangle . \tag{5.155}$$

This is the form of the lattice W–T identities for flavour currents which we shall mainly use in what follows.

The continuum limit of the vector and axial vector currents in (5.149), (5.151)–(5.153) can be investigated in lattice perturbation theory. In order to illustrate the structure of current insertions in lattice Feynman graphs, let us consider the vector current $V_{Sx\mu}$ in (5.153). Replacing the link variables by their perturbative expansion (5.50), and using the Fourier-transformed fields in (8.38), (8.39), one obtains with $\bar{p} = p - k$, $p + p_1 - k$,

$p + p_1 + p_2 - k$, respectively, in the three terms on the right hand side,

$$\tilde{V}_{Sk\mu} \equiv \sum_x e^{-ik\cdot x} V_{Sx\mu} = \int_p \left(\tilde{\bar{\psi}}_{\bar{p}} \gamma_\mu \frac{\lambda_S}{2} \tilde{\psi}_p \right) \frac{1}{2} \left[e^{-i\bar{p}_\mu} + e^{ip_\mu} \right]$$

$$+ig \int_p \int_{p_1} \left(\tilde{\bar{\psi}}_{\bar{p}} \gamma_\mu \frac{\lambda_S}{2} \frac{\lambda_r}{2} \tilde{A}_\mu^r(p_1) \tilde{\psi}_p \right) \frac{1}{2} \left[e^{-i(\bar{p}_\mu - p_{1\mu}/2)} - e^{i(p_\mu + p_{1\mu}/2)} \right]$$

$$-\frac{g^2}{4} \int_p \int_{p_1} \int_{p_2} \left(\tilde{\bar{\psi}}_{\bar{p}} \gamma_\mu \frac{\lambda_S}{2} \left[\frac{1}{3} \tilde{A}_\mu^r(p_1) \tilde{A}_\mu^r(p_2) + d_{rst} \frac{\lambda_t}{2} \tilde{A}_\mu^r(p_1) \tilde{A}_\mu^s(p_2) \right] \tilde{\psi}_p \right)$$

$$\cdot \frac{1}{2} \left[e^{-i(\bar{p}_\mu - p_{1\mu}/2 - p_{2\mu}/2)} + e^{i(p_\mu + p_{1\mu}/2 + p_{2\mu}/2)} \right] + O(g^3) . \tag{5.156}$$

A graphical representation of these terms is given in fig. 5.13. The Feynman rules for these insertion vertices are (with no summation over μ):

$$\text{fig. 5.13(a)} = \delta_{\bar{c}c} (\gamma_\mu)_{\bar{\alpha}\alpha} \left(\frac{1}{2}\lambda_S \right)_{\bar{q}q} \frac{1}{2} \left[e^{-i\bar{p}_\mu} + e^{ip_\mu} \right] ,$$

$$\text{fig. 5.13(b)} = ig \delta_{\mu\mu_1} (\gamma_\mu)_{\bar{\alpha}\alpha} \left(\frac{1}{2}\lambda_S \right)_{\bar{q}q} \left(\frac{1}{2}\lambda_r \right)_{\bar{c}c}$$

$$\cdot \frac{1}{2} \left[e^{-i(\bar{p}_\mu - p_{1\mu}/2)} - e^{i(p_\mu + p_{1\mu}/2)} \right] ,$$

$$\text{fig. 5.13(c)} = -\frac{g^2}{2} (\gamma_\mu)_{\bar{\alpha}\alpha} \left(\frac{1}{2}\lambda_S \right)_{\bar{q}q} \delta_{\mu_1\mu_2} \left(\frac{1}{3}\delta_{r_1 r_2} + d_{r_1 r_2 t} \frac{1}{2}\lambda_t \right)_{\bar{c}c}$$

$$\cdot \frac{1}{2} \left[e^{-i(\bar{p}_\mu - p_{1\mu}/2 - p_{2\mu}/2)} + e^{i(p_\mu + p_{1\mu}/2 + p_{2\mu}/2)} \right] . \tag{5.157}$$

In the case of the axial vector current $A_{Sx\mu}$ in (5.152), γ_μ is everywhere replaced by $\gamma_\mu \gamma_5$.

In the continuum limit the currents are multiplicatively renormalized and also mixed with each other and with other composite operators, similarly to any other composite operator [A3,A8]. The normalization of currents is, however, restricted by the above W–T identities. In fact, for exactly conserved currents the normalization is uniquely determined. For instance, if in (5.150) one takes $\mathcal{O} = \bar{\psi}_y \psi_z$, and assumes exact flavour symmetry (which implies $[am, \lambda_S] = 0$), then one obtains

$$\left\langle \Delta_\mu^{(x)b} V_{Sx\mu}^{con} \bar{\psi}_y \psi_z \right\rangle = (\delta_{xy} - \delta_{xz}) \left\langle \bar{\psi}_y \frac{\lambda_S}{2} \psi_z \right\rangle . \tag{5.158}$$

In the continuum limit $a \to 0$ we have

$$\langle \cdots \rangle \to \langle 0| T\{\cdots\} |0 \rangle , \quad \Delta_\mu^{(x)b} \to a\frac{\partial}{\partial x_\mu} , \quad V_{Sx\mu}^{con} \to a^3 V(x)_{S\mu} ,$$

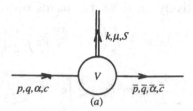

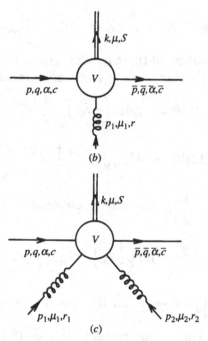

Fig. 5.13. Different terms in perturbation theory for the insertion of the flavour vector current $V_{Sx\mu}$: (a) zeroth order in the gauge coupling, (b) first order, (c) second order. Quarks are denoted by full lines, gluons by curly lines and the current by a double line.

$$\overline{\psi}_y \to \overline{\psi}(y)a^{3/2}Z_\psi^{1/2} , \quad \psi_z \to a^{3/2}Z_\psi^{1/2}\psi(z) , \quad \delta_{xy} \to a^4\delta^4(x-y) , \quad (5.159)$$

therefore the continuum limit of (5.158) is

$$\frac{\partial}{\partial x_\mu}\langle 0|T\{V(x)_{S\mu}\overline{\psi}(y)\psi(z)\}|0\rangle$$

$$= [\delta^4(x-y) - \delta^4(x-z)]\langle 0|T\left\{\overline{\psi}(y)\frac{\lambda_S}{2}\psi(z)\right\}|0\rangle . \quad (5.160)$$

Comparing with (5.145), this is in agreement with the equal time commutators

$$\delta(x_0 - y_0)[V(x)_{S0}, \overline{\psi}_{qac}(y)] = \delta^4(x - y)\overline{\psi}_{q'ac}(y)\left(\frac{\lambda_S}{2}\right)_{q'q},$$

$$\delta(x_0 - z_0)[V(x)_{S0}, \psi_{qac}(z)] = -\delta^4(x - z)\left(\frac{\lambda_S}{2}\right)_{qq'}\psi_{q'ac}(z). \quad (5.161)$$

Note that (5.160) only refers to the vacuum expectation value of this, but one can also deduce

$$\frac{\partial}{\partial x_\mu}\langle A|T\{V(x)_{S\mu}\overline{\psi}(y)\psi(z)\}|B\rangle$$

$$= [\delta^4(x - y) - \delta^4(x - z)]\langle A|T\left\{\overline{\psi}(y)\frac{\lambda_S}{2}\psi(z)\right\}|B\rangle, \quad (5.162)$$

for any on mass shell (physical) states $\langle A|$ and $|B\rangle$. In order to obtain (5.162) one has to start in (5.150) with $\mathcal{O} = A_u\overline{\psi}_y\psi_z B_v$, where A_u and B_v are interpolating fields for the states $\langle A|$ and $|B\rangle$, respectively. Since possible additional terms in (5.150) are then proportional to δ_{ux} or δ_{vx}, they do not contribute to the pole residuum, which gives the on mass shell matrix element. (See the reduction formulas in (8.63), (8.64), and remember that after the continuation to Euclidean space–time the particle poles are reproduced in the Laplace transforms by exponential contributions at large Euclidean time separations.)

The equal time commutation relations in (5.161) express the fact that the space integrals of the current densities (5.138) are the generators of the SU(3) flavour symmetry group. Note that this uniquely fixes the normalization of the four-vector current $V(x)_{S\mu}$ (the normalization of the fields $\overline{\psi}(y)$ and $\psi(z)$, of course, cancels in (5.161)). In other words, as a consequence of the W–T identity (5.150),

$$\lim_{a\to 0} a^{-3}V_{Sx\mu}^{con} = V(x)_{S\mu} \quad (5.163)$$

is the correctly normalized conserved flavour vector current operator. This can also be stated by saying that in the continuum limit $V_{Sx\mu}^{con}$ 'is not renormalized'.

Up to now equal quark masses (exact flavour symmetry) were assumed. In nature the u- and d-quark masses are both very small and close to each other, but the s-quark mass is larger. The explicit flavour symmetry breaking due to the quark mass differences appears also in the W–T identities (5.150) or (5.155). In the case of $m_d = m_u$, the bare quark mass matrix is

$$m = \begin{pmatrix} m_u & 0 & 0 \\ 0 & m_u & 0 \\ 0 & 0 & m_s \end{pmatrix} = \frac{1}{3}(2m_u + m_s) + \frac{\lambda_8}{\sqrt{3}}(m_u - m_s) , \qquad (5.164)$$

therefore we have from (8.16) and (8.21)

$$\left[m, \frac{\lambda_S}{2} \right] = \frac{\mathrm{i}}{\sqrt{3}}(m_u - m_s) f_{8SR}\lambda_R ,$$

$$\left\{ m, \frac{\lambda_S}{2} \right\} = \frac{1}{3}(2m_u + m_s)\lambda_S + \frac{1}{\sqrt{3}}(m_u - m_s)\left(\frac{2}{3}\delta_{8S} + d_{8SR}\lambda_R \right) . \quad (5.165)$$

These are the combinations appearing on the right hand side of the W–T identities, which express the explicit breaking of the chiral flavour symmetry due to the quark masses. For instance, if $m_u \neq m_s$, then the vector current $V^{con}_{Sx\mu}$ for $S = 4, 5, 6, 7$ is not exactly conserved, and the definition of the continuum vector current in (5.163) for these S-values becomes more subtle. Without going into details here, let us just remark that the procedure is in principle similar to the definition of the partially conserved axial vector current in the next subsection.

5.3.2   PCAC and the quark mass

The PCAC relation (5.140) tells us that in the limit of exactly zero quark mass, when the pion is a true Goldstone boson with exactly zero mass, the divergence of the axial vector current vanishes. Since the axial symmetry of the lattice action is explicitly broken, even for zero bare quark mass ($m = 0$) by the Wilson term proportional to r, the construction of the correctly normalized conserved axial vector current in the continuum limit is more involved than the definition of the conserved vector current in (5.163). In the present section, for simplicity, only the equal quark mass case $m_u = m_d = m_s \equiv m$ will be considered. For some aspects of the flavour symmetry breaking due to quark mass differences see [5.66, 5.67].

If in the continuum limit one defines the operators

$$\lim_{a \to 0} a^{-3} A_{Sx\mu} \equiv A(x)_{S\mu} , \qquad \lim_{a \to 0} a^{-4} X_{ASx} \equiv X(x)_{AS} ,$$

$$P_{Sx} \equiv \left(\overline{\psi}_x \gamma_5 \frac{\lambda_S}{2} \psi_x \right) , \qquad \lim_{a \to 0} a^{-3} P_{Sx} \equiv P(x)_S , \qquad (5.166)$$

and takes $\mathcal{O}$ in the axial vector W–T identity (5.155) to be the product of appropriate interpolating fields then, similarly to (5.162), for the on mass shell matrix elements one obtains the relation

$$\langle A | \frac{\partial}{\partial x_\mu} A(x)_{S\mu} | B \rangle = \langle A | 2mP(x)_S + X(x)_{AS} | B \rangle . \qquad (5.167)$$

Owing to the appearance of $X(x)_{AS}$ on the right hand side this is not the correct PCAC relation, because the axial vector current is not conserved at $m = 0$. In the tree graph approximation, however, X_{ASx} has mass dimension 5 (it is proportional to the fifth power of the lattice spacing a), therefore it seems that in the continuum limit $X(x)_{AS} = 0$. For instance, the matrix element of X_{ASx} between free quark states is given by

$$\langle \mathbf{p}_2, \sigma_2 | X_{ASx} | \mathbf{p}_1, \sigma_1 \rangle = -ra^5 m^2 e^{ix \cdot (p_2 - p_1)}$$

$$\cdot \bar{u}(\mathbf{p}_2, \sigma_2) \gamma_5 \frac{\lambda_S}{2} u(\mathbf{p}_1, \sigma_1) + O(a^7) . \tag{5.168}$$

This is in Minkowski space, and the quark states are normalized according to (8.51)–(8.57). In higher orders of the gauge coupling, however, the vanishing of $\lim_{a \to 0} a^{-4} X_{ASx}$ can be compensated by the ultraviolet divergence of loop integrations. This is the potential source of anomalies in the continuum limit of the divergence of axial vector currents, as discussed for an external electromagnetic field at the end of section 4.4 (for the anomaly of the flavour singlet axial vector current in QCD see the discussion near (5.209)).

In the case of the flavour non-singlet axial vector current $A(x)_{S\mu}$ there is no anomaly but, as a consequence of ultraviolet divergences in loops, X_{ASx} mixes with the dimension 4 operator $\partial/\partial x_\mu A(x)_{S\mu}$ and the dimension 3 operator $P(x)_S$, that is [5.66]

$$X_{ASx} = \overline{X}_{ASx} - 2a\bar{m} P_{Sx} - (Z_A - 1)\Delta_\mu^b A_{Sx\mu} . \tag{5.169}$$

Here, by definition, $\overline{X}_{ASx}$ does not contribute to (5.167) because

$$\lim_{a \to 0} a^{-4} \langle A | \overline{X}_{ASx} | B \rangle = 0 . \tag{5.170}$$

Using the decomposition (5.169), the axial vector W–T identity in (5.155) in the flavour symmetric case can be written as

$$\left\langle \mathcal{O} \Delta_\mu^b A_{Sx\mu}^{ren} \right\rangle + \left\langle \left(\frac{\mathcal{O} \overleftarrow{\partial}}{\partial \psi_x} \gamma_5 \frac{\lambda_S}{2} \psi_x \right) + \left(\overline{\psi}_x \gamma_5 \frac{\lambda_S}{2} \frac{\overrightarrow{\partial} \mathcal{O}}{\partial \overline{\psi}_x} \right) \right\rangle$$

$$= \left\langle \mathcal{O} 2a(m - \bar{m}) P_{Sx} + \mathcal{O} \overline{X}_{ASx} \right\rangle . \tag{5.171}$$

Here $A_{Sx\mu}^{ren}$ is the *renormalized axial vector current* defined by

$$A_{Sx\mu}^{ren} \equiv Z_A A_{Sx\mu} . \tag{5.172}$$

The condition (5.170) does not yet determine $\overline{X}_{ASx}$ uniquely. A sufficient condition can be obtained from the requirement that in the continuum limit the renormalized axial vector current has to satisfy the correct equal time commutation relations with the quark field. For the conserved vector current the analogous relations were obtained from (5.158), therefore let us consider in (5.171) $\mathcal{O} = \overline{\psi}_y \psi_z$, which gives

$$\left\langle \Delta_\mu^{(x)b} A_{Sx\mu}^{ren} \overline{\psi}_y \psi_z \right\rangle = \left\langle 2a(m - \overline{m}) P_{Sx} \overline{\psi}_y \psi_z + \overline{X}_{ASx} \overline{\psi}_y \psi_z \right\rangle$$

$$- (\delta_{xy} + \delta_{xz}) \left\langle \overline{\psi}_y \gamma_5 \frac{\lambda_S}{2} \psi_z \right\rangle . \tag{5.173}$$

In the continuum limit $a \to 0$ this has to be consistent with the equal time commutation relations (analogous to (5.161)):

$$\delta(x_0 - y_0)[A(x)_{S0}, \overline{\psi}_{q\alpha c}(y)] = -\delta^4(x - y)\overline{\psi}_{q'\alpha'c}(y)(\gamma_5)_{\alpha'\alpha} \left(\frac{\lambda_S}{2} \right)_{q'q} ,$$

$$\delta(x_0 - z_0)[A(x)_{S0}, \psi_{q\alpha c}(z)] = -\delta^4(x - z)(\gamma_5)_{\alpha\alpha'} \left(\frac{\lambda_S}{2} \right)_{qq'} \psi_{q'\alpha'c}(z) . \tag{5.174}$$

Consistency is achieved for

$$A(x)_{S\mu} \equiv \lim_{a \to 0} a^{-3} A_{Sx\mu}^{ren} = \lim_{a \to 0} a^{-3} Z_A A_{Sx\mu} \tag{5.175}$$

if, as in (5.162), for any pair $\langle A|, |B \rangle$ of physical states on mass shell we have

$$\lim_{a \to 0} a^{-7} \langle A | T\{\overline{X}_{ASx} \overline{\psi}_y \psi_z\} | B \rangle = 0 . \tag{5.176}$$

This is a stronger condition than (5.170).

The axial vector W–T identity (5.171), and its special case (5.173), show that the conservation of the axial vector current is achieved at $m = \overline{m}$. This is the *chiral limit* of lattice QCD with Wilson quarks. Since $a\overline{m} = a\overline{m}(am, g ; r)$ is a function of the bare parameters am, g (and r), the chiral limit is defined by the solution of the equation

$$am_{cr} = a\overline{m}(am_{cr}, g ; r) . \tag{5.177}$$

For fixed r the solution $am_{cr} \equiv \frac{1}{2} K_{cr}^{-1} - 4r$ can be considered as a function of the bare gauge coupling g. This defines the *critical line* for zero quark mass in terms of the *critical hopping parameter* K_{cr}.

In lattice perturbation theory $a\overline{m}$ and Z_A can be calculated as functions of the bare parameters from the condition (5.176) with $A = B = 0$. In 1-loop order one has, for instance, [5.66]

$$a\overline{m} = -\frac{2}{3}rg^2 \int_q \left\{ \frac{1}{\hat{q}^2} \left[3 + \frac{\hat{q}^2(4 - \frac{1}{4}\hat{q}^2) - \overline{q}^2}{\overline{q}^2 + \frac{1}{4}r^2(\hat{q}^2)^2} \right] \right.$$

$$\left. - amr\frac{\hat{q}^2(4 - \frac{1}{4}\hat{q}^2) - \overline{q}^2}{[\overline{q}^2 + \frac{1}{4}r^2(\hat{q}^2)^2]^2} \right\} + O(g^4)$$

$$= -\delta(am) + \frac{2}{3}amr^2g^2 \int_q \frac{\hat{q}^2(4 - \frac{1}{4}\hat{q}^2) - \bar{q}^2}{[\bar{q}^2 + \frac{1}{4}r^2(\hat{q}^2)^2]^2} + O(g^4) , \quad (5.178)$$

where $\delta(am)$ is given by (5.57) with $N = 3$. Since the perturbative solution of (5.177) for am_{cr} is of order g^2, we have from (5.177) and (5.178) $am_{cr} = -\delta(am) + O(g^4)$, therefore the lowest order g^2-dependence of am_{cr} and K_{cr} agrees with (5.57)–(5.59).

Remember that in (5.59) the critical line for zero quark mass was determined by requiring the linearly divergent contribution to the renormalized quark mass to vanish. Here we see that, to lowest order in g^2, this is the same as the critical line given by (5.177), which is the condition for the conservation of the axial vector current. If the continuum limit at $g^2 = am = 0$ is approached along that critical line, then the existence of a conserved axial charge operator follows from the conservation of the axial vector current operator $A(x)_{S\mu}$ in (5.175). This together with the assumption that the axial symmetry is spontaneously broken by the non-zero vacuum expectation value of $(\bar{\psi}_x\psi_x)$ imply, by the Goldstone theorem, that the pseudoscalar boson mass is zero.

The axial vector current renormalization factor Z_A can also be calculated in lattice perturbation theory from the condition (5.176) with $A = B = 0$ [5.66]. The result is, similarly to $a\overline{m}$ in (5.178), a power series in the bare coupling squared g^2 with finite coefficients. In fact, it is possible to show to all orders of perturbation theory that Z_A (as well as Z_V) is a function of g and of the Wilson parameter r only [5.68]. Since in the continuum limit $g^2 \to 0$, the axial vector current renormalization factor Z_A tends to 1. Therefore Z_A in (5.175) can also be omitted from the definition of the continuum axial vector current operator $A(x)_{S\mu}$ (the definition in (5.166) is also correct). In numerical calculations, however, which are performed at non-zero g^2, it is better to use $A^{ren}_{S\mu}$ as defined in (5.172), because Z_A is still quite different from 1 [5.69].

Instead of $A_{S\mu}$, one can also consider the local axial vector current $A^{loc}_{S\mu}$ in (5.151), and define a renormalization factor Z^{loc}_A, such that

$$A^{ren}_{S\mu} \equiv Z_A A_{S\mu} = Z^{loc}_A A^{loc}_{S\mu} + O(a^4) . \quad (5.179)$$

For the vector currents in (5.153) a similar definition is

$$V^{ren}_{S\mu} \equiv V^{con}_{S\mu} = Z_V V_{S\mu} + O(a^4) = Z^{loc}_V V^{loc}_{S\mu} + O(a^4) . \quad (5.180)$$

The finite renormalization factors Z_A, Z^{loc}_A, Z_V and Z^{loc}_V, which all tend to 1 in the continuum limit, can be determined numerically. In the case of the vector currents the simplest procedure is to consider the ratios of their two-point functions to the two-point function of $V^{con}_{S\mu}$ [5.69]. For the axial vector currents it is better to use some three-point A–A–V current expectation values in connection with the current algebra relations (see next subsection).

The form of the axial vector W–T identities in (5.171) or (5.173) suggests that we define, for fixed g^2 (and r),

$$\overline{m}_0 \equiv m - \overline{m} \tag{5.181}$$

as the bare quark mass parameter [5.66]. Since $a\overline{m}$ is a function of am, too, and $(am - am_{cr})$ is small in the continuum limit, one can expand at $am = am_{cr}$ and obtain

$$a\overline{m}_0 \simeq (am - am_{cr})Z_m \; ; \quad Z_m \equiv 1 - \left.\frac{\partial(a\overline{m})}{\partial(am)}\right|_{m=m_{cr}} . \tag{5.182}$$

This means that

$$\overline{m}_0 \simeq \frac{Z_m}{2a}\left(\frac{1}{K} - \frac{1}{K_{cr}}\right) . \tag{5.183}$$

Compared with the definition of m_0 in (5.56), (5.60), this contains an extra finite renormalization factor Z_m, which tends to 1 in the continuum limit. For non-zero g^2, however, where the numerical calculations are done, it is better to use the definition (5.181), because Z_m can still substantially deviate from 1 [5.69]. In 1-loop lattice perturbation theory Z_m can be obtained from (5.182) and (5.178) as

$$Z_m = 1 - \frac{2}{3}r^2 g^2 \int_q \frac{\hat{q}^2(4 - \frac{1}{4}\hat{q}^2) - \overline{q}^2}{[\overline{q}^2 + \frac{1}{4}r^2(\hat{q}^2)^2]^2} + O(g^4) . \tag{5.184}$$

Note that if Z_A is known, it is easy to determine $\overline{m}_0$. This is because for $a \to 0$ the W–T identity (5.171), together with (5.172) and (5.170), imply that, for any pair of on mass shell states $\langle A|$, $|B\rangle$, we have

$$Z_A\langle A|\Delta_\mu^b A_{Sx\mu}|B\rangle = 2a\overline{m}_0\langle A|P_{Sx}|B\rangle + O(a^5) . \tag{5.185}$$

Here the $O(a^5)$ term is negligible for small a, because the rest is $O(a^4)$, therefore

$$\tilde{m}_0 \equiv \frac{\overline{m}_0}{Z_A} = \frac{\langle 0|\Delta_\mu^b A_{Sx\mu}|\pi\rangle}{2a\langle 0|P_{Sx}|\pi\rangle} + O(a) , \tag{5.186}$$

where, for definiteness, the matrix elements between the vacuum and the lowest pseudoscalar meson state $|\pi\rangle$ were taken. These matrix elements can easily be obtained from the two-point functions $\langle \Delta_\mu^{(x)b} A_{Sx\mu}\Delta_\nu^{(y)b} A_{Sy\nu}\rangle$ and $\langle P_{Sx}P_{Sy}\rangle$, respectively, using the behaviour at large $|x_0 - y_0|$ (see (5.94) and (5.95)).

5.3.3   Current algebra

Using the correctly normalized vector and axial vector currents (5.180) and (5.179), respectively, the equal time commutators in (5.139) can be verified in lattice perturbation theory. In order to extract the equal time commutators appearing, for instance, in the relations (5.143) and (5.145), one has to consider the continuum limit of the W–T identities (5.150), (5.155) or (5.171). The perturbative calculations were performed in the 1-loop approximation in [5.66] (see also [5.29, 5.20, 5.30, 5.39, 5.70]), and showed that the current algebra relations are indeed satisfied.

Let us consider here, as an illustration, the axial vector–axial vector commutator in the case of equal mass (flavour symmetric) quarks. In the W–T identity (5.171) let us take $\mathcal{O} = A^{ren}_{Tyv} = Z^{loc}_A A^{loc}_{Tyv} + O(a^4)$, which gives with (5.181)

$$\langle \Delta^{(x)b}_\mu A^{ren}_{Sx\mu} A^{ren}_{Tyv} \rangle = 2a\overline{m}_0 \langle P_{Sx} A^{ren}_{Tyv} \rangle + \langle \overline{X}_{ASx} A^{ren}_{Tyv} \rangle$$

$$+ \frac{Z^{loc}_A}{Z^{loc}_V} \delta_{xy} \mathrm{if}_{STU} \langle V^{ren}_{Uyv} \rangle + O(a^8) . \tag{5.187}$$

In the chiral limit with zero quark mass ($\overline{m}_0 = 0$) for $a \to 0$, explicit evaluation in 1-loop perturbation theory shows [5.66] that

$$\Delta^{(x)b}_\mu \langle A|T\{A^{ren}_{Sx\mu} A^{ren}_{Tyv}\}|B\rangle = \delta_{xy} \mathrm{if}_{STU} \langle A|V^{ren}_{Uyv}|B\rangle + O(a^8) . \tag{5.188}$$

Here $\langle A|$ and $|B\rangle$ are an arbitrary pair of states on mass shell. The $O(a^8)$ piece is negligible for small a, because the rest is $O(a^7)$. The important statement in (5.188) is that the $Z_{A,V}$ factors cancel, which differ from 1 by $O(g^2)$ terms. This happens because in $\langle A|T\{\overline{X}_{ASx} A^{ren}_{Tyv}\}|B\rangle$ the appropriate local contact term (proportional to δ_{xy}) is contained. (Remember that $\overline{X}_{ASx}$ is uniquely defined, including the contact terms, by (5.176). In fact, the difference between (5.170) and (5.176) is just that the latter fixes the contact terms, too.)

In the continuum limit (5.188) gives

$$\frac{\partial}{\partial x_\mu} \langle A|T\{A(x)_{S\mu} A(y)_{Tv}\}|B\rangle = \delta^4(x - y) \mathrm{if}_{STU} \langle A|V(x)_{Uv}|B\rangle , \tag{5.189}$$

which is consistent with the equal time commutator of two axial vector currents in (5.139). Note that, in perturbation theory, confinement is not taken into account, therefore the on mass shell states $\langle A|$ and $|B\rangle$ are constructed from free quarks and gluons. The current algebra hypothesis in the framework of lattice QCD is equivalent to the assumption that the relations like (5.189) are valid for physical on mass shell states respecting confinement.

The cancellation of the Z-factors of current renormalization in the W–T identities like (5.188) can be used for the numerical determination of, for instance, Z_A or Z^{loc}_A [5.69]. For this purpose one can consider an A–A–V current expectation value. In the zero quark mass continuum limit one has to recover the W–T identity

$$\frac{\partial}{\partial x_\mu} \langle A | T \{ A(x)_{S\mu} A(y)_{Tv} V(z)_{U\rho} \} | B \rangle$$

$$= \delta^4(x - y) \mathrm{i} f_{STV} \langle A | T \{ V(y)_{Vv} V(z)_{U\rho} | B \rangle$$

$$+ \delta^4(x - z) \mathrm{i} f_{SUV} \langle A | T \{ A(y)_{Tv} A(z)_{V\rho} | B \rangle \ . \qquad (5.190)$$

Similarly to (5.187), the W–T identity in (5.171) with

$$\mathcal{O} = A_{Tyv}^{ren} V_{Uz\rho}^{ren} = Z_A^{loc} Z_V^{loc} A_{Tyv}^{loc} V_{Uz\rho}^{loc} + O(a^4)$$

gives

$$\langle \Delta_\mu^{(x)b} A_{Sx\mu}^{ren} A_{Tyv}^{ren} V_{Uz\rho}^{ren} \rangle = 2a\bar{m}_0 \langle P_{Sx} A_{Tyv}^{ren} V_{Uz\rho}^{ren} \rangle + \langle \overline{X}_{ASx} A_{Tyv}^{ren} V_{Uz\rho}^{ren} \rangle$$

$$+ \frac{Z_A^{loc}}{Z_V^{loc}} \delta_{xy} \mathrm{i} f_{STV} \langle V_{Vyv}^{ren} V_{Uz\rho}^{ren} \rangle + \frac{Z_V^{loc}}{Z_A^{loc}} \delta_{xz} \mathrm{i} f_{SUV} \langle A_{Tyv}^{ren} A_{Vz\rho}^{ren} \rangle + \bar{O}(a) \ . \quad (5.191)$$

Here $\bar{O}(a)$ stands for correction terms of higher order in a, which are negligible in the continuum limit. The Z-factors in front of the equal time commutator terms disappear, if the contact terms in $\overline{X}_{ASx}$ are such that

$$\langle \overline{X}_{ASx} A_{Tyv}^{loc} V_{Uz\rho}^{loc} \rangle = \left(\frac{Z_V^{loc}}{Z_A^{loc}} - 1 \right) \delta_{xy} \mathrm{i} f_{STV} \langle V_{Vyv}^{loc} V_{Uz\rho}^{loc} \rangle$$

$$+ \left(\frac{Z_A^{loc}}{Z_V^{loc}} - 1 \right) \delta_{xz} \mathrm{i} f_{SUV} \langle A_{Tyv}^{loc} A_{Vz\rho}^{loc} \rangle + \bar{O}(a) \ . \qquad (5.192)$$

From the definition in (5.169), and using (5.181) and (5.186), one can write $\overline{X}_{ASx}$ as

$$\overline{X}_{ASx} = -\eta_{Sx} + Z_A \xi_{Sx} \ ;$$

$$\eta_{Sx} \equiv \Delta_\mu^b A_{Sx\mu} - 2am P_{Sx} - X_{ASx} \ , \quad \xi_{Sx} \equiv \Delta_\mu^b A_{Sx\mu} - 2a\tilde{m}_0 P_{Sx} \ . \quad (5.193)$$

Here η_{Sx} is defined in such a way that its contribution in (5.192), due to the W–T identity (5.155), just reproduces the terms proportional to -1 on the right hand side. Therefore the condition (5.192) can also be formulated as

$$\langle \xi_{Sx} A_{Tyv}^{loc} V_{Uz\rho}^{loc} \rangle = \frac{Z_V^{loc}}{Z_A Z_A^{loc}} \delta_{xy} \mathrm{i} f_{STV} \langle V_{Vyv}^{loc} V_{Uz\rho}^{loc} \rangle$$

$$+ \frac{Z_A^{loc}}{Z_A Z_V^{loc}} \delta_{xz} \mathrm{i} f_{SUV} \langle A_{Tyv}^{loc} A_{Vz\rho}^{loc} \rangle + \bar{O}(a) \ . \qquad (5.194)$$

This formula can be used for the numerical determination of the axial

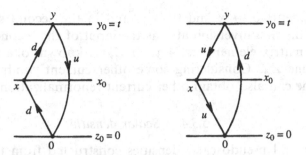

Fig. 5.14. The quark line graphs contributing to the three-point expectation value of the $\langle \xi AV \rangle$ three-point function.

vector current renormalization factors. By using the relations (5.179) and (5.180) one can write this also in terms of the other lattice currents, for instance,

$$\langle \xi_{Sx} A_{Ty\nu} V_{Uz\rho} \rangle = \frac{Z_V}{Z_A^2} \delta_{xy} \mathrm{i} f_{STV} \langle V_{Vy\nu} V_{Uz\rho} \rangle$$

$$+ \frac{1}{Z_V} \delta_{xz} \mathrm{i} f_{SUV} \langle A_{Ty\nu} A_{Vz\rho} \rangle + \bar{O}(a) \ . \tag{5.195}$$

The numerical evaluation of the three-point expectation value on the left hand side of (5.195) is not difficult, if appropriate quantum numbers are chosen. For instance, for $S = 3$, $T = 1 + \mathrm{i}2 \equiv \pi^-$ and $U = 1 - \mathrm{i}2 \equiv \pi^+$, because of $f_{STU} \to \epsilon_{STU}$, we have

$$\langle \xi_{3x} A_{\pi^- y\nu} V_{\pi^+ z\rho} \rangle = \frac{Z_V}{Z_A^2} \delta_{xy} \langle V_{\pi^- y\nu} V_{\pi^+ z\rho} \rangle$$

$$- \frac{1}{Z_V} \delta_{xz} \langle A_{\pi^- y\nu} A_{\pi^+ z\rho} \rangle + \bar{O}(a) \ . \tag{5.196}$$

Putting $v = \rho = n = \{1, 2, 3\}$, and the point z to the origin $z = 0$, and defining $y \equiv (\mathbf{y}t)$, one gets

$$\sum_x \sum_{\mathbf{y},n} \langle \xi_{3x} A_{\pi^- (\mathbf{y}t)n} V_{\pi^+ 0n} \rangle = \frac{Z_V}{Z_A^2} \sum_{\mathbf{y},n} \langle V_{\pi^- (\mathbf{y}t)n} V_{\pi^+ 0n} \rangle$$

$$- \frac{1}{Z_V} \sum_{\mathbf{y},n} \langle A_{\pi^- (\mathbf{y}t)n} A_{\pi^+ 0n} \rangle + \bar{O}(a) \ . \tag{5.197}$$

The left hand side here can easily be evaluted, because only the simple quark line graphs depicted in fig. 5.14 contribute. Since these only contain a single quark line loop connecting three points, one can calculate the necessary propagator matrix elements by two matrix inversions per spin and colour indices. One can start from the origin 0 and calculate the matrix elements $0 \to x$ and $0 \to y$, and in the second step, taking the output of the first inversion at x as the input of the second inversion, the necessary matrix elements $x \to y$. If Z_V is known, one can use (5.197) to determine Z_A. Considering some other current combinations, like in (5.194), one can also obtain other current renormalization factors [5.69].

5.3.4   *Scalar densities*

The scalar and pseudoscalar densities constructed from the quark fields play an important rôle in the broken chiral symmetry of low energy QCD. For instance, the non-zero vacuum expectation value of the flavour singlet scalar density S_x drives the spontaneous breaking of the chiral symmetry at zero quark masses. The eighth component of the octet scalar density S_{8x} is proportional to the $SU(3)_V$ flavour symmetry breaking Hamiltonian, which describes the effects of the mass difference of s- and (u, d)-quarks. The pseudoscalar octet densities P_{Sx} ($S = 1, 2, \ldots, 8$) can serve as interpolating fields for the quasi-Goldstone bosons in the pseudoscalar meson octet.

The definition of these densities is

$$S_x \equiv (\overline{\psi}_x \psi_x), \qquad S_{Sx} \equiv \left(\overline{\psi}_x \frac{\lambda_S}{2} \psi_x \right),$$

$$P_x \equiv (\overline{\psi}_x \gamma_5 \psi_x), \qquad P_{Sx} \equiv \left(\overline{\psi}_x \gamma_5 \frac{\lambda_S}{2} \psi_x \right). \tag{5.198}$$

In the flavour symmetric continuum limit the singlets and octets are renormalized, in general, with different factors:

$$S_x^{ren} \equiv Z_{S0} S_x, \qquad S(x) \equiv \lim_{a \to 0} a^{-3} S_x^{ren},$$

$$S_{Sx}^{ren} \equiv Z_S S_{Sx}, \qquad S(x)_S \equiv \lim_{a \to 0} a^{-3} S_{Sx}^{ren},$$

$$P_x^{ren} \equiv Z_{P0} P_x, \qquad P(x) \equiv \lim_{a \to 0} a^{-3} P_x^{ren},$$

$$P_{Sx}^{ren} \equiv Z_P P_{Sx}, \qquad P(x)_S \equiv \lim_{a \to 0} a^{-3} P_{Sx}^{ren}. \tag{5.199}$$

In fact, the renormalization conditions can be chosen to a large extent arbitrarily.

There is, however, a restriction for the ratio of the pseudoscalar to scalar densities, which is the consequence of the fact that scalar densities appear in the lattice action (5.48). This implies that the derivative of an arbitrary expectation value with respect to am is

$$\frac{\partial}{\partial(am)} \langle \mathcal{O} \rangle = - \left\langle \mathcal{O} \sum_y S_y \right\rangle + \langle \mathcal{O} \rangle \left\langle \sum_y S_y \right\rangle. \tag{5.200}$$

(The second term on the right hand side is due to the partition function, which provides the correct normalization.) Taking $\mathcal{O} = A_u \mathcal{O}_x B_v$, where A_u and B_v are interpolating fields for the on mass shell states $\langle A|$ and $|B\rangle$, respectively, and using the reduction formulas in (8.63) and (8.64), one obtains after rearrangement

$$\langle A|T\left\{\mathcal{O}_x \sum_y S_y\right\}|B\rangle = \langle A|\mathcal{O}_x|B\rangle\langle 0|\sum_y S_y|0\rangle - \frac{\partial}{\partial(am)}\langle A|\mathcal{O}_x|B\rangle . \quad (5.201)$$

As a special case let us consider the lattice divergence of the renormalized axial vector current: $\mathcal{O}_x = \Delta_\mu^{(x)b} A_{Sx\mu}^{ren} \equiv \sum_{\mu=1}^4 (A_{Sx\mu}^{ren} - A_{Sx-\hat\mu\mu}^{ren})$. Using (5.179), (5.181), (5.182) and (5.185) one can show that

$$\frac{\partial}{\partial(am)}\langle A|\Delta_\mu^b A_{Sx\mu}^{ren}|B\rangle = 2Z_m\langle A|P_{Sx}|B\rangle$$

$$+ 2a\bar{m}_0 \frac{\partial}{\partial(am)}\langle A|P_{Sx}|B\rangle + O(a^4) . \quad (5.202)$$

In the chiral limit with zero quark mass (5.201) and (5.202) imply

$$\Delta_\mu^{(x)b}\langle A|T\{A_{Sx\mu}^{ren} \sum_y S_y\}|B\rangle = -2Z_m\langle A|P_{Sx}|B\rangle + O(a^4) . \quad (5.203)$$

In the continuum limit $a \to 0$ this has to reproduce the W–T identity

$$\frac{\partial}{\partial x_\mu}\langle A|T\{A(x)_{S\mu}S(y)\}|B\rangle = -2\delta^4(x-y)\langle A|P(y)_S|B\rangle . \quad (5.204)$$

Therefore, according to (5.199), the ratio of the renormalization factors has to satisfy [5.67]

$$\frac{Z_P}{Z_{S0}} = Z_m . \quad (5.205)$$

This can also be directly checked in 1-loop lattice perturbation theory [5.66], where the result for the ratio Z_P/Z_{S0} agrees with Z_m in (5.184). The non-perturbative determination of the ratios of renormalization factors for the pseudoscalar and scalar densities can be based on the W–T identity for the three-point function A_μ-S-P [5.67]. The procedure is quite similar to the one applied for Z_A in the previous subsection.

The non-perturbative definition of the scalar singlet vacuum expectation value $\langle S_x \rangle$ is also based on the appropriate W–T identity. $\langle S_x \rangle$ is called the *'quark condensate'*, and signals the spontaneous breaking of the chiral symmetry. For its definition let us consider (5.171) with $\mathcal{O} = P_{Ty}$, which gives with (5.181) and (8.21)

$$\langle \Delta_\mu^{(x)b} A_{Sx\mu}^{ren} P_{Ty} \rangle = 2a\bar{m}_0 \langle P_{Sx} P_{Ty} \rangle$$

$$+ \langle \overline{X}_{ASx} P_{Ty} \rangle - \delta_{xy} \left\langle \frac{1}{3} \delta_{ST} S_y + d_{STU} S_{Uy} \right\rangle . \qquad (5.206)$$

In the case of an exact SU(3)$_V$ flavour symmetry (with equal quark masses) the last term vanishes: $\langle S_{Uy} \rangle = 0$. As was also discussed in previous subsections, in the continuum limit the term containing $\overline{X}_{ASx}$ provides some local contact (Schwinger) terms, which are proportional to the delta function $\delta^4(x - y)$ and its derivatives. Some of these terms vanish after summation over x. The rest can be added to $\langle S_y \rangle$, which gives, by definition, the *subtracted expectation value* of the scalar singlet density $\langle S_y \rangle_{sub}$. Therefore the sum of (5.206) over x, in the limit of massless quarks, is

$$\frac{1}{3} \delta_{ST} \langle S_y \rangle_{sub} = \lim_{\overline{m}_0 \searrow 0} \left\{ 2a\overline{m}_0 \sum_x \langle P_{Sx} P_{Ty} \rangle \right\} . \qquad (5.207)$$

Here it is important that the right hand side is evaluated at non-zero (positive) quark mass ($\overline{m}_0 > 0$), and the limit $\overline{m}_0 \to 0$ is taken by extrapolation. For small positive $\overline{m}_0$ the pion mass is still non-zero but small. In fact, in (5.207) for $\overline{m}_0 \searrow 0$ only the pion contributes. To every order of perturbation theory $\langle S_y \rangle_{sub}$ is zero by construction. The non-perturbative contributions can be numerically obtained from (5.207) [5.69].

5.3.5   *The U(1) problem*

In previous subsections the continuum limit of W–T identities for the flavour octet vector and axial vector currents was investigated. The associated charges in (5.138) are the generators of the broken SU(3)$_V$ $\otimes$ SU(3)$_A$ chiral flavour symmetry. In fact, the symmetry of the classical continuum QCD action with three massless quarks is larger than SU(3)$_V$ $\otimes$ SU(3)$_A$, namely U(3)$_V$ $\otimes$ U(3)$_A$ = U(1)$_V$ $\otimes$ U(1)$_A$ $\otimes$ SU(3)$_V$ $\otimes$ SU(3)$_A$. The continuum current operators associated to U(1)$_V$ and U(1)$_A$ are, respectively, $V(x)_\mu$ and $A(x)_\mu$ defined as

$$V(x)_\mu \equiv \overline{\psi}(x)\gamma_\mu\psi(x) , \quad A(x)_\mu \equiv \overline{\psi}(x)\gamma_\mu\gamma_5\psi(x) , \qquad (5.208)$$

where a sum over flavour (q), Dirac (α) and colour (c) indices of the quark field $\psi(x) \equiv \psi(x)_{q\alpha c}$ is understood.

The flavour singlet vector current $V(x)_\mu$ is exactly conserved even for different quark masses. The corresponding conserved current on the lattice $V_{x\mu}^{con}$ can be defined by the replacement $\lambda_S/2 \to 1$ in (5.149), and the associated W–T identity is the same as (5.150) with $\lambda_S/2 \to 1$. This expresses the exact conservation of the number of quarks minus the number of antiquarks.

In the case of the flavour singlet axial vector current $A(x)_\mu$, the U(1)$_A$ symmetry is explicitly broken by the Wilson term in the lattice action, similarly to the flavour octet axial vector current $A(x)_{Sx\mu}$. The difference is that whereas the (partial) conservation of $A(x)_{S\mu}$ is restored in the continuum limit (see (5.167)–(5.170)), the presence of an Adler–Bell–Jackiw anomaly in the divergence of $A(x)_\mu$ prevents the restoration of the $U(1)_A$ symmetry, even in the massless continuum limit, because the divergence of $A(x)_\mu$ is given by

$$\frac{\partial}{\partial x_\mu} A(x)_\mu = \frac{N_f g^2}{32\pi^2} \epsilon_{\mu\nu\rho\sigma} F^a(x)_{\mu\nu} F^a(x)_{\rho\sigma} . \tag{5.209}$$

Here $N_f = 3$ is the number of quark flavours, g^2 the colour gauge coupling, and $F^a(x)_{\mu\nu}$ $(a = 1, 2, \ldots, 8)$ the colour field strength tensor. The anomalous term on the right hand side arises in lattice QCD with Wilson fermions in an analogous way to the case of an external electromagnetic field, studied at the end of section 4.4 (see also [5.71]).

The $U(1)_A$ axial anomaly in (5.209) is proportional to the *topological charge density*

$$q(x) \equiv \frac{g^2}{64\pi^2} \epsilon_{\mu\nu\rho\sigma} F^a(x)_{\mu\nu} F^a(x)_{\rho\sigma} . \tag{5.210}$$

Non-zero contributions to $q(x)$ are given by instantons and other topologically non-trivial gauge field configurations [A3]. The *topological charge* of a Euclidean colour gauge field configuration is

$$Q \equiv \int d^4x \, q(x) = \frac{g^2}{64\pi^2} \epsilon_{\mu\nu\rho\sigma} \int d^4x \, F^a(x)_{\mu\nu} F^a(x)_{\rho\sigma} . \tag{5.211}$$

An important consequence of the Atiyah–Singer index theorem [5.72] is that the difference of the number of positive chirality zero modes (n_+) and negative chirality zero modes (n_-) of the Dirac operator in an external colour gauge field is equal to the topological charge:

$$Q = n_+ - n_- . \tag{5.212}$$

The physical consequence of the $U(1)_A$ anomaly in (5.209) is that the $U(1)_A$ symmetry of the classical continuum action is not realized in the quantum theory. The true continuous chiral symmetry is $U(1)_V \otimes SU(3)_V \otimes SU(3)_A$. Its spontaneous breaking to $U(1)_V \otimes SU(3)_V$ at small u-, d- and s-quark masses implies the existence of eight low-mass quasi-Goldstone bosons. The ninth pseudoscalar meson, the η', does not become a Goldstone boson if the quark masses are put equal to zero. In nature the η' meson is substantially heavier than the π-, K- and η-mesons, in agreement with the observations. This resolution of the $U(1)_A$ problem is made quantitative by the Witten–Veneziano formula [5.73]

$$\frac{\langle Q^2 \rangle_{\text{quenched}}}{\Omega} \equiv \chi_t^{\text{quenched}}$$

$$\simeq \frac{f_\pi^2}{4N_f}(m_{\eta'}^2 + m_\eta^2 - 2m_K^2) \simeq (180 \text{ MeV})^4 , \qquad (5.213)$$

which connects the pseudoscalar masses to the topological susceptibility in the quenched approximation χ_t^{quenched}. The quenched approximation arises because (5.213) can be derived in the limit of a very large number of colours ($N_c \to \infty$), when the internal quark loops are suppressed. This means that χ_t^{quenched} has to be computed in pure gauge theory, which is much easier than to compute expectation values in full QCD with dynamical fermions.

The difficulty in the numerical evaluation of the topological susceptibility lies in the lattice definition of the topological charge Q (for a review see [5.74]). The straightforward transcription of (5.211) into the lattice language suffers from large renormalizations [5.75], which are difficult to control. The geometrical method [5.76], which is based on an interpolation of the colour gauge field obtained from the link variables, is plagued by the presence of dislocations [5.77]. The dislocations are short distance lattice artifacts, which can eventually completely spoil the continuum limit, not to mention the results of realistic numerical simulations in the available range of the gauge coupling. The cooling method [5.78] is devised to overcome the problem of lattice artifacts in the geometrical method by smoothing out short-range fluctuations before a topological charge measurement is performed. In fact, at present this seems to be the most robust and reliable way of calculating the topological susceptibility. Nevertheless, the relaxation steps of the cooling algorithms introduce a non-locality in the lattice approximants of the local continuum density, which is questionable until the physical scale is not well separated from the scale of the lattice spacing. Finally the fermionic method [5.79] determines the topological charge of a gauge configuration by counting on the lattice ($n_+ - n_-$) in the continuum index theorem (5.212). It is encouraging that the results of this method are close to the results obtained by cooling [5.80].

5.3.6   Current matrix elements

The vector and axial vector currents of the chiral flavour symmetry were defined in the Wilson formulation of lattice QCD through the Ward–Takahashi identities. Their matrix elements play an important rôle in the electroweak interactions of hadrons. For instance, the electromagnetic current of the light (u-, d- and s-) quarks is

$$J_{em}(x)_\mu = V(x)_{3\mu} + \frac{1}{\sqrt{3}}V(x)_{8\mu} , \qquad (5.214)$$

and the charged weak current, which is coupled to the W-boson, is

$$J_W(x)_\mu = \frac{1}{2} \cos \Theta \left[V(x)_{1\mu} + iV(x)_{2\mu} - A(x)_{1\mu} - iA(x)_{2\mu} \right]$$

$$+ \frac{1}{2} \sin \Theta \left[V(x)_{4\mu} + iV(x)_{5\mu} - A(x)_{4\mu} - iA(x)_{5\mu} \right] . \tag{5.215}$$

(Θ is the Cabibbo angle.) A similar expression holds also for the neutral weak current of hadrons coupled to the Z-boson.

The simplest matrix elements to calculate are those between the vacuum and single hadron states. These determine the leptonic decays of vector and pseudoscalar mesons. They can be extracted from the asymptotic behaviour of two-point current–current correlation functions at large Euclidean time separations, by using the formulas (5.94) and (5.95).

The matrix elements of the vector and axial vector currents between two single hadron states define the electromagnetic and weak form factors, structure functions, semileptonic decay form factors etc. These can be obtained from the ratios of three-point to two-point functions of currents. For instance, the matrix element $\langle B|J(0)|A \rangle$ of the current component $J(0)$ appears in the asymptotic behaviour of the Euclidean three-point function

$$\langle 0|T\{B^+(y_4)J(0)A(-x_4)\}|0\rangle = \text{ for } y_4 > 0 > -x_4$$

$$= \lim_{T \to \infty} \left(\text{Tr} \, e^{-TH} \right)^{-1} \text{Tr} \left\{ B^+(0) e^{-y_4 H} J(0) e^{-x_4 H} A(0) e^{-(T-y_4-x_4)H} \right\} . \tag{5.216}$$

The operators $B^+(y_4)$ at Euclidean time y_4, and $A(-x_4)$ at Euclidean time $-x_4$, are some appropriate interpolating fields for the states $\langle B|$ and $|A\rangle$, respectively. (T denotes, as usual, the time extension of the lattice.) Assuming that in the given channels the lowest energy states are $\langle B|$ and $|A\rangle$, one obtains, similarly to (5.94) and (5.95),

$$\langle B|J(0)|A \rangle = \langle B|B(0)|0 \rangle \langle 0|A^+(0)|A \rangle$$

$$\cdot \lim_{x_4, y_4 \to \infty} \lim_{T \to \infty} \frac{\langle B^+(y_4)J(0)A(-x_4) \rangle}{\langle B^+(y_4)B(0) \rangle \langle A^+(0)A(-x_4) \rangle} . \tag{5.217}$$

Here the first two factors on the right hand side can be determined from the asymptotic behaviour of two-point functions.

From the point of view of numerical simulations the expectation values, which are easiest to obtain, are the ones corresponding to connected quark line graphs. For instance, in the case of a flavour non-singlet current the nucleon–nucleon matrix element is given by connected quark line graphs as in fig. 5.15(a). For the flavour singlet currents there is an additional disconnected contribution, too (see fig. 5.15(b)). In contrast to fig. 5.15(b), the necessary propagator matrix elements for fig. 5.15(a) can be calculated

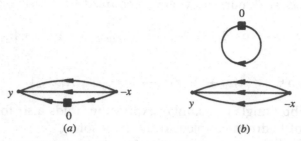

Fig. 5.15. Connected (*a*) and disconnected (*b*) quark line graphs contributing to the nucleon–nucleon matrix elements of currents. The current insertion is at the origin, the nucleon operators are at $-x$ and y.

by two matrix inversions per spin and colour indices: one can start from the origin (0) and calculate the matrix elements $0 \to -x$ and $0 \to y$ and then, taking the output of the first inversion at $-x$ as the input of the second inversion, the necessary matrix elements $-x \to y$. (This holds also if there is a summation over $-x$ and y, say, over timeslices. The direction of the quark lines can be reversed, if necessary, by using the relation (5.15).) For summaries of current matrix element calculations see [5.81, 5.82, 5.83, 5.84].

Besides the hadronic matrix elements of single currents, another important type of matrix elements are those of the low energy effective Hamiltonian of non-leptonic decays [5.85] (for reviews see also [5.86, 5.87]). The non-leptonic decays of hadrons are triggered by a short distance W-boson exchange between two currents $J_W(x)_\mu$ in (5.215). At the long distance (hadronic) scale the effect of the hadron structure is taken into account by the hadronic matrix elements of some local operator products. In the continuum formulation the renormalization between the two (short distance and long distance) scales can be described in the framework of operator product expansions [5.88]. The renormalized effective Hamiltonian is written as

$$H_{eff} \propto \sum_k C^{(k)}(\mu, M_W) \mathcal{O}^{(k)}(\mu) , \qquad (5.218)$$

where the Wilson coefficients $C^{(k)}$ describe the dependence on the renormalization scale μ, and $\mathcal{O}^{(k)}(\mu)$ ($k = 1, 2, \ldots$) is an operator basis at the scale μ (expressed by polynomials of the quark and gluon fields). On the lattice the operators $\mathcal{O}^{(k)}(\mu)$ have to be related to the corresponding lattice operators $\mathcal{O}^{(k)}_{latt}(a)$ defined at the momentum scale $\mu \simeq a^{-1}$. The relation between the two types of operators is calculable in perturbation theory [5.87]. The basic approach is to calculate free quark matrix elements on the lattice and in the continuum, and compare the results. Such a 'perturbative matching' leads to relations of the form

$$\mathcal{O}^{(k)}(\mu) = \sum_l Z_{kl}(a\mu, g)\mathcal{O}^{(l)}_{latt}(a) \ . \tag{5.219}$$

The difficulty in evaluating the non-leptonic decay matrix elements is that the operator mixing in (5.219) (the sum over l) brings in not only operators with the same or higher mass dimension as $\mathcal{O}^{(k)}$, but in general also operators with lower dimensions. The coefficients Z_{kl} with l refering to lower dimension operators are proportional to some inverse power of the lattice spacing $a^{-\Delta_{kl}}$, and hence contain power divergences in the continuum limit (Δ_{kl} is given by the difference of the naive power counting dimensions). Such coefficients cannot be reliably calculated in perturbation theory, therefore either they have to be avoided by exploiting exact lattice symmetries, or have to be subtracted non-perturbatively [5.85, 5.87]. The coefficients relating $\mathcal{O}^{(k)}$ to operators of the same or higher dimensions are given to a good approximation by perturbation theory.

Although the coefficients Z_{kl} of the operators with higher dimensions are, in the continuum limit, proportional to positive powers of the lattice spacing, for finite lattice spacings such contributions appear as disturbing 'lattice artifacts'. Therefore it is worth trying to eliminate them partially in the framework of the improvement program [5.26]. This means for instance that by tree level improvement the coefficients of order $O(g^0 a\mu)$ can be eliminated, leaving terms of order $O(g^2 a\mu)$ and $O(g^0(a\mu)^2)$ only. This can be achieved by choosing an 'improved lattice action', and at the same time introducing an 'improved operator basis' [5.89].

Numerical results on non-leptonic decay matrix elements are collected and discussed in the reviews [5.81, 5.82, 5.83].

5.4 Hadron thermodynamics

It is a long standing expectation that strongly interacting hadronic matter has a qualitatively new behaviour at temperatures and/or densities which are comparable to hadronic scales. It has even been argued that copious resonance production leads to a singularity of the thermodynamic functions, which would correspond to the existence of a highest limiting temperature [5.90]. However, due to the internal structure of hadrons, at temperatures of the order of the pion mass, or densities a few times that of ordinary nuclear matter, the hadrons start to overlap, and the interaction among the constituents becomes important. On the basis of asymptotic freedom one can expect that at asymptotically large temperatures (and, maybe, densities) the hadronic matter appears as an asymptotically free gas of quarks and gluons [5.91]. The qualitative differences between this

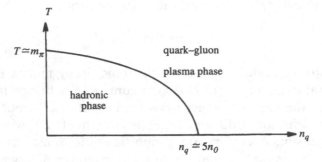

Fig. 5.16. The schematic phase structure of strongly interacting matter in the (n_q, T)-plane (n_q = quark number density = 3 baryon number density, T = temperature). n_0 is the normal nuclear matter density, m_π the pion mass.

high energy *quark–gluon plasma* and the low energy hadronic matter, with confinement and spontaneous symmetry breaking of the chiral symmetry, presumably imply the existence of a phase transition between the two regimes (see fig. 5.16). Indeed, in the strong coupling limit of lattice QCD such a phase transition could be shown to exist [5.92]. The existence and the properties of this phase transition have an important impact on the experimental search for the quark–gluon plasma in high energy heavy ion collisions, as well as on the theoretical understanding of the early universe, because during the early stages of its expansion the universe cooled down from the quark–gluon phase to the hadronic phase.

The properties of the high energy quark–gluon plasma may be guessed on the basis of the perturbative expansion of the thermodynamic potential. It turns out, however, that in order $O(g^6)$ serious infrared divergences appear, which can only be cured by some partial resummation, and then one has to take into account all higher orders [5.93]. Therefore perturbation theory has only a limited predictive power for the asymptotic behaviour of the quark–gluon plasma. Its range of validity is uncertain, and the appearance of infrared singularities calls for the dynamical emergence of chromoelectric and chromomagnetic mass scales lower than $O(T)$ for $T \to \infty$. This prevents the perturbative treatment of long distance properties. In particular, it cannot be expected that the perturbative approach is well suited for the study of inherently non-perturbative aspects of the phase transition separating the high temperature plasma phase from the low temperature hadronic phase.

This shows that hadronic thermodynamics is mainly a domain of the non-perturbative lattice approach. The Monte Carlo simulations can provide unique quantitative results for the phase transition temperature, the order of the phase transition, the equation of state and other physical quantities, which are of immediate interest for theory and experiment. In fact, our understanding of the thermodynamics at hadronic scales is growing in parallel with our knowledge about lattice QCD (for summaries of the lattice results see [5.94, 5.95, 5.96, 5.97]).

5.4.1   *Hadron thermodynamics on the lattice*

The thermodynamic quantities can be obtained as derivatives of the logarithm of the partition function (Z). They are defined in the *thermodynamic limit*, when the spatial volume tends to infinity ($V \to \infty$). The temperature (T) is determined by the extension of the lattice in the Euclidean time direction ($L_t \equiv L_4$) (see section 1.8). (Note that in the present section the notation conventions summarized in the appendix are partly changed.) Denoting the lattice spacing in the Euclidean time direction by a_t, the inverse temperature is given by

$$T^{-1} = a_t L_t .\qquad (5.220)$$

It will be convenient to choose in the spatial directions a different lattice spacing, denoted by a_s, therefore the physical volume is

$$V = (a_s L_s)^3 .\qquad (5.221)$$

Here, for simplicity, the lattice extensions in the three spatial directions are assumed to be equal: $L_1 = L_2 = L_3 \equiv L_s$. The ratio of the two lattice spacings will be denoted by

$$\xi \equiv \frac{a_s}{a_t} .\qquad (5.222)$$

Important thermodynamic quantities are the *energy density*

$$\epsilon \equiv -\frac{\partial \ln Z}{V \partial T^{-1}} = \frac{T^2}{V} \frac{\partial \ln Z}{\partial T} ,\qquad (5.223)$$

and *pressure*

$$P \equiv T \frac{\partial \ln Z}{\partial V} .\qquad (5.224)$$

These formulas show why is it important to have a variable lattice spacing ratio ξ, namely, in order to be able to perform derivatives with respect to T^{-1} in (5.220) and V in (5.221). Another advantage of using a finer lattice spacing in the temporal direction ($\xi > 1$) is to retain enough Matsubara frequencies for high temperatures (i. e. to keep L_t in (5.220) large enough even for large T).

The lattice spacings a_s, a_t, or in the isotropic case $a \equiv a_s = a_t$, do not explicitly appear in the lattice action. They are introduced in the continuum limit, together with the lattice Λ-parameter, as functions of the bare couplings (see (5.78) in section 5.1 for the isotropic case). An *anisotropic lattice* can be obtained by introducing anisotropic bare couplings. For instance, in the case of an SU(N) lattice gauge theory the coupling g_t for the time-like plaquettes is different from the coupling g_s for the space-like plaquettes. Therefore, the action corresponding to (5.2) can be written as

$$S_g[U] = \beta_s \sum_{\text{sp}} \left(1 - \frac{1}{N}\text{Re Tr } U_{\text{sp}}\right) + \beta_t \sum_{\text{tp}} \left(1 - \frac{1}{N}\text{Re Tr } U_{\text{tp}}\right) . \quad (5.225)$$

where U_{sp} and U_{tp} are plaquette variables for space-like and time-like pla-quettes, respectively. The bare parameters $\beta_{s,t}$ are conveniently introduced as

$$\beta_s \equiv \frac{2N}{g_s^2 \xi} , \qquad \beta_t \equiv \frac{2N\xi}{g_t^2} . \quad (5.226)$$

The fermionic part of the QCD lattice action for a single flavour of Wilson fermions is, instead of (5.5),

$$S_q[U, \psi, \overline{\psi}] = \sum_x \left\{ (\overline{\psi}_x \psi_x) - K_s \sum_{\mu=\pm 1}^{\pm 3} (\overline{\psi}_{x+\hat{\mu}}[1 + \gamma_\mu] U_{x\mu}\psi_x) \right.$$

$$\left. - K_t \sum_{\mu=\pm 4} (\overline{\psi}_{x+\hat{\mu}}[1 + \gamma_\mu] U_{x\mu}\psi_x) \right\} , \quad (5.227)$$

where K_s and K_t are the hopping parameters on space-like and time-like links, respectively, and for the Wilson parameter $r = 1$ was taken. For staggered fermions on an anisotropic lattice (5.7) becomes

$$S_q[U, \psi, \overline{\psi}] = \sum_x \left\{ am(\overline{\psi}_x \psi_x) \right.$$

$$+ \frac{1}{2} \sum_{\mu=1}^{3} \alpha_{x\mu} \left[(\overline{\psi}_x U_{x\mu}^+ \psi_{x+\hat{\mu}}) - (\overline{\psi}_{x+\hat{\mu}} U_{x\mu}\psi_x)\right]$$

$$\left. + \frac{\gamma_t}{2} \alpha_{x4} \left[(\overline{\psi}_x U_{x4}^+ \psi_{x+\hat{4}}) - (\overline{\psi}_{x+\hat{4}} U_{x4}\psi_x)\right] \right\} . \quad (5.228)$$

Here am is the fermion mass in lattice units and γ_t is the anisotropy parameter analogous to K_t/K_s for Wilson fermions.

The parameters of the lattice action $S = S_g + S_q$ can be considered on the anisotropic lattice as functions of the two lattice spacings a_s and a_t. On a given lattice (fixed L_s and L_t) the derivatives in the thermodynamic quantities like (5.223) or (5.224) can be expressed by derivatives with respect to the variables a_s and a_t [5.98, 5.99]. Using (5.220)–(5.222) one obtains

$$\epsilon = T^4 \left(\frac{L_t}{\xi L_s}\right)^3 \left\langle a_t \frac{\partial S}{\partial a_t}\right\rangle,$$

$$P = -\frac{T^4}{3}\left(\frac{L_t}{\xi L_s}\right)^3 \left\langle a_s \frac{\partial S}{\partial a_s}\right\rangle. \tag{5.229}$$

Let us first consider the thermodynamic quantities in pure gauge theory. Denoting the values of space-like and time-like plaquettes as

$$R_{\rm sp} \equiv 1 - \frac{1}{N}\operatorname{Re}\operatorname{Tr} U_{\rm sp}, \quad R_{\rm tp} \equiv 1 - \frac{1}{N}\operatorname{Re}\operatorname{Tr} U_{\rm tp}, \tag{5.230}$$

we have

$$a_t \frac{\partial S_g}{\partial a_t} = \left(1 + g_s^2 a_t \frac{\partial g_s^{-2}}{\partial a_t}\right)\beta_s \sum_{\rm sp} R_{\rm sp} + \left(-1 + g_t^2 a_t \frac{\partial g_t^{-2}}{\partial a_t}\right)\beta_t \sum_{\rm tp} R_{\rm tp},$$

$$a_s \frac{\partial S_g}{\partial a_s} = \left(-1 + g_s^2 a_s \frac{\partial g_s^{-2}}{\partial a_s}\right)\beta_s \sum_{\rm sp} R_{\rm sp} + \left(1 + g_t^2 a_s \frac{\partial g_t^{-2}}{\partial a_s}\right)\beta_t \sum_{\rm tp} R_{\rm tp}. \tag{5.231}$$

The derivatives of the couplings $g_{s,t}$, which appear here, can be calculated numerically. Before describing a suitable numerical procedure, it is useful to study this question in lattice perturbation theory [5.100]. For a fixed ratio of the lattice spacings ξ in (5.222) one can compare the theory to a corresponding theory with isotropic coupling g, which has the same isotropic lattice spacing as the spatial lattice spacing of the anisotropic lattice a_s. The relation between a_s and g is given by the lattice β-function $\beta_{LAT}(g)$ according to (5.78):

$$a_s \Lambda_{LAT} \equiv (\beta_0 g^2)^{-\beta_1/(2\beta_0^2)} e^{-1/(2\beta_0 g^2)}$$

$$\cdot \exp\left\{-\int_0^g dh\left[\frac{1}{\beta_{LAT}(h)} + \frac{1}{\beta_0 h^3} - \frac{\beta_1}{\beta_0^2 h}\right]\right\}. \tag{5.232}$$

(This implies that $dg(a_s)/d\log a_s = -\beta_{LAT}(g)$.) With this definition one can consider the couplings $g_{s,t}$ as functions of g and ξ. Using the method of background fields [5.37], one requires that the effective action as a function of g be independent of the asymmetry ξ. For small g this gives the behaviour [5.101, 5.100]

$$g_s^{-2} \equiv g^{-2} + C_s(g^2, \xi) = g^{-2} + c_s(\xi) + O(g^2),$$

$$g_s^2 = g^2[1 - c_s(\xi)g^2 + O(g^4)];$$

$$g_t^{-2} \equiv g^{-2} + C_t(g^2, \xi) = g^{-2} + c_t(\xi) + O(g^2) \,,$$

$$g_t^2 = g^2[1 - c_t(\xi)g^2 + O(g^4)] \,. \tag{5.233}$$

For the ratio of the lattice spacings this implies

$$\xi = (\beta_t/\beta_s)^{1/2} \left\{ 1 + \frac{1}{2}[c_s(\xi) - c_t(\xi)]g^2 + O(g^4) \right\} \,. \tag{5.234}$$

This means that for small couplings the ratio of the lattice spacings is approximately equal to the naive ratio given by the asymmetry of the couplings in the lattice action. The renormalization of ξ, due to the gauge interaction, becomes small in this limit.

From (5.229), (5.231) and (5.233) one obtains the energy density of the gauge field ϵ_g as

$$\epsilon_g = T^4 \frac{3L_t^4}{\xi^3} \left\langle \left(1 - g_s^2 \xi \frac{\partial C_s(g^2, \xi)}{\partial \xi} \right) \beta_s R_{\mathrm{sp}} \right.$$

$$\left. + \left(-1 - g_t^2 \xi \frac{\partial C_t(g^2, \xi)}{\partial \xi} \right) \beta_t R_{\mathrm{tp}} \right\rangle \,. \tag{5.235}$$

Similarly, the pressure of the gauge field P_g is

$$P_g = \frac{1}{3}\epsilon_g - T^4 \frac{2L_t^4 \beta_{LAT}(g)}{\xi^3 g^3}$$

$$\cdot \left\langle \left(1 - g^4 \frac{\partial C_s(g^2, \xi)}{\partial g^2} \right) g_s^2 \beta_s R_{\mathrm{sp}} + \left(1 - g^4 \frac{\partial C_t(g^2, \xi)}{\partial g^2} \right) g_t^2 \beta_t R_{\mathrm{tp}} \right\rangle \,. \tag{5.236}$$

In the small gauge coupling limit one can use the perturbative result (5.233), and for the lattice β-function the first universal term of the expansion (5.66), which gives $\beta_{LAT}(g)/g^3 \to -\beta_0$. For instance, for the gauge field the quantity

$$\delta \equiv \frac{1}{3}\epsilon - P \tag{5.237}$$

becomes

$$\delta_g = -T^4 \frac{L_t^4}{\xi^3} [2\beta_0 g^2 + O(g^4)]\langle \beta_s R_{\mathrm{sp}} + \beta_t R_{\mathrm{tp}} \rangle \,. \tag{5.238}$$

The quantity δ is characteristic for the interactions, because for an ideal relativistic Bose gas $\delta = 0$ [A6]. $3\delta = \epsilon - 3P$ is the trace of the energy–momentum tensor, which is related to the breaking of conformal symmetry.

All the thermodynamic quantities defined up to now still contain vacuum contributions. This can be seen from the limit of the dimensionless ratios like ϵ/T^4 or P/T^4, which tend to finite values for $T \to 0$. The physical values of the thermodynamic quantities can be obtained by subtracting the vacuum contributions. Another possibility is to consider only differences for two non-zero values of the temperature. For the purpose of illustration let us consider δ_g in the special case of $\xi = 1$ ($g_s = g_t \equiv g$), when (5.236) and $C_s(g^2, 1) = C_t(g^2, 1) = 0$ imply

$$\delta_g(T)T^{-4} = L_t^4 \frac{4N\beta_{LAT}(g)}{g^3} \langle R_{sp} + R_{tp} \rangle_{g,L_t} . \qquad (5.239)$$

The expectation value is evaluated here, at given g and L_t, in the thermodynamic limit $L_s \to \infty$. Keeping g fixed, one has $TL_t = T'L'_t$, therefore

$$\left[\delta_g(T) - \delta_g\left(T\frac{L_t}{L'_t}\right) \right] T^{-4} = L_t^4 \frac{4N\beta_{LAT}(g)}{g^3}$$

$$\cdot \left\{ \langle R_{sp} + R_{tp} \rangle_{g,L_t} - \langle R_{sp} + R_{tp} \rangle_{g,L'_t} \right\} . \qquad (5.240)$$

In the limit $L'_t \to \infty$ the second temperature TL_t/L'_t tends to zero, and we get

$$[\delta_g(T) - \delta_g(0)]T^{-4} = L_t^4 \frac{4N\beta_{LAT}(g)}{g^3}$$

$$\cdot \left\{ \langle R_{sp} + R_{tp} \rangle_{g,L_t} - \langle R_{sp} + R_{tp} \rangle_{g,\infty} \right\} . \qquad (5.241)$$

In practice, when L_s is finite and large, one can take $L'_t = L_s$ to represent zero temperature.

As stated before, $c_s(\xi)$ and $c_t(\xi)$ in (5.233) can be calculated in lattice parturbation theory. In the thermodynamic quantities, for instance in (5.235) and (5.236), their derivatives appear. At $\xi = 1$ we have [5.100]

$$\frac{dc_s(\xi)}{d\xi}\bigg|_{\xi=1} = 4N\left[\frac{N^2-1}{32N^2} 0.586844\ldots + 0.000499\ldots \right]$$

$$-\frac{N_f}{2} \cdot \begin{cases} 0.00794\ldots & \text{(Wilson)} \\ 0.00062\ldots & \text{(staggered)} \end{cases} ,$$

$$\frac{dc_t(\xi)}{d\xi}\bigg|_{\xi=1} = 4N\left[-\frac{N^2-1}{32N^2} 0.586844\ldots + 0.005306\ldots \right]$$

$$-\frac{N_f}{2} \cdot \begin{cases} 0.00052\ldots & \text{(Wilson)} \\ 0.00782\ldots & \text{(staggered)} \end{cases} . \qquad (5.242)$$

Here also the contributions of N_f flavours of fermions ('quarks') in the fundamental representation of the SU(N) gauge group are included, either in the Wilson (5.227) or in the staggered (5.228) formulation [5.102, 5.103].

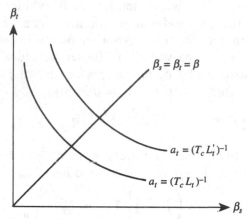

Fig. 5.17.   The curves with constant a_t in the (β_s, β_t)-plane for the anisotropic action, which can be obtained by determining the critical temperature T_c on lattices with different time extension L_t.

The perturbative formulas for the anisotropy parameter ξ become only reliable at small g^2. This is difficult to achieve in real simulations, therefore it is better to apply a non-perturbative numerical procedure [5.99, 5.104]. An important tool to fix to absolute scale of a_t is to measure the critical temperature T_c of the phase transition between the hadronic and quark–gluon plasma phase. Since T_c can be determined on lattices with any number of sites in the time direction (L_t), the curves with constant a_t for fixed $T = T_c$ and any L_t can be easily obtained (see fig. 5.17). For a simulation on a lattice with fixed L_t this gives a_t, in principle, only for the discrete set of temperatures $T = T_c L'_t / L_t$, with an arbitrary L'_t. In the continuum limit, however, when $L_t \to \infty$, this discrete set of points becomes dense in T. The dependence of the ratio $\xi = a_s/a_t$ on the bare parameters $\beta_{s,t}$ in the action (5.225) can be numerically determined by matching Wilson loop expectation values to the corresponding ones on isotropic lattices. Knowing a_t and ξ (and hence also a_s) in a dense set of points in the (β_s, β_t)-plane is enough for the approximate determination of the derivatives $\partial g_{s,t}^{-2} / \partial a_{s,t}$ appearing in (5.231).

In order to find the connection to the perturbative formulas it is, in fact, convenient to compare to an isotropic theory with coupling $\bar{g}$, which has a lattice spacing equal to a_t. $\bar{g}$ is determined by an equation like (5.232), where a_s is replaced by a_t and g by $\bar{g}$. Then, similarly to (5.233), we have

$$g_s^{-2} \equiv \bar{g}^{-2} + \bar{C}_s(\bar{g}^2, \xi) = \bar{g}^{-2} + \bar{c}_s(\xi) + O(\bar{g}^2) \,,$$

$$g_t^{-2} \equiv \bar{g}^{-2} + \bar{C}_t(\bar{g}^2, \xi) = \bar{g}^{-2} + \bar{c}_t(\xi) + O(\bar{g}^2) \,. \tag{5.243}$$

In the small coupling limit it follows from the definition of $\bar{g}$ and (5.232) that

$$\bar{g}^{-2} = g^{-2} + 2\beta_0 \log \xi + O(g^2) , \qquad (5.244)$$

therefore

$$\bar{c}_{s,t}(\xi) = c_{s,t}(\xi) - 2\beta_0 \log \xi . \qquad (5.245)$$

Using the information in fig. 5.17, it is convenient to estimate the derivatives with respect to ξ for fixed $\bar{g}$. (Note that for small couplings one can use (5.234) for ξ, therefore the numerical measurement of ξ is not necessary.) Of course, analogous expressions for the energy density and pressure in (5.235) and (5.236) can also be derived in terms of $\bar{C}_{s,t}(\bar{g}^2, \xi)$, instead of $C_{s,t}(g^2, \xi)$.

Another possibility for the non-perturbative calculation of the pressure is to use the connection to the Helmholz free energy density f [5.105], which is defined as

$$f \equiv \epsilon - Ts = -\frac{T}{V} \ln Z , \qquad (5.246)$$

with s as the entropy density. To be specific, let us consider here pure gauge theory on isotropic lattices with $\beta_s = \beta_t \equiv \beta = 2Ng^{-2}$ in the action (5.225). In this case the derivative of the logarithm of the partition function with respect to β is

$$\frac{\partial \ln Z_g}{\partial \beta} = \left\langle -\frac{\partial S_g}{\partial \beta} \right\rangle = -3L_s^3 L_t \langle R_{\mathrm{sp}} + R_{\mathrm{tp}} \rangle . \qquad (5.247)$$

Therefore the free energy density of the gauge field f_g can be determined, up to an integration constant, by integrating over β. Removing the vacuum contribution in the same way as in (5.241), one obtains

$$[f_g(T) - f_g(0)]T^{-4} = 3L_t^4$$

$$\cdot \int_0^\beta d\bar{\beta} \left\{ \langle R_{\mathrm{sp}} + R_{\mathrm{tp}} \rangle_{\bar{\beta}, L_t} - \langle R_{\mathrm{sp}} + R_{\mathrm{tp}} \rangle_{\bar{\beta}, \infty} \right\} . \qquad (5.248)$$

The lower limit of integration was specified here at $\beta = 0$, which corresponds to infinite lattice spacing ($a = \infty$), and hence to zero temperature, where the left hand side vanishes. Other thermodynamic quantities can be obtained from the Helmholz free energy density, if one assumes the homogenity of the system, when the pressure is given by $P = -f$. Homogenity means that the increase of the volume of the system by a multiplicative factor implies the increase of the extensive thermodynamic quantities by the same factor. This general property can usually be expected to hold in a pure phase of a very large system. This latter condition is, of course, not easy to achieve in numerical simulations.

The thermodynamic quantities for quarks (fermions in the fundamental representation of SU(N)) can also be obtained from (5.229), if in the lattice action $S = S_g + S_q$ the quark part S_q is also taken into account. (In this way every thermodynamic quantity is given by a sum of gluonic and fermionic contributions as, for instance, $\epsilon = \epsilon_g + \epsilon_q$ etc.) The formulas analogous to (5.235) and (5.236), as well as the perturbative derivatives of the bare couplings with respect to the anisotropy parameter ξ, were worked out for the anisotropic actions with Wilson, respectively, staggered quarks (5.227) and (5.228) in [5.102, 5.103]. Here we only consider the special case of isotropic lattices ($\xi = 1$), where the formulas corresponding to (5.241) and (5.248) can be derived. Let us start with Wilson fermions with the isotropic hopping parameter $K \equiv K_s = K_t$ in (5.227). For definiteness, in the continuum limit we fix the ratio of the RGI quark mass to the Λ-parameter $\lambda \equiv M/\Lambda$ in (5.85). Therefore the hopping parameter $K = K(g, \lambda)$ will be considered as a function of the bare gauge coupling g and λ. For vanishing quark mass $K(g, 0) = K_{cr}(g)$ is the critical line, which is approximately given for small g^2 by (5.62). Since for $\xi = 1$ (5.229) implies

$$\delta \equiv \frac{1}{3}\epsilon - P = \frac{T^4}{3}\left(\frac{L_t}{L_s}\right)^3 \left\langle a\frac{\partial S}{\partial a}\right\rangle, \qquad (5.249)$$

the quark contribution coming from S_q is

$$\delta_q = -\frac{2T^4}{3}L_t^4 \beta_{LAT}(g, \lambda)\frac{\partial K(g, \lambda)}{\partial g}\langle 3R_{sl} + R_{tl}\rangle. \qquad (5.250)$$

Here $\beta_{LAT}(g, \lambda)$ is the lattice β-function, which appears in the single variable RGE for fixed λ, similar to (5.77). (Actually (5.77) refers to zero quark mass $\lambda = 0$, but a similar RGE holds also for any fixed value of λ.) The real link expectation values R_{sl} and R_{tl} are defined by

$$R_{sl} \equiv \operatorname{Re}\operatorname{Tr}_{cs}\left\{Q_{x,x+\hat{n}}^{-1}(1 + \gamma_n)U_{xn}\right\},$$

$$R_{tl} \equiv \operatorname{Re}\operatorname{Tr}_{cs}\left\{Q_{x,x+\hat{4}}^{-1}(1 + \gamma_4)U_{x4}\right\}, \qquad (5.251)$$

with $n = 1, 2, 3$, and Tr_{cs} denoting a trace over colour and spinor indices. (In deriving (5.250) the relations (5.15) and (5.1) were used.) Note that at fixed RGI quark mass λ, in the gauge field contribution δ_g (5.239), the β-function $\beta_{LAT}(g)$ is also replaced by $\beta_{LAT}(g, \lambda)$.

For the derivation of the Helmholz free energy density f one needs, besides (5.247), the derivative

$$\frac{\partial \ln Z}{\partial K} = \left\langle -\frac{\partial S}{\partial K}\right\rangle = -2L_s^3 L_t\langle 3R_{sl} + R_{tl}\rangle. \qquad (5.252)$$

In order to obtain $[f(T) - f(0)]T^{-4}$ in the point with bare parameters β, K, on a lattice with time-like extension L_t, one has to start at the point $(\beta = K = 0)$, corresponding to infinite lattice spacing and infinite quark mass, and integrate, say, first $\partial \ln Z / \partial \beta$ over β at fixed $K = 0$, and then $\partial \ln Z / \partial K$ over K at fixed β. The result is

$$[f(T) - f(0)]T^{-4} =$$

$$3L_t^4 \int_0^\beta d\bar\beta \left\{ \langle R_{\mathrm{sp}} + R_{\mathrm{tp}} \rangle_{\bar\beta, K=0, L_t} - \langle R_{\mathrm{sp}} + R_{\mathrm{tp}} \rangle_{\bar\beta, K=0, \infty} \right\}$$

$$+ 2L_t^4 \int_0^K d\bar K \left\{ \langle 3R_{\mathrm{sl}} + R_{\mathrm{tl}} \rangle_{\beta, \bar K, L_t} - \langle 3R_{\mathrm{sl}} + R_{\mathrm{tl}} \rangle_{\beta, \bar K, \infty} \right\} . \tag{5.253}$$

Since at $K = 0$ the action describes the pure gauge theory, in this way the free energy is obtained as a sum of the gluonic contribution in pure gauge theory (5.248), plus the fermion contribution at fixed β.

For staggered fermions the derivation of (5.250) and (5.253) is rather similar to the case of Wilson fermions. Only the following replacements have to be performed:

$$K \longrightarrow am , \quad \langle 6R_{\mathrm{sl}} + 2R_{\mathrm{tl}} \rangle \longrightarrow \langle (\overline\psi_x \psi_x) \rangle = -\langle \mathrm{Tr}_c \, Q_{xx}^{-1} \rangle . \tag{5.254}$$

A consequence is that in (5.250) $\partial(am)/\partial g$ appears. Therefore, since in the case of staggered fermions the critical line for zero quark mass is $am = 0$, independently of g, the quark contribution δ_q vanishes for zero quark mass. In this case δ is entirely given by the gluonic contribution δ_g.

5.4.2 Deconfinement and chiral symmetry restoration

At high temperatures and/or densities the hadronic matter is expected to undergo a phase transition into a quark–gluon plasma phase, where the characteristic low energy features, like confinement and spontaneous chiral symmetry breaking are lost, and the short distance behaviour of matter is dominated by the asymptotic freedom of QCD. The phase transition or rapid crossover separating the low energy hadronic phase from the high energy quark–gluon plasma phase is expected to occur at some temperature $T_c \simeq m_\pi$, roughly equal to the pion mass, where copious thermal production of pions sets in. The thermodynamic properties near the phase transition are determined by the three light quarks (u, d and s), because the heavier quarks are exponentially suppressed by $\exp(-m_{c,b,t}/T_c)$. As a consequence, the phase transition can be described without the heavy quarks, apart from their passive rôle as indicators of the plasma phase [5.106].

In a first approximation one can neglect the mass differences among light quarks, and consider QCD with $N_f = 3$ degenerate light quark

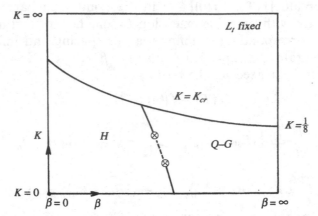

Fig. 5.18. The expected phase structure of hadronic matter for three Wilson quark flavours at fixed temporal lattice extension L_t in the (β, K)-plane. The line $K = K_{cr}$ corresponds to zero quark mass. The hadronic (H) and quark–gluon plasma (Q–G) phase is separated in part by first order phase transitions (full lines ending in a $\otimes$) or by a rapid crossover (dashed line).

flavours. Taking isotropic lattice spacings ($a_t = a_s \equiv a$) and Wilson quarks, and fixing for the moment the temporal extension of the lattice (L_t), the temperature in (5.220) can be considered as a function of the bare parameters $\beta \equiv 6g^{-2}$ and K, because the lattice spacing is $a = a(\beta, K)$ (see at the end of section 5.1). The qualitative phase structure in the (β, K)-plane, for fixed L_t, is shown by fig. 5.18.

The $K = 0$ boundary in fig. 5.18 corresponds to infinite quark mass, when QCD is reduced to pure SU(3) gauge theory. The gluonic action S_g in (5.2) has a global Z_3 symmetry, which is spontaneously broken at the deconfining phase transition. The Z_3 transformations are defined on all time-like gauge fields originating at the points of a given temporal hyperplane:

$$U_{\mathbf{x}, x_4, 4} \longrightarrow z U_{\mathbf{x}, x_4, 4} \quad (x_4 \text{ fixed}) , \quad z \in Z_3 . \qquad (5.255)$$

Since the elements of Z_3, which define the centre of SU(3), commute with every $U \in$ SU(3), the lattice action S_g is left invariant by these Z_3 transformations. Similarly, the product of the link variables ($U_{x\mu}$) along any closed curve, which crosses the $x_4 = $ const. plane the same number of times in the positive as in the negative direction, is also invariant. However, the product of link variables along closed curves, which go around the torus in the time direction (closed by the periodic boundary conditions), might change under the transformation in (5.255). For instance, the

Polyakov loops defined by

$$L_{\mathbf{x}} \equiv \mathrm{Tr} \left\{ \prod_{x_4=1}^{L_t} U_{\mathbf{x},x_4,4} \right\}, \qquad (5.256)$$

transform according to

$$L_{\mathbf{x}} \longrightarrow z L_{\mathbf{x}}. \qquad (5.257)$$

The expectation values of Polyakov loops probe the screening properties of a static colour triplet test charge in the surrounding gluonic medium. It measures the excess free energy $F_q(T)$ induced by the presence of the static test charge [5.107]:

$$e^{-F_q(T)/T} \propto |\langle L \rangle| \equiv \left| \left\langle \frac{1}{L_s^3} \sum_{\mathbf{x}} L_{\mathbf{x}} \right\rangle \right|. \qquad (5.258)$$

In the absence of dynamical quarks (which is the case for infinitely heavy quarks) a single colour triplet charge cannot be screened in the confined phase, therefore $F_q(T)$ is infinite and the expectation value of the Polyakov loop is zero. In the deconfined phase F_q is finite and $\langle L \rangle \neq 0$. Therefore, in pure gauge theory L is an order parameter for the deconfinement phase transition. If instead of the physical colour SU(3) one considers SU(N), then the centre is Z_N, and in the deconfinement phase the non-zero expectation value of the Polyakov loop breaks the global Z_N symmetry.

The products of link variables appearing in the definition of the Polyakov loops $L_{\mathbf{x}}$ in (5.256) are important degrees of freedom characterizing the deconfinement phase transition. Integrating out all other degrees of freedom one obtains a three-dimensional Z_N spin system (in the case of SU(N) gauge theory), which describes the critical behaviour of the deconfinement phase transition. If one assumes that in this effective spin system the short-range ferromagnetic couplings dominate, then one can predict the order of the phase transition [5.108]. For instance, for SU(2) gauge theory the prediction is that the deconfinement phase transition is of second order, for the physically important SU(3) case, in agreement with numerical simulations [5.109, 5.95, 5.96], of first order.

In the presence of dynamical fermions (e. g. $K > 0$ for Wilson quarks), the Polyakov loop is not an order parameter, because $\langle L \rangle$ is not exactly zero in the hadronic phase. The dynamical quarks can screen the external colour triplet charge, and the free energy excess F_q is finite. On the boundary at zero quark mass ($K = K_{cr}$) the dominant feature of the transition between hadronic and quark–gluon phases is the restoration of the chiral symmetry. The non-zero vacuum expectation value of the quark condensate $\langle \overline{\psi}_x \psi_x \rangle$ becomes zero in the quark–gluon plasma phase. The order of this chiral phase transition can be analysed in a three-dimensional effective chiral model with N_f^2 scalar fields [5.110]. The renormalization group analysis in this effective model suggests that the chiral transition at zero quark mass is of first order for $N_f \geq 3$. For $N_f = 2$ the prediction is uncertain, because it depends on the relative magnitude of the effective coupling parameters. For $N_f = 1$ there is no chiral phase transition at all, at most a rapid crossover. The numerical simulations with $N_f = 4$ definitely support a first order phase transition in the chiral limit. For $N_f = 3$ there could still be a first order phase transition, but at $N_f = 2$ it seems that there is only a rapid crossover between the hadronic and quark–gluon plasma phase [5.97]. This means that on fig. 5.18 with $N_f = 3$ degenerate flavours there is indeed a first order line near $K = K_{cr}$. Nevertheless, in nature the strange quark is heavier than the up and down quarks, therefore the real situation could be closer to the $N_f = 2$ case. In the real world the transition from the hadronic to the quark–gluon phase could happen along a line which passes through the dashed gap between the two first order phase boundaries for very heavy and very light quarks.

An important observation for Wilson quarks is that it is favourable to use $\tilde{m}_0$ in (5.186) as the quark mass parameter, because its value is almost independent of whether the system is in the high temperature phase or in the low temperature one [5.111].

Although the first order phase transition at small quark masses can be mainly characterized by the restoration of the chiral symmetry, the numerical simulations show that nearly at the same values of bare parameters, where the vacuum expectation value of the quark condensate $\langle \overline{\psi}_x \psi_x \rangle$ becomes zero, also the expectation value of the Polyakov loop shows a steep rise. In this sense the restoration of the chiral symmetry is accompanied by a simultaneous deconfinement of the colour charges.

The deconfinement in the quark–gluon plasma phase is manifested by the qualitative behaviour of the heavy quark–antiquark potential $V_{q\bar{q}}$, which can also be defined by the expectation value of two Polyakov loops as

$$\exp\left\{-V_{q\bar{q}}(a\mathbf{x}, T)/T\right\} \equiv \frac{\langle \text{Tr}\, L_{\mathbf{x}_0}\, \text{Tr}\, L_{\mathbf{x}_0+\mathbf{x}}^+ \rangle}{|\langle L \rangle|^2}. \tag{5.259}$$

The normalization with $|\langle L \rangle|^2$ is necessary, in order to eliminate the divergent self-energy contributions. The heavy quark potential signals the deconfinement phase transition by a qualitative change. In pure gauge theory the qualitative behaviour in the two phases can be parametrized by

$$V_{q\bar{q}}(\mathbf{r}, T) \simeq \begin{cases} \alpha(T)/|\mathbf{r}| + \sigma(T)|\mathbf{r}| & \text{for } T < T_c \\ \alpha(T)/|\mathbf{r}|\, e^{-\mu(T)|\mathbf{r}|} & \text{for } T > T_c \end{cases}. \tag{5.260}$$

Here $\sigma(T)$ denotes the string tension, and $\mu(T)$ the Debye screening mass. The short distance behaviour can be characterized in both phases by the temperature-dependent coupling $\alpha(T)$. The short and intermediate distance properties of $V_{q\bar{q}}$ are very important for the understanding of the heavy quark bound state spectrum, which can be a valuable indicator of the quark–gluon plasma phase in high energy heavy ion collisions [5.106, 5.94].

The characterization of the quark–gluon plasma as a weakly interacting gas of quarks and gluons is based on asymptotic freedom and perturbation theory. At asymptotically high temperatures the free relativistic gas of SU(N) gluons and N_f quark flavours (fermions in the fundamental representation of SU(N)) is characterized by the energy density

$$\epsilon = \frac{\pi^2}{15} \left(N^2 - 1 + \frac{7}{4} N N_f \right) T^4 , \tag{5.261}$$

and pressure $P = \frac{1}{3}\epsilon$ ($\delta = 0$). This can be conjectured to occur in the high temperature quark–gluon plasma. Nevertheless, the serious infrared problems in high temperature perturbation theory [5.93], which cannot be dealt with in the standard perturbative framework, already show that asymptotic freedom may only be valid for short distances and short time scales of the order T^{-1}. (For a reformulation of high temperature perturbation theory see, for instance, [5.112, 5.113].) It seems plausible that at scales larger than $[g(T)^2 T]^{-1}$, where $g(T)^2$ is the colour gauge coupling running with the temperature, the plasma exhibits confining features [5.114]. Since $g(T)^2$ goes to zero logarithmically for $T \to \infty$, at sufficiently high temperatures the deconfined and confined scales are well separated. The confinement acts only in long range, low frequency modes of the plasma. Thus the high temperature plasma is, to a good approximation, a gas of quasifree quarks and gluons. However, quanta with momenta of the order of $g(T)^2 T$ are subject to non-perturbative confining effects. As a consequence of this, it may very well be that only colour singlet modes produce poles and branch points in linear response functions. This could have a great deal of influence on the equation of state, the rate of entropy production upon a phase change, plasma transport properties etc.

The expected restoration of the chiral SU(3) $\otimes$ SU(3) symmetry suggests that in the chiral limit in the quark–gluon plasma phase the pseudoscalar mesons appear in a massive degenerate $\pi - \sigma$ multiplet, the vector mesons in a $\rho - a_1$ multiplet, and the baryons in parity doublets. The first numerical investigations of the hadronic screening lengths, measuring the long distance behaviour of spatial correlations of hadronic operators, support these expectations [5.115].

5.4.3   *Non-zero quark number density*

Up to now the thermodynamics of hadronic matter was discussed at high temperature and vanishing net quark number density. (In the low energy hadronic phase non-zero net quark number density appears as non-zero net baryon number density, which is a third of the quark number density.) Non-vanishing quark number density ($n_q \neq 0$) can be described by a non-zero *quark chemical potential* ($\mu_q \neq 0$) in the grand canonical partition function $Z(T,\mu_q) \equiv \mathrm{Tr}\{\exp[-(H - \mu_q N_q)/T]\}$, with the Hamiltonian H and quark number operator N_q. n_q can be calculated from

$$n_q = \frac{T}{V}\frac{\partial \ln Z}{\partial \mu_q} = \frac{a^{-3}}{L_s^3 L_t}\frac{\partial \ln Z}{\partial(a\mu_q)}\,, \tag{5.262}$$

where in the second form an isotropic lattice spacing (a) is assumed.

Zero or small net quark number densities occur in the early universe or in the central rapidity region of high energy heavy ion collisions. Under such physical circumstances the phase transition between the hadronic and quark–gluon phase occurs at high temperatures, near the $n_q = 0$ axis in fig. 5.16. In the baryon-rich fragmentation region of heavy ion collisions, or in the interior of large neutron stars, however, the net quark number density is large and the temperature relatively low. In this case the relevant part of the phase transition curve in fig. 5.16 is near the $T = 0$ axis, where the transition is driven by the very high pressure squeezing the baryons very close together in a small volume.

The introduction of the quark chemical potetial μ_q in the QCD lattice action has to satisfy the trivial requirement that for free quarks (without colour gauge interaction) the well-known results for the ideal relativistic Fermi gas [A6] have to be reproduced in the continuum limit. The quark matrix Q in (5.13), (5.14), (5.5) for Wilson quarks with non-zero chemical potential, on an isotropic lattice, can be written in general [5.116, 5.117, 5.118] as

$$Q_{yx} = \delta_{yx} - K \sum_{n=\pm 1}^{\pm 3} \delta_{y,x+\hat{n}}(r + \gamma_n)U_{xn}$$

$$- K\, F(a\mu_q)\delta_{y,x+\hat{4}}(r + \gamma_4)U_{x4} - K\, G(a\mu_q)\delta_{y+\hat{4},x}(r - \gamma_4)U_{y4}^+ \,. \tag{5.263}$$

Here F and G are, for the moment, arbitrary functions of the chemical potential in lattice units ($a\mu_q$), to be determined later. For staggered quarks in (5.7) one can similarly define

$$Q_{yx} = am\delta_{yx} + \frac{1}{2}\sum_{n=1}^{3}\alpha_{xn}\left[\delta_{y+\hat{n},x}U_{yn}^+ - \delta_{y,x+\hat{n}}U_{xn}\right]$$

$$+ \frac{\alpha_{x4}}{2} \left[G(a\mu_q)\delta_{y+\hat{4},x} U^+_{y4} - F(a\mu_q)\delta_{y,x+\hat{4}} U_{x4} \right] . \tag{5.264}$$

In order to constrain the functions $F(a\mu_q)$ and $G(a\mu_q)$ let us consider the free energy density f (5.246) in the absence of gauge interactions $(U_{x\mu} = 1)$:

$$f = -\frac{T}{V} \ln Z = -T^4 \frac{L_t^3}{L_s^3} \ln Z . \tag{5.265}$$

After performing the Grassmannian path integral, according to (5.12) one obtains

$$f T^{-4} = -\frac{L_t^3}{L_s^3} \log \det Q = -\frac{L_t^3}{L_s^3} \log \det \tilde{Q} . \tag{5.266}$$

Here in the second step a unitary transformation into momentum space was performed, which is defined by

$$\tilde{Q}_{lk} \equiv \frac{1}{L_s^3 L_t} \sum_{xy} e^{-iy \cdot l + ix \cdot k} Q_{yx} . \tag{5.267}$$

This gives

$$f T^{-4} = -\frac{L_t^3}{L_s^3} \sum_k \log \det_{sc} \left\{ 1 - K \left[\sum_{n=1}^{3} (2r \cos k_n - 2i\gamma_n \sin k_n) \right. \right.$$

$$\left. \left. + 2rR \cos (k_4 + i\theta) - 2iR\gamma_4 \sin (k_4 + i\theta) \right] \right\} , \tag{5.268}$$

where R and θ are defined by

$$R \equiv (FG)^{1/2} , \qquad \tanh \theta \equiv \frac{F - G}{F + G} ;$$

$$\frac{1}{2}(F + G) = R \cosh \theta , \qquad \frac{1}{2}(F - G) = R \sinh \theta . \tag{5.269}$$

(5.268) still contains the vacuum contribution, which can be removed by subtracting $f T^{-4}$ at $\mu_q = 0$. Calculating the determinant in spinor–colour indices $\det_{sc}$ gives

$$[f(T, \mu_q) - f(T, 0)] T^{-4}$$

$$= -L_t^4 \frac{6}{L_s^3 L_t} \sum_k \left\{ \log \left[\left(1 - 2rK \sum_{n=1}^{3} \cos k_n - 2rKR \cos (k_4 + i\theta) \right)^2 \right. \right.$$

$$\left. \left. + 4K^2 \sum_{n=1}^{3} \sin^2 k_n + 4K^2 R^2 \sin^2 (k_4 + i\theta) \right]$$

$$-\log\left[\left(1-2rK\sum_{\mu=1}^{4}\cos k_\mu\right)^2+4K^2\sum_{\mu=1}^{4}\sin^2 k_\mu\right]\right\}. \qquad (5.270)$$

In the continuum limit $a \to 0$ the momentum is $k_\mu = ap_\mu$, and in the limit $L_s, L_t \to \infty$ the momentum sums become integrals by (8.41). It turns out that the correct continuum limit is obtained only if

$$R \equiv 1, \qquad \theta(a\mu_q) = a\mu_q + O(a\mu_q)^2. \qquad (5.271)$$

Otherwise, at the critical value of the hopping parameter (see (4.125)), the free energy density and other thermodynamic quantities diverge. This means that the simplest choice [5.116, 5.117]

$$F(a\mu_q) = \frac{1}{G(a\mu_q)} = e^{a\mu_q}, \qquad (5.272)$$

is essentially unique, at least near the continuum limit. In this case (5.270) gives

$$f(T,\mu_q) - f(T,0) \longrightarrow$$

$$-\frac{6}{(2\pi)^4}\int_{-\infty}^{\infty}d^4p\,\log\frac{m^2+\sum_{n=1}^{3}p_n^2+(p_4+i\mu_q)^2}{m^2+p^2}. \qquad (5.273)$$

It can also be shown that in the continuum limit the choice (5.272) gives the correct energy and number density of an ideal relativistic Fermi gas [5.116, 5.118].

The numerical simulation of QCD at non-zero quark chemical potential $\mu_q \neq 0$ is difficult, because the relation (5.16) does not hold. The quark determinant is complex, therefore the effective gauge action S_{eff} in (5.17) and the Boltzmann factor $\exp(-S_{eff})$ also become complex. This means that $\exp(-S_{eff})$ cannot be used for the transition probability in an updating process. In principle one can define the transition probability by the absolute value of the quark determinant:

$$\det Q \equiv e^{i\varphi_Q}|\det Q| = e^{i\varphi_Q}[\det(QQ^+)]^{1/2},$$

$$S_{eff}[U] \equiv S_g[U] - \frac{1}{2}\log\det(QQ^+) = S_g[U] - \frac{1}{2}\mathrm{Tr}\,\log(QQ^+), \quad (5.274)$$

and include its phase in the measurement. For instance, for a pure gluonic quantity $A[U]$, instead of (5.18), one can use

$$\langle A[U]\rangle = \frac{\langle A[U]e^{i\varphi_Q}\rangle_{S_{eff}}}{\langle e^{i\varphi_Q}\rangle_{S_{eff}}}. \qquad (5.275)$$

The problem with this mathematically correct representation is that, in practice, it usually does not work. Namely, $e^{i\varphi_Q}$ has a very strong fluctuation on the typical configurations in an updating process, which is generated by S_{eff} in (5.274). Such an updating process most of the time produces 'wrong' configurations with the wrong values of the measurable quantities. These contributions have to be cancelled by a strongly fluctuating phase. Therefore (5.275) can only be used on small lattices and for small chemical potentials [5.119].

A technique for calculating the value of the quark determinant is based on the reduction of the quark matrix to the size of a timeslice [5.119]. For simplicity, let us consider here only the case of staggered quarks. Wilson quarks can be treated similarly. Using the freedom of the choice of the staggered fermion factor $\alpha_{x\mu}$ (see section 4.3), let us define $\alpha_{x4} \equiv 1$. An example for such a choice is

$$\alpha_{x4} = 1, \qquad \alpha_{xn} = (-1)^{x_4 + x_1 + \cdots + x_{n-1}} \quad (n = 1, 2, 3) . \tag{5.276}$$

In this case the block structure of the quark matrix, in the temporal gauge and with (5.272), is

$$Q_{yx} = \begin{pmatrix} B_1 & \frac{1}{2}e^{-a\mu_q} & 0 & \cdots & 0 & \frac{1}{2}e^{a\mu_q}U \\ -\frac{1}{2}e^{a\mu_q} & B_2 & \frac{1}{2}e^{-a\mu_q} & \cdots & 0 & 0 \\ 0 & -\frac{1}{2}e^{a\mu_q} & B_3 & \cdots & 0 & 0 \\ \cdots & \cdots & \cdots & \cdots & \cdots & \cdots \\ -\frac{1}{2}e^{-a\mu_q}U^+ & 0 & 0 & \cdots & -\frac{1}{2}e^{a\mu_q} & B_{L_t} \end{pmatrix} . \tag{5.277}$$

The blocks here are of size $3L_s^3 \otimes 3L_s^3$, and belong to given timeslices (the factors 3 take into account colour indices). Owing to the temporal gauge $U_{x4} = 0$ ($x_4 = 1, 2, \ldots, L_t - 1$), the products of the gauge variables along time-like links, which are denoted by U, appear only at $x_4 = L_t$. The blocks containing U and U^+ have an extra factor -1 because of the antiperiodic boundary conditions for fermions in the time direction. The blocks $B_{1,2,\ldots,L_t}$ stand for the mass term and spatial hopping terms.

The chemical potential can be moved to the last timeslice by multiplying the jth column of Q by $e^{ja\mu_q}$ and the jth row by $e^{-ja\mu_q}$. The leftmost column can be moved over to the right without changing the determinant (det Q = det Q'):

$$Q' = \begin{pmatrix} \frac{1}{2} & 0 & \cdots & \frac{1}{2}e^{\mu_q/T}U & B_1 \\ B_2 & \frac{1}{2} & \cdots & 0 & -\frac{1}{2} \\ -\frac{1}{2} & B_3 & \cdots & 0 & 0 \\ 0 & -\frac{1}{2} & \cdots & 0 & 0 \\ \cdots & \cdots & \cdots & \cdots & \cdots \\ 0 & 0 & \cdots & \frac{1}{2} & 0 \\ 0 & 0 & \cdots & B_{L_t} & -\frac{1}{2}e^{-\mu_q/T}U^+ \end{pmatrix} . \tag{5.278}$$

The extension of Q' can be reduced to a single block by Gaussian elimination: subtract $2B_2$ times the first row from the second row and add the first row to the third row. Repeat this procedure until only the last two rows remain, and proceed in the last step correspondingly. Since the B_j are sparse matrices, the necessary multiplications by them are easy. At the end one has to calculate the determinant of a $3L_s^3 \otimes 3L_s^3$ matrix, which requires asymptotically $\propto (L_s^3)^3 = L_s^9$ operations.

Another possibility for simulation at non-zero chemical potential is to use the polymer representation of the fermion determinant as a sum over polymers (see section 4.1). The problem of the fluctuating phases remains, however, and the procedure is only successful in special cases, like for staggered fermions at infinite gauge coupling [5.24]. For another attempt to simulate $\mu_q \neq 0$ on small lattices see [5.120].

Useful information about the behaviour in the region of small chemical potential can also be obtained by studying the response of the quark number density to infinitesimal changes in the chemical potential. This is measured by the *quark number susceptibility*

$$\chi_q \equiv \frac{\partial n_q}{\partial \mu_q} = \frac{T}{V} \frac{\partial^2 \ln Z}{\partial \mu_q^2} = \frac{a^{-2}}{L_s^3 L_t} \frac{\partial^2 \ln Z}{\partial (a\mu_q)^2} . \tag{5.279}$$

The quark number density defined in (5.262) can be expressed in the effective gauge theory at $\mu_q = 0$ as

$$n_q a^3 = (L_s^3 L_t)^{-1} \left\langle \frac{\partial}{\partial (a\mu_q)} \mathrm{Tr} \log Q \right\rangle$$

$$= (L_s^3 L_t)^{-1} \left\langle \mathrm{Tr} \left\{ Q^{-1} \frac{\partial Q}{\partial (a\mu_q)} \right\} \right\rangle . \tag{5.280}$$

Similarly, the quark number susceptibility is

$$\chi_q a^2 = (L_s^3 L_t)^{-1} \left[\left\langle \mathrm{Tr} \left\{ Q^{-1} \frac{\partial Q}{\partial (a\mu_q)} \right\} \mathrm{Tr} \left\{ Q^{-1} \frac{\partial Q}{\partial (a\mu_q)} \right\} \right. \right.$$

$$+ \mathrm{Tr} \left\{ Q^{-1} \frac{\partial^2 Q}{\partial (a\mu_q)^2} - Q^{-1} \frac{\partial Q}{\partial (a\mu_q)} Q^{-1} \frac{\partial Q}{\partial (a\mu_q)} \right\} \right\rangle$$

$$\left. - \left\langle \mathrm{Tr} \left\{ Q^{-1} \frac{\partial Q}{\partial (a\mu_q)} \right\} \right\rangle^2 \right] . \tag{5.281}$$

These expectation values can be evaluated in the $\mu_q = 0$ limit [5.121]. In order to avoid the full inversion of Q, which would be prohibitively expensive, the traces in (5.281) can be approximated by unbiased estimators. These can be obtained by introducing N_η vectors of complex Gaussian random numbers $\eta^{(v)}$ ($v = 1, 2, \ldots, N_\eta$), of dimension equal to the dimension of Q. If the components of $\eta^{(v)}$ are distributed according to $\exp\{-(\eta^{(v)}, \eta^{(v)})\} = \exp\{-\sum_x (\eta_x^{(v)+} \eta_x^{(v)})\}$, one can easily prove that, for instance,

$$\lim_{N_\eta \to \infty} \frac{1}{N_\eta} \sum_{v=1}^{N_\eta} \left(\eta^{(v)}, Q^{-1} \frac{\partial Q}{\partial(a\mu_q)} \eta^{(v)} \right) = \mathrm{Tr} \left\{ Q^{-1} \frac{\partial Q}{\partial(a\mu_q)} \right\}. \qquad (5.282)$$

In other words, the diagonal matrix element with respect to the random vector is an unbiased estimator for the trace. Using such estimators in (5.281), the quark number susceptibility can be obtained at $a\mu_q = 0$ as the average over random vectors and gauge configurations.

6

Higgs and Yukawa models

The electroweak interactions in the Standard Model are described by a chiral $SU(2)_L \otimes U(1)_Y$ symmetric gauge theory. The chiral symmetry is spontaneously broken by the vacuum expectation value of the scalar Higgs boson field. The consequence of the spontaneous symmetry breaking is that most of the elementary particles (weak vector bosons, quarks and leptons) become massive through their coupling to the Higgs boson field [A3,A4]. The study of non-perturbative spontaneous symmetry breaking is one of the main motivations for the lattice investigations of the Higgs sector in the Standard Model.

Another important aspect is the mathematically rigorous definition of the chiral $SU(2)_L \otimes U(1)_Y$ symmetric quantum field theory by means of the continuum limit of a suitably discretized lattice theory. In QCD the coupling is asymptotically free, therefore the continuum limit is defined near the ultraviolet stable Gaussian fixed point at zero coupling (see, for instance, in fig. 5.7). The couplings in the electroweak sector, with the exception of the $SU(2)_L$ gauge coupling, are not asymptotically free, therefore the Gaussian fixed point at zero couplings is ultraviolet unstable. This means that the lines of constant physics do not converge to the Gaussian fixed point, as in fig. 5.7. The simplest assumption which makes the definition of a continuum limit possible is that there is another, non-trivial (non-Gaussian) fixed point, somewhere else in the bare parameter space, where the lines of constant physics converge. In fact, in renormalized perturbation theory, where the renormalized couplings are considered as free parameters, one implicitly assumes the existence of such a non-trivial fixed point. An important goal of the non-perturbative lattice studies of the electroweak sector of the Standard Model is the search for such fixed points, suitable for the definition of a continuum limit.

The weak and electromagnetic gauge couplings are small in nature and can, therefore, be treated by perturbation theory. However, the couplings of the Higgs scalar field (the quartic self-coupling and the Yukawa couplings to fermions) could, in principle, be strong if some presently unknown (scalar or fermion) matter particles were heavy relative to the scale of the scalar vacuum expectation value. The existence of such strong couplings would require a non-perturbative treatment.

This discussion shows that the non-perturbative lattice studies of the electroweak sector, and in particular the embedded Higgs sector, is mainly motivated by the necessity of a better understanding of the theoretical framework of the Standard Model, or its possible extensions, rather than by the practical need for a numerical determination of some measurable parameters. In the present chapter the basic ingredients of the lattice Higgs and Yukawa models will be briefly summarized. For reviews of the extended literature on this subject see, for instance, [6.1, 6.2, 6.3, 6.4, 6.5].

6.1   Lattice Higgs models

The spontaneous symmetry breaking in the minimal Standard Model is a consequence of the non-vanishing vacuum expectation value of an $SU(2)_L$ doublet complex scalar Higgs field Φ_{Ax} ($A = 1, 2$) with hypercharge $Y = +1$. Higgs models describe the self-interaction of the Higgs field and its coupling to the $SU(2)_L$ and/or $U(1)_Y$ gauge fields. The gauge coupling of $SU(2)_L$ and $U(1)_Y$ is usually denoted, respectively, by g and $\frac{1}{2}g'$. Since the ratio of g' and g is related to the Weinberg angle θ_W by $g'/g = \tan\theta_W$, and θ_W is relatively small ($\sin^2\theta_W \simeq 0.23$), in a first approximation the coupling to the $U(1)_Y$ gauge field can be neglected. Therefore the simplest prototype Higgs model is the $SU(2)$ Higgs model with a complex scalar doublet field.

The fermions (three families of leptons and quarks) are coupled to the Higgs field through Yukawa couplings, which are in most cases very small (the only exception is the top quark with an expected Yukawa coupling of about 0.5). Therefore, if these small Yukawa couplings are neglected, the fermions communicate to the gauge-scalar sector only through the weakly coupled chiral gauge fields. For instance, $SU(2)_L$ is coupled to the left-handed chiral components of the fermion doublets, as the notation tells. Therefore, the $SU(2)$ ($\equiv SU(2)_L$) Higgs model describing the interaction of a complex scalar doublet field with the $SU(2)$ gauge field is a reasonable first approximation to the Higgs sector of the Standard Model. Switching on the $U(1) \equiv U(1)_Y$ gauge coupling of the scalar field gives the $SU(2) \otimes U(1)$ Higgs model.

The lattice discretization of the Higgs models essentially follows from the lattice formulation of the pure scalar ϕ^4 model (chapter 2), and the pure gauge model (chapter 3 and section 4.5). An important new aspect is the interplay between spontaneous symmetry breaking and gauge invariance.

6.1.1　Lattice actions

The complex doublet scalar field $\{\Phi_{Ax},\ A = 1, 2\}$ in the Higgs sector of the Standard Model can be described in different ways. Another doublet field with opposite hypercharge ($Y = -1$) is often introduced by

$$\tilde{\Phi}_{Ax} \equiv \epsilon_{AB}\Phi^*_{Bx} = i\tau_{2,AB}\Phi^*_{Bx}\ , \tag{6.1}$$

where ϵ_{AB} is the antisymmetric unit tensor, and τ_r ($r = 1, 2, 3$) denote isospin Pauli matrices. In this chapter automatic summation over repeated internal symmetry indices (like, for instance, A or r) is usually assumed. Combining Φ_{Ax} and $\tilde{\Phi}_{Ax}$ into a $2 \otimes 2$ matrix field gives

$$\varphi_x \equiv \begin{pmatrix} \tilde{\Phi}_{1x} & \Phi_{1x} \\ \tilde{\Phi}_{2x} & \Phi_{2x} \end{pmatrix} \equiv \mathbf{1}\sigma(x) + i\tau_r\pi_r(x)$$

$$= \begin{pmatrix} \sigma(x) + i\pi_3(x) & i\pi_1(x) + \pi_2(x) \\ i\pi_1(x) - \pi_2(x) & \sigma(x) - i\pi_3(x) \end{pmatrix}\ , \tag{6.2}$$

that is

$$\Phi_{Ax} = \varphi_{A2,x}\ , \qquad \tilde{\Phi}_{Ax} = \varphi_{A1,x}\ , \tag{6.3}$$

and the four real fields $\phi_S(x)$ ($S = 0, 1, 2, 3$) are given by

$$\phi_0(x) \equiv \sigma(x) = \frac{1}{2}(\tilde{\Phi}_{1x} + \Phi_{2x}) = \frac{1}{2}(\Phi_{2x} + \Phi^*_{2x})\ ,$$

$$\phi_1(x) \equiv \pi_1(x) = \frac{-i}{2}(\tilde{\Phi}_{2x} + \Phi_{1x}) = \frac{-i}{2}(\Phi_{1x} - \Phi^*_{1x})\ ,$$

$$\phi_2(x) \equiv \pi_2(x) = \frac{1}{2}(-\tilde{\Phi}_{2x} + \Phi_{1x}) = \frac{1}{2}(\Phi_{1x} + \Phi^*_{1x})\ ,$$

$$\phi_3(x) \equiv \pi_3(x) = \frac{i}{2}(-\tilde{\Phi}_{1x} + \Phi_{2x}) = \frac{i}{2}(\Phi_{2x} - \Phi^*_{2x})\ . \tag{6.4}$$

The realness of $\phi_S(x)$ can be expressed in terms of the matrix field φ_x by the relation

$$\varphi^+_x = \epsilon^{-1}\varphi^T_x\epsilon = \tau_2\varphi^T_x\tau_2\ . \tag{6.5}$$

The conventional notations $\sigma(x), \pi_r(x)$ express the equivalence of the pure scalar part of the SU(2) Higgs sector to the four-component O(4) symmetric ϕ^4 model or 'sigma model' (see chapter 2). The O(4) symmetry transformations are defined by the equivalence of the O(4) group to $SU(2)_L \otimes SU(2)_R$:

$$\varphi'_x = U_L^{-1} \varphi_x U_R , \qquad U_{L,R} \in SU(2)_{L,R} . \tag{6.6}$$

The realness condition (6.5) is preserved by this transformation, and because of

$$\det \varphi'_x = \det \varphi_x = \phi_S(x) \phi_S(x) , \tag{6.7}$$

it is clear that (6.6) defines an orthogonal transformation of the real components $\phi_S(x)$. Note that according to (6.3) the $SU(2)_L$ transformations act on Φ_{Ax} correctly as on an $SU(2)_L$ doublet. The definition of $SU(2)_L \otimes SU(2)_R$ in (6.6) can, therefore, be extended to the fermions in such a way, that $SU(2)_{L(R)}$ acts on the $L(R)$-handed chiral components of the fermion doublets (see section 6.2).

The lattice action of the $SU(2)$ ($\equiv SU(2)_L$) Higgs model consists of the pure gauge part $S_g[U]$ plus the gauged scalar part $S_\phi[U, \phi]$:

$$S[U, \phi] = S_g[U] + S_\phi[U, \phi] . \tag{6.8}$$

$S_g[U]$ is given, as usual, by a sum over plaquettes of the link gauge variables $U_{x\mu} \in SU(2)$ (see chapter 3):

$$S_g[U] = \beta \sum_p \left(1 - \frac{1}{2} \text{Tr} \, U_p \right) , \tag{6.9}$$

where $\beta = 4/g^2$ gives the bare $SU(2)$ gauge coupling. The scalar part of the lattice action in terms of the doublet field ($\Phi_x \equiv$ column (Φ_{1x}, Φ_{2x})) is

$$S_\phi[U, \Phi] = \sum_x \left\{ \mu_\phi (\Phi_x^+ \Phi_x) + \lambda (\Phi_x^+ \Phi_x)^2 - \kappa \sum_{\mu=\pm 1}^{\pm 4} (\Phi_{x+\hat{\mu}}^+ U_{x\mu} \Phi_x) \right\} . \tag{6.10}$$

Here the normalization of the scalar field is left general. The number of physically relevant bare parameters is only two. In lattice perturbation theory (see next subsection) it is convenient to fix the normalization freedom by $\kappa = \frac{1}{2}$, and then the bare mass squared in lattice units is $\mu_0^2 = 2\mu_\phi - 8$. In numerical simulations the best choice [6.6] is

$$\mu_\phi = 1 - 2\lambda , \tag{6.11}$$

and then, by adding a constant,

$$S_\phi[U, \Phi] = \sum_x \left\{ (\Phi_x^+ \Phi_x) + \lambda [(\Phi_x^+ \Phi_x) - 1]^2 - \kappa \sum_{\mu=\pm 1}^{\pm 4} (\Phi_{x+\hat{\mu}}^+ U_{x\mu} \Phi_x) \right\} . \tag{6.12}$$

This form is well suited also for the numerical study of the limit of infinitely strong bare quartic self-coupling $\lambda \to \infty$, when the length of the

Higgs field is frozen to unity:

$$(\Phi_x^+ \Phi_x) \equiv \Phi_{Ax}^* \Phi_{Ax} = \phi_S(x)\phi_S(x) = 1 .\tag{6.13}$$

In terms of the $2 \otimes 2$ matrix field φ_x in (6.2) we have

$$S_\phi[U,\varphi] = \sum_x \left\{ \frac{1}{2} \text{Tr}\,(\varphi_x^+ \varphi_x) + \lambda \left[\frac{1}{2} \text{Tr}\,(\varphi_x^+ \varphi_x) - 1 \right]^2 \right.$$

$$\left. -\kappa \sum_{\mu=1}^{4} \text{Tr}\,(\varphi_{x+\hat\mu}^+ U_{x\mu} \varphi_x) \right\} ,\tag{6.14}$$

where in the last term a relation like (6.5) was used for the gauge field variable $U_{x\mu}$. From (6.2) it follows that

$$\varphi_x^+ \varphi_x = (\sigma(x)^2 + \pi_r(x)\pi_r(x))\mathbf{1} = \phi_S(x)\phi_S(x)\mathbf{1} \equiv \rho_x^2 \mathbf{1} .\tag{6.15}$$

Therefore φ_x can be represented as

$$\varphi_x \equiv \rho_x \alpha_x , \qquad \rho_x \geq 0 , \qquad \alpha_x \in SU(2) .\tag{6.16}$$

ρ_x is the length of the Higgs field, and α_x collects the angular variables in an SU(2) element.

The integration measure in the path integral in terms of the real components $\phi_S(x)$ is

$$\prod_{x,\mu} \int_{SU(2)} dU_{x\mu} \cdot \prod_{x,S} \int_{-\infty}^{+\infty} d\phi_S(x) ,\tag{6.17}$$

where $dU_{x\mu}$ denotes the invariant Haar measure in SU(2). The integration over the scalar field can also be performed over the length and angular variables as

$$\prod_S \int_{-\infty}^{+\infty} d\phi_S(x) \propto \int_0^\infty \rho_x^3 \, d\rho_x \int_{SU(2)} d\alpha_x .\tag{6.18}$$

The second form is useful if the scalar part of the action is represented by the variables (ρ_x, α_x):

$$S_\phi[U,\rho,\alpha] = \sum_x \left\{ \rho_x^2 + \lambda(\rho_x^2 - 1)^2 - \kappa \sum_{\mu=1}^{4} \rho_{x+\hat\mu}\rho_x \, \text{Tr}\,(\alpha_{x+\hat\mu}^+ U_{x\mu}\alpha_x) \right\} .$$

$$\tag{6.19}$$

This shows that the lattice action can be formulated entirely in terms of gauge invariant variables: the Higgs field length ρ_x, and the gauge invariant link variable

$$V_{x\mu} \equiv \alpha_{x+\hat\mu}^+ U_{x\mu}\alpha_x .\tag{6.20}$$

As a function of $V_{x\mu}$ and ρ_x we have

$$S[V,\rho] = S_g[V] + \sum_x \left\{ \rho_x^2 + \lambda(\rho_x^2 - 1)^2 - \kappa \sum_{\mu=1}^4 \rho_{x+\hat{\mu}}\rho_x \operatorname{Tr} V_{x\mu} \right\} . \quad (6.21)$$

Here we used the gauge invariance of the pure gauge action, which implies $S_g[U] = S_g[V]$. Since the angular part of the Higgs field α_x does not appear in the action $S[V,\rho]$ at all, one can trivially perform the path integral over it. The integrations left over are

$$\prod_{x,\mu} \int_{SU(2)} dV_{x\mu} \cdot \prod_x \int_0^\infty \rho_x^3 \, d\rho_x . \quad (6.22)$$

Here the gauge invariance of the SU(2) integration was also used, which implies $dU_{x\mu} = dV_{x\mu}$.

The action in the form (6.21) can also be considered as the action in the *unitary gauge*. Namely, a local gauge transformation

$$U'_{x\mu} = \Lambda_{x+\hat{\mu}}^{-1} U_{x\mu} \Lambda_x , \qquad \alpha'_x = \Lambda_x^{-1}\alpha_x , \quad (6.23)$$

can be defined in such a way that

$$\alpha'_x = 1 , \qquad \varphi'_x = \begin{pmatrix} \rho_x & 0 \\ 0 & \rho_x \end{pmatrix} , \qquad \Phi'_x = \begin{pmatrix} 0 \\ \rho_x \end{pmatrix} . \quad (6.24)$$

Since the appropriate gauge transformation is defined just by $\Lambda_x = \alpha_x$, the transformed gauge field is $U'_{x\mu} = V_{x\mu}$. The gauge defined by (6.24) is called the 'unitary gauge', because it eliminates the unphysical gauge degrees of freedom.

The physical variables are gauge invariant, therefore in numerical simulations it is not necessary to keep the angular part of the Higgs field α_x, which varies with a gauge transformation according to (6.23). In other words one can use the unitary gauge action (6.21). This action has a global (x-independent) SU(2)$_W$ *weak isospin invariance* defined by the transformations ($U \in$ SU(2)$_W$)

$$\rho'_x = \rho_x , \qquad V'_{x\mu} = U^{-1}V_{x\mu}U . \quad (6.25)$$

In terms of the variables $(\varphi_x, U_{x\mu})$ this corresponds to the transformations

$$\varphi'_x = U^{-1}\varphi_x U , \qquad U'_{x\mu} = U^{-1}U_{x\mu}U , \quad (6.26)$$

which is the same as (6.6) with $U = U_L = U_R$. This means that the weak isospin is defined by the diagonal subgroup of SU(2)$_L \otimes$ SU(2)$_R$. The physical spectrum, of course, respects this global symmetry. The lowest lying particle states are: the isoscalar spin-0 Higgs boson represented, for instance, by the variable ρ_x, and the spin-1 W-boson with isospin 1. This latter can be represented by the real fields

$$V_{rx\mu} \equiv -i \operatorname{Tr} \{\tau_r V_{x\mu}\} . \quad (6.27)$$

The masses of the Higgs boson (m_H) and W-boson (m_W) can be determined, as usual, by the asymptotic behaviour of the two-point functions of ρ_x, respectively, $V_{rx\mu}$ at large distances [6.7].

Another useful item of physical information is obtained by measuring the static energy $E(r)$ of an external SU(2) doublet charge pair at distance r. This can be determined on the lattice by the expectation values of the rectangular Wilson loops $W_{R,T} \equiv \frac{1}{2} \mathrm{Tr}\, U_{R,T}$, with time elongation T and space distance R, as

$$aE(aR) = -\lim_{T \to \infty} \frac{1}{T} \log \langle W_{R,T} \rangle . \tag{6.28}$$

In the Higgs phase with broken SU(2) symmetry the static energy can be parametrized by a screened Yukawa potential [6.7, 6.8]

$$aE(aR) \simeq \mathrm{const.} - \frac{3\alpha_W}{4R} e^{-am_W R} . \tag{6.29}$$

This gives another possibility of defining the W-boson mass m_W. In addition, the coefficient α_W defines a renormalized gauge coupling.

The inclusion of the U(1)$_Y$ gauge interaction into the SU(2)$_L$ Higgs model [6.9] is achieved by defining the gauge link variable $U_{x\mu}$ as the product of $U_{Lx\mu} \in$ SU(2)$_L$ times a U(1)$_Y$ element $U_{Yx\mu} \equiv \exp(i\theta_{x\mu})$:

$$U_{x\mu} \equiv U_{Lx\mu} U_{Yx\mu} \in \mathrm{SU(2)}_L \otimes \mathrm{U(1)}_Y . \tag{6.30}$$

For the U(1)$_Y$ gauge part we take here the compact formulation equivalent to S_W in (4.279), and denote the bare coupling conventionally by $\frac{1}{2}g'$ (and $\beta' \equiv 4g'^{-2}$). The lattice action

$$S[U_L, U_Y, \Phi] = S_g[U_L, U_Y] + S_\phi[U, \Phi] \tag{6.31}$$

consists again of the pure gauge part S_g and scalar part S_ϕ. The gauge part is

$$S_g[U_L, U_Y] = \beta \sum_p \left(1 - \frac{1}{2} \mathrm{Tr}\, U_{Lp} \right) + \beta' \sum_p (1 - \cos\theta_p) , \tag{6.32}$$

where the U(1)$_Y$ plaquette variable is

$$\theta_p \equiv \theta_{x\mu\nu} \equiv \theta_{x+\hat{\mu},\nu} - \theta_{x\nu} - \theta_{x+\hat{\nu},\mu} + \theta_{x\mu} \quad (1 \leq \mu < \nu \leq 4) . \tag{6.33}$$

The scalar part of the action, in the normalization corresponding to (6.11), has the same form as (6.12):

$$S_\phi[U, \Phi] = \sum_x \left\{ (\Phi_x^+ \Phi_x) + \lambda[(\Phi_x^+ \Phi_x) - 1]^2 - \kappa \sum_{\mu=\pm 1}^{\pm 4} (\Phi_{x+\hat{\mu}}^+ U_{x\mu} \Phi_x) \right\} . \tag{6.34}$$

The $2 \otimes 2$ matrix field in (6.2) can also be introduced. The link variables in negative directions are now given by

$$U_{x,-\mu} \equiv U^+_{x-\hat{\mu},\mu} = U^+_{Lx-\hat{\mu},\mu} U^*_{Yx-\hat{\mu},\mu} \, . \tag{6.35}$$

Owing to

$$(\Phi^+_{x-\hat{\mu}} U_{x,-\mu} \Phi_x) = (\tilde{\Phi}^+_x U_{Lx-\hat{\mu},\mu} \tilde{\Phi}_{x-\hat{\mu}}) U^*_{Yx-\hat{\mu},\mu} \, , \tag{6.36}$$

we have

$$(\tilde{\Phi}^+_{x+\hat{\mu}} U_{Lx\mu} \tilde{\Phi}_x) e^{-i\theta_{x\mu}} + (\Phi^+_{x+\hat{\mu}} U_{Lx\mu} \Phi_x) e^{i\theta_{x\mu}}$$

$$= \text{Tr} \left(\varphi^+_{x+\hat{\mu}} U_{Lx\mu} \varphi_x e^{-i\tau_3 \theta_{x\mu}} \right) \, . \tag{6.37}$$

Therefore the $SU(2)_L \otimes U(1)_Y$ gauged scalar action in (6.34) can also be written as

$$S_\phi[U, \varphi] = \sum_x \left\{ \frac{1}{2} \text{Tr} (\varphi^+_x \varphi_x) + \lambda \left[\frac{1}{2} \text{Tr} (\varphi^+_x \varphi_x) - 1 \right]^2 \right.$$

$$\left. -\kappa \sum_{\mu=1}^{4} \text{Tr} \left(\varphi^+_{x+\hat{\mu}} U_{Lx\mu} \varphi_x e^{-i\tau_3 \theta_{x\mu}} \right) \right\} \, . \tag{6.38}$$

The form of the gauge invariant hopping term (proportional to κ) shows that in the scalar Higgs sector the $U(1)_Y$ group is embedded in $SU(2)_R$. Namely, a complete gauging of the global $SU(2)_L \otimes SU(2)_R$ symmetry in (6.6) would result in the hopping term

$$-\kappa \sum_{\mu=1}^{4} \text{Tr} \left(\varphi^+_{x+\hat{\mu}} U_{Lx\mu} \varphi_x U^+_{Rx\mu} \right) \, . \tag{6.39}$$

The $SU(2)_L \otimes U(1)_Y$ local gauge transformation of the matrix Higgs field is

$$\varphi'_x = \Lambda^{-1}_{Lx} \varphi_x e^{i\tau_3 b_x} \qquad (\Lambda_{Lx} \in SU(2)_L, \ 0 \le b_x < 2\pi) \, . \tag{6.40}$$

The transformation to the unitary gauge defined by (6.24) is, therefore, not unique because

$$\varphi'_x = \begin{pmatrix} \rho_x & 0 \\ 0 & \rho_x \end{pmatrix} = e^{-(i/2)\tau_3 a_x} \begin{pmatrix} \rho_x & 0 \\ 0 & \rho_x \end{pmatrix} e^{(i/2)\tau_3 a_x} \qquad (0 \le a_x < 4\pi) \, . \tag{6.41}$$

This remaining local gauge invariance corresponds to the electromagnetic $U(1)_Q$ subgroup of $SU(2)_L \otimes U(1)_Y$, which is not broken by a non-zero vacuum expectation value of the Higgs field. The definition of the $U(1)_Q$

gauge transformation in (6.41) is obviously consistent with the general definition of the electric charge as

$$Q = T_{L3} + T_{R3} + \frac{1}{2}(B - L) \equiv T_{L3} + \frac{1}{2}Y \,, \qquad (6.42)$$

where $T_{L(R)3}$ is the third generator of SU(2)$_{L(R)}$, and B is the baryon number, L the lepton number. (Of course, for the Higgs field $B = L = 0$.)

The hopping term L_ϕ of the scalar action in the unitary gauge is

$$L_\phi = -\kappa \sum_{\mu=1}^{4} \rho_{x+\hat{\mu}} \rho_x \, \mathrm{Tr} \left(U'_{Lx\mu} \mathrm{e}^{-i\tau_3 \theta'_{x\mu}} \right) \,. \qquad (6.43)$$

Representing the gauge fields in the unitary gauge by

$$U'_{Lx\mu} \equiv \exp \left\{ \mathrm{ig} \sum_{r=1}^{3} \frac{\tau_r}{2} W'_{r\mu}(x) \right\} \,, \qquad \theta'_{x\mu} \equiv \frac{1}{2} g' B'_\mu(x) \,, \qquad (6.44)$$

the small gauge coupling expansion of L_ϕ becomes

$$L_\phi = -\kappa \sum_{\mu=1}^{4} \rho_{x+\hat{\mu}} \rho_x \left\{ 2 - \sum_{r=1}^{2} \frac{1}{4}(g W'_{r\mu}(x))^2 \right.$$

$$\left. -\frac{1}{4}(g' B'_\mu(x) - g W'_{3\mu}(x))^2 + \cdots \right\} \,. \qquad (6.45)$$

This shows that the massive gauge bosons are: the charged W-bosons $W_{1,2\mu}(x)$ and the Z-boson, which is the combination defined by the *Weinberg angle* θ_W:

$$Z'_\mu(x) \equiv \sin \theta_W \, B'_\mu(x) - \cos \theta_W \, W'_{3\mu}(x) \quad (\tan \theta_W \equiv g'/g) \,. \qquad (6.46)$$

The orthogonal combination of $B'_\mu(x)$ and $W'_{3\mu}(x)$ is the massless photon field

$$A'_\mu(x) \equiv \sin \theta_W \, W'_{3\mu}(x) + \cos \theta_W \, B'_\mu(x) \,. \qquad (6.47)$$

6.1.2   Lattice perturbation theory

The formulation of perturbation theory in lattice Higgs models combines the ϕ^4 model (see chapter 2) with pure gauge theory (chapter 3). The new feature is the interplay between spontaneous symmetry breaking and gauge invariance. For definiteness, we shall consider here the SU(2) Higgs model with the action (6.14), following [6.10]. The extension to SU(2)$_L \otimes$ U(1)$_Y$ with the action (6.38), or to other gauge groups and/or scalar field representations, is straightforward.

The SU(2) gauge field in lattice perturbation theory can be described by the real link variables $A_\mu^r(x)$ $(r = 1, 2, 3)$ as

$$U_{x\mu} = \exp\left\{ig\frac{\tau_r}{2}A_\mu^r(x)\right\} \equiv 1a_\mu^0(x) + i\tau_r a_\mu^r(x) \ . \tag{6.48}$$

Introducing the length of the isovector gauge variable by

$$|A_\mu(x)| \equiv \left(A_\mu^r(x)A_\mu^r(x)\right)^{1/2} \ , \tag{6.49}$$

the four real variables $a_\mu^S(x)$ $(S = 0, 1, 2, 3)$ can be expressed as

$$a_\mu^0(x) = \cos\left(\frac{g}{2}|A_\mu(x)|\right) = 1 - \frac{g^2}{8}A_\mu^r(x)A_\mu^r(x) + O(g^4) \ ,$$

$$a_\mu^r(x) = \frac{A_\mu^r(x)}{|A_\mu(x)|}\sin\left(\frac{g}{2}|A_\mu(x)|\right)$$

$$= \frac{g}{2}A_\mu^r(x) - \frac{g^3}{48}A_\mu^r(x)A_\mu^s(x)A_\mu^s(x) + O(g^5) \ . \tag{6.50}$$

The invariant Haar measure in terms of $a_\mu^S(x)$ is

$$\int dU_{x\mu} = \frac{1}{\pi^2}\int d^4 a_\mu^S(x)\,\delta(a_\mu^T(x)a_\mu^T(x) - 1)$$

$$= \frac{1}{2\pi^2}\int d^4 a_\mu^S(x)\exp\left\{\sum_{j=1}^{\infty}\frac{(a_\mu^s(x)a_\mu^s(x))^j}{2j}\right\}$$

$$\cdot\theta(1 - a_\mu^r(x)a_\mu^r(x))\left[\delta\left(a_\mu^0(x) - \left(1 - a_\mu^s(x)a_\mu^s(x)\right)^{1/2}\right)\right.$$

$$\left. +\delta\left(a_\mu^0(x) + \left(1 - a_\mu^s(x)a_\mu^s(x)\right)^{1/2}\right)\right] \ . \tag{6.51}$$

For small g the zeroth component $a_\mu^0(x)$ is near $+1$, therefore only the first δ-function is important. In terms of $A_\mu^r(x)$ we have

$$\int dU_{x\mu} \propto \int d^3 A_\mu^r(x)\,\theta\left(\frac{2\pi}{g} - |A_\mu(x)|\right)\frac{\sin^2\left(\frac{1}{2}g|A_\mu(x)|\right)}{\frac{1}{4}g^2|A_\mu(x)|^2} \ . \tag{6.52}$$

The measure factor can be included in the Boltzmann factor $\exp(-S)$ by adding to the action S the measure term

$$S_m[A] \equiv -\sum_x\sum_{\mu=1}^{4}\log\left\{\frac{\sin^2\left(\frac{1}{2}g|A_\mu(x)|\right)}{\frac{1}{4}g^2|A_\mu(x)|^2}\right\}$$

$$= \sum_x \sum_{\mu=1}^{4} \frac{g^2}{12} A_\mu^r(x) A_\mu^r(x) + O(g^4) \; . \qquad (6.53)$$

The whole action contains, besides the pure gauge part $S_g[A] \equiv S_g[U]$ given by (6.9), and the measure term $S_m[A]$, also the gauge fixing term $S_{gf}[A,\varphi]$, the Faddeev–Popov term $S_{FP}[A,\varphi,\eta,\overline{\eta}]$ and, of course, the scalar part $S_\phi[A,\varphi]$ containing the scalar–gauge interaction:

$$S[A,\varphi,\eta,\overline{\eta}] = S_g[A] + S_m[A] + S_{gf}[A,\varphi] + S_{FP}[A,\varphi,\eta,\overline{\eta}] + S_\phi[A,\varphi] \; . \quad (6.54)$$

The pure gauge part can be expanded in powers of $A_\mu^r(x)$ (or $a_\mu^r(x)$ [6.10]) as in the general SU(N) case (see chapter 3). The gauge fixing function f_{rx}, which determines S_{gf} and S_{FP}, can be chosen, for instance, as

$$f_{rx}[A,\pi] \equiv \sum_{\mu=1}^{4} (a_\mu^r(x) - a_\mu^r(x-\hat{\mu})) - \frac{\alpha g^2}{4} v \pi_r(x) \; . \qquad (6.55)$$

Here α is an arbitrary gauge parameter, assumed to be positive ($\alpha > 0$), $a_\mu^r(x)$ is given as a function of $A_\mu^r(x)$ by (6.50), and v is the expectation value of the scalar field discussed below. The gauge defined by (6.55) is the lattice analogue of the 't Hooft gauge [6.11], which has several merits. It diagonalizes the quadratic form in (A,π), therefore there is no $A - \pi$ mixing term in the propagators. In addition, in this gauge both the would-be Goldstone bosons $\pi_r(x)$ and the Faddeev–Popov ghost fields $\eta_{rx}, \overline{\eta}_{rx}$ acquire a non-zero mass.

The gauge fixing term and the Faddeev–Popov term in the action can be identified by the usual procedure [A3]. A local SU(2) gauge transformation of the field variables

$$U_{(\Lambda)x\mu} = \Lambda_{x+\hat{\mu}}^{-1} U_{x\mu} \Lambda_x \; , \quad \varphi_{(\Lambda)x} = \Lambda_x^{-1} \varphi_x \; , \qquad (6.56)$$

can be parametrized by

$$\Lambda_x \equiv \mathbf{1}\Lambda_{0x} + i\tau_s \Lambda_{sx} \; \in \text{SU}(2) \; . \qquad (6.57)$$

Let us consider the path integral of an arbitrary gauge invariant function $\Sigma[U,\varphi]$, which can be written as

$$\mathscr{I} \equiv \int [\mathrm{d}U \, \mathrm{d}\varphi] \Sigma[U,\varphi]$$

$$= \int [\mathrm{d}U \, \mathrm{d}\varphi] \Sigma[U,\varphi] \Delta_f[A,\varphi] \prod_{rx} \delta \left(f_{rx}[A,\pi] - g_{rx} \right) \; . \qquad (6.58)$$

The second form is obtained by inserting

$$1 = \Delta_f[A,\varphi] \int \prod_y \mathrm{d}\Lambda_y \cdot \prod_{rx} \delta \left(f_{rx}[A_{(\Lambda)}, \pi_{(\Lambda)}] - g_{rx} \right) \; , \qquad (6.59)$$

and then performing the integral over Λ_y using gauge invariance (see section 3.3). Here Δ_f is the *Faddeev–Popov determinant*

$$\Delta_f[A, \varphi] \equiv \det \left(\frac{\partial f_{rx}[A_{(\Lambda)}, \pi_{(\Lambda)}]}{\partial \Lambda_{sy}} \right)_{\Lambda_{sy}=0}$$

$$\equiv \det \mathcal{M}_{sy,rx} = \Delta_f[A_{(\Lambda)}, \varphi_{(\Lambda)}] , \tag{6.60}$$

and $A_{(\Lambda)}$, $\pi_{(\Lambda)}$ and $\varphi_{(\Lambda)}$ denote the gauge transformed variables A, π and φ, respectively. The function g_{rx} of rx in (6.58), (6.59) is, for the moment, completely arbitrary.

With the help of the pairs of Grassmann variables $(\eta_{rx}, \bar{\eta}_{rx})$, called Faddeev–Popov ghost fields, the Faddeev–Popov determinant Δ_f can be written as (see section 3.3 and (4.22) in section 4.1)

$$\Delta_f = \det \mathcal{M} = \int [d\bar{\eta}\, d\eta] e^{-\sum_{xy} \bar{\eta}_{sy} \mathcal{M}_{sy,rx} \eta_{rx}}$$

$$\equiv \int [d\bar{\eta}\, d\eta] e^{-S_{FP}[A,\varphi,\eta,\bar{\eta}]} . \tag{6.61}$$

The freedom of choosing g_{rx} in (6.58) can be used to multiply by a weight factor $\exp\{-2/(\alpha g^2) \sum_x g_{rx} g_{rx}\}$ and integrate over g_{rx}. Apart from an uninteresting constant factor, this gives for the path integral $\mathscr{I}$

$$\mathscr{I} \propto \int [dU\, d\varphi\, d\bar{\eta}\, d\eta] \Sigma[U, \varphi] e^{-S_{gf}[A,\varphi] - S_{FP}[A,\varphi,\eta,\bar{\eta}]} , \tag{6.62}$$

where the gauge fixing term S_{gf} is

$$S_{gf}[A, \varphi] \equiv \frac{2}{\alpha g^2} \sum_x f_{rx}[A, \pi] f_{rx}[A, \pi]$$

$$= \frac{2}{\alpha g^2} \sum_x \sum_{\mu,\nu=1}^{4} (a_\mu^r(x) - a_\mu^r(x - \hat{\mu}))(a_\nu^r(x) - a_\nu^r(x - \hat{\nu}))$$

$$+ \frac{\alpha g^2}{8} v^2 \sum_x \pi_r(x)\pi_r(x) - v \sum_x \sum_{\mu=1}^{4} \pi_r(x)(a_\mu^r(x) - a_\mu^r(x - \hat{\mu})) . \tag{6.63}$$

In the total action (6.54) the last term ($A - \pi$ mixing) is cancelled by a corresponding term with opposite sign in the scalar part S_ϕ.

The explicit form of the Faddeev–Popov term S_{FP} can be easily obtained by performing the derivatives in the definition (6.60). The result is

$$S_{FP}[A, \varphi, \eta, \bar{\eta}] = \sum_{xy} \bar{\eta}_{sy} \mathcal{M}_{sy,rx} \eta_{rx}$$

$$= \sum_x \left\{ \sum_{\mu=1}^{4} \left[a_\mu^0(x) \left(\bar{\eta}_{rx} - \bar{\eta}_{rx+\hat{\mu}} \right) \left(\eta_{rx} - \eta_{rx+\hat{\mu}} \right) \right. \right.$$

$$\left. + \epsilon_{rst} \left(\bar{\eta}_{rx} + \bar{\eta}_{rx+\hat{\mu}} \right) a_\mu^s(x) \left(\eta_{tx} - \eta_{tx+\hat{\mu}} \right) \right]$$

$$\left. + \frac{\alpha g^2}{4} v^2 \bar{\eta}_{rx} \eta_{rx} + \frac{\alpha g^2}{4} v \bar{\eta}_{rx} (\delta_{rs}\sigma(x) + \epsilon_{rst}\pi_t(x)) \eta_{sx} \right\} . \qquad (6.64)$$

In order to obtain expressions in terms of $A_\mu^r(x)$, both here and in (6.63), one has to use the relations given by (6.50).

The remaining scalar part of the action S_ϕ will be given here in the phase with spontaneous symmetry breaking, where the Higgs field has a non-zero expectation value $\langle \varphi_x \rangle \neq 0$. This is conventionally transformed to the σ-direction, and the notations are

$$\varphi_x \equiv \mathbf{1}\sigma_0(x) + \mathrm{i}\tau_r\pi_r(x) , \qquad \sigma_0(x) \equiv \sigma(x) + v ; \qquad \langle \sigma(x) \rangle = 0 . \quad (6.65)$$

Note that a non-zero expectation value $v = \langle \sigma_0(x) \rangle \neq 0$ can only occur in a fixed gauge. If the gauge is not fixed, and in the path integral one is also integrating over the gauge degrees of freedom, the expectation value of $\sigma_0(x)$ is exactly zero [6.12]. However, if the gauge is fixed, $\langle \sigma_0(x) \rangle \neq 0$ is possible because the transcription of the path integral $\mathscr{I}$ in (6.62) is only valid for a gauge invariant function $\Sigma[U, \varphi]$. For a gauge variant function, like φ_x, the gauge fixing can change the expectation value. In fact, $\langle \sigma_0(x) \rangle$ can be used in the SU(2) Higgs model as an order parameter for the Higgs phase.

In terms of the variables $\sigma_0(x)$, $\pi_r(x)$ and $a_\mu^S(x)$ the scalar part of the action is, according to (6.14),

$$S_\phi[A, \varphi] = \sum_x \left\{ \frac{1}{2}(\mu_0^2 + 8)(\sigma_0(x)^2 + \pi_r(x)\pi_r(x)) \right.$$

$$+ \lambda_0(\sigma_0(x)^2 + \pi_r(x)\pi_r(x))^2 - \frac{1}{2} \sum_{\mu=1}^{4} \mathrm{Tr} \left[(\sigma_0(x + \hat{\mu}) - \mathrm{i}\tau_r\pi_r(x + \hat{\mu})) \right.$$

$$\left. \left. \cdot (a_\mu^0(x) + \mathrm{i}\tau_s a_\mu^s(x))(\sigma_0(x) + \mathrm{i}\tau_t\pi_t(x)) \right] \right\} . \qquad (6.66)$$

Here the general normalization of the scalar field in (6.10) is fixed by the condition $\kappa = \frac{1}{2}$. In this case $\mu_\phi = \frac{1}{2}(\mu_0^2 + 8)$, and in order to distinguish from the normalization convention in (6.12) and (6.14), the bare quartic coupling is denoted by λ_0. Working out the different terms in S_ϕ as a function of v and $\sigma(x)$, and omitting an inessential constant, one obtains

$$S_\phi[A, \varphi] = \sum_x \left\{ \sigma(x)v(\mu_0^2 + 4v^2\lambda_0) + \sigma(x)^2 \left(\frac{\mu_0^2}{2} + 6v^2\lambda_0 \right) \right.$$

$$+\pi_r(x)\pi_r(x) \left(\frac{\mu_0^2}{2} + 2v^2\lambda_0 \right) + 4v\lambda_0\sigma(x)(\sigma(x)^2 + \pi_r(x)\pi_r(x))$$

$$+\lambda_0(\sigma(x)^2 + \pi_r(x)\pi_r(x))^2 + \sum_{\mu=1}^{4} \left[(\sigma(x) - \sigma(x + \hat{\mu}))\sigma(x) \right.$$

$$+(\pi_r(x) - \pi_r(x + \hat{\mu}))\pi_r(x) + (1 - a_\mu^0(x))v^2$$

$$+va_\mu^r(x)(\pi_r(x) - \pi_r(x + \hat{\mu})) + (1 - a_\mu^0(x))v(\sigma(x + \hat{\mu}) + \sigma(x))$$

$$+(1 - a_\mu^0(x))(\sigma(x + \hat{\mu})\sigma(x) + \pi_r(x + \hat{\mu})\pi_r(x))$$

$$\left. \left. +a_\mu^r(x)(\sigma(x + \hat{\mu})\pi_r(x) - \pi_r(x + \hat{\mu})\sigma(x)) + \epsilon_{rst}\pi_r(x + \hat{\mu})a_\mu^s(x)\pi_t(x) \right] \right\} .$$

$$(6.67)$$

The vacuum expectation value of the scalar field in lattice units v was implicitly introduced in (6.65), as an unknown function of the bare parameters $v = v(g^2, \mu_0^2, \lambda_0)$ or $v = v(g^2, \kappa, \lambda)$. This function is determined by the condition $\langle \sigma(x) \rangle = 0$. If the interactions are neglected, this implies that the coefficient of the term linear in $\sigma(x)$ is zero, that is

$$v(\mu_0^2 + 4\lambda_0 v^2) = 0 \longrightarrow v_1 = 0 , \qquad v_2 = \left(\frac{-\mu_0^2}{4\lambda_0} \right)^{1/2} . \qquad (6.68)$$

The first solution $v = v_1$ corresponds to the symmetric phase, the second one $v = v_2$ to the broken phase occuring at $\mu_0^2 < 0$. If the interaction is taken into account by perturbation theory, the condition $\langle \sigma(x) \rangle = 0$ means that the sum of the tadpole graphs, with a single external σ-line, has to be zero. In the broken phase this gives v as a power series in g^2 and λ_0:

$$v = \left(\frac{-\mu_0^2}{4\lambda_0} \right)^{1/2} + O(g^2, \lambda_0) . \qquad (6.69)$$

The Feynman rules can be derived by expanding the action S in (6.54) in power series of $A_\mu^r(x)$, $\sigma(x)$, $\pi_r(x)$, η_{rx} and $\bar{\eta}_{rx}$. (In order to express $a_\mu^S(x)$ by $A_\mu^r(x)$ one has to use (6.50).) The second order terms define the propagators. In momentum space the propagator of the gauge field $A_\mu^r(x)$ turns out to be

$$\widetilde{\Delta}^{(\alpha)}(k)_{sv,r\mu} = \frac{\delta_{sr}}{(am_W)^2 + \hat{k}^2} \left[\delta_{v\mu} - (1 - \alpha) \frac{\hat{k}_v \hat{k}_\mu}{\alpha(am_W)^2 + \hat{k}^2} \right] . \qquad (6.70)$$

The notations in (8.40) are used, and the W-mass squared in lattice units is defined (see remark below) as

$$(am_W)^2 \equiv \left(\frac{1}{2}gv \right)^2 . \qquad (6.71)$$

The propagator of the Faddeev–Popov ghost field is

$$\widetilde{\Delta}^{FP}(k)_{sr} = \frac{\delta_{sr}}{\alpha(am_W)^2 + \hat{k}^2} , \qquad (6.72)$$

and those of the scalar fields are

$$\widetilde{\Delta}^\sigma(k) = \frac{1}{(am_\sigma)^2 + \hat{k}^2} ,$$

$$\widetilde{\Delta}^\pi(k) = \frac{1}{(am_\pi)^2 + \alpha(am_W)^2 + \hat{k}^2} , \qquad (6.73)$$

with

$$(am_\sigma)^2 \equiv \mu_0^2 + 12v^2\lambda_0 , \qquad (am_\pi)^2 \equiv \mu_0^2 + 4v^2\lambda_0 . \qquad (6.74)$$

The squared masses $(am_W)^2$, $(am_\sigma)^2$ and $(am_\pi)^2$ appearing in the propagators can be arbitrarily shifted according to $(am)^2 \to (am)^2 + \delta(am)^2$. In order to compensate for the shift $\delta(am)^2$, which is assumed to be a power series in g^2 and λ_0, one has to take into acount also two-point interaction vertices ('insertions') proportional to $-\delta(am)^2$. (For an analogous procedure in the case of the quark propagator in QCD see section 5.1.) A good choice of $\delta(am)^2$ is such that in the continuum limit $(am)^2 + \delta(am)^2$ vanishes for every physical particle. The freedom of choosing the propagator mass was, in fact, already exploited in (6.71), where a piece $g^2/6$ coming from the measure term S_m was not included.

The cubic, quartic and higher order interaction vertices are given by the corresponding terms in the expansion of the action in powers of $A_\mu^r(x)$, $\sigma(x)$, $\pi_r(x)$, η_{rx} and $\bar{\eta}_{rx}$. Their form is easily derived, for instance, from (6.64) and (6.67). The pure gauge vertices arise from the expansion of $S_g[A]$ and $S_m[A]$ in (6.53) (see chapter 3). As usual in lattice perturbation theory with gauge fields, besides the 16 vertices with mass dimension four or lower, there are also an infinite number of higher dimensional vertices produced, for instance, by the expansion of $a_\mu^S(x)$ in powers of $A_\mu^r(x)$.

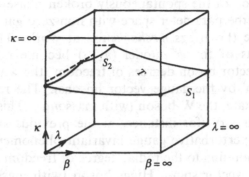

Fig. 6.1. The schematic phase structure of the SU(2) Higgs model with scalar doublet field. $\beta \equiv 4/g^2$ defines the bare SU(2) gauge coupling, λ is the bare scalar quartic self-coupling, and κ is the scalar hopping parameter. Above the phase transition surface (T) there is the Higgs phase, below it the confinement phase. On the pieces of the boundary denoted by S_1 and S_2 the phase transition is of second order.

6.1.3   *Phase structure and symmetry restoration*

The phase structure of the SU(2) Higgs model of a Higgs scalar doublet with lattice action (6.8), (6.9) and (6.12) or (6.14) is characterized by a phase transition between the *confinement phase* at small values of the scalar hopping parameter κ, and the *Higgs phase* at larger κ (see fig. 6.1). The numerical simulations, which have shown metastability by hystereses [6.13] and double peaked distributions [6.14], suggest that the phase transition surface (T) is everywhere of first order, except for the boundaries at $\beta = \infty$ (S_1), and at large enough λ and small β (S_2). The strength of the first order transition grows for decreasing λ and, if β is not too small, also for decreasing β.

At $\beta = \infty$ $(g^2 = 0)$ the gauge field decouples from the scalar field, and we have a pure scalar O(4) symmetric ϕ^4 model. In this limit the boundary S_1 of the phase transition surface is the well known second order phase transition line separating the symmetric phase at small κ from the phase with spontaneously broken symmetry at larger κ (see chapter 2). At non-zero gauge coupling the symmetric phase of the ϕ^4 model is continued by the confinement phase. This phase resembles in some sense QCD: the SU(2) 'colour' gauge field confines the scalar matter particles ('scalar quarks') into massive colour singlet bound states. On the $\kappa = 0$ boundary the scalar field is decoupled, because it becomes infinitely heavy, and one is left with a pure SU(2) gauge theory. With increasing κ the scalars become lighter, and the glueball states of pure gauge theory are increasingly mixed with the bound states of scalar constituents.

The continuation of the spontaneously broken phase of pure ϕ^4 theory to the interior of the parameter space with non-zero gauge coupling leads to a phase where the *Higgs mechanism* is at work. The three massless Goldstone bosons of the ϕ^4 model ($\pi_{1,2,3}$) become massive by mixing with the gauge vector boson degrees of freedom (the would-be Goldstone bosons are 'eaten' by the gauge vector bosons). The result is a massive isovector spin-1 state, the W-boson (with mass m_W). This can also be seen in perturbation theory (for instance, in the previous subsection), but it is, in fact, a non-perturbative gauge invariant phenomenon. The massive σ-particle corresponding to the radial degree of freedom of the scalar field is the physical isovector spin-0 Higgs boson (with mass m_H). For weak gauge couplings the W-bosons are still light, therefore the mass ratio

$$R_{HW} \equiv \frac{m_H}{m_W} \tag{6.75}$$

is larger than 2. As a consequence, the physical Higgs boson is not stable. It is a resonance decaying into two or more W-bosons. At small λ and/or β, however, R_{HW} can be smaller than 2, and then the Higgs boson is stable. Concerning the λ-dependence, remember that in the pure ϕ^4 limit ($\beta = \infty$) m_H is connected to the renormalized quartic coupling λ_R by

$$am_H = v_R(8\lambda_R)^{1/2} \equiv v_R(g_R/3)^{1/2} , \tag{6.76}$$

where v_R is the renormalized vacuum expectation value of the scalar field in lattice units ($g_R \equiv 4!\lambda_R$ is the renormalized quartic coupling in the conventional normalization introduced in chapter 2). Across the phase transition surface there is a characteristic and dramatic change in the mass ratio R_{HW}. For instance, at large bare quartic coupling R_{HW} is small in the confinement phase and large in the phase with Higgs mechanism [6.15, 6.8, 6.16].

At strong gauge coupling (small β) and larger values of κ the Higgs model resembles neither the pure ϕ^4 model nor the pure gauge model. In this region the distinction between 'confinement' and the 'Higgs mechanism' loses its meaning [6.17]. This is shown by the analytic connection of the two regions beyond the boundary S_2 in fig. 6.1. The existence of such a connection can also be proven rigorously [6.18]. It implies that, at least near the boundary S_2, the particle spectrum and other physical properties below and above the phase transition surface are qualitatively similar. In the confinement region, below the surface T, the Higgs boson and W-boson states are confined bound states of scalar constituents, with the appropriate quantum numbers. Of course, at large β, far away from the boundary S_2, the two phases can become physically quite different, especially above the scale of the W- and Higgs boson masses. However, at low energies, much below these masses, where at present most of our phenomenological observations are concentrated, there might be corresponding points in the two phases also at weak SU(2) gauge couplings, where the phenomenology is quite similar. This is the basis of the speculations [6.19] about the possible relevance of the confinement phase for the physics in the Standard Model. This 'strongly coupled Standard Model' may be a viable alternative to the usual scenario of the Standard Model, although there might be some problems with chirality, when the light fermions are included. In any case, at energies comparable to the scale of the renormalized vacuum expectation value ($v_R \simeq 250$ GeV) essential deviations in the qualitative behaviour are expected [6.19].

In view of the existence of an analytic connection between the confinement and Higgs region, they represent two different regions within a single phase. Nevertheless, because of the quantitative differences, below and above the surface T in fig. 6.1, it has become a custom to call the two regions two different phases. It is also possible to construct non-local, gauge invariant order parameters, like for instance the 'vacuum overlap order parameter' (VOOP), which are useful for the characterization of the two situations [6.20].

Besides the ratio of the Higgs boson mass to the W-boson mass R_{HW}, the other dimensionless parameter in the SU(2) Higgs model is the renormalized SU(2) gauge coupling denoted by α_W. In nature α_W is small. At the scale of m_W the experimental value is

$$\alpha_W \simeq 0.04 . \tag{6.77}$$

This corresponds to a renormalized SU(2) coupling $g_W^2 \equiv 4\pi\alpha_W \simeq 0.5$. Therefore, with $g^2 \simeq g_W^2$, which is a good first approximation for small g^2 and $am_W = O(1)$, the relevant region in $\beta = 4g^{-2}$ is near $\beta \simeq 8$. Based on the experience in pure gauge theory and QCD, it is expected that at $\beta \simeq 8$ in the confinement phase the numerical simulation is impossible, because the lowest physical mass in lattice units is extraordinarily small. (According to the perturbative β-function in (5.66), the SU(2) glueball masses at $\beta \simeq 8$ could be as small as $\simeq 10^{-7}$.) However, in the Higgs region the situation is different: the W-mass and Higgs boson mass in lattice units can be tuned to $am_{W,H} \simeq O(0.1) - O(1)$, and the numerical simulation is feasible [6.8].

The phase structure of the $SU(2)_L \otimes U(1)_Y$ Higgs model can be characterized similarly, but it is more complicated than the phase structure of the simple SU(2) Higgs model, because new phase transitions are induced by a strong U(1)-coupling [6.9].

At high temperatures ($T \gg m_W$) it is expected on general physical grounds, and on the basis of approximate perturbative calculations, that the spontaneous symmetry breaking disappears. The symmetry gets 'restored' [6.21, 6.22]. Therefore, in the early universe the spontaneous

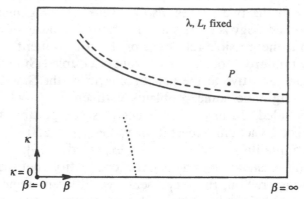

Fig. 6.2. The upwards shift of the confinement–Higgs phase transition (from the full line to the dashed line) on a $L_t \cdot L_s^3$ lattice ($L_s \gg L_t$), describing finite temperatures. The position of the deconfinement phase transition (probably only a rapid crossover) is also shown by a dotted line. P indicates a possible physical point at zero temperature, with weak gauge coupling in the Higgs phase.

electroweak symmetry breaking appears only after some time, when the temperature is cooled down below the phase transition temperature T_{sr} of the *electroweak symmetry restoration*.

On the lattice the (inverse) temperature is controlled by the temporal extension of the lattice (see section 5.4). The symmetry restoration is exhibited by a small shift of the confinement–Higgs phase transition surface towards larger values of the scalar hopping parameter κ [6.23, 6.24]. This is illustrated in the (β, κ)-plane (for fixed λ) by fig. 6.2, where the continuation of the deconfining phase transition of pure SU(2) gauge theory is also shown in the interior of the parameter space (non-zero κ). Since for SU(2) the deconfinement phase transition is of second order, this continuation is probably not a real phase transition, but only a rapid crossover, which disappears if κ is sufficiently large (compare to the analogous fig. 5.18 in QCD with Wilson quarks). In addition, even this crossover occurs at strong gauge coupling, in the confinement region of the phase diagram, hence it is not relevant for the physical region (near point P in fig. 6.2).

The physical value of the symmetry restoration temperature (T_{sr}) can be obtained as follows: suppose that on an asymmetric lattice ($L_t \ll L_s$) the phase transition occurs at some point P of the parameter space, where on the zero temperature (symmetric) lattice ($L_t = L_s =$ *large*) the renormalized gauge coupling α_W is equal to the physical value in (6.77), the Higgs–W mass ratio R_{HW} in (6.75) has some known value, and the

W-mass in lattice units is am_W. Then we have, for the given α_W and R_{HW},

$$\frac{T_{sr}}{m_W} = (am_W L_t)^{-1}.$$ (6.78)

The order of the symmetry restoring phase transition, as a function of the temperature, is not known at present. However, at zero temperature the confinement–Higgs phase transition surface (see fig. 6.1) is probably of first order everywhere in the bare parameter space. It is plausible that on an asymmetric lattice this phase transition surface is only shifted to larger κ values, but the order is unchanged. This would mean that the symmetry restoring phase transition is always of first order. Assuming that the qualitative behaviour at $T = 0$ is preserved, one can also conclude that the first order transition weakens if β and/or λ is increasing. If β is tuned in such a way that the renormalized gauge coupling α_W is equal to its physical value (6.77), then this dependence on λ implies that the first order symmetry restoring phase transition becomes weaker for increasing Higgs–W mass ratio R_{HW}.

In order to obtain thermodynamic quantities, like energy density (ϵ) or pressure (P), and for an easier control of the high temperature behaviour, one can introduce different lattice spacings in time direction (a_t) and in space directions (a_s). (See the more detailed discussion on finite temperature field theory in sections 1.8 and 5.4.) This is achieved by choosing different coupling strengths in the action for time-like and space-like directions. Therefore, instead of the lattice action (6.8), (6.9), (6.14) we have

$$S[U, \varphi] = \beta_s \sum_{sp} \left(1 - \frac{1}{2} \mathrm{Tr}\, U_{sp}\right) + \beta_t \sum_{tp} \left(1 - \frac{1}{2} \mathrm{Tr}\, U_{tp}\right)$$

$$+ \sum_x \left\{ \frac{1}{2} \mathrm{Tr}\, (\varphi_x^+ \varphi_x) + \lambda \left[\frac{1}{2} \mathrm{Tr}\, (\varphi_x^+ \varphi_x) - 1 \right]^2 \right.$$

$$\left. - \kappa_s \sum_{\mu=1}^3 \mathrm{Tr}\, (\varphi_{x+\hat\mu}^+ U_{x\mu} \varphi_x) - \kappa_t \mathrm{Tr}\, (\varphi_{x+\hat4}^+ U_{x4} \varphi_x) \right\}.$$ (6.79)

The asymmetry of the lattice spacings $\xi \equiv a_s/a_t$ is a function of the ratios [6.25]

$$\gamma_g \equiv \left(\frac{\beta_t}{\beta_s}\right)^{1/2}, \qquad \gamma_\phi \equiv \left(\frac{\kappa_t}{\kappa_s}\right)^{1/2}.$$ (6.80)

Neglecting quantum correction (i. e. considering only the propagators without interactions) we have $\xi = \gamma_g = \gamma_\phi$. In the general case, however, the true relations $\gamma_g = \gamma_g(\xi)$, $\gamma_\phi = \gamma_\phi(\xi)$ have to be determined non-perturbatively (see also section 5.4 for the analogous problem in QCD). For this one can use, for instance, the requirement that the two-point correlation functions of the W-boson variables in (6.27) have to be isotropic at large physical Euclidean distances. If the functions $\gamma_{g,\phi}(\xi)$ are known, the thermodynamic quantities can be calculated similarly as in section 5.4 for QCD.

In order to illustrate this procedure, let us consider the quantity in (5.249)

$$\delta \equiv \frac{1}{3}\epsilon - P = \frac{T^4}{3}\left(\frac{L_t}{L_s}\right)^3 \left\langle \frac{\partial S}{\partial \ln a} \right\rangle$$

$$= \frac{T^4}{3}\left(\frac{L_t}{L_s}\right)^3 \left\langle \frac{\partial \kappa}{\partial \ln a}\frac{\partial S}{\partial \kappa} + \frac{\partial \lambda}{\partial \ln a}\frac{\partial S}{\partial \lambda} + \frac{\partial \beta}{\partial \ln a}\frac{\partial S}{\partial \beta} \right\rangle . \qquad (6.81)$$

This can be calculated on lattices with isotropic lattice spacings $a_s = a_t = a$. The partial derivatives of the bare parameters have to be evaluated here for fixed physical parameters α_W and R_{HW}. That is, one has to know the lines of constant physics in the bare parameter space, where α_W and R_{HW} are constant, and only the lattice spacing a (and hence am_W) is changing. These curves can be parametrized by the functions

$$\kappa = \kappa(am_W, \alpha_W, R_{HW}) , \quad \lambda = \lambda(am_W, \alpha_W, R_{HW}) ,$$

$$\beta = \beta(am_W, \alpha_W, R_{HW}) , \qquad (6.82)$$

and then from (6.81) we get

$$\delta = \frac{T^4}{3}L_t^4 \left\langle -\frac{\partial \kappa(am_W, \alpha_W, R_{HW})}{\partial \ln(am_W)} \sum_{\mu=1}^{4} \text{Tr}\,(\varphi_{x+\hat{\mu}}^+ U_{x\mu}\varphi_x) \right.$$

$$+ \frac{\partial \lambda(am_W, \alpha_W, R_{HW})}{\partial \ln(am_W)}\left[\frac{1}{2}\text{Tr}\,(\varphi_x^+ \varphi_x) - 1\right]^2$$

$$\left. + \frac{\partial \beta(am_W, \alpha_W, R_{HW})}{\partial \ln(am_W)}\left[3\left(1 - \frac{1}{2}\text{Tr}\,U_{\text{sp}}\right) + 3\left(1 - \frac{1}{2}\text{Tr}\,U_{\text{tp}}\right)\right] \right\rangle . \quad (6.83)$$

The difficult task here is to determine numerically the partial derivatives. For a general set-up of this problem, and for a discussion of the qualitative behaviour of the lines of constant physics in the SU(2) Higgs model, see the next subsection.

6.1.4   Triviality upper bound

As discussed in chapter 2, the pure ϕ^4 limit of the Higgs sector of the Standard Model has a trivial continuum limit. This means that for the renormalized quartic coupling g_R, which is related to the Higgs boson mass by (6.76), there is a cut-off dependent upper bound. This upper bound tends to zero for infinite cut-off, therefore in the continuum limit the renormalized quartic coupling is necessarily zero.

In the SU(2) Higgs model the question is, how does the SU(2) gauge coupling influence the triviality of the continuum limit. Can the asymptotically free gauge coupling induce a non-zero renormalized quartic coupling in the continuum? This question can be simply formulated in terms of the *lines of constant physics (LCP)*, which are curves in the bare parameter space, where the dimensionless physical quantities, like the renormalized gauge coupling α_W and the Higgs–W mass ratio R_{HW}, are constant, and only the lattice spacing a changes. A non-trivial continuum limit means that there are some LCPs with non-zero α_W and R_{HW}, which tend for $a \to 0$ to a second order phase transition point, say, on the boundary S_1 of the phase transition surface T in fig. 6.1.

Before discussing the qualitative behaviour of LCPs in the SU(2) Higgs model let us consider the equations for the LCPs in general [6.10], which are useful in many respects, for instance, for the calculation of thermodynamic quantities as in (6.83). Let us consider a general lattice quantum field theory with n relevant bare couplings $g_1, g_2, \ldots, g_n$. This means that in order to specify the LCPs one has to keep $(n-1)$ independent dimensionless physical quantities $F_2, F_3, \ldots, F_n$ constant:

$$F_j(g_1, g_2, \ldots, g_n) = F_{j0} = \text{const.} \quad (j = 2, 3, \ldots, n) . \qquad (6.84)$$

The LCPs are characterized by the constant values F_{j0}. The points of a specific LCP can be parametrized, for instance, by the first bare coupling g_1: $g_j = g_j(g_1; F_2, F_3, \ldots, F_n)$ $(j = 2, 3, \ldots, n)$. In this case we have from $\mathrm{d}F_j = 0$

$$\frac{\mathrm{d}g_j(g_1)}{\mathrm{d}g_1} = \frac{\det_{n-1}^{[1,j]}(\partial F/\partial g)}{\det_{n-1}^{[1,1]}(\partial F/\partial g)} . \qquad (6.85)$$

Here $\det_{n-1}^{[i,k]}(\partial F/\partial g)$ denotes the $(n-1) \otimes (n-1)$ subdeterminant of the $n \otimes n$ derivative matrix $\partial F/\partial g$, which belongs to the matrix element $(\partial F/\partial g)_{ik} = \partial F_i/\partial g_k$.

Another possibility is to parametrize the points of an LCP by the value of some reference physical quantity F_1. For instance, it is reasonable to take F_1 as some physical mass in lattice units, or its logarithm:

$$F_1 \equiv am \quad \text{or} \quad \log(am) . \qquad (6.86)$$

In this case the differential equations for $g_i(F_1)$ ($i = 1, 2, \ldots, n$) are

$$\frac{\mathrm{d} g_i(F_1)}{\mathrm{d} F_1} = \frac{\det_{n-1}^{[1,i]} (\partial F/\partial g)}{\det_n (\partial F/\partial g)} \equiv \beta_{g_i/F_1}(F_1, g_2, g_3, \ldots, g_n) . \tag{6.87}$$

On the right hand side β_{g_i/F_1} is a generalized Callan–Symanzik β-function, which is considered here as a function of the reference quantity F_1 and bare parameters $g_2, g_3, \ldots, g_n$. In this way (6.87) is a system of first order differential equations for the LCPs.

The generalized β-functions β_{g_i/F_1} can, in principle, be determined by numerical simulations. A direct way to obtain $\partial F_i/\partial g_k$ is the following: let us assume that F_i is a known function of the expectation values $\langle q_1 \rangle, \ldots, \langle q_R \rangle$. Then

$$-\frac{\partial \langle q_r \rangle}{\partial g_k} = -\frac{\partial}{\partial g_k} \left(\frac{\int [\mathrm{d}\phi] e^{-S[\phi]} q_r}{\int [\mathrm{d}\phi] e^{-S[\phi]}} \right) = \left\langle q_r \frac{\partial S}{\partial g_k} \right\rangle - \langle q_r \rangle \left\langle \frac{\partial S}{\partial g_k} \right\rangle \tag{6.88}$$

implies that

$$\frac{\partial F_i}{\partial g_k} = -\sum_{r=1}^{R} \frac{\partial F_i}{\partial \langle q_r \rangle} \left\{ \left\langle q_r \frac{\partial S}{\partial g_k} \right\rangle - \langle q_r \rangle \left\langle \frac{\partial S}{\partial g_k} \right\rangle \right\} . \tag{6.89}$$

Therefore the numerical problem is to determine the correlation of the quantities q_r and the partial derivatives of the action $\partial S/\partial g_k$.

As an example, let us imagine that $F_1 = am$ is extracted from the ratio of some timeslice correlations at time separation t and $(t + 1)$:

$$F_1 \equiv am_{(t,t+1)} \equiv \log \frac{\langle s_0 s_t \rangle - \langle s_0 \rangle^2}{\langle s_0 s_{t+1} \rangle - \langle s_0 \rangle^2} . \tag{6.90}$$

In this case (6.89) gives

$$\frac{\partial F_1}{\partial g_k} = \left[\langle s_0 s_{t+1} \rangle - \langle s_0 \rangle^2 \right]^{-1}$$

$$\cdot \left\{ \left\langle s_0 s_{t+1} \frac{\partial S}{\partial g_k} \right\rangle - \langle s_0 s_{t+1} \rangle \left\langle \frac{\partial S}{\partial g_k} \right\rangle - 2 \langle s_0 \rangle \left\langle s_0 \frac{\partial S}{\partial g_k} \right\rangle + 2 \langle s_0 \rangle^2 \left\langle \frac{\partial S}{\partial g_k} \right\rangle \right\}$$

$$- (s_{t+1} \longrightarrow s_t) . \tag{6.91}$$

Substituting back $\partial F_i/\partial g_k$ into (6.85) gives the direction of the LCPs in the point $(g_1, g_2, \ldots, g_n)$, whereas (6.87) gives the generalized β-functions needed, for instance, in thermodynamic quantities as in (6.83).

Sometimes it is also useful to consider curves in subspaces of the bare parameter space, which belong to constant values of an appropriately smaller number of physical quantities. These 'lines of partially constant physics (LPCP)' are defined, in general, by fixing $(n-k)$ physical quantities $F_2, F_3, \ldots, F_{n-k+1}$ and $(k - 1)$ bare parameters $g_{n-k+2}, g_{n-k+3}, \ldots, g_n$. The differential equations for the LPCPs have similar forms as (6.85) and (6.87). For simplicity, let us consider here only the case with $n = 3$, as we have in the SU(2) Higgs model. (There we can take, for instance, $g_1 = g^2$, $g_2 = \kappa$, $g_3 = \lambda$ and $F_1 = am_W$, $F_2 = \alpha_W$, $F_3 = R_{HW}$.) In the plane with $g_3 = \text{const.}$ let us keep $F_2(g_1, g_2, g_3) = F_{20}$ constant. Then we have

$$\frac{dg_2}{dg_1} = -\frac{\partial F_2}{\partial g_1} \left(\frac{\partial F_2}{\partial g_2}\right)^{-1},$$

$$\frac{dg_2(F_1)}{dF_1} = -\frac{\partial F_2}{\partial g_1} \left(\frac{\partial F_1}{\partial g_1}\frac{\partial F_2}{\partial g_2} - \frac{\partial F_1}{\partial g_2}\frac{\partial F_2}{\partial g_1}\right)^{-1}$$

$$= -\frac{\partial F_2}{\partial g_1} \left[\det_2^{[3,3]}(\partial F/\partial g)\right]^{-1},$$

$$\frac{dF_3(F_1)}{dF_1} = -\det_2^{[1,3]}(\partial F/\partial g) \left[\det_2^{[3,3]}(\partial F/\partial g)\right]^{-1}. \tag{6.92}$$

The last equation describes the change of a third physical quantity F_3 along the LPCPs in the $g_3 = \text{const.}$ planes.

The equations like (6.85), (6.87) and (6.92) are valid under quite general circumstances. For instance, the isotropy of the lattice is not assumed. In fact, such equations can be very useful in the cases if the temporal (a_t) and spatial ($a_s \equiv \xi a_t$) lattice spacings are different as, for instance, in some thermodynamic applications. The asymmetry of the lattice spacings can be included in an enlarged set of dimensionless physical quantities. As an example, one can take the anisotropic action of the SU(2) Higgs model in (6.79), with five parameters ($\lambda, \beta_s, \beta_t, \kappa_s, \kappa_t$). In this case the masses extracted from the asymptotic behaviour of two-point correlations are different in the time-like and space-like directions: $a_t m_{W,H}$, respectively, $a_s m_{W,H}$. For the reference quantity one can take, for instance, $F_1 = a_t m_W$, and the other four independent dimensionless quantities can be: $F_2 = \alpha_W$, $F_3 = R_{HW}$, $F_4 = a_s m_W / a_t m_W$ and $F_5 = a_s m_H / a_t m_H$. In order to obtain a Euclidean invariant continuum limit one requires, of course, that $F_4 = F_5 \equiv \xi$.

In the region of small bare couplings one can use bare lattice perturbation theory to find the LCPs (or LPCPs). From the perturbative expressions one can, for instance, derive the differential equations in (6.87). In the SU(2) Higgs model one can use, for instance, the perturbative 1-loop effective potential to derive [6.10]

$$\frac{d\lambda_0(\tau)}{d\tau} = \frac{1}{16\pi^2}\left[96\lambda_0^2 + \frac{9}{32}g^4 - 9\lambda_0 g^2 + \cdots\right],$$

$$\frac{\mathrm{d}g^2(\tau)}{\mathrm{d}\tau} = \frac{1}{16\pi^2}\left[-\frac{43}{3}g^4 + \cdots\right] . \qquad (6.93)$$

Here $\tau \equiv \log{(am_R)^{-1}}$ is the logarithm of the inverse renormalized scalar mass in lattice units, and the normalization of bare parameters is chosen according to (6.66). The dots stand for terms of higher order $O(\lambda_0^3, \lambda_0^2 g^2, \lambda_0 g^4, g^6)$.

The behaviour of the solutions of the differential equations in (6.93) shows that, at least in the region of small bare couplings, where the higher order corrections can be neglected, the lines of constant physics do not reach the second order phase transition line S_1 at $g^2 = 0$. Therefore, a non-trivial continuum limit cannot be defined for small couplings. The reason is, that for large τ (corresponding to large cut-offs) $g^2 \to 0$ according to the second equation above, and consequently the first $O(\lambda_0^2)$ term in the first equation dominates, which pushes λ_0 towards large values. Therefore, the Gaussian fixed point at $g^2 = \lambda_0 = 0$ is an *infrared stable fixed point*, which repels the LCPs, in the same way as in pure ϕ^4 theory.

The perturbative treatment is, of course, not valid at large bare quartic couplings $\lambda_0 \gg 1$. Therefore the LCPs still could, in principle, be recollected at some point of the boundary S_1 at a large value of λ_0. The weak gauge coupling expansion near the $g^2 = 0$ critical line suggests, however, that this does not actually happen [6.10] (see the next subsection). The reason is the following. If instead of the bare quartic coupling λ_0 the renormalized quartic coupling λ_r in a corresponding point of pure ϕ^4 theory, at the same (κ, λ), is taken as the parameter along the LCPs, then the differential equations (6.93) become

$$\frac{\mathrm{d}\lambda_r(\tau)}{\mathrm{d}\tau} = \frac{1}{16\pi^2}\left[-9\lambda_r g^2 + \cdots\right] ,$$

$$\frac{\mathrm{d}g^2(\tau)}{\mathrm{d}\tau} = \frac{1}{16\pi^2}\left[-\frac{43}{3}g^4 + \cdots\right] . \qquad (6.94)$$

Note that λ_r is not equal to the renormalized quartic coupling λ_R at non-zero g^2. For small g^2 we have, however, $\lambda_R = \lambda_r + O(g^2)$. Omitting higher order terms, the equations (6.94) have the solutions

$$g^2(\tau) = \left[g_0^{-2} + \frac{43}{48\pi^2}(\tau - \tau_0)\right]^{-1} ,$$

$$\lambda_r(\tau) = \lambda_{r0}\left[1 + \frac{43g_0^2}{48\pi^2}(\tau - \tau_0)\right]^{-27/43} , \qquad (6.95)$$

with the initial values g_0^2, λ_{r0} at $\tau = \tau_0$. In pure ϕ^4 theory there is a cut-off dependent upper bound $\lambda_r(\tau)_{max}$ on the renormalized coupling λ_r, which

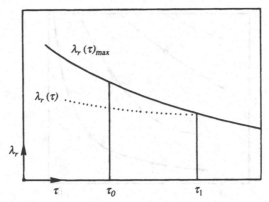

Fig. 6.3. The change of λ_r along the LCPs (dotted line). The full line is the upper bound on λ_r, which is reached in the ϕ^4 theory at infinite bare quartic coupling $\lambda = \infty$.

behaves for large τ (see chapter 2) as

$$\lambda_r(\tau)_{max} \simeq \frac{\pi^2}{6\tau}. \tag{6.96}$$

Since according to (6.95) $\lambda_r(\tau)$ goes asymptotically slower to zero than $\lambda_r(\tau)_{max}$, every LCP will reach the upper bound at some finite $\tau = \tau_1$ (see fig. 6.3).

The expected qualitative behaviour of the projections of the LCPs on to the (β, λ) plane is illustrated by fig. 6.4. As it is shown by the figure, the non-trivial LCPs all end on the boundary of the parameter space at $\lambda = \infty$, in the same way as in pure ϕ^4 theories. Therefore, the continuum limit of the SU(2) Higgs model at small gauge coupling is trivial.

The effect of the gauge coupling makes the upper bound on the renormalized quartic coupling somewhat higher than in pure ϕ^4 theory, especially if the cut-off is high. This can qualitatively be seen already in (6.93), because of the negative $O(\lambda_0 g^2)$ contribution in the first equation. The quantitative change for small g^2 can be obtained from the ratio $\lambda_{r0}/\lambda_r(\tau)$ in (6.95). Namely, if the physical scale and the cut-off scale are close to each other, the renormalized couplings are approximately equal to the bare couplings. Therefore $g_0^2 \simeq 0.5$ corresponds to the physical value of the SU(2) gauge coupling, and at $(\tau - \tau_0) \simeq 40$, which refers to a cut-off at the Planck mass, we get $\lambda_{r0}/\lambda_r(\tau)_{max} = \lambda_{r0}/\lambda_r(\tau) \simeq 1.9$. Neglecting the small difference between λ_R at the physical scale and λ_{r0}, we see that the upper bound on the renormalized quartic coupling is increased by this factor, due to the cumulative effect of the gauge interaction between the physical scale and the cut-off scale.

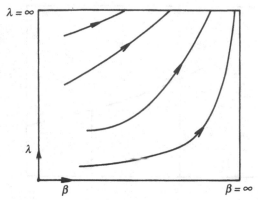

Fig. 6.4. The qualitative picture of LCPs in the SU(2) Higgs model projected on to the (β, λ) plane, as expected on the basis of the weak gauge coupling expansion. The arrows point in the direction of decreasing lattice spacings. Note that in reality there is a two-parameter family of LCPs, but here only a one-parameter subset is shown for simplicity.

This discussion shows that the cut-off dependent upper bound on the Higgs boson mass is reached in the SU(2) Higgs model at infinite bare quartic coupling $\lambda = \infty$, where it can be numerically determined. The advantage of the numerical simulations at finite β is that the W-bosons are massive, and therefore no difficulties arise due to the infrared singularities caused by massless Goldstone bosons. (In the pure ϕ^4 limit an external field has to be introduced, in order to make the Goldstone bosons massive.) However, at small gauge couplings corresponding to the physical value (6.77), the Higgs–W mass ratio is of the order of 10, therefore large lattices are needed. The first studies on lattices of size up to 16^4 [6.26, 6.27] showed that the upper bound at the smallest possible cut-off is

$$R_{HW} < 9 \, . \tag{6.97}$$

According to the next subsection, this agrees well with the upper bound deduced in the pure ϕ^4 limit (see chapter 2). But for controlling both finite volume effects and lattice artifacts in the SU(2) Higgs model, one would need lattices whith linear extensions larger by at least a factor of two.

6.1.5 *Weak gauge coupling limit*

Both the $U(1)_Y$ and $SU(2)_L$ gauge couplings are weak in nature, therefore in the Higgs sector of the Standard Model one can take into account their effects as small perturbations in an O(4) symmetric ϕ^4 theory. In particular, in the SU(2) Higgs model, which is mainly discussed in this chapter, this means to consider the limit of weak SU(2) gauge coupling: $g^2 \to 0$.

An important effect of a non-zero gauge coupling in the physically relevant phase with broken symmetry is that the massless Goldstone bosons and gauge vector bosons mix with each other, to form a massive W-boson (the 'Higgs mechanism'). At the tree level of perturbation theory, appropriate for small λ_0 and g^2, the W-mass in lattice units am_W is given by (6.71). The generalization to an arbitrary quartic scalar coupling λ_0 (and small g^2) is the *Dashen–Neuberger formula* [6.28]:

$$(am_W)^2 = \left(\frac{1}{2}gF\right)^2 + O(g^4, a^4) = \left(\frac{1}{2}gv_R\right)^2 + O(g^4, a^4) . \tag{6.98}$$

The quantity F is defined by the π-boson to vacuum matrix element

$$\langle 0|J^{(A)}_{rx\mu}|\pi_s(\mathbf{p})\rangle = i\delta_{rs}p_\mu F e^{-ix \bullet p} \tag{6.99}$$

of the axial vector current

$$J^{(A)}_{rx\mu} = \pi_r(x + \hat{\mu})\sigma_0(x) - \sigma_0(x + \hat{\mu})\pi_r(x)$$

$$= \sigma_0(x)\Delta^{(x)f}_\mu \pi_r(x) - \pi_r(x)\Delta^{(x)f}_\mu \sigma_0(x) . \tag{6.100}$$

In this definition the unshifted σ-field $\sigma_0(x)$ defined by (6.65) is used. (Note that the normalization of F in (6.99) is related to the QCD pion decay constant defined in (5.141) by $F\sqrt{2} \to f_\pi$.) Other currents belonging to the SU(2)$_L \otimes$ SU(2)$_R$ symmetry are the vector current

$$J^{(V)}_{rx\mu} = \epsilon_{rst}\pi_s(x)\pi_t(x + \hat{\mu}) = \epsilon_{rst}\pi_s(x)\Delta^{(x)f}_\mu \pi_t(x) , \tag{6.101}$$

and the chiral currents

$$J^{(R)}_{rx\mu} = \frac{1}{2}\left(J^{(V)}_{rx\mu} + J^{(A)}_{rx\mu}\right) , \qquad J^{(L)}_{rx\mu} = \frac{1}{2}\left(J^{(V)}_{rx\mu} - J^{(A)}_{rx\mu}\right) . \tag{6.102}$$

The Dashen–Neuberger formula (6.98) is a consequence of the SU(2)$_L \otimes$ SU(2)$_R$ Ward identities in ϕ^4 theory, which among others imply the relation [6.29]

$$F = v_R \equiv Z_\pi^{-1/2}v . \tag{6.103}$$

Here v_R is the renormalized vacuum expectation value of the scalar field, and $Z_\pi^{1/2}$ is the renormalization factor of the π-field, defined as usual (see (8.50)) by

$$\langle 0|\pi_r(x)|\pi_s(\mathbf{p})\rangle = \delta_{rs}Z_\pi^{1/2}e^{-ix \bullet p} . \tag{6.104}$$

Another consequence of the non-zero gauge coupling is that the phase transition between the symmetric and broken phases becomes first order.

For small quartic coupling this can be deduced from renormalized perturbation theory, applied to the effective potential $V_{eff}(\sigma)$ [6.30]. In lattice perturbation theory the 1-loop effective action for the SU(2) Higgs model can be written as [A3]

$$\Gamma_{1-\text{loop}}[A, \sigma, \pi] = S[A, \sigma, \pi]$$

$$+ \frac{1}{2}\text{Tr} \log \{\mathscr{D}[A, \sigma, \pi]\Delta\} - \text{Tr} \log \{\mathscr{M}[A, \sigma, \pi]\Delta^{FP}\} . \tag{6.105}$$

Here $S[A, \sigma, \pi] \equiv S[A, \sigma, \pi, \eta_x = 0, \overline{\eta}_x = 0]$ is the lattice action (6.54) without the Faddeev–Popov term, $\mathscr{D}$ is the second derivative matrix of $S[A, \sigma, \pi]$ with respect to the bosonic variables $[A, \sigma, \pi]$, Δ is the bosonic propagator matrix, $\mathscr{M}$ is the Faddeev–Popov ghost matrix in (6.64), and Δ^{FP} is the corresponding propagator. The effective potential of a spatially constant σ-field $V_{eff}(\sigma)$ is usually considered in the 't Hooft–Landau gauge, when the gauge fixing parameter tends to zero: $\alpha \to 0$. Evaluating (6.105) in lattice perturbation theory, according to (6.54)–(6.74), one obtains

$$V_{eff}^{1-\text{loop}}(\sigma) \equiv \frac{1}{\Omega}\Gamma_{1-\text{loop}}[A(x) = 0, \sigma(x) = \sigma, \pi(x) = 0]$$

$$= \frac{\mu_0^2}{2}[(\sigma + v)^2 - v^2] + \lambda_0[(\sigma + v)^4 - v^4]$$

$$+ \int_k \left\{ \frac{1}{2} \log \left[1 + \frac{12\lambda_0}{(am_\sigma)^2 + \hat{k}^2}\left((\sigma + v)^2 - v^2\right) \right] \right.$$

$$+ \frac{3}{2} \log \left[1 + \frac{4\lambda_0}{(am_\pi)^2 + \hat{k}^2}\left((\sigma + v)^2 - v^2\right) \right]$$

$$\left. + \frac{9}{2} \log \left[1 + \frac{g^2/4}{(am_W)^2 + \hat{k}^2}\left((\sigma + v)^2 - v^2\right) \right] \right\} . \tag{6.106}$$

A constant term was added here in such a way that $V_{eff}(\sigma)$ vanishes at $\sigma = 0$.

The vacuum expectation value v can be determined from the requirement that the absolute minimum of $V_{eff}(\sigma)$ is at $\sigma = 0$, hence

$$\left. \frac{dV_{eff}(\sigma)}{d\sigma} \right|_{\sigma=0} = 0 . \tag{6.107}$$

In terms of the more natural variable

$$w_0 \equiv v(4\lambda_0)^{1/2} , \tag{6.108}$$

(6.106), (6.107) and (6.74) give

$$0 = w_0 \left\{ w_0^2 + \mu_0^2 + 12\lambda_0 \mathscr{I}_1(\mu_0^2 + 3w_0^2) \right.$$

$$\left. +12\lambda_0 \mathscr{I}_1(\mu_0^2 + w_0^2) + \frac{9}{4}g^2 \mathscr{I}_1\left(\frac{g^2}{16\lambda_0}w_0^2\right) \right\} . \tag{6.109}$$

The lattice sums (or integrals) $\mathscr{I}_n$ are defined, in general, by

$$\mathscr{I}_n(\mu^2) \equiv \int_k (\mu^2 + \hat{k}^2)^{-n} \equiv J_n(\mu) . \tag{6.110}$$

The solutions of (6.109) are $w_0 = 0$, and $w_0 \neq 0$ such that the contents of the curly brackets vanish. The first solution is the absolute minimum of the effective potential in the symmetric (confinement) phase, the second one in the broken (Higgs) phase. The first order phase transition between the two phases occurs on the surface, where two solutions belong to equally deep minima of the effective potential.

It turns out that for weak gauge coupling, where (6.106) is relevant, the first order phase transition is weak, and at the phase transition the lattice spacing a is rather small (the correlation lengths are large), especially for larger values of the quartic coupling λ_0. In this case the bare perturbation theory is not reliable because of the appearance of logarithmic singularities of the loop integrals. Therefore (6.106) and (6.109) are not directly applicable. But if one considers the renormalized effective potential in a renormalization scheme, where the renormalized quantities are defined at zero four-momentum, and some non-zero constant value of the σ-field $\sigma(x) = a\bar{\sigma}$ (with lattice spacing a), one can improve the perturbative expansion of the effective potential by using the renormalization group [6.30]. In this way the perturbative expansion becomes applicable also at small lattice spacings (large cut-offs) near the phase transition. In particular, the definitions of renormalized scalar mass μ_R, renormalized quartic coupling λ_R, and scalar field renormalization factor Z_R are

$$\frac{\mu_R^2(a\bar{\sigma})}{Z_R(a\bar{\sigma})} \equiv \frac{\mathrm{d}^2 V_{eff}(\sigma = a\bar{\sigma})}{\mathrm{d}\sigma^2} , \qquad \frac{\lambda_R(a\bar{\sigma})}{Z_R(a\bar{\sigma})^2} \equiv \frac{\mathrm{d}^4 V_{eff}(\sigma = a\bar{\sigma})}{4! \, \mathrm{d}\sigma^4} ,$$

$$Z_R(a\bar{\sigma}) \equiv \frac{1}{8} \sum_{\mu=1}^{4} \frac{\partial^2}{\partial k_\mu \, \partial k_\mu}\bigg|_{k=0} \frac{1}{\Omega} \sum_{xy} e^{ik\cdot(x-y)}$$

$$\cdot \frac{\partial^2 \Gamma[A(x) = 0, \sigma(x) = a\bar{\sigma}, \pi(x) = 0]}{\partial\sigma(x) \, \partial\sigma(y)} . \tag{6.111}$$

The notations $\mu_R^2(a\bar{\sigma}), \ldots$ express the fact that these renormalized quantities depend on the scale introduced by the constant field $\bar{\sigma}$. The renormalization group equations applied by Coleman and Weinberg [6.30] describe the invariance of the effective potential under the change of $\bar{\sigma}$. The Callan–Symanzik functions that appear can be approximated for small couplings by the universal perturbative terms. (For instance, for the β-functions see the right hand sides in (6.93).)

If the renormalized effective potential is written in terms of the renormalized quantities $\mu_R^2(a\bar{\sigma})$, $\lambda_R(a\bar{\sigma})$, and some suitably defined scale dependent renormalized gauge coupling squared $g_W^2(a\bar{\sigma}) \equiv 4\pi\alpha_W(a\bar{\sigma})$, then one can apply the renormalization group equations, starting from the high scale where $a\bar{\sigma} = O(1)$, down to the physical scale $\bar{\sigma}/m_W = O(1)$. At the high scale the renormalized quantities are not much different from the bare ones (for instance, $\lambda_R = \lambda_0 + O(\lambda_0^2, g^2\lambda_0, g^4)$) and, as long as the renormalized couplings remain small, one can use the perturbative Callan–Symanzik functions. In this way one can obtain the effective potential at the physical (long distance, infrared) scale, which determines the properties of the phase transition. It turns out [6.31], that the first order phase transition is on the surface where

$$g_R \equiv 4!\lambda_R = g_{Rmin} = -\frac{27}{256\pi^2}g_W^4 \,, \qquad (6.112)$$

and in the Higgs phase we have $g_R \geq g_{Rmin}$. Since perturbation theory gives for small g_R and g_W^2 [6.31]

$$R_{HW}^2 \equiv \frac{m_H^2}{m_W^2} = \frac{4g_R}{3g_W^2} + \frac{9}{32\pi^2}g_W^2 + O(g_R^2g_W^{-2}, g_R, g_Rg_W^2, g_W^4) \,, \qquad (6.113)$$

(6.112) implies a minimum of the Higgs–W mass ratio [6.32], which is reached on the Higgs phase side of the phase transition:

$$R_{HW} \geq R_{HWmin} = \frac{3}{8\pi}g_W \simeq 0.08 \,. \qquad (6.114)$$

In a region where the W-mass in lattice units is not very small, and therefore $g^2 = g_W^2 + O(g_W^4)$ and $\lambda_R = \lambda_r + O(\lambda_r^2, \lambda_r g^2)$, the combination of (6.76) and the Dashen–Neuberger relation (6.98) gives

$$R_{HW}^2 = \frac{32\lambda_r}{g_W^2} + O(\lambda_r^2g_W^{-2}, \lambda_r, g_W^2) \,. \qquad (6.115)$$

Here λ_r is the renormalized quartic coupling in the $g^2 = 0$ limit. Taking for the renormalized gauge coupling the physical value $g_W^2 \simeq 0.5$, for low cut-offs one can use (6.115) for the non-perturbative determination of the cut-off dependent upper bound on R_{HW} [6.28]. The results obtained in the $O(4)$-symmetric ϕ^4 model confirm, to a good approximation, the upper bound (6.97) (see chapter 2).

The Dashen–Neuberger formula (6.98) is an important special case of more general relations, which can be derived from the *weak gauge coupling expansions* of the SU(2) Higgs model. Such expansions can be formulated both with gauge fixing [6.10], and in a gauge invariant way using the unitary gauge [6.33]. Here let us briefly discuss the former case. One starts by relating the lattice actions in the points with bare parameters (β, κ, λ) and $(\beta = \infty, \kappa, \lambda)$. For the scalar part of the action we take here the normalization convention in (6.14). The gauge fixing is performed as in pure gauge theory, that is the gauge fixing function is given by (6.55) without the term proportional to v. The complete gauge action can then be written as

$$S_g^{(0)}[A, \eta, \overline{\eta}] \equiv S_g[A] + S_m[A] + S_{gf}[A] + S_{FP}[A, \eta, \overline{\eta}] \ . \tag{6.116}$$

In this case we have

$$S[A, \sigma, \pi, \eta, \overline{\eta}] = S_\phi^{(0)}[\sigma, \pi] + S_g^{(0)}[A, \eta, \overline{\eta}] + 2\kappa \sum_x \sum_{\mu=1}^4 \left\{ J_{rx\mu}^{(L)} 2a_\mu^r(x) \right.$$

$$\left. + (\sigma_0(x + \hat{\mu})\sigma_0(x) + \pi_r(x + \hat{\mu})\pi_r(x))(1 - a_\mu^0(x)) \right\} \ , \tag{6.117}$$

where $S_\phi^{(0)}$ is the pure scalar action at the bare parameters (κ, λ), $J_{rx\mu}^{(L)}$ is the SU(2)$_L$ current defined by (6.100)–(6.102), and the notations in (6.48) are also used.

Our aim is to calculate the generating function of the connected Green functions

$$W[h, i] \equiv \log \frac{I[h, i]}{I[0, 0]} \ , \tag{6.118}$$

where

$$I[h, i] \equiv \int [\mathrm{d}A \, \mathrm{d}\sigma \, \mathrm{d}\pi \, \mathrm{d}\overline{\eta} \, \mathrm{d}\eta] \exp \left\{ -S[A, \sigma, \pi, \eta, \overline{\eta}] \right.$$

$$\left. + \sum_x \left[h_0(x)\sigma_0(x) + h_r(x)\pi_r(x) + \sum_{\mu=1}^4 i_\mu^r(x)A_\mu^r(x) \right] \right\} \ , \tag{6.119}$$

with the external sources $[h, i]$. Using the decomposition (6.117), this can obviously be written as

$$I[h, i] \equiv \int [\mathrm{d}\sigma \, \mathrm{d}\pi] \exp \left\{ -S_\phi^{(0)}[\sigma, \pi] + \sum_x [h_0(x)\sigma_0(x) + h_r(x)\pi_r(x)] \right\}$$

$$\cdot \int [\mathrm{d}A\,\mathrm{d}\bar{\eta}\,\mathrm{d}\eta]\exp\left\{-S_g^{(0)}[A,\eta,\bar{\eta}] + \sum_x \sum_{\mu=1}^4 \left[A_\mu^r(x)\left(i_\mu^r(x) - 2\kappa g J_{rx\mu}^{(L)}\right)\right.\right.$$

$$\left.\left. -\frac{\kappa g^2}{4}A_\mu^s(x)A_\mu^s(x)(\sigma_0(x+\hat{\mu})\sigma_0(x) + \pi_r(x+\hat{\mu})\pi_r(x)) + \cdots\right]\right\}. \quad (6.120)$$

The dots stand for terms of higher order in $gA_\mu^r(x)$, which arise from the expansions in (6.50). Performing the inner integrals over the gauge variables, one obtains an expression in terms of some connected Green functions in pure gauge theory at the given bare parameter β. Finally, the integral over the scalar degrees of freedom results in Green functions in pure ϕ^4 theory at the bare parameters (κ, λ), which contain, besides the elementary fields σ, π, also some composite currents like $J_{rx\mu}^{(L)}$ etc. [6.10]. These latter can be expressed in terms of the renormalized Green functions and field renormalization factors in ϕ^4. In this way one can explore our knowledge about the behaviour of the Green functions of the ϕ^4 theory, for instance, the triviality of the continuum limit.

A simple example of the application of the weak gauge coupling expansion is to determine the $O(g^2)$ correction to the critical hopping parameter, on the boundary separating the symmetric and broken phases. The result is [6.10]

$$\kappa_{cr}(\lambda, g^2) = \kappa_{cr}(\lambda, 0)\left[1 + \frac{9}{32}g^2 J_1(0) + \cdots\right]$$

$$= \kappa_{cr}(\lambda, 0)[1 + g^2 \cdot 0.04358\ldots + \cdots], \quad (6.121)$$

with J_1 defined in (6.110) (see also in section 2.2). This is valid for arbitrarily large λ, too. If $\lambda_0 \equiv \lambda/(4\kappa^2)$ is small, then from ordinary lattice perturbation theory we have the double expansion

$$\kappa_{cr}(\lambda_0, g^2) = \frac{1}{8} + \frac{\lambda_0}{64}[24J_1(0) - 1] + \frac{9}{256}g^2 J_1(0) + \cdots$$

$$= \frac{1}{8} + \lambda_0 \cdot 0.04247\ldots + g^2 \cdot 0.005446\ldots + \cdots. \quad (6.122)$$

6.2   Lattice Yukawa models

The masses of fermions (quarks and leptons) originate from the spontaneous symmetry breaking in the Higgs sector of the Standard Model. They are proportional to the vacuum expectation value of the scalar field, and the proportionality factors are the Yukawa couplings. Most quarks and leptons are very light on the scale of the vacuum expectation value

($v_R \simeq 250$ GeV), therefore the corresponding Yukawa couplings are very weak. An exception is the top quark, which has a Yukawa coupling of order one, and perhaps there exist some further heavy fermions with even stronger Yukawa couplings.

Strong Yukawa couplings can be studied non-perturbatively on the lattice. They are interesting on their own, but in addition, their feedback on the scalar sector can influence the triviality bound for the Higgs boson mass, and perhaps even the whole triviality issue. In fact, as we shall discuss below, in the case of a trivial continuum limit the existence of strong Yukawa couplings also implies a cut-off dependent lower bound on the Higgs boson mass.

6.2.1   Lattice actions

The theoretical challenge for the lattice formulation of the fermionic part of the electroweak sector is to implement the $SU(2)_L \otimes U(1)_Y$ local gauge symmetry. This is a chiral symmetry in the sense that the left-handed and right-handed components of the fermion fields transform differently, and the representation on the known particle states is not real. The difficulty arises due to the Nielsen–Ninomiya theorem (see section 4.4), which essentially tells us that in the fermion propagator there is an equal number of left-handed and right-handed states for every set of quantum numbers. In other words, on the lattice the spectrum of propagating states consists of fermion–mirror-fermion pairs (a parity transformation brings a left-handed fermion component into a right-handed one and vice versa).

In the naive discretization of the Dirac fermion action the mirror pairs are represented by the lattice fermion doublers: there are eight pairs of fermion states with opposite chirality. In QCD the 15 superfluous doubler states can be kept at the cut-off scale by adding the Wilson term to the action (see section 4.2), and in the continuum limit an appropriate tuning of the hopping parameter is enough for the description of the spontaneously broken chiral symmetry at zero quark masses (see section 5.3). At finite lattice spacings, however, the chiral symmetry of the lattice action is explicitly broken by the Wilson term. Therefore the Wilson fermions are not well suited for a formulation of the electroweak interactions, where the chiral symmetry is to be gauged, and should therefore not be broken. The same is true also for staggered fermions: according to section 4.3, at finite lattice spacings the exact symmetry of the massless action is $U(1)_o \otimes U(1)_e$. Two of the four fermion flavours described by the staggered fermion action transform non-trivially under $U(1)_o$ and trivially under $U(1)_e$, another two have opposite transformation properties. With respect to the chiral $U(1)$ subgroup these pairs are, however, mirror fermion pairs, because of the opposite eigenvalues of γ_5

(compensated by opposite eigenvalues of t_5 acting in flavour space, as shown by (4.179)). In addition, $U(1)_o \otimes U(1)_e$ can only be identified with some subgroup of $SU(2)_L \otimes U(1)_Y$, therefore the rest of the physical chiral symmetry is explicitly broken at finite lattice spacings also in the case of staggered fermions.

For the lattice formulation of chiral Yukawa models one can try to modify the Wilson fermion action in such a way that chiral invariance is restored by the insertion of the Higgs scalar field at appropriate places [6.34, 6.35]. However, there is both analytical and numerical evidence that such models do not have an appropriate continuum limit (for a review see [6.5]).

Another, more radical, possibility to generalize the Wilson fermion action for chiral interactions is to introduce, besides the fermion field $(\psi_x, \overline{\psi}_x)$, also a *mirror fermion field* $(\chi_x, \overline{\chi}_x)$, and write the Wilson term as an off-diagonal ψ–χ mixing term. With a general field normalization the $SU(2)_L \otimes SU(2)_R$ symmetric lattice action becomes [6.36]

$$S[\varphi, \psi, \overline{\psi}, \chi, \overline{\chi}] = S_\varphi[\varphi] + S_{\varphi\psi\chi}[\varphi, \psi, \overline{\psi}, \chi, \overline{\chi}] \,,$$

$$S_\varphi[\varphi] = \sum_x \left\{ \frac{\mu_\phi}{2} \mathrm{Tr}\,(\varphi_x^+ \varphi_x) + \lambda \left[\frac{1}{2} \mathrm{Tr}\,(\varphi_x^+ \varphi_x) \right]^2 - \kappa \sum_{\mu=1}^{4} \mathrm{Tr}\,(\varphi_{x+\hat{\mu}}^+ \varphi_x) \right\} \,,$$

$$\tag{6.123}$$

where the fermionic part is

$$S_{\varphi\psi\chi}[\varphi, \psi, \overline{\psi}, \chi, \overline{\chi}] = \sum_x \left\{ \mu_{\psi\chi} \left[(\overline{\chi}_x \psi_x) + (\overline{\psi}_x \chi_x) \right] \right.$$

$$- \sum_{\mu=\pm 1}^{\pm 4} \left[K_\psi (\overline{\psi}_{x+\hat{\mu}} \gamma_\mu \psi_x) + K_\chi (\overline{\chi}_{x+\hat{\mu}} \gamma_\mu \chi_x) \right]$$

$$+ K_r \sum_{\mu=\pm 1}^{\pm 4} \left[(\overline{\chi}_x \psi_x) - (\overline{\chi}_{x+\hat{\mu}} \psi_x) + (\overline{\psi}_x \chi_x) - (\overline{\psi}_{x+\hat{\mu}} \chi_x) \right]$$

$$+ G_\psi \left[(\overline{\psi}_{Rx} \varphi_x^+ \psi_{Lx}) + (\overline{\psi}_{Lx} \varphi_x \psi_{Rx}) \right]$$

$$\left. + G_\chi \left[(\overline{\chi}_{Rx} \varphi_x \chi_{Lx}) + (\overline{\chi}_{Lx} \varphi_x^+ \chi_{Rx}) \right] \right\} \,.$$

$$\tag{6.124}$$

With respect to $SU(2)_L \otimes SU(2)_R$ the scalar field is transformed according to (6.6), and the fermion field as

$$\psi'_{Lx} = U_L^{-1} \psi_{Lx} \,, \quad \psi'_{Rx} = U_R^{-1} \psi_{Rx} \,,$$

$$\overline{\psi}'_{Lx} = \overline{\psi}_{Lx} U_L , \quad \overline{\psi}'_{Rx} = \overline{\psi}_{Rx} U_R . \tag{6.125}$$

The $SU(2)_L \otimes SU(2)_R$ transformations of the mirror fermion fields are

$$\chi'_{Lx} = U_R^{-1} \chi_{Lx} , \quad \chi'_{Rx} = U_L^{-1} \chi_{Rx} ,$$

$$\overline{\chi}'_{Lx} = \overline{\chi}_{Lx} U_R , \quad \overline{\chi}'_{Rx} = \overline{\chi}_{Rx} U_L . \tag{6.126}$$

This implies that the action (6.124) is exactly chiral symmetric.

The parameter $\mu_{\psi\chi}$ in the $SU(2)_L \otimes SU(2)_R$ symmetric action (6.124) for a mirror pair of fermion doublets is an off-diagonal ψ–χ mixing mass, which is allowed by the symmetry. K_ψ and K_χ are the hopping parameters for the ψ- (fermion), respectively, χ- (mirror-fermion) field, K_r is the bare parameter multiplying the off-diagonal Wilson term, and G_ψ and G_χ are the chiral Yukawa couplings of the fermion, respectively, mirror fermion. A traditional normalization in lattice perturbation theory is to fix

$$K_\psi = K_\chi = \frac{1}{2} , \qquad K_r \equiv \frac{r}{2} , \tag{6.127}$$

whereas in numerical simulations it is convenient to take

$$K_\psi = K_\chi \equiv K , \quad K_r \equiv rK , \quad \bar{\mu} \equiv \mu_{\psi\chi} + 8rK = 1 . \tag{6.128}$$

An important property of the action (6.124) is that by the extension of the proofs for pure scalar ϕ^4 models (chapter 2), and for free Wilson fermions (section 4.2), reflection positivity can be proven in some parts of the parameter space, in particular for non-negative scalar hopping parameters $\kappa \geq 0$, with respect to link-reflection [6.37]. This ensures the unitarity of the corresponding quantum field theory in Minkowski space–time.

A simplified version of the lattice action (6.124) is obtained if, instead of the fermion doublets, only singlets are considered. In this case the chiral symmetry is $U(1)_L \otimes U(1)_R$, and the scalar field has only two real (one complex) components. The action has the same form as (6.124), only the Yukawa coupling terms have to be replaced by [6.38, 6.39, 6.37]

$$\cdots + G_\psi(\overline{\psi}_x[\phi_{1x} + i\gamma_5\phi_{2x}]\psi_x) + G_\chi(\overline{\chi}_x[\phi_{1x} - i\gamma_5\phi_{2x}]\chi_x) . \tag{6.129}$$

The idea of explicitly introducing the mirror fermion fields into the lattice action is only justified if, by tuning the parameters one can, at least approximately, decouple the superfluous degrees of freedom, and describe the electroweak Standard Model to a good approximation. It is, however, plausible that if the bare parameters of the mirror fermions are introduced explicitly, then this tuning can be done more directly. Otherwise it is presumably more difficult to control the mirror doublers, which appear not only kinematically in the propagator, but can also be dynamically produced as composite mirror fermion states by a strong Yukawa coupling.

Such a dynamical fermion doubling occurs, for instance, in the lattice σ-model with (non-chiral) Wilson fermions at infinitely strong bare Yukawa coupling, as shown by the hopping parameter expansion [6.40]. Another example of the dynamical generation of composite mirror fermion states was observed in the model of [6.34], in numerical simulations of the symmetric phase at strong Yukawa couplings [6.41].

A first insight into the $SU(2)_L \otimes SU(2)_R$ or $U(1)_L \otimes U(1)_R$ symmetric chiral Yukawa model defined by the actions (6.124), (6.129) can be obtained at small couplings in lattice perturbation theory [6.39, 6.37, 6.42]. In the symmetric phase the momentum space fermion propagator is, in the notations of (8.40),

$$\tilde{\Delta}_p = (\bar{p}^2 + \mu_p^2)^{-1} \begin{pmatrix} -i\gamma \cdot \bar{p} & \mu_p \\ \mu_p & -i\gamma \cdot \bar{p} \end{pmatrix}, \tag{6.130}$$

where the $2 \otimes 2$ matrix acts in the ψ–χ space, and the momentum dependent mass in lattice units μ_p is, with $K \equiv K_\psi = K_\chi$,

$$\mu_p \equiv \frac{1}{2K}(\mu_{\psi\chi} + K_r \hat{p}^2) \longrightarrow \mu_{\psi\chi} + \frac{r}{2}\hat{p}^2. \tag{6.131}$$

The second form is valid in the perturbative normalization (6.127). Since μ_p does not vanish at the corners of the Brillouin zone different from $p = 0$, in the continuum limit the doublers of both the fermion and the mirror fermion are heavy. The light fermion spectrum consists of a single mirror pair of fermions, corresponding to a minimal doubling of the physical fermion states.

The fermion mass matrix M can be defined as the inverse propagator at zero four-momentum, therefore in the symmetric phase (6.130) gives

$$M = \begin{pmatrix} 0 & \mu_{\psi\chi} \\ \mu_{\psi\chi} & 0 \end{pmatrix}. \tag{6.132}$$

This can be diagonalized by an orthogonal transformation

$$\bar{O}^+ M O = \begin{pmatrix} \mu_{\psi\chi} & 0 \\ 0 & \mu_{\psi\chi} \end{pmatrix},$$

$$O = \frac{1}{\sqrt{2}} \begin{pmatrix} 1 & \gamma_5 \\ 1 & -\gamma_5 \end{pmatrix}, \qquad \bar{O}^+ = \frac{1}{\sqrt{2}} \begin{pmatrix} 1 & 1 \\ -\gamma_5 & \gamma_5 \end{pmatrix}. \tag{6.133}$$

This shows that in the symmetric phase the masses of the fermion and mirror fermion are equal. The degeneracy implies that the diagonalization of the fermion mass matrix is not unique. Another interesting way of diagonalization can be achieved by

$$\bar{U}^+ M U = \begin{pmatrix} \mu_{\psi\chi} & 0 \\ 0 & \mu_{\psi\chi} \end{pmatrix},$$

$$U = \begin{pmatrix} P_L & P_R \\ P_R & P_L \end{pmatrix}, \qquad \bar{U}^+ = \begin{pmatrix} P_R & P_L \\ P_L & P_R \end{pmatrix}. \tag{6.134}$$

Here $P_{L,R}$ are the chiral projectors in (8.11).

This way of diagonalization means that the left- and right-handed components of ψ and χ are reshuffled, in order to obtain a vector-like kinetic part. Sometimes it is advantageous to do this directly in the action, especially for applications in the symmetric phase (for instance, in the hopping parameter expansion [6.43].) The new fields are

$$\psi_{Ax} \equiv \psi_{Lx} + \chi_{Rx}, \qquad \overline{\psi}_{Ax} \equiv \overline{\psi}_{Lx} + \overline{\chi}_{Rx},$$

$$\psi_{Bx} \equiv \chi_{Lx} + \psi_{Rx}, \qquad \overline{\psi}_{Bx} \equiv \overline{\chi}_{Lx} + \overline{\psi}_{Rx}. \tag{6.135}$$

Denoting the (ψ_A, ψ_B)-pair by $\Psi_x \equiv (\psi_{Ax}, \psi_{Bx})$, and using the field normalization condition in (6.128), $\bar{\mu} = 1$, the fermion matrix Q in the fermionic part of the action

$$S_{\varphi\psi\chi} = \sum_{x,y} \overline{\Psi}_y Q(\varphi)_{yx} \Psi_x, \tag{6.136}$$

becomes

$$Q(\varphi)_{yx} = \begin{pmatrix} \delta_{yx} - K \sum_\mu \delta_{y,x+\hat{\mu}}(\gamma_\mu + r) & \delta_{yx}\varphi_x(G_\alpha + \gamma_5 G_\beta) \\ \delta_{yx}\varphi_x^+(G_\alpha - \gamma_5 G_\beta) & \delta_{yx} - K \sum_\mu \delta_{y,x+\hat{\mu}}(\gamma_\mu + r) \end{pmatrix}. \tag{6.137}$$

The coupling constant combinations $G_{\alpha,\beta}$ are defined as

$$G_\alpha \equiv \frac{1}{2}(G_\psi + G_\chi), \qquad G_\beta \equiv \frac{1}{2}(G_\psi - G_\chi). \tag{6.138}$$

The result of the transformation (6.135) is that the fermion action without interaction becomes diagonal in ψ_A, ψ_B, and equals the free Wilson action. The dependence on chirality is only manifested in the Yukawa couplings. The main interest in Yukawa models with explicit mirror fermions is, however, in the phase with broken symmetry, where the two fermion masses are generally different, and the unique way of fermion mass diagonalization is a generalization of (6.133). Therefore, in the broken phase the original form of the action in (6.124) is more adequate.

In the broken phase the momentum space fermion propagator is, in the perturbative normalization (6.127),

$$\tilde{\Delta}_p = \tilde{D}(p)^{-1} \begin{pmatrix} (\bar{p}^2 + \mu_\chi^2)\mu_\psi - \mu_p^2\mu_\chi & (\bar{p}^2 + \mu_p^2 - \mu_\psi\mu_\chi)\mu_p \\ (\bar{p}^2 + \mu_p^2 - \mu_\psi\mu_\chi)\mu_p & (\bar{p}^2 + \mu_\psi^2)\mu_\chi - \mu_p^2\mu_\psi \end{pmatrix}$$

$$- i\gamma \cdot \bar{p}\tilde{D}(p)^{-1} \begin{pmatrix} \bar{p}^2 + \mu_p^2 + \mu_\chi^2 & -(\mu_\psi + \mu_\chi)\mu_p \\ -(\mu_\psi + \mu_\chi)\mu_p & \bar{p}^2 + \mu_p^2 + \mu_\psi^2 \end{pmatrix}, \tag{6.139}$$

with

$$\tilde{D}(p) = (\bar{p}^2 + \mu_p^2 - \mu_\psi \mu_\chi)^2 + \bar{p}^2 (\mu_\psi + \mu_\chi)^2 , \qquad (6.140)$$

and

$$\mu_\psi \equiv G_\psi v , \qquad\qquad \mu_\chi \equiv G_\chi v , \qquad\qquad (6.141)$$

where v is the vacuum expectation value of the scalar field in lattice units.

The mass matrix, that is the inverse propagator at zero four momentum, is in the broken phase

$$M = \begin{pmatrix} \mu_\psi & \mu_{\psi\chi} \\ \mu_{\psi\chi} & \mu_\chi \end{pmatrix} . \qquad (6.142)$$

This can be diagonalized by a rotation in the ψ–χ space with an angle α satisfying

$$\sin\alpha = \mu_{\psi\chi}\sqrt{2}\Big\{ (\mu_\psi - \mu_\chi)^2 + 4\mu_{\psi\chi}^2$$

$$+ |\mu_\psi - \mu_\chi| \left[(\mu_\psi - \mu_\chi)^2 + 4\mu_{\psi\chi}^2 \right]^{1/2} \Big\}^{-1/2} , \qquad (6.143)$$

and the mass eigenvalues are

$$\mu_{1,2} = \frac{1}{2} \left\{ \mu_\psi + \mu_\chi \pm \left[(\mu_\psi - \mu_\chi)^2 + 4\mu_{\psi\chi}^2 \right]^{1/2} \right\} . \qquad (6.144)$$

Therefore, in the broken phase the ψ- and χ-states are mixed in general, and there are two different fermion masses. In the special case $\mu_\chi \gg \mu_\psi, \mu_{\psi\chi}$ the mixing is very small, and the mirror fermion is much heavier than the fermion. Such a situation is consistent with the present phenomenological knowledge about electroweak interactions [6.44], therefore it can also be considered a sufficiently good approximation of a minimal Standard Model without mirror fermions.

Of course, in order to describe the Standard Model really one has to introduce mass splittings within the doublets, the $SU(2)_L \otimes U(1)_Y$ symmetry has to be gauged and, finally, the fermions have to come in a multiplicity corresponding to (at least) three generations. In each fermion generation there is a lepton doublet and three equal mass quark doublets corresponding to three colours.

The mass splitting within doublets can be easily achieved, if in the SU(2) doublet space the Yukawa couplings are represented by diagonal matrices like, for instance,

$$G_\psi = \begin{pmatrix} G_{\psi u} & 0 \\ 0 & G_{\psi d} \end{pmatrix} . \qquad (6.145)$$

The corresponding Yukawa coupling term in (6.124) becomes

$$\cdots + \left[(\overline{\psi}_{Rx} G_\psi \varphi_x^+ \psi_{Lx}) + (\overline{\psi}_{Lx} \varphi_x G_\psi \psi_{Rx}) \right] \cdots . \qquad (6.146)$$

In this case the global symmetry is reduced from $SU(2)_L \otimes SU(2)_R$ to $SU(2)_L \otimes U(1)_{R3}$, where $U(1)_{R3}$ is generated by the third component of the right-handed isospin. (Note that, besides the Yukawa couplings, also some other bare parameters can depend on the doublet indices, provided that the $SU(2)_L \otimes U(1)_{R3}$ symmetry is preserved.) In addition to $SU(2)_L \otimes U(1)_{R3}$, there is always a vector-like, isospin singlet $U(1)_F$ symmetry for fermion number conservation, which corresponds to the global symmetry transformations

$$\varphi'_x = \varphi_x , \quad \psi'_x = e^{-i\alpha}\psi_x , \quad \chi'_x = e^{-i\alpha}\chi_x ,$$

$$\overline{\psi}'_x = \overline{\psi}_x e^{i\alpha} , \quad \overline{\chi}'_x = \overline{\chi}_x e^{i\alpha} . \tag{6.147}$$

The physical $U(1)_Y$ symmetry is a combination of $U(1)_{R3}$ and $U(1)_F$ given, in accordance with (6.42), by the definition of the *weak hypercharge*

$$Y = B - L + 2T_{R3} . \tag{6.148}$$

The $(B-L)$ quantum number is proportional to the fermion number. For quark doublets it is $1/3$, and for lepton doublets -1.

The gauging of the exact global $SU(2)_L \otimes SU(2)_R$ symmetry in the lattice action (6.124) is straightforward. One has only to insert the gauge link variables $U_{Lx\mu} \in SU(2)_L$ and $U_{Rx\mu} \in SU(2)_R$ if necessary, namely if the scalar or fermion fields are not at the same lattice points. For instance, the scalar hopping term becomes (6.39), and the naive fermion kinetic term in (6.124) has to be replaced by

$$\cdots - K_\psi \left[(\overline{\psi}_{Lx+\hat{\mu}}\gamma_\mu U_{Lx\mu}\psi_{Lx}) + (\overline{\psi}_{Rx+\hat{\mu}}\gamma_\mu U_{Rx\mu}\psi_{Rx}) \right]$$

$$- K_\chi \left[(\overline{\chi}_{Lx+\hat{\mu}}\gamma_\mu U_{Rx\mu}\chi_{Lx}) + (\overline{\chi}_{Rx+\hat{\mu}}\gamma_\mu U_{Lx\mu}\chi_{Rx}) \right] \cdots . \tag{6.149}$$

The gauging of $SU(2)_L \otimes U(1)_Y$ can be obtained from this on the basis of the definition of the weak hypercharge Y in (6.148). This means that $U_{L,Rx\mu}$ has to be replaced everywhere by

$$U_{Lx\mu} \longrightarrow U_{Lx\mu}U_{Yx\mu} , \qquad U_{Rx\mu} \longrightarrow U_{Yx\mu} . \tag{6.150}$$

Here the $U(1)_Y$ link variable $U_{Yx\mu}$ is defined by

$$U_{Yx\mu} \equiv e^{+iY\theta_{x\mu}} , \tag{6.151}$$

where Y is understood as a $2 \otimes 2$ diagonal matrix in the isospin (doublet) space, with elements given by (6.148).

Up to now only gauge invariant lattice formulations of the $SU(2)_L \otimes U(1)_Y$ symmetric electroweak interactions were considered. If explicit gauge invariance is abandoned, one can also formulate chiral lattice gauge theories without introducing mirror fermion fields [6.45]. The consistency of this approach is, however, not known at present.

6.2.2 *The Golterman–Petcher theorem*

The lattice formulation of the electroweak interactions of lepton doublets is particularly subtle. The right-handed components of the neutrinos are singlets with respect to the $SU(3) \otimes SU(2)_L \otimes U(1)_Y$ gauge interactions, therefore they can be coupled to some other physical particle only through their masses and Yukawa couplings. In the case of a massless neutrino the right-handed component is totally decoupled, and in the continuum can also be completely omitted from a minimal Standard Model. On the lattice the decoupling of massless right-handed neutrinos can be achieved in the continuum limit, if the premises of the Golterman–Petcher theorem [6.46] are fulfilled.

In order to formulate the Golterman–Petcher theorem, let us first consider in more detail the lattice action of a lepton doublet (v, l) (v is the neutrino, $l = e^-, \mu^-$ or τ^- is the corresponding negative charged lepton). As discussed in the previous subsection, besides the lepton doublet field $\psi_x \equiv (\psi_{vx}, \psi_{lx})$ one also needs the mirror doublet field $\chi_x \equiv (\chi_{vx}, \chi_{lx})$, which is assumed to become very heavy in the continuum limit. We shall work here in the normalization convention (6.127), and on the basis of (6.124), (6.146) write the fermionic part of the lattice action as

$$S_{\varphi\psi\chi} \equiv S^{(1)}_{\varphi\psi\chi} + S^{(2)}_{\varphi\psi\chi} \,,$$

$$S^{(1)}_{\varphi\psi\chi} = \sum_x \sum_{\mu=\pm 1}^{\pm 4} \left\{ -\frac{1}{2} \left[(\overline{\psi}_{x+\hat{\mu}} \gamma_\mu \psi_x) + (\overline{\chi}_{x+\hat{\mu}} \gamma_\mu \chi_x) \right] \right.$$

$$\left. + \frac{r}{2} \left[(\overline{\chi}_x \psi_x) - (\overline{\chi}_{x+\hat{\mu}} \psi_x) + (\overline{\psi}_x \chi_x) - (\overline{\psi}_{x+\hat{\mu}} \chi_x) \right] \right\} \,,$$

$$S^{(2)}_{\varphi\psi\chi} = \sum_x \left\{ G_{\psi v} \left[\left(\overline{\psi}_{Rvx} (\tilde{\Phi}^+_x \psi_{Lx}) \right) + \left((\overline{\psi}_{Lx} \tilde{\Phi}_x) \psi_{Rvx} \right) \right] \right.$$

$$+ G_{\psi l} \left[\left(\overline{\psi}_{Rlx} (\Phi^+_x \psi_{Lx}) \right) + \left((\overline{\psi}_{Lx} \Phi_x) \psi_{Rlx} \right) \right]$$

$$+ G_{\chi v} \left[\left(\overline{\chi}_{Lvx} (\tilde{\Phi}^+_x \chi_{Rx}) \right) + \left((\overline{\chi}_{Rx} \tilde{\Phi}_x) \chi_{Lvx} \right) \right]$$

$$+ G_{\chi l} \left[\left(\overline{\chi}_{Llx} (\Phi^+_x \chi_{Rx}) \right) + \left((\overline{\chi}_{Rx} \Phi_x) \chi_{Llx} \right) \right]$$

$$+ \mu_L \left[(\overline{\chi}_{Rvx} \psi_{Lvx}) + (\overline{\psi}_{Lvx} \chi_{Rvx}) + (\overline{\chi}_{Rlx} \psi_{Llx}) + (\overline{\psi}_{Llx} \chi_{Rlx}) \right]$$

$$+ \mu_{Rv} \left[(\overline{\chi}_{Lvx} \psi_{Rvx}) + (\overline{\psi}_{Rvx} \chi_{Lvx}) \right] + \mu_{Rl} \left[(\overline{\chi}_{Llx} \psi_{Rlx}) + (\overline{\psi}_{Rlx} \chi_{Llx}) \right] \right\} \,. \quad (6.152)$$

Here in the second part containing the Yukawa couplings and the fermion–mirror-fermion mixing mass terms the chiral components, as defined in

(8.12), are written out separately, and the Higgs doublet fields Φ_x and $\tilde{\Phi}_x$ defined in (6.2) are used. The Yukawa couplings for v and l are different, as in (6.145), and a splitting within the doublet, which respects the $SU(2)_L \otimes U(1)_Y$ symmetry, is also allowed in the mixing mass terms.

The Golterman–Petcher theorem for the decoupling of v_R holds, if the lattice action is symmetric with respect to the x-independent fermionic shift symmetry

$$\psi'_{Rvx} = \psi_{Rvx} - \epsilon \, , \qquad \overline{\psi}'_{Rvx} = \overline{\psi}_{Rvx} - \overline{\epsilon} \, , \qquad (6.153)$$

where ϵ and $\overline{\epsilon}$ are Grassmann variables. In (6.152) this is true for

$$G_{\psi v} = \mu_{Rv} = 0 \, , \qquad (6.154)$$

when the fields ψ_{Rvx} and $\overline{\psi}_{Rvx}$ only appear with lattice derivatives acting on them.

In order to describe the consequences of the *Golterman–Petcher shift symmetry* (6.153), let us consider the generating function

$$Z[\eta, \overline{\eta}, \zeta, \overline{\zeta}, j^+] \equiv Z[J] \equiv \int [d\Phi \, d\overline{\psi} \, d\psi \, d\overline{\chi} \, d\chi] \exp \left\{ - S[\Phi, \psi, \overline{\psi}, \chi, \overline{\chi}] \right.$$

$$\left. + \sum_x \left[(\overline{\eta}_x \psi_x) - (\overline{\psi}_x \eta_x) + (\overline{\zeta}_x \chi_x) - (\overline{\chi}_x \zeta_x) + (j_x^+ \Phi_x) \right] \right\} \, , \qquad (6.155)$$

and vary it, similarly to the derivation of Ward–Takahashi identities, by making the shift parameters space–time dependent: $\epsilon \to \epsilon_x$, $\overline{\epsilon} \to \overline{\epsilon}_x$. It is easy to see that, according to the definitions in section 4.1, the Grassmann integrals are invariant with respect to such shifts, therefore from the term linear in ϵ_x one obtains

$$0 = \int [d\Phi \, d\overline{\psi} \, d\psi \, d\overline{\chi} \, d\chi] \exp \left\{ -S + \sum_x \left[(\overline{\eta}_x \psi_x) - \cdots + (j_x^+ \Phi_x) \right] \right\}$$

$$\cdot \left[\overline{\eta}_{Lvx} + \frac{1}{2} \sum_{\mu=1}^4 (\Delta_\mu^f + \Delta_\mu^b) \overline{\psi}_{Rvx} \gamma_\mu + \frac{r}{2} \Delta \overline{\chi}_{Lvx} \right] \, . \qquad (6.156)$$

Similarly, the terms linear in $\overline{\epsilon}_x$ give

$$0 = \int [d\Phi \, d\overline{\psi} \, d\psi \, d\overline{\chi} \, d\chi] \exp \left\{ -S + \sum_x \left[(\overline{\eta}_x \psi_x) - \cdots + (j_x^+ \Phi_x) \right] \right\}$$

$$\cdot \left[-\eta_{Lvx} - \frac{1}{2} \sum_{\mu=1}^4 \gamma_\mu (\Delta_\mu^f + \Delta_\mu^b) \psi_{Rvx} + \frac{r}{2} \Delta \chi_{Lvx} \right] \, . \qquad (6.157)$$

One defines the generating function of one particle irreducible vertices as the Legendre transform of $W \equiv \log Z$:

$$\Gamma[\Phi_e, \psi_e, \overline{\psi}_e, \chi_e, \overline{\chi}_e] \equiv -W[\eta, \overline{\eta}, \zeta, \overline{\zeta}, j^+]$$

$$+ \sum_x \left[(\overline{\eta}_x \psi_{ex}) - (\overline{\psi}_{ex} \eta_x) + (\overline{\zeta}_x \chi_{ex}) - (\overline{\chi}_{ex} \zeta_x) + (j_x^+ \Phi_{ex}) \right] . \qquad (6.158)$$

Here on the right hand side the 'external currents' $J \equiv (\eta, \overline{\eta}, \zeta, \overline{\zeta}, j^+)$ are considered as functions of the 'field expectation values' $(\Phi_e, \psi_e, \overline{\psi}_e, \chi_e, \overline{\chi}_e)$. The relation $J = J(\Phi_e, \psi_e, \overline{\psi}_e, \chi_e, \overline{\chi}_e)$ is determined by the following set of equations:

$$\Phi_{ex} = \frac{\partial W}{\partial j_x^+} , \qquad \psi_{ex} = \frac{\partial W}{\partial \overline{\eta}_x} , \qquad \overline{\psi}_{ex} = \frac{\partial W}{\partial \eta_x} ,$$

$$\chi_{ex} = \frac{\partial W}{\partial \overline{\zeta}_x} , \qquad \overline{\chi}_{ex} = \frac{\partial W}{\partial \zeta_x} . \qquad (6.159)$$

Taking care of the anticommutativity of the Grassmann variables, the relations (6.156) and (6.157) become in terms of Γ

$$0 = -\frac{\partial \Gamma}{\partial \psi_{eRvx}} + \frac{1}{2} \sum_{\mu=1}^{4} (\Delta_\mu^f + \Delta_\mu^b) \overline{\psi}_{eRvx} \gamma_\mu + \frac{r}{2} \Delta \overline{\chi}_{eLvx} ,$$

$$0 = \frac{\partial \Gamma}{\partial \overline{\psi}_{eRvx}} - \frac{1}{2} \sum_{\mu=1}^{4} \gamma_\mu (\Delta_\mu^f + \Delta_\mu^b) \psi_{eRvx} + \frac{r}{2} \Delta \chi_{eLvx} . \qquad (6.160)$$

These remarkable identities express the consequences of the invariance of the action with respect to the Golterman–Petcher shift symmetry (6.153), which holds if (6.154) is satisfied. For the inverse propagator of ν_R (6.160) implies, with (8.40),

$$\sum_y e^{-ik \cdot (y-x)} \frac{\partial^2 \Gamma}{\partial \psi_{eRvx} \partial \overline{\psi}_{eRvy}} = i\gamma \cdot \overline{k} ,$$

$$\sum_y e^{-ik \cdot (y-x)} \frac{\partial^2 \Gamma}{\partial \chi_{eLvx} \partial \overline{\psi}_{eRvy}}$$

$$= \sum_y e^{-ik \cdot (y-x)} \frac{\partial^2 \Gamma}{\partial \psi_{eRvx} \partial \overline{\chi}_{eLvy}} = \frac{r}{2} \hat{k}^2 . \qquad (6.161)$$

This means that the ν_R-components of the inverse propagator are equal to the corresponding components of the free lattice inverse propagator. In the continuum limit $a \to 0$, when $\overline{k}_\mu, \hat{k}_\mu \to a p_\mu$, the ψ_{Rv}-$\overline{\psi}_{Rv}$ component tends to a free massless inverse propagator, and the $\chi_{Lv}-\overline{\psi}_{Rv}$ and $\psi_{Rv}-\overline{\chi}_{Lv}$ components become negligible, because they are of the order a^2. Moreover, taking higher derivatives of (6.160), it follows that every one-particle irreducible vertex of order three or higher, which involves ψ_{Rv} and/or $\overline{\psi}_{Rv}$, vanishes identically. Therefore in the continuum limit the fields ψ_{Rv} and $\overline{\psi}_{Rv}$ describe a free massless v_R state, which completely decouples from the rest of the physical fields.

It is easy to see that the above derivation is not affected by the inclusion of the $SU(2)_L \otimes U(1)_Y$ gauge interaction. Namely, ψ_{Rv} and $\overline{\psi}_{Rv}$ are singlets with respect to $SU(2)_L \otimes U(1)_Y$, therefore the derivative terms in $S^{(1)}_{\varphi\psi\chi}$ in (6.152), which contain ψ_{Rv} and/or $\overline{\psi}_{Rv}$, are unchanged after switching on the gauge interactions, and the identities in (6.160) remain valid. This is the *Golterman–Petcher theorem: a lattice action satisfying the global fermionic shift symmetry, with respect to the transformations (6.153), describes in the continuum limit a massless right-handed neutrino, which completely decouples from the interacting physical states.*

It is clear that the decoupling holds for every fermion field component which satisfies a shift symmetry like (6.153). Without gauge interactions this can also be used for a complete decoupling of the mirror fermion states. For instance, the lattice action (6.124) describing the Yukawa interaction of a mirror pair of fermion doublets with the Higgs boson field φ_x is symmetric with respect to

$$\chi'_x = \chi_x - \epsilon, \qquad \overline{\chi}'_x = \overline{\chi}_x - \overline{\epsilon}, \qquad (6.162)$$

if

$$\mu_{\psi\chi} = G_\chi = 0. \qquad (6.163)$$

Therefore, in this case in the continuum limit the massless mirror fermions described by $\chi_x, \overline{\chi}_x$ are completely decoupled from the fermions described by $\psi_x, \overline{\psi}_x$ [6.4, 6.42]. Of course, this decoupling does not work if the $SU(2)_L \otimes U(1)_Y$ gauge interactions are switched on, because the members of the mirror doublet are not singlets with respect to $SU(2)_L \otimes U(1)_Y$.

Let us now briefly discuss the neutrino, if it has a small non-zero mass. In this case the right-handed component is also physical, and the neutrino states can mix a little with the heavy mirror neutrino. The non-zero neutrino masses are produced by $\mu_L, \mu_{Rv} \neq 0$ and, in the phase with spontaneously broken symmetry, by

$$\mu_\psi \equiv G_{\psi v} v, \qquad \mu_\chi \equiv G_{\chi v} v, \qquad (6.164)$$

with $v = \langle \Phi_{2x} \rangle = \langle \widetilde{\Phi}_{1x} \rangle$ being the vacuum expectation value of the scalar field. The neutrino mass terms in (6.152) are

$$\cdots + \mu_\psi \left[(\overline{\psi}_{Rvx} \psi_{Lvx}) + (\overline{\psi}_{Lvx} \psi_{Rvx}) \right] + \mu_\chi \left[(\overline{\chi}_{Rvx} \chi_{Lvx}) + (\overline{\chi}_{Lvx} \chi_{Rvx}) \right]$$

$$+ \mu_L \left[(\bar{\chi}_{Rvx}\psi_{Lvx}) + (\bar{\psi}_{Lvx}\chi_{Rvx})\right] + \mu_{Rv}\left[(\bar{\chi}_{Lvx}\psi_{Rvx}) + (\bar{\psi}_{Rvx}\chi_{Lvx})\right] . \quad (6.165)$$

Therefore the neutrino mass matrix on the $(\bar{\psi}_{Rv}, \bar{\psi}_{Lv}, \bar{\chi}_{Rv}, \bar{\chi}_{Lv}) \otimes$
$(\psi_{Lv}, \psi_{Rv}, \chi_{Lv}, \chi_{Rv})$ basis is

$$M_v = \begin{pmatrix} \mu_\psi & 0 & \mu_{Rv} & 0 \\ 0 & \mu_\psi & 0 & \mu_L \\ \mu_L & 0 & \mu_\chi & 0 \\ 0 & \mu_{Rv} & 0 & \mu_\chi \end{pmatrix} . \quad (6.166)$$

For $\mu_{Rv} \neq \mu_L$ this is not symmetric, hence one has to diagonalize $M_v^T M_v$
by $O^T M_v^T M_v O$, and $M_v M_v^T$ by $\bar{O}^T M_v M_v^T \bar{O}$, where

$$O = \begin{pmatrix} \cos\alpha_L & 0 & \sin\alpha_L & 0 \\ 0 & \cos\alpha_R & 0 & \sin\alpha_R \\ -\sin\alpha_L & 0 & \cos\alpha_L & 0 \\ 0 & -\sin\alpha_R & 0 & \cos\alpha_R \end{pmatrix} ,$$

$$\bar{O} = \begin{pmatrix} \cos\alpha_R & 0 & \sin\alpha_R & 0 \\ 0 & \cos\alpha_L & 0 & \sin\alpha_L \\ -\sin\alpha_R & 0 & \cos\alpha_R & 0 \\ 0 & -\sin\alpha_L & 0 & \cos\alpha_L \end{pmatrix} . \quad (6.167)$$

The rotation angles of the left-handed and right-handed components
satisfy

$$\tan(2\alpha_L) = \frac{2(\mu_\chi\mu_L + \mu_\psi\mu_{Rv})}{\mu_\chi^2 + \mu_{Rv}^2 - \mu_\psi^2 - \mu_L^2} ,$$

$$\tan(2\alpha_R) = \frac{2(\mu_\chi\mu_{Rv} + \mu_\psi\mu_L)}{\mu_\chi^2 + \mu_L^2 - \mu_\psi^2 - \mu_{Rv}^2} , \quad (6.168)$$

respectively, and the two (positive) mass-squared eigenvalues are given by

$$\mu_{(v)1,2}^2 = \frac{1}{2}\left\{ \mu_\chi^2 + \mu_\psi^2 + \mu_L^2 + \mu_{Rv}^2 \mp \left[(\mu_\chi^2 - \mu_\psi^2)^2 + (\mu_L^2 - \mu_{Rv}^2)^2 + \right.\right.$$

$$\left.\left. 2(\mu_\chi^2 + \mu_\psi^2)(\mu_L^2 + \mu_{Rv}^2) + 8\mu_\chi\mu_\psi\mu_L\mu_{Rv}\right]^{1/2} \right\} . \quad (6.169)$$

It can also be proven that the mass matrix itself is diagonalized by

$$\bar{O}^T M_v O = O^T M_v^T \bar{O} = \begin{pmatrix} \mu_{(v)1} & 0 & 0 & 0 \\ 0 & \mu_{(v)1} & 0 & 0 \\ 0 & 0 & \mu_{(v)2} & 0 \\ 0 & 0 & 0 & \mu_{(v)2} \end{pmatrix} . \quad (6.170)$$

This shows that for $\mu_\psi, \mu_L, \mu_{Rv} \ll \mu_\chi$ there is a light state with mass $\mu_{(v)1}$
and a heavy state with mass $\mu_{(v)2}$. According to (6.168), for $\mu_L \neq \mu_{Rv}$ the

fermion–mirror-fermion mixing angle in the left-handed sector is different from the one in the right-handed sector.

An alternative way to diagonalize the mass matrix is to write the mass terms with left-handed fields only, by introducing, instead of the right-handed components, the charge conjugated left-handed fields

$$\psi_{cL} \equiv C \overline{\psi}_R^T , \qquad \overline{\psi}_{cL} \equiv \psi_R^T C ,$$ (6.171)

where the charge conjugation matrix C is defined in section 5.2 by (5.107). The neutrino mass terms are then

$$\cdots + \mu_\psi \left[(\psi_{cLvx}^T C \psi_{Lvx}) + (\overline{\psi}_{Lvx} C \overline{\psi}_{cLvx}^T) \right]$$

$$+ \mu_\chi \left[(\chi_{cLvx}^T C \chi_{Lvx}) + (\overline{\chi}_{Lvx} C \overline{\chi}_{cLvx}^T) \right]$$

$$+ \mu_L \left[(\chi_{cLvx}^T C \psi_{Lvx}) + (\overline{\psi}_{Lvx} C \overline{\chi}_{cLvx}^T) \right]$$

$$+ \mu_{Rv} \left[(\psi_{cLvx}^T C \chi_{Lvx}) + (\overline{\chi}_{Lvx} C \overline{\psi}_{cLvx}^T) \right] .$$ (6.172)

The corresponding symmetrized mass matrix on the $(\psi_{Lv}^T C, \psi_{cLv}^T C, \chi_{Lv}^T C, \chi_{cLv}^T C) \otimes (\psi_{Lv}, \psi_{cLv}, \chi_{Lv}, \chi_{cLv})$ basis is

$$M_v^{(L)} = \frac{1}{2} \begin{pmatrix} 0 & \mu_\psi & 0 & \mu_L \\ \mu_\psi & 0 & \mu_{Rv} & 0 \\ 0 & \mu_{Rv} & 0 & \mu_\chi \\ \mu_L & 0 & \mu_\chi & 0 \end{pmatrix} .$$ (6.173)

This symmetrical matrix can be diagonalized by an O(4) rotation. It has two pairs of eigenvalues with equal absolute values and opposite signs: $\pm \mu_{(v)1}$, respectively, $\pm \mu_{(v)2}$, which on the left-handed basis correspond to two fermion states with Dirac masses $\mu_{(v)1}$ and $\mu_{(v)2}$.

The diagonalization of the mass matrices for charged leptons and quarks can be achieved on the fermion–mirror-fermion basis in a similar way [6.47].

6.2.3 *Numerical simulations, phase structure*

Numerical simulations of Yukawa models can reveal the non-perturbative behaviour of a coupled scalar–fermion system at strong couplings. At small Yukawa coupling the scalar field is not much affected by the feedback of the fermion interactions, therefore the system is qualitatively similar to a pure ϕ^4 model (see chapter 2). If, however, the bare Yukawa coupling is strong, some new non-perturbative phenomena, like new phases and phase transitions, can occur. In fact, the primary goal of the first numerical studies of some simple Yukawa models was to map out the phase structure at intermediate and large bare Yukawa couplings.

The Monte Carlo simulations can be carried out by the unbiased hybrid Monte Carlo algorithm for dynamical fermions (see section 7.6), if a flavour doubling of the fermion spectrum is introduced. If the fermion matrix of the original (first) fermion flavour is denoted by Q, then the fermion matrix of the second flavour is Q^+, which has opposite chiralities with respect to Q (see, for instance, the Yukawa coupling terms in the action (6.124), where $\varphi_x \to \varphi_x^+$ means a transition from fermion to mirror fermion). This flavour doubling leads to an extension of the global symmetry. For instance, in the $U(1)_L \otimes U(1)_R$ symmetric model with a mirror pair of fermion fields the chiral symmetry becomes $U(1)_L \otimes U(1)_R \otimes U(1)_{1-2}$ [6.37], where $U(1)_{1-2}$ is a vector-like symmetry corresponding to the conservation of the difference of the number of first and second flavours. If the model is described in terms of left-handed fields (the right-handed fields are represented by the left-handed components of the charge-conjugate fields according to (6.171)), then the quantum numbers of the eight fermion fields and the Higgs field are

$$
\begin{array}{cccc@{\qquad}cccc}
 & U(1)_L & U(1)_R & U(1)_{1-2} & & U(1)_L & U(1)_R & U(1)_{1-2} \\[4pt]
\psi_L^{(1)}: & 1 & 0 & 1 & \psi_L^{(2)}: & 0 & 1 & -1 \\
\psi_{cL}^{(1)}: & 0 & -1 & -1 & \psi_{cL}^{(2)}: & -1 & 0 & 1 \\
\chi_L^{(1)}: & 0 & 1 & 1 & \chi_L^{(2)}: & 1 & 0 & -1 \\
\chi_{cL}^{(1)}: & -1 & 0 & -1 & \chi_{cL}^{(2)}: & 0 & -1 & 1 \\
\phi: & 1 & -1 & 0 & & & &
\end{array}
$$

$$(6.174)$$

Concerning chirality it is instructive to consider the possible fermion mass terms in this flavour doubled model. The mass terms allowed by the chiral symmetry $U(1)_L \otimes U(1)_R \otimes U(1)_{1-2}$ are those connecting $\psi^{(1)}$ with $\chi^{(1)}$ or $\psi^{(2)}$ with $\chi^{(2)}$ but not $\psi^{(1)}$ with $\psi^{(2)}$ and $\chi^{(1)}$ with $\chi^{(2)}$ (these latter are forbidden by $U(1)_{1-2}$). The vacuum expectation value of the scalar field breaks $U(1)_L \otimes U(1)_R$ to its diagonal subgroup, but $U(1)_{1-2}$ is not spontaneously broken. A *'chiral' set of fields* can be defined by the requirement that no mass term is allowed by the symmetry and/or generated by spontaneous symmetry breaking. In this sense, for instance, the subset $\{\psi_L^{(1)}, \psi_{cL}^{(1)}\}$ is 'chiral' and 'anomalous' (it has a non-zero U(1)-anomaly). A subset like $\{\psi_L^{(1)}, \psi_{cL}^{(1)}, \chi_L^{(1)}, \chi_{cL}^{(1)}\}$ is 'non-chiral' and 'non-anomalous'. Finally, the subset $\{\psi_L^{(1)}, \psi_{cL}^{(1)}, \psi_L^{(2)}, \psi_{cL}^{(2)}\}$, which is obtained after decoupling the mirror (χ-) fields, is 'chiral' and 'non-anomalous'. (With respect to the subgroup $U(1)_L \otimes U(1)_R$ this latter subset is 'non-chiral', but there is no reason why only this subgroup

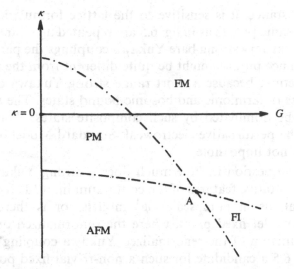

Fig. 6.5. The typical phase structure of Yukawa models in the plane of scalar hopping parameter (κ) and bare Yukawa coupling (G). FM = ferromagnetic, PM = paramagnetic, AFM = antiferromagnetic, FI = ferrimagnetic phase.

should be considered.)

The doubling of the fermion spectrum can be avoided if in the Monte Carlo integration measure $|\det Q|$ is taken. This can be done, for instance, by using the hybrid classical–Langevin algorithm discussed in section 7.5. The result is, of course, only reliable if the effect of the phase of the fermion determinant is small. Then either one can neglect it completely, or take it into account as a small correction in the measured averages.

The typical phase structure of a Yukawa model in the region of small and intermediate Yukawa couplings, as obtained in non-perturbative studies, is illustrated by fig. 6.5. Such a picture emerges both in simple non-chiral [6.48, 6.49, 6.50, 6.51], as well as in chiral models [6.52, 6.53]. PM is the *paramagnetic or symmetric phase* with vanishing scalar vacuum expectation values. The other phases are characterized by non-zero expectation values of the scalar field $\langle \phi_x \rangle \neq 0$ and/or the *staggered scalar field* $\langle \hat{\phi}_x \rangle \neq 0$, where

$$\hat{\phi}_x \equiv (-1)^{x_1+x_2+x_3+x_4} \phi_x = e^{i\pi(x_1+x_2+x_3+x_4)} \phi_x . \qquad (6.175)$$

In the *ferromagnetic phase* (FM) we have $\langle \phi_x \rangle \neq 0$ and $\langle \hat{\phi}_x \rangle = 0$, in the *antiferromagnetic phase* (AFM) $\langle \phi_x \rangle = 0$ and $\langle \hat{\phi}_x \rangle \neq 0$, whereas in the *ferrimagnetic phase* both the scalar and the staggered scalar expectation values are non-zero: $\langle \phi_x \rangle \neq 0$, $\langle \hat{\phi}_x \rangle \neq 0$. An interesting point of the phase diagram is point 'A' in fig. 6.5, where all four phases meet.

At still stronger bare Yukawa couplings the phase structure is not universal, for instance, it is sensitive to the lattice formulation [6.49]. In many cases the same phases as in fig. 6.5 are repeated again in the opposite order. However, at very strong bare Yukawa couplings the particle content of the long distance physics might be quite different from the field content of the lattice action, because a short range strong Yukawa coupling can produce all sorts of fermionic and bosonic bound states. The light particle spectrum may be dominated by such composite states. This makes the connection to the perturbative electroweak Standard Model in this region cumbersome, if not impossible.

An important question is, how much does a strong Yukawa coupling influence the qualitative features of the continuum limit? Is the continuum limit still trivial, as in pure scalar ϕ^4 models, or is there perhaps a non-trivial ultraviolet fixed point, where the renormalized quartic scalar coupling is a function of the renormalized Yukawa coupling? Is perhaps point 'A' in fig. 6.5 a candidate for such a non-trivial fixed point [6.50]?

In order to obtain numerical hints for answering these questions, the renormalized quantities in the fermionic sector have to be computed. As a simple example, let us consider here the fermionic renormalized quantities in the symmetric (PM) phase of the $U(1)_L \otimes U(1)_R$ symmetric Yukawa model with mirror pairs of fermion fields, defined by the lattice action (6.129).

Since in the symmetric phase both fermions and bosons are massive, the renormalization can be defined at zero four-momentum without a danger of infrared singularities. Let us denote the mirror pair of fermion fields by $\Psi \equiv (\psi, \chi)$. The behaviour of the fermion propagator matrix in momentum space for $p \to 0$ is

$$\tilde{\Delta}_\Psi(p) \equiv \sum_x e^{-ip \cdot (y-x)} \Delta_{yx}^\Psi = A - ip \cdot \gamma B + O(p^2) . \qquad (6.176)$$

The inverse propagator in the same limit is

$$\tilde{\Delta}_\Psi(p)^{-1} = M + ip \cdot \gamma N + \mathcal{O}(p^2) , \qquad (6.177)$$

where the matrices M and N are given by

$$M = A^{-1}, \qquad N = A^{-1}BA^{-1} . \qquad (6.178)$$

The matrices A and B can be determined in a numerical simulation from the expectation values of fermionic timeslices. Assuming antiperiodic boundary conditions for the fermionic fields in the time direction, and periodic ones otherwise, possible definitions on the lattice are:

$$A = \frac{1}{2}\left[\tilde{\Delta}_\Psi\left(0,0,0,+\frac{\pi}{T}\right) + \tilde{\Delta}_\Psi\left(0,0,0,-\frac{\pi}{T}\right)\right]$$

$$= \sum_x \cos\left[\frac{\pi}{T}(y_4 - x_4)\right] \Delta^{\Psi}_{yx} = \sum_x \cos\left[\frac{\pi}{T}(y_4 - x_4)\right] \langle \Psi_y \bar{\Psi}_x \rangle \,,$$

$$B\gamma_4 = \frac{i}{2\sin(\pi/T)}\left[\tilde{\Delta}_{\Psi}\left(0,0,0,+\frac{\pi}{T}\right) - \tilde{\Delta}_{\Psi}\left(0,0,0,-\frac{\pi}{T}\right)\right]$$

$$= \sum_x \frac{\sin\left[(\pi/T)(y_4 - x_4)\right]}{\sin(\pi/T)}\Delta^{\Psi}_{yx} = \sum_x \frac{\sin\left[(\pi/T)(y_4 - x_4)\right]}{\sin(\pi/T)}\langle \Psi_y \bar{\Psi}_x \rangle \,.$$

$$(6.179)$$

Here, as usual, $\langle \ldots \rangle$ denotes an expectation value with respect to the Boltzmann factor e^{-S} and, for instance, x_4 is the time-coordinate of the lattice point $x \equiv (x_1, x_2, x_3, x_4)$. The $O(p^2)$ corrections in (6.176) are small, if the smallest time-like momentum π/T is sufficiently small relative to the fermion mass in lattice units. This can be achieved if the lattice is long enough in the time direction.

In the symmetric phase, due to chiral symmetry, the matrices A, B have in the (ψ, χ)-space the general form

$$A = \begin{pmatrix} 0 & A_{\psi\chi} \\ A_{\psi\chi} & 0 \end{pmatrix}, \qquad B = \begin{pmatrix} B_{\psi\psi} & 0 \\ 0 & B_{\chi\chi} \end{pmatrix}. \qquad (6.180)$$

Therefore, M and N are given by

$$M = \begin{pmatrix} 0 & A_{\psi\chi}^{-1} \\ A_{\psi\chi}^{-1} & 0 \end{pmatrix}, \qquad N = \begin{pmatrix} B_{\chi\chi}A_{\psi\chi}^{-2} & 0 \\ 0 & B_{\psi\psi}A_{\psi\chi}^{-2} \end{pmatrix}. \qquad (6.181)$$

The wave function renormalization has to transform N to the unit matrix. Let us define the renormalized fermion fields by

$$\Psi_R \equiv Z_{\Psi}^{-1/2}\Psi, \qquad \bar{\Psi}_R \equiv \bar{\Psi}Z_{\Psi}^{-1/2} \,. \qquad (6.182)$$

According to (6.181) the matrix $Z_{\Psi}^{1/2}$ is given by

$$Z_{\Psi}^{1/2} \equiv \begin{pmatrix} \sqrt{Z_{\psi}} & 0 \\ 0 & \sqrt{Z_{\chi}} \end{pmatrix} = \begin{pmatrix} A_{\psi\chi}B_{\chi\chi}^{-1/2} & 0 \\ 0 & A_{\psi\chi}B_{\psi\psi}^{-1/2} \end{pmatrix}. \qquad (6.183)$$

Multiplying the unrenormalized mass matrix M by $Z_{\Psi}^{1/2}$ from left and right gives the renormalized fermion mass matrix M_R and renormalized fermion mass μ_R:

$$M_R \equiv Z_{\Psi}^{1/2}MZ_{\Psi}^{1/2} = \begin{pmatrix} 0 & A_{\psi\chi}(B_{\psi\psi}B_{\chi\chi})^{-1/2} \\ A_{\psi\chi}(B_{\psi\psi}B_{\chi\chi})^{-1/2} & 0 \end{pmatrix}$$

$$\equiv \begin{pmatrix} 0 & \mu_R \\ \mu_R & 0 \end{pmatrix}. \qquad (6.184)$$

In order to define the renormalized Yukawa couplings in terms of the expectation values on the lattice, let us introduce some shorthand notations. For instance, the type of bilinear fermionic expectation values occuring in (6.179) can be denoted as

$$\langle \psi_L \bar{\chi}_R \rangle_0 \delta_{\rho\sigma} \equiv \frac{1}{\Omega} \sum_{x,y} e^{-i(\pi/T)(y_4 - x_4)} \langle \psi_{L\rho y} \bar{\chi}_{R\sigma x} \rangle \; . \tag{6.185}$$

Here ρ, σ are the spinor indices of the chiral components of fermion fields. Similarly, the fermion–fermion–scalar expectation values for the renormalized Yukawa couplings are introduced as, for instance,

$$\langle \bar{\psi}_R \phi_1 \psi_L \rangle_0 \delta_{\rho\sigma} \equiv \frac{1}{\Omega} \sum_{x,y,z} e^{-i(\pi/T)(y_4 - x_4)} \langle \bar{\psi}_{R\sigma x} \phi_{1z} \psi_{L\rho y} \rangle \; . \tag{6.186}$$

The diagonal matrix built from these expectation values is

$$\langle \bar{\Psi}\Phi\Psi \rangle_0 \equiv \mathrm{diag}\,\{ \langle \bar{\psi}_R \phi_1 \psi_L \rangle_0, \langle \bar{\psi}_L \phi_1 \psi_R \rangle_0, \langle \bar{\chi}_R \phi_1 \chi_L \rangle_0, \langle \bar{\chi}_L \phi_1 \chi_R \rangle_0 \} \; . \tag{6.187}$$

Using these notations the renormalized Yukawa coupling matrix in the space of chiral components $(\psi_L, \psi_R, \chi_L, \chi_R)$ is defined as

$$G_R \equiv \frac{m_R^2}{(Z_\phi)^{1/2}} M_R Z_\Psi^{-1/2} \langle \bar{\Psi}\Phi\Psi \rangle_0 Z_\Psi^{-1/2} M_R \; . \tag{6.188}$$

This is given directly in terms of the expectation values, but one can easily see that it is equivalent to the usual definition in terms of the one-particle irreducible vertex functions. G_R is a diagonal matrix $G_R = \mathrm{diag}\,\{ G_{R\psi}, G_{R\psi}, G_{R\chi}, G_{R\chi} \}$, where

$$G_{R\psi} = \frac{m_R^2 \mu_R^2}{Z_\chi (4Z_\phi)^{1/2}} \langle \bar{\chi}_R \phi \chi_L \rangle_0 = \frac{m_R^2 \mu_R^2}{Z_\chi (4Z_\phi)^{1/2}} \langle \bar{\chi}_L \phi^+ \chi_R \rangle_0 \; ,$$

$$G_{R\chi} = \frac{m_R^2 \mu_R^2}{Z_\psi (4Z_\phi)^{1/2}} \langle \bar{\psi}_R \phi^+ \psi_L \rangle_0 = \frac{m_R^2 \mu_R^2}{Z_\psi (4Z_\phi)^{1/2}} \langle \bar{\psi}_L \phi \psi_R \rangle_0 \; . \tag{6.189}$$

Here m_R is the renormalized boson mass, and Z_ϕ the wave function renormalization factor of the scalar field. (The renormalized quantities in the scalar sector are defined in the same way as for the ϕ^4 model in chapter 2.) Equivalent expressions useful for the numerical determination are

$$G_{R\psi} = \frac{Z_\psi (Z_\phi)^{1/2} \langle \bar{\chi}_R \phi \chi_L \rangle_0}{\langle \phi^+ \phi \rangle_0 \langle \chi_L \bar{\psi}_R \rangle_0 \langle \psi_L \bar{\chi}_R \rangle_0} = \frac{Z_\psi (Z_\phi)^{1/2} \langle \bar{\chi}_L \phi^+ \chi_R \rangle_0}{\langle \phi^+ \phi \rangle_0 \langle \chi_R \bar{\psi}_L \rangle_0 \langle \psi_R \bar{\chi}_L \rangle_0} \; ,$$

$$G_{R\chi} = \frac{Z_\chi (Z_\phi)^{1/2} \langle \bar{\psi}_R \phi^+ \psi_L \rangle_0}{\langle \phi^+ \phi \rangle_0 \langle \psi_L \bar{\chi}_R \rangle_0 \langle \chi_L \bar{\psi}_R \rangle_0} = \frac{Z_\chi (Z_\phi)^{1/2} \langle \bar{\psi}_L \phi \psi_R \rangle_0}{\langle \phi^+ \phi \rangle_0 \langle \psi_R \bar{\chi}_L \rangle_0 \langle \chi_R \bar{\psi}_L \rangle_0} \; . \tag{6.190}$$

In hybrid Monte Carlo simulations the expectation values bilinear in the fermionic variables, which appear in the above formulas, can be effectively determined by 'noisy estimators' in terms of the pseudofermion field (see (7.150) in section 7.4). The definition of the fermionic renormalized quantities in the broken phase can be formulated similarly [6.37].

6.2.4 *Vacuum stability lower bound*

In this subsection it will be assumed that the continuum limit of the Yukawa models is trivial, in the same way as for pure scalar ϕ^4 models (see chapter 2). This is suggested by the 1-loop perturbative β-functions which have, besides the infrared stable fixed point at vanishing couplings, no other fixed point, in particular no non-trivial ultraviolet stable fixed point, which would be necessary for a non-trivial continuum limit.

The qualitative behaviour of the perturbative β-functions is similar in a large class of models. To be specific, let us consider an $SU(2)_L \otimes SU(2)_R$ symmetric Yukawa model with N_f mirror pairs of degenerate fermion doublets with equal Yukawa couplings. (The lattice action for $N_f = 1$ is given by (6.124).) In this case the 1-loop β-functions for the quartic scalar coupling (g_R) and the Yukawa couplings of the fermion and mirror fermion ($G_{R\psi}$, respectively, $G_{R\chi}$) are [6.36, 6.42]:

$$\frac{dg_R}{d\log\mu} = \frac{1}{16\pi^2} \left[4g_R^2 + 16N_f g_R(G_{R\psi}^2 + G_{R\chi}^2) - 96N_f(G_{R\psi}^4 + G_{R\chi}^4) \right] ,$$

$$\frac{dG_{R\psi}}{d\log\mu} = \frac{1}{16\pi^2} \cdot 4N_f G_{R\psi}(G_{R\psi}^2 + G_{R\chi}^2) ,$$

$$\frac{dG_{R\chi}}{d\log\mu} = \frac{1}{16\pi^2} \cdot 4N_f G_{R\chi}(G_{R\psi}^2 + G_{R\chi}^2) . \tag{6.191}$$

Here the renormalization group equations are given, which describe the change of the renormalized couplings for fixed bare couplings as functions of the renormalization scale μ. The same 1-loop β-functions also apply along a 'line of constant physics' in the bare parameter space, as functions of the lattice spacing a.

Note that for $G_{R\psi} = \pm G_{R\chi} \equiv G_R$ these are identical to the 1-loop β-functions in a Yukawa model of $2N_f$ degenerate doublets with equal Yukawa couplings. Namely, the Yukawa couplings of fermion and mirror fermion doublets are equivalent by charge conjugation. For instance, the Yukawa coupling of the mirror fermion doublet in (6.124) can be written, with the help of the charge conjugated fields in (6.171), as

$$(\overline{\chi}_{Rx}\varphi_x\chi_{Lx}) + (\overline{\chi}_{Lx}\varphi_x^+\chi_{Rx})$$

$$= (\bar{\chi}_{cRx}\epsilon^{-1}\varphi_x^+\epsilon\chi_{cLx}) + (\bar{\chi}_{cLx}\epsilon^{-1}\varphi_x\epsilon\chi_{cRx}) . \qquad (6.192)$$

Here the relation (6.5) was also used. This shows that in terms of the new fields $\chi'_x \equiv \epsilon\chi_{cx}$, $\bar{\chi}'_x \equiv \bar{\chi}_{cx}\epsilon^{-1}$ the Yukawa coupling is the same as for a fermion doublet. In other words, a mirror doublet is equivalent to a doublet, as long as the $SU(3) \otimes U(1)_Y$ gauge interactions are neglected. (The $\psi-\chi$ mixing mass term appears on the new basis as an off-diagonal Majorana mass similar to (6.172).)

The three β-functions in (6.191) simultaneously vanish only at the Gaussian infrared stable fixed point $g_R = G_{R\psi} = G_{R\chi} = 0$. This corresponds to a trivial continuum limit, and implies a cut-off dependent upper bound on the renormalized quartic and Yukawa couplings. For low cut-offs the upper bound on the renormalized Yukawa couplings is expected roughly near the *tree unitarity bound* [6.54, 6.42]

$$G_{R\psi}^2, \ G_{R\chi}^2, \ G_{R\psi}G_{R\chi} \le \frac{4\pi}{N_f} , \qquad (6.193)$$

where the tree level fermion–fermion scattering amplitudes saturate unitarity. The upper bound on the renormalized quartic scalar coupling is expected to be of the same order of magnitude as the upper bound in the $O(4)$ symmetric ϕ^4 model without Yukawa couplings (see in chapter 2), although the exact bound can, of course, show some dependence on $G_{R\psi}$ and $G_{R\chi}$.

Besides the infrared stable fixed point at vanishing couplings, the other important characteristics of the β-functions in (6.191) can be seen on the corresponding equations for the ratios

$$\frac{d(g_R/G_{R\psi}^2)}{d\log\mu} = \frac{4G_{R\psi}^2}{16\pi^2}\left[\frac{g_R^2}{G_{R\psi}^4}\right.$$

$$\left.+2N_f\left(1+\frac{G_{R\chi}^2}{G_{R\psi}^2}\right)\frac{g_R}{G_{R\psi}^2} - 24N_f\left(1+\frac{G_{R\chi}^4}{G_{R\psi}^4}\right)\right] ,$$

$$\frac{d(G_{R\chi}/G_{R\psi})}{d\log\mu} = 0 . \qquad (6.194)$$

This shows that the ratio $G_{R\chi}/G_{R\psi}$ remains stable at an arbitrary value, and there is a particular ratio $g_R/G_{R\psi}^2 > 0$, where the square bracket on the right hand side of the first equation vanishes, which is also stable. The qualitative behaviour of the renormalization group flow for fixed $G_{R\chi}/G_{R\psi}$ is shown in fig. 6.6. As one can see from the figure, if one integrates the renormalization group equations towards larger renormalization scales μ from a point below the separating line of the flow (S), at some point the quartic coupling becomes negative. This property of the β-functions

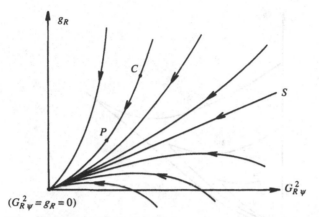

Fig. 6.6. The qualitative picture of the renormalization group flow of the renormalized quartic coupling (g_R) and Yukawa coupling squared ($G^2_{R\psi}$), for fixed $G_{R\chi}/G_{R\psi}$ and decreasing renormalization scale $\mu \rightarrow 0$, as implied by the 1-loop β-functions. C is a point at the cut-off scale, P is a corresponding point on the same trajectory at the physical scale.

implies the existence of a cut-off dependent lower bound on the renormalized quartic coupling, which is called in the literature the *vacuum stability lower bound* [6.55].

In order to see how this bound arises in a lattice formulation, let us discuss the effective potential of a constant scalar field ϕ [6.37]. The perturbative expansion of the effective potential breaks down for large values of ϕ. This is due to the appearance of logarithms of the field in the coefficient of the quartic term. In the literature this situation has been dealt with by a summation of the leading logs by means of the renormalization group [6.30, 6.55]. The coefficient of the ϕ^4 term is then equal to the running quartic coupling at a scale μ, which is given by the value of the field. Given some values of the renormalized couplings g_R, $G_{R\psi}$ and $G_{R\chi}$ at the physical scale set by the physical masses, the renormalization group equations can be integrated to some high scale to yield the corresponding quartic coupling. Since the β-function of the quartic scalar coupling is in some region negative, namely at small couplings and large ratios $G_{R\psi}/g_R$ and/or $G_{R\chi}/g_R$, it may happen that for large fields one ends up with a running coupling which is negative, and the effective potential appears to bend over to large negative values. This situation has been called 'vacuum instability', and the corresponding values of the renormalized couplings, for which it occurs, are excluded from the allowed region.

Now the question poses itself, how can this happen in a non-perturbative lattice formulation, where the effective potential is a well-defined quantity, which is known to be convex due to a theorem of Symanzik

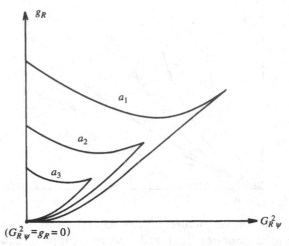

Fig. 6.7. The allowed region of the renormalized quartic and Yukawa coupling for different lattice spacings, in the case of a trivial continuum limit corresponding to the 1-loop β-functions. For decreasing lattice spacings $a_1 > a_2 > a_3$ the allowed region shrinks to the trivial point $g_R = G_{R\psi}^2 = 0$.

[6.29]? Consequently it cannot possibly bend over to negative values for large fields, because it is finite (e.g. zero by convention) at its minima.

The first possibility is that at strong bare Yukawa couplings there is a first order phase transition to another, deeper minimum of the effective potential, and the convexity is provided by the Maxwell construction.

In the case of absence of a first order phase transition, which is suggested by fig. 6.5, the resolution of this apparent paradox is related to the fact that the running coupling at the cut-off scale $\mu \simeq \Lambda \simeq a^{-1}$ is essentially equal to the bare coupling. In order that the path integral be well defined, the bare coupling λ has to be positive. The renormalization group flow may now be followed from the cut-off scale to the physical scale, where the renormalized couplings are defined. Those values of the renormalized couplings that can be reached starting from any positive value of the bare coupling form the physical admissible region. Those outside would not correspond to any positive bare λ, and cannot be realized for the given cut-off. In particular the boundary corresponding to $\lambda = 0$ yields the *vacuum stability bound*.

To conclude, in absence of a first order phase transition the non-perturbative formulation of the *vacuum stability bound* is the requirement of a positive bare quartic coupling λ. The problem is not primarily the large field behaviour of the effective potential. Furthermore the cut-off plays an important rôle, which is not always emphasized in the literature.

The exact effective potential and exact β-functions are, of course, not

known. Therefore one has to rely on some approximations, like perturbation theory or numerical simulations. Without knowing the qualitative
behaviour of the β-functions it is impossible to derive the vacuum stability
bound. In particular, the qualitative discussion is different in the case of
a trivial continuum limit, which is qualitatively represented by the 1-loop
β-functions, or if a non-trivial fixed point at non-zero couplings exist, as
suggested by the qualitative features of the 2-loop approximation [6.37].
(For the explicit expressions of the β-functions up to 2-loop order in the
model considered in (6.191) see [6.42].) Since the qualitative behaviour
of the β-functions is not yet known, for the moment the vacuum stability bound can only be discussed if some scenario is assumed for the
β-functions. Here we assume the absence of a non-trivial fixed point,
corresponding to the qualitative behaviour of the 1-loop β-functions.

As discussed above, in this case the renormalization flow of the couplings
is as shown by fig. 6.6. The renormalized couplings at the physical scale
(e.g. point P on the figure) are connected to the bare couplings along a
flow line (point C on the figure). The length of the flow line connecting C
and P depends, of course, on the scale ratio at C and P. For a given scale
ratio, i. e. for a given physical mass in lattice units, one can put the point
C on the positive $G_{R\psi}$ axis where $g_R = 0$. The set of corresponding points
P defines a curve, which is the *vacuum stability bound* for the given ratio
of cut-off to physical mass. Namely, if one wanted to go with P along
the flow line closer to the $g_R = 0$ axis, then the corresponding point C
would have a negative bare quartic coupling, and the lattice action would
become unstable. Saying it differently, since the 1-loop contribution of the
fermion loop to the quartic coupling (at given bare couplings) is positive,
it can happen that the bare quartic coupling needed for some combination
of renormalized couplings is negative. (Note that here we refer to the true
1-loop contribution, not to its leading log at large fields.)

It is clear from the figure, that for very large scale differences the lower
limit implied by the requirement of stability tends to the separating line S
of the flow from $g_R = \infty$ and $g_R = -\infty$. Of course, in the case of a trivial
continuum limit one has to take into account also the cut-off dependent
upper limit on the renormalized couplings. The allowed region in the
$(G_{R\psi}^2, g_R)$-plane is bounded by the upper limit for g_R, and by the vacuum
stability bound (see fig. 6.7).

Note that if the effect of weak gauge couplings is also included, then
there is an additional small excluded pocket near zero couplings. This is
the consequence of the Weinberg–Linde bound [6.32] (see section 6.1).

7
Simulation algorithms

7.1   Numerical simulation and Markov processes

The goal of numerical simulations in lattice quantum field theory is to estimate the expectation values of some functions $A[\varphi]$ of the field variables $[\varphi] \equiv \{\varphi_{x\alpha}\}$ ($\varphi_{x\alpha}$ denotes a real field component with index α at lattice site x). This is given by path integrals as

$$\langle A \rangle = Z^{-1} \int [\mathrm{d}\varphi] \mathrm{e}^{-S[\varphi]} A[\varphi] , \qquad Z = \int [\mathrm{d}\varphi] \mathrm{e}^{-S[\varphi]} . \qquad (7.1)$$

$S[\varphi]$ is the lattice action, which is assumed to be a real function of the field variables. For the moment we only consider bosonic path integrals. This means that the Grassmann variables for fermions are already integrated out, resulting in the fermion determinant

$$\det Q[\varphi] \equiv \mathrm{e}^{-S_F[\varphi]} . \qquad (7.2)$$

Therefore, if $S_B[\varphi]$ denotes the pure bosonic action, we have

$$S[\varphi] \equiv S_B[\varphi] + S_F[\varphi] . \qquad (7.3)$$

The assumption of a real $S[\varphi]$ means a positive fermion determinant, which is a restriction on the class of allowed fermionic theories.

In the case of lattice field systems of interest, the number of integration variables in

$$[\mathrm{d}\varphi] \equiv \prod_{x,\alpha} \mathrm{d}\varphi_{x\alpha} \qquad (7.4)$$

is very large (typically larger than 10^4, and sometimes also exceeding 10^6). For so many variables, apart from trivial systems, the only possibility is to try a Monte Carlo integration. In the simplest case this would mean to generate *field configurations* $[\varphi_n]$ randomly in the space of field variables. This would, however, be very inefficient, because for large lattice volumes

Ω the integrand of path integrals is sharply peaked at some specific configurations. This can be explained by the analogy of lattice quantum field theory to a classical statistical mechanics system in four dimensions (see chapter 1). Typical lattice actions are, for instance, the scalar ϕ^4 action with $\varphi \to \phi$ (see chapter 2):

$$S = \sum_x \left\{ \phi_x^2 + \lambda(\phi_x^2 - 1)^2 - \kappa \sum_\mu \phi_{x+\hat{\mu}} \phi_x \right\} , \qquad (7.5)$$

or the Wilson action of a U(N) lattice gauge field theory with $\varphi \to U$ (see chapter 3):

$$S[U] = \beta \sum_p \left(1 - \frac{1}{N} \mathrm{Re\,Tr}\, U_p \right) . \qquad (7.6)$$

As one can see from this, a typical lattice action contains a summation over the lattice. In order to display the essential statistical mechanics features of the lattice field system, let us write the action as

$$S[\varphi] \equiv \beta E[\varphi] \equiv \Omega\beta\epsilon[\varphi] . \qquad (7.7)$$

In statistical mechanics $\beta = 1/(kT)$ is the inverse temperature. In lattice field theory it is some overall parameter in the action, like $\beta = 2N/g^2$ for gauge fields, or $\beta = \kappa$ in scalar ϕ^4 models at $\lambda = \infty$. $E[\varphi]$ is analogous to the 'energy', $\epsilon[\varphi]$ to the 'energy density' given by the average plaquette for gauge fields, and by the average link for scalar fields. It is also useful to define the 'density of states' (or 'spectral density') $D(\epsilon)$ by

$$D(\epsilon) \equiv e^{\Omega s(\epsilon)} \equiv \int [\mathrm{d}\varphi] \delta \left(\epsilon - \frac{S[\varphi]}{\Omega\beta} \right) . \qquad (7.8)$$

Here $s(\epsilon)$ is analogous to the 'entropy density'. In this case the 'partition function' $Z \equiv Z_\beta$ is given by

$$Z_\beta = \int \mathrm{d}\epsilon\, D(\epsilon) e^{-\Omega\beta\epsilon} = \int \mathrm{d}\epsilon\, e^{\Omega[s(\epsilon) - \beta\epsilon]} . \qquad (7.9)$$

This shows that the probability distribution in ϵ is

$$\rho_\beta(\epsilon) \equiv \frac{D(\epsilon)}{Z_\beta} e^{-\Omega\beta\epsilon} = Z_\beta^{-1} e^{\Omega[s(\epsilon) - \beta\epsilon]} . \qquad (7.10)$$

As in statistical mechanics, one can call $f(\epsilon) \equiv \beta\epsilon - s(\epsilon)$ the 'free energy density'.

Since the number of lattice points Ω is large, in the path integral only a small vicinity of the minimum of the free energy density will substantially contribute. An efficient Monte Carlo integration has to take this into account. What is required is an *importance sampling* of the field configurations with the weight $\exp(-S[\varphi])$. This means that the distribution of configurations in the sample generated during the Monte Carlo integration has to follow the *Boltzmann factor* $\exp(-S[\varphi])$. In this context, according to the statistical mechanics analogy, it is useful to introduce the concept of an *ensemble of configurations*: *an ensemble consists of an infinite number of field configurations, with a density $W[\varphi]$ defined on the measure $[d\varphi]$*. In the *canonical ensemble* (or *equilibrium ensemble*) the density W_c is proportional to the Boltzmann factor:

$$W_c[\varphi] \propto e^{-S[\varphi]} . \tag{7.11}$$

The task in numerical simulations is to generate *samples* consisting of a large number N of configurations $\{[\varphi_n], n = 1, 2, \ldots, N\}$, in such a way that the distribution within a sample approximates the desired distribution in the canonical ensemble. The *sample average* of the quantity A is

$$\overline{A} \equiv \frac{1}{N} \sum_{n=1}^{N} A[\varphi_n] . \tag{7.12}$$

This is an *estimator* of the *ensemble average* $\overline{\overline{A}}$, which is equal to the expectation value:

$$\overline{\overline{A}} = \langle A \rangle . \tag{7.13}$$

This distinction between sample average and ensemble average is important for the discussion of error estimates (see below).

7.1.1   *Updating processes*

The configuration sample approximating the canonical ensemble is generated in numerical simulations as a sequence $\{[\varphi_n], 1 \leq n \leq N\}$. This sequence of field configurations is prepared by repeated application of an algorithm which creates from a configuration the next one: $[\varphi_n] \to [\varphi_{n+1}]$. This procedure of creating a sequence of configuration is called *updating*. The updating is a stochastic process in the sense that the transition $[\varphi] \to [\varphi']$ happens with a given *transition probability* $P([\varphi'] \leftarrow [\varphi])$. Of course, during realization of the updating by a computer program, the transition from one configuration to the next is usually deterministic. The transition probability refers to a large number of independent updatings. (The updating program is in this sense similar to a pseudo-random number generator. In fact, the updating is usually built on the use of a pseudo-random number sequence.)

Another, simpler formulation of the transition probability P uses the above introduced concept of configuration ensembles. According to this, an updating step changes the ensemble density W as

$$W'[\varphi'] = \sum_{[\varphi]} P\left([\varphi'] \leftarrow [\varphi]\right) W[\varphi] \equiv \int [\mathrm{d}\varphi] P\left([\varphi'] \leftarrow [\varphi]\right) W[\varphi] \,. \quad (7.14)$$

In a short matrix notation this can be written as $W' = PW$, if the densities are considered as vectors, and the transition probability as a square matrix. (Summation or integration over $[\varphi]$ can be alternatively used here.) The transition probability P obviously has to satisfy

$$\sum_{[\varphi']} P\left([\varphi'] \leftarrow [\varphi]\right) \equiv \int [\mathrm{d}\varphi'] P\left([\varphi'] \leftarrow [\varphi]\right) = 1 \,. \quad (7.15)$$

In addition, it is assumed that for any pairs of configurations

$$P\left([\varphi'] \leftarrow [\varphi]\right) > 0 \,. \quad (7.16)$$

This condition is called *strong ergodicity*, because it tells us that every configuration can be reached with a finite probability from any other one. In concrete applications this can be achieved, if an updating 'step' is defined by a sufficiently large number of repetitions of some smaller, elementary steps. In other words, the transition probability for a 'step' is some power of simpler transition probabilities: $P \equiv P_{\mathrm{simple}}^k$. For the density W it is usual to assume the normalization condition

$$\sum_{[\varphi]} W[\varphi] \equiv \int [\mathrm{d}\varphi] W[\varphi] = 1 \,. \quad (7.17)$$

A stochastic process satisfying (7.15)–(7.17) is called a *Markov process* (the sequence of configurations is a *Markov chain*). In our case of updating we can also call it *updating process*.

An important condition on the updating process is that given some 'reasonable' initial ensemble of configurations with density W_0, the repeated application of updating steps brings this to the canonical (equilibrium) ensemble:

$$\lim_{k \to \infty} P^k W_0 = W_c \,. \quad (7.18)$$

This means that one can start the updating from any 'reasonable' configuration (where the meaning of 'reasonable' will be defined below). Considered as an ensemble, this has a density one at the given configuration and zero elsewhere. After a while the repeated application of updating steps takes this initial ensemble in the vicinity of the desired equilibrium ensemble. From (7.18) it follows that the canonical ensemble is a fixed point of the transition probability:

$$P W_c = W_c \,. \quad (7.19)$$

We require that this fixed point is unique. Otherwise the result of the numerical simulation would be non-unique. As a matrix equation, (7.19) tells us that the density of the canonical ensemble belongs to an eigenvalue 1 of the transition probability matrix.

A sufficient (but not necessary) condition for (7.19) is *detailed balance*:

$$P\left([\varphi'] \leftarrow [\varphi]\right) W_c[\varphi] = P\left([\varphi] \leftarrow [\varphi']\right) W_c[\varphi'] . \tag{7.20}$$

This and (7.15) implies that

$$\sum_{[\varphi]} P\left([\varphi'] \leftarrow [\varphi]\right) W_c[\varphi] = \sum_{[\varphi]} P\left([\varphi] \leftarrow [\varphi']\right) W_c[\varphi'] = W_c[\varphi'] , \tag{7.21}$$

which is the same as (7.19). The requirement in (7.18) is also satisfied: it follows from the theorem of Perron on positive matrices (see, for instance, in [7.1]), that the eigenvalue 1 of the matrix P is non-degenerate, and every other eigenvalue is less than 1. Therefore, repeated application of P on W damps out all orthogonal components, exept for the component parallel to the eigenvector with eigenvalue 1. The condition a 'reasonable' intial ensemble density W_0 in (7.18) has to satisfy is that it has to have a non-zero overlap with the canonical ensemble. Owing to strong ergodicity (7.16), this is always true for any single initial configuration. Examples of transition probabilities satisfying the detailed balance condition (7.20) are provided by the Metropolis and heatbath algorithms discussed below.

The function $A[\varphi]$ defining the measured quantity can, in principle, be any integrable function of the field configuration $[\varphi]$. In practice, however, there are some requirements which A has to satisfy, in order that the sample average $\overline{A}$ in a not unreasonably large sample be close to the canonical ensemble average $\overline{\overline{A}} = \langle A \rangle$. In order to see this, one has to remember that the space of field configurations has a very large number of dimensions. Correspondingly, there is an extremely large number of different configurations. Even in the simple case of the Ising model, where the field variables have only two values ± 1, on a lattice with Ω points the number of different field configurations is 2^Ω. There is no chance to visit all possible configurations in any updating sequence of reasonable length. In fact, only a very small portion of configurations can be generated. This means that, for instance, a simple function defined to be 1 on a particular configuration, and zero everywhere else, can practically never be measured numerically. Another, more important, example is if the function has a strongly oscillating sign (or phase) on 'typical' configurations occuring often in the updating process. Such functions are not well suited for numerical determination in a simulation.

Another important class of functions difficult for numerical measurement is obtained, if in an updating process at action parameter β one wants to determine expectation values at a different β'. It follows from the definition in (7.1) and from (7.9), that an expectation value $\langle A \rangle_{\beta'}$ at β' can be expressed by an expectation value at β as

$$\langle A \rangle_{\beta'} = Z_{\beta'}^{-1} \int [d\varphi] e^{-\Omega\beta'\epsilon[\varphi]} A[\varphi] = \left\langle A \frac{Z_\beta}{Z_{\beta'}} e^{\Omega(\beta-\beta')\epsilon} \right\rangle_\beta . \tag{7.22}$$

The expression in $\langle \cdots \rangle_\beta$ is a strongly varying function, if the shift in β is $\Delta\beta \equiv \beta' - \beta = O(1)$. In this case the generated sample of configurations is peaked at the wrong value of ϵ, and configurations where the function multiplying A is large practically never occur. Therefore, for $\Delta\beta = O(1)$ the relation (7.22) cannot be used. The only possibility is to choose $\Delta\beta$ so small that $\Omega\Delta\beta = O(1)$. Such a small interval in β is usually not interesting. An exception is the vicinity of phase transitions, where things are rapidly changing, exactly at a such fine scale in β.

Under these conditions the relation in (7.22) is useful. This is the basis of the *histogram method* (or *spectral density method*) [7.2, 7.3]. In order to illustrate this method, let us consider the (analogue) energy $E[\varphi]$ defined in (7.7), and some other physical quantity $A[\varphi]$. The interesting range of values of the pair (E, A) has to be divided into an appropriate number of bins (typically a few hundred to thousand). During the updating process one counts the number of occurences $N(E_i, A_j)$ of the values (E_i, A_j) in the bin (i, j), and defines the *frequency function*

$$f_{ij} \equiv \frac{N(E_i, A_j)}{\sum_{ij} N(E_i, A_j)} \simeq \rho_\beta(E_i, A_j) = \frac{D(E_i, A_j)}{Z_\beta} e^{-\beta E_i} . \tag{7.23}$$

As it is shown here, f_{ij} is an estimator of the probability distribution ρ_β, which is proportional to the density of states ('spectral density') $D(E, A)$, similarly to (7.10). The probability distribution at another β' is obviously given by

$$\rho_{\beta'}(E_i, A_j) = \frac{f_{ij} e^{(\beta-\beta')E_i}}{\sum_{ij} f_{ij} e^{(\beta-\beta')E_i}} . \tag{7.24}$$

Knowledge of $\rho_{\beta'}$ is, of course, enough to calculate the expectation value of any function of (E, A) at the action parameter β', as long as $\rho_{\beta'}$ is well determined in the relevant bins. The advantage of the histogram method is that in the vicinity of phase transitions, where some physical quantities are rapidly changing, one can obtain detailed information in terms of curves (as functions of β') from a single simulation. Note that the frequency function f_{ij} in (7.23) can also be used for the determination of the density of states $D(E, A)$, which is proportional to $f_{ij} \exp(\beta E_i)$. The sorting of the values of (E, A) into bins introduces some error in the calculation of expectation values. This can be avoided if the full history of the run, with all obtained values of (E, A), is recorded.

7.1.2   Updating with constraints

Besides the straightforward updating process, it is sometimes also useful to consider updating with constraints in the range of some functions of field variables. An example of the application of updating with constraints is the *spectral density method* [7.4], which is a way to obtain detailed information about the partition function by calculating the density of states $D(\epsilon)$ in (7.8). Such information is particularly useful near phase transitions. An important question is, for instance, the behaviour of the complex zeros of the partition function, which come close to the real axis in the limit of infinite volumes [A6]. One possibility for studying the density of states in unconstrained updating was mentioned at the end of the previous subsection in connection with the histogram method. The obtained frequency function of bins f_{ij} summed over j is proportional to the product $D(\epsilon)\exp(-\Omega\beta\epsilon)$. Similarly, in an unconstrained updating process corresponding to a simple Monte Carlo integration (no importance sampling), when the updating steps are done with uniform distribution with respect to the path integral measure $[\mathrm{d}\varphi]$, the frequency function of the bins in ϵ gives directly $D(\epsilon)$.

More generally, if in the updating process with importance sampling, the action $S[\varphi] \equiv \Omega\beta\epsilon$ is replaced by $\Omega s_W(\epsilon)$ (corresponding to a weight $\exp(-\Omega s_W(\epsilon))$), then the frequency function of the bins is proportional to

$$D(\epsilon)e^{-\Omega s_W(\epsilon)} = e^{\Omega[s(\epsilon) - s_W(\epsilon)]} . \qquad (7.25)$$

The above special cases correspond, respectively, to $s_W = \beta\epsilon$ and $s_W = 0$. From the numerical point of view the best situation is, when the distribution over the bins is nearly uniform. Therefore, the best choice is to take an s_W close to s [7.5]. The approximation to s can be improved step by step, for instance, by starting on small lattices and using the obtained s as a first estimate on the larger lattice.

To obtain a good estimate for $s(\epsilon)$ in the whole range of the (analogue) energy density ϵ is, however, rather difficult. Therefore, an improvement can be achieved, if the ϵ-range is divided into several parts, with a few overlapping bins at the edges [7.4]. In this case $D\exp(-\Omega s_W)$ is obtained within each subset of bins by a random walk constrained to remain in the subset. The relative normalization of the frequency functions in adjacent sets of bins is determined by comparing the common bins. In detail, the updating process inside the chosen set of bins is the same as normal updating with the action $\Omega s_W(\epsilon)$. Changes leading out of the subset are not accepted, the configuration remains in the same bin where it was, nevertheless the frequency function of the bin is increased by 1. In order to illustrate the necessity of increasing the frequency function of the bin in cases for which the random walk would lead out of the subset, let us consider a simple example with three bins in total, with probability $\frac{1}{3}$ each. The unconstrained random walk moves with probability $\frac{1}{2}$ to the left and right. If the random walk is in the middle bin, this correctly gives for the probability after the next step $P(2)' = \frac{1}{2}P(1) + \frac{1}{2}P(3) = \frac{1}{3}$. In the bins at the edges one of these contributions is missing, therefore it has to be replaced by the contribution of the events when the random walk stays in the bin. We have then, for instance, $P(1)' = \frac{1}{2}P(2) + \frac{1}{2}P(1) = \frac{1}{3}$. The general case can be treated similarly (see [7.5]).

The spectral density method can obviously be extended to more general cases, for instance to calculate the density of states in two variables, as $D(E, A)$ in (7.23). Other kinds of updatings with constraints are useful for the determination of the *constraint effective potential* in scalar field theories [7.6] (see also in chapter 2). The constraint effective potential requires keeping the sum of the field over the lattice constant. This can be achieved either by always changing simultaneously pairs of sites with opposite amounts [7.6], or by going to momentum space and keeping the zero momentum mode unchanged [7.7].

7.1.3   Error estimates

In numerical simulations the updating process creates a configuration sample $[\varphi_n]$, $(n = 1, 2, \ldots, N)$. The task is to determine the expectation value of different quantities. The simplest kind of quantities are defined by a function of the field variables $A[\varphi]$. An estimator of their expectation values is given by the sample average $\overline{A}$ defined in (7.12). Averaging over an infinite number of samples or over an infinitely large sample ($N \to \infty$) gives the ensemble average equal to the expectation value: $\overline{\overline{A}} = \langle A \rangle$. Such quantities, which can be directly obtained as averages, may be called *primary quantities*, in distinction to the *secondary quantities* determined by functions of the averages.

A simple example of secondary quantities is the correlation of two primary quantities $A[\varphi], B[\varphi]$, which is defined as

$$(AB) \equiv \langle AB \rangle - \langle A \rangle \langle B \rangle \; . \tag{7.26}$$

A specific kind of correlation often occurs in connection with two-point functions. Namely, if $S_t[\varphi]$ is a *timeslice field* depending only on field variables at $x_4 = t$, then

$$(S_{t_1} S_{t_2}) = (S_{t_1 - t_2 + t_0} S_{t_0}) = \langle S_{t_1} S_{t_2} \rangle - \langle S_{t_1} \rangle \langle S_{t_2} \rangle \; . \tag{7.27}$$

The first equality here is the consequence of time translation invariance. For instance, in case of a scalar field ϕ, the simplest timeslice field is just the average over the timeslice, which corresponds to projecting out zero spatial momentum:

$$S_t \equiv \frac{1}{V} \sum_{\mathbf{x}} \phi_{\mathbf{x}t} \ . \tag{7.28}$$

The timeslice correlations are often used to determine the *masses* (see, for instance, in (5.96)). The mass in lattice units m can be extracted from the large time distance behaviour of zero spatial momentum timeslice correlations:

$$\langle S_{t+t_0} S_{t_0} \rangle \simeq C_0 + C_1(e^{-mt} + e^{-m(T-t)}) \ . \tag{7.29}$$

The constant C_0, which is usually proportional to e^{-mT}, occurs only if $\langle S_t \rangle \neq 0$. Since (7.29) is only the asymptotic behaviour for $t, T \to \infty$, the exact value of the mass m is not easy to determine (one has to specify some well-defined numerical limiting procedure). It is conceptually simpler to consider *effective masses*, which are defined at a given lattice time extension T and fixed timeslice pair t_1, t_2. Let us consider first the case $C_0 = 0$, and define C_1 and $m(t_1, t_2, T)$ by the solution of the system of two equations

$$\langle S_{t_1+t_0} S_{t_0} \rangle = C_1(e^{-m(t_1,t_2,T)t_1} + e^{-m(t_1,t_2,T)(T-t_1)}) \ ,$$

$$\langle S_{t_2+t_0} S_{t_0} \rangle = C_1(e^{-m(t_1,t_2,T)t_2} + e^{-m(t_1,t_2,T)(T-t_2)}) \ . \tag{7.30}$$

In this case the true mass is given by

$$m = \lim_{t_1,t_2,T \to \infty} m(t_1, t_2, T) \ . \tag{7.31}$$

The effective mass $m(t_1, t_2, T)$ can be numerically obtained [7.8] from the ratio

$$r_{12} \equiv \frac{\langle S_{t_1+t_0} S_{t_0} \rangle}{\langle S_{t_2+t_0} S_{t_0} \rangle} = \frac{e^{-m(t_1,t_2,T)t_1} + e^{-m(t_1,t_2,T)(T-t_1)}}{e^{-m(t_1,t_2,T)t_2} + e^{-m(t_1,t_2,T)(T-t_2)}} \ . \tag{7.32}$$

With the notations

$$\tau_i \equiv \left(\frac{1}{2}T - t_i \right) \ , \qquad x \equiv e^{-m(t_1,t_2,T)} \ , \tag{7.33}$$

this implies

$$r_{12}(x^{\tau_2} + x^{-\tau_2}) = (x^{\tau_1} + x^{-\tau_1}) \ . \tag{7.34}$$

This can be easily solved numerically for x, to give $m(t_1, t_2, T) = -\log x$. In the special case $\tau_2 = 0$, $r_{12} \to r_1$ we have the analytic expression

$$m(t_1, T/2, T) = \tau_1^{-1} \log \left[r_1 + (r_1^2 - 1)^{1/2} \right] \ . \tag{7.35}$$

If in (7.29) there is a non-zero constant C_0, it can be cancelled by subtracting the value of the timeslice correlation at some third time t_3.

One can take, for instance, $t_3 = \frac{1}{2}(t_1 + t_2)$. In this case the numerical equation to be solved is

$$(r_{13} - 1)(x^{\tau_2} + x^{-\tau_2} - x^{\tau_3} - x^{-\tau_3}) = (r_{23} - 1)(x^{\tau_1} + x^{-\tau_1} - x^{\tau_3} - x^{-\tau_3}) . \quad (7.36)$$

The effective masses $m(t_1, t_2, T)$, as secondary quantities, are still relatively simply defined functions of the sample averages. More complicated functions do also occur, for instance, if the mass $m(t_1, t_2, T)$ is defined by fitting the behaviour of the timeslice correlations by the form in (7.29), in an interval $[t_1, t_2]$ on a lattice with time extension T. The fit is usually defined by the minimum of χ^2, which also depends on the errors of the correlations (see below, at (7.46)). More generally, an important class of secondary quantities is given by the parameters of fits of some primary or secondary quantities.

An important general question is, how good are the estimates obtained from a finite (but hopefully large) sample of configurations? In order to formulate this question mathematically, it is useful to consider the canonical (equilibrium) ensemble not only as an infinite collection of configurations, but also as an infinite collection of samples of configurations (for instance, produced by an infinity of updating processes). Sometimes also the number of steps in the updating process N has to be considered large (going to infinity).

In the ideal case, when the configurations contained in a sample are all statistically independent, the sample average $\overline{A}$ is normally distributed around the mean value $\overline{\overline{A}}$, with variance

$$\sigma_{\overline{A}}^2 = \frac{\overline{A^2} - \overline{A}^2}{N - 1} = \frac{\overline{(A - \overline{A})^2}}{N - 1} . \quad (7.37)$$

This is the consequence of the central limit theorem [7.9], and is true under some rather mild assumptions on the distribution of the numerically measured values of the primary quantity $A_n \equiv A[\varphi_n]$ ($n = 1, 2, \ldots, N$). In this case the quoted error on the sample average would be $\overline{\overline{A}} = \overline{A} \pm \sigma_{\overline{A}}$.

The error estimate in (7.37) is usually too optimistic, because during the updating process the subsequent configurations have lots of similarities, they are by no means independent. (Quite often one step in the updating consists only of a single local change of the configuration, or of a 'sweep' of local changes over the whole lattice.) This correlation in the sequence of generated configurations is called *autocorrelation*. For a primary quantity A the autocorrelation is defined as

$$(A_n A_{n+\tau}) \equiv \langle A_n A_{n+\tau} \rangle - \langle A_n \rangle \langle A_{n+\tau} \rangle$$

$$= \langle A_n A_{n+\tau} \rangle - \langle A \rangle^2 = \langle (A_n - \overline{A})(A_{n+\tau} - \overline{A}) \rangle . \quad (7.38)$$

In the different steps here we used $\langle A_n \rangle = \langle \overline{A} \rangle = \langle A \rangle$. (Remember that the ensemble is an infinite collection of samples!) For time independent updating processes and infinitely large samples the autocorrelation $(A_n A_{n+\tau})$ depends only on the time difference τ. (For finite samples there is some dependence on the initial configuration, too.) Note, that here we use 'time' in the sense of the 'computer time' labelling the configurations, and not as the Euclidean time on the lattice ($t \equiv x_4$).

In terms of the autocorrelation the true variance of $\overline{A}$ is

$$\sigma_{\overline{A}}^2 = \left\langle \left[\frac{1}{N} \sum_{n=1}^{N} (A_n - \langle A \rangle) \right]^2 \right\rangle = \sum_{\tau=-N}^{N} \frac{N - |\tau|}{N^2} (A_n A_{n+\tau})$$

$$\xrightarrow{N \to \infty} (AA) \frac{2\tau_{int,A}}{N} \simeq (\overline{A^2} - \overline{A}^2) \frac{2\tau_{int,A}}{N} , \qquad (7.39)$$

where the *integrated* (or *effective*) *autocorrelation time* $\tau_{int,A}$ is defined as

$$\tau_{int,A} \equiv \frac{1}{2} \sum_{\tau=-\infty}^{+\infty} \frac{(A_n A_{n+\tau})}{(AA)} . \qquad (7.40)$$

Comparing (7.39) with (7.37), one can see that due to the autocorrelation the effective number of independent measurements is $N/(2\tau_{int,A})$. That is, if the numerical measurement of A takes a substantial computer time, it is better to skip about $2\tau_{int,A}$ configurations between the measurements.

As it was discussed after (7.21), due to Perron's theorem, the relaxation of the configuration ensemble towards equilibrium is dominated by the second largest eigenvalue $\lambda_2 < 1$ of the transition probability matrix. Since in n updating steps this implies a damping λ_2^n, the effective autocorrelation time $\tau_{int,A}$ is typically of the order of the *exponential autocorrelation time* τ_{exp}, which is defined by

$$\tau_{exp} \equiv \frac{-1}{\log \lambda_2} \simeq (1 - \lambda_2)^{-1} . \qquad (7.41)$$

This is, however, only an estimate for $\tau_{int,A}$ which may, for instance, also depend on the measured quantity A. Note that $\tau_{int,A}$ and τ_{exp} play different rôles: $\tau_{int,A}$ tells us how often the measurement of the quantity A has to be performed for an effective use of computer time. τ_{exp} is characterizing the asymptotic decrease of the autocorrelation. From the practical point of view, for the measurement of A, of course, only $\tau_{int,A}$ matters. The exponential autocorrelation time τ_{exp} has the advantage, that it depends only on the updating process.

In practice one can numerically determine the autocorrelation, and using (7.40) with a truncation in the sum over time differences, it is also possible to calculate the effective autocorrelation time $\tau_{int,A}$. Another practical way of obtaining correct error estimates from an updating sequence with autocorrelations is *binning*. Assuming that the sequence is long enough, one can build blocks of subsequent configurations, called *bins*, and average the primary quantities first in the bins. The obtained bin averages themselves can then be considered as results of single measurements, and can be used to estimate the variance according to (7.37). If the bins are large enough (larger than the autocorrelation time), then the average values in different bins are practically uncorrelated, and the obtained error estimate $\sigma_{\overline{A}}$ in (7.37) is correct. Therefore, if the bins are further increased, beyond the autocorrelation time, the error estimates remain constant. (This can also be a practical way to observe the autocorrelation time.) Of course, access to computer time is limited, therefore the number of very large bins is usually small. In this case the error of the error estimate becomes large, and a further increase of the bin size is useless.

The continuum limit of lattice quantum field theories is always defined in the vicinity of *critical points*, that is near second order phase transitions in the bare parameter space. This causes a difficulty in autocorrelations, because the correlation length $\xi = m^{-1}$ diverges near the critical points. This implies a divergence of the autocorrelation time according to the *dynamical scaling law*

$$\tau_{int,A} \propto \xi^{z(A)}. \tag{7.42}$$

Here $z(A)$ is the *dynamical critical exponent*, which is characteristic to the updating process, and may also depend on the observed quantity A. (Note that a physical quantity is usually sensitive to regions with typical extensions similar to the correlation length.) In local updatings we usually have $z \simeq 2$, which can be understood as due to the 'diffusion' of the changes induced by the local updating. Obviously, for $z(A) \neq 0$ (7.42) implies a deterioration of the effectivity of the updating process near critical points. This is called *critical slowing down*.

Good updating algorithms avoid critical slowing down as much as possible. One can try to fight it by improving the algorithm, for instance, by including some appropriate global changes. In fact, in some special cases, it is also possible to achieve $z(A) \simeq 0$ (see the cluster algorithms in section 7.7). Nevertheless, one has to keep in mind, that for constant physical lattice volumes the number of lattice sites in d dimensions has to increase as ξ^d. Therefore, the relevant increase in computation time in the continuum limit is at least ξ^{d+z}, which is bad enough for $d = 4$, even with $z = 0$.

Up to now we mainly concentrated on the error estimates of the primary quantities, which are directly obtained as sample averages. In principle, the secondary quantities can be treated similarly. Since, however, the functions of the averages have to be evaluated on large samples, some care has to be taken. In order to obtain correct error estimates, sometimes one has to consider very large (mathematically speaking, infinite) samples, a condition which is difficult to fulfil in practice. Also the attitude towards considering blocks of subsequent configurations (which we called up to now 'bins') is different for secondary quantities, because their values are defined on samples. Therefore, dividing the whole sample into smaller parts means considering *subsamples*. If the subsamples are large enough, one can determine any primary or secondary quantity in them, and use (7.37) as an error estimate for the average of the subsample averages, with N as the number of subsamples. In practice this condition is only seldomly fulfilled: large bins are not much different from small subsamples, and neither of them are sufficiently large. Sometimes even the computation of some quantities takes a lot of computer time, and therefore one has to try to estimate the errors from a relatively small number of measurements. We shall return to this problem at the end of this subsection.

Now let us consider another interesting problem, namely the error estimate of fit parameters. An important aspect here is the correlation between the fitted quantities. The problem is quite similar to the error estimates of fit parameters in real (non-numerical) experiments, therefore the discussion here will be kept short. (For more details see [7.9] and [7.10].) The general task is to estimate the parameters λ_i, $(i = 1, 2, \ldots, i_{max})$, to find their statistical errors and to see how well the theory describes the data obtained in the numerical experiment. (There is almost always some approximation involved in the form of the fit because, for instance, some higher excited states are omitted or finite lattice volume effects are neglected etc.)

To be specific let us consider a sample of independent measurements of some primary or secondary quantities y_t, $(t = 1, 2, \ldots, t_{max})$. (The autocorrelation of the configurations in the updating process is assumed to be eliminated, for instance, by binning subsequent configurations into large enough bins.) The sample averages are defined as

$$\overline{y_t} \equiv \frac{1}{N} \sum_{n=1}^{N} y_{tn} \ . \tag{7.43}$$

The *correlation* between the measured quantities is characterized by the *covariance matrix*

$$C_{tu} \equiv \frac{1}{N(N-1)} \sum_{n=1}^{N} (y_{tn} - \overline{y_t})(y_{un} - \overline{y_u})$$

$$= \frac{1}{N-1} (\overline{y_t y_u} - \overline{y_t}\ \overline{y_u}) \ . \tag{7.44}$$

The symmetrical matrix $C_{tu} = C_{ut}$ can be determined numerically as a secondary quantity. If there were no correlations among the quantities, the error estimate for $\overline{y_t}$ would be given by the diagonal element C_{tt}, as $\overline{\overline{y_t}} = \overline{y_t} \pm \sqrt{C_{tt}}$. Let us assume that the multivariable distribution of the sample averages is Gaussian. (For this it is not necessary that the distribution of the individual measured values y_{tn} be Gaussian. It is enough, for instance, that y_{tn} have finite variance.) This means that the probability distribution of $\overline{y_t}$ is

$$P(\overline{y}) \propto \exp\left\{-\frac{1}{2}\sum_{tu}(\overline{y_t} - \overline{\overline{y_t}})C_{tu}^{-1}(\overline{y_u} - \overline{\overline{y_u}})\right\}. \tag{7.45}$$

The theoretical formula $f_t(\lambda)$, to be fitted to $\overline{y_t}$ for $t = 1, 2, \ldots, t_{max}$, depends on the parameters λ_i ($i = 1, 2, \ldots, i_{max}$). The best fit is defined by the 'maximum likelihood' corresponding to the 'least squares', to the minimum of χ^2 defined as

$$\chi^2 \equiv \sum_{tu}(\overline{y_t} - f_t(\lambda))C_{tu}^{-1}(\overline{y_u} - f_u(\lambda)). \tag{7.46}$$

The minimum χ^2 occurs at $\lambda_i \equiv \overline{\lambda_i}$. If the measurement sample and the eventual subsamples defining secondary quantities are infinitely large (and if the theory is good), the ensemble average of λ_i gives the ensemble average of y_t by

$$\overline{\overline{y_t}} = f_t(\overline{\overline{\lambda}}). \tag{7.47}$$

The condition of the χ^2 minimum is

$$0 = \frac{\partial\chi^2}{\partial\lambda_i}\bigg|_{\lambda=\overline{\lambda}} = 2\sum_{tu}\frac{\partial f_t(\overline{\lambda})}{\partial\overline{\lambda_i}}C_{tu}^{-1}\left[f_u(\overline{\lambda}) - \overline{y_u}\right]. \tag{7.48}$$

The question is, how do $\overline{\lambda_i}$ fluctuate around $\overline{\overline{\lambda_i}}$, as $\overline{y_t}$ fluctuate around $\overline{\overline{y_t}}$? In other words, what are the errors of $\overline{\lambda_i}$? Clearly, the probability distribution of $\overline{y_t}$ determines the probability distribution of $\overline{\lambda_i}$. The covariance matrix of the fit parameters is given by

$$\Delta_{ij} = \int[d\overline{y}]P(\overline{y})(\overline{\lambda_i} - \overline{\overline{\lambda_i}})(\overline{\lambda_j} - \overline{\overline{\lambda_j}}). \tag{7.49}$$

Δ_{ij} can be numerically measured by considering large subsamples, if $\overline{\overline{\lambda_i}}$ is approximated by the best fit parameters in the total sample.

It is convenient to introduce the *Hessian matrix* by

$$H_{ij} \equiv \frac{\partial^2\chi^2}{\partial\lambda_i\,\partial\lambda_j}\bigg|_{\lambda=\overline{\lambda}}$$

$$= 2\sum_{tu}\left\{\frac{\partial^2 f_t(\overline{\lambda})}{\partial\overline{\lambda_i}\,\partial\overline{\lambda_j}}C_{tu}^{-1}\left[f_u(\overline{\lambda}) - \overline{y_u}\right] + \frac{\partial f_t(\overline{\lambda})}{\partial\overline{\lambda_i}}C_{tu}^{-1}\frac{\partial f_u(\overline{\lambda})}{\partial\overline{\lambda_j}}\right\}. \tag{7.50}$$

In the linear approximation

$$\overline{\lambda}_i = \overline{\overline{\lambda}}_i + \sum_t \frac{\partial \overline{\lambda}_i}{\partial \overline{y}_t} (\overline{y}_t - \overline{\overline{y}}_t) + O(\overline{y} - \overline{\overline{y}})^2 \; , \tag{7.51}$$

we have

$$\Delta_{ij} = \sum_{tu} \frac{\partial \overline{\lambda}_i}{\partial \overline{y}_t} C_{tu} \frac{\partial \overline{\lambda}_j}{\partial \overline{y}_u} \; . \tag{7.52}$$

Taking the derivative of the minimum condition (7.48) with respect to $\overline{y}_u$, it follows

$$\frac{\partial \overline{\lambda}_i}{\partial \overline{y}_u} = 2 \sum_{j,t} H_{ij}^{-1} \frac{\partial f_t(\overline{\lambda})}{\partial \overline{\lambda}_j} C_{tu}^{-1} \; . \tag{7.53}$$

This gives for the covariance matrix of the fit parameters

$$\Delta_{ij} = 4 \sum_{lm,tu} H_{il}^{-1} \frac{\partial f_t(\overline{\lambda})}{\partial \overline{\lambda}_l} C_{tu}^{-1} \frac{\partial f_u(\overline{\lambda})}{\partial \overline{\lambda}_m} H_{mj}^{-1} \; . \tag{7.54}$$

If the samples are large and the fits are good, in (7.50) one can neglect the term proportional to $f(\overline{\lambda}) - \overline{y}$, and then

$$\Delta_{ij} = 2H_{ij}^{-1} + O\left(f(\overline{\lambda}) - \overline{y}\right) \; . \tag{7.55}$$

The distribution of the best fit parameters $\overline{\lambda}_i$ is also Gaussian around the ensemble averages $\overline{\overline{\lambda}}_i$, that is

$$P(\overline{\lambda}) \propto \exp\left\{ -\frac{1}{2} \sum_{ij} (\overline{\lambda}_i - \overline{\overline{\lambda}}_i) \Delta_{ij}^{-1} (\overline{\lambda}_j - \overline{\overline{\lambda}}_j) \right\} \; . \tag{7.56}$$

In particular, if one is interested only in the error of a single fit parameter λ_i, no matter what the other parameters do, we can quote as an error

$$\overline{\overline{\lambda}}_i = \overline{\lambda}_i \pm \sqrt{\Delta_{ii}} \; . \tag{7.57}$$

More generally, the errors of any subset of parameters are given by the corresponding submatrix of Δ_{ij}. The distribution of the parameters can be characterized by ellipsoids, with axes given by the eigenvalues of the submatrix.

Assuming that the measurement sample is large enough, one can consider many subsamples, which are themselves still large enough, and obtain estimates for the error of any primary or secondary quantity from the observed fluctuations. In practice, however, the statistics is often not so large. In this case the secondary quantities, which are functions of the primary averages, have to be considered with a particular care. Namely, taking the average and constructing the functions are usually non-commutative.

The best estimate of the value of the secondary quantities can be obtained from the averages over the total sample, not from the average of the function values on subsamples. The distribution of the values of the functions in subsamples can be used for the estimate of the errors, if there are enough large subsamples.

If this is not the case, one can use the *jackknife analysis* for a more reliable estimate of the error of secondary quantities [7.11, 7.12, 7.13]: Consider a not very large sample of independent measurements of a primary quantity A. The measured values are $A_1, A_2, \ldots, A_{N_s}$, and the sample average is

$$\overline{A} \equiv \frac{1}{N_s} \sum_{s=1}^{N_s} A_s . \tag{7.58}$$

The best estimate of a secondary quantity is $\overline{y} = y(\overline{A})$ (not $\overline{y(A)}$). A stable error estimator for $\overline{y}$ can be derived from the jackknife averages obtained by omitting a single measurement from the sample in all possible ways:

$$A_{(J)s} \equiv \frac{1}{N_s - 1} \sum_{r \neq s} A_r . \tag{7.59}$$

The corresponding values of the secondary quantity are the *jackknife estimators* $y_{(J)s} \equiv y(A_{(J)s})$, with an average

$$\overline{y_{(J)}} \equiv \frac{1}{N_s} \sum_{s=1}^{N_s} y_{(J)s} . \tag{7.60}$$

The variance of the jackknife estimators can be obtained as

$$\sigma_{(J)\overline{y}}^2 \equiv \frac{N_s - 1}{N_s} \sum_{s=1}^{N_s} \left(y_{(J)s} - \overline{y_{(J)}} \right)^2 . \tag{7.61}$$

For primary quantities this is equivalent to the simple variance in (7.37): $\sigma_{(J)\overline{A}} = \sigma_{\overline{A}}$. For secondary quantities, however, the jackknife variance estimator is usually more reliable, which gives as an error estimate $\overline{y} = \overline{y} \pm \sigma_{(J)\overline{y}}$.

7.1.4 Improved estimators

In a numerical simulation one is usually interested in measuring a set of physical quantities with as small statistical errors as possible. The statistical errors can also be minimized by a clever choice of the measured quantities. Namely, it is possible that, in order to measure the expectation value $\langle A \rangle$, one can choose another quantity A_I with the same mean value

$$\langle A_I \rangle = \langle A \rangle , \tag{7.62}$$

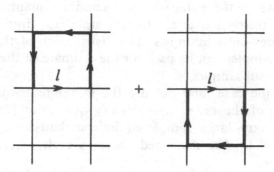

Fig. 7.1. Two of the contributions to S_l. The link is denoted by l, and the rest of the plaquettes (the 'staples') are drawn with double lines.

but with

$$\langle A_I^2 \rangle < \langle A^2 \rangle \,. \tag{7.63}$$

According to (7.37) the variance of $\overline{A_I}$ is smaller than the variance of $\overline{A}$. Since for infinite statistics both $\overline{A_I}$ and $\overline{A}$ approach the same mean value $\overline{\overline{A_I}} = \overline{\overline{A}}$, A_I is an *improved estimator* for the expectation value $\langle A \rangle = \overline{\overline{A}}$.

As an example, let us consider [7.14] pure SU(N) lattice gauge theory with the Wilson lattice action. An important class of measurable quantities depends linearly on the link variable $U_l \in SU(N)$, and has the form

$$A[U] \equiv \mathrm{Tr}\,\{ U_l R[\check{U}_l] \} \,. \tag{7.64}$$

Here $\check{U}_l$ stands for all other link variables besides U_l. Let us separate the Wilson action (7.6) into a component depending on U_l and the rest:

$$S[U] \equiv -\frac{\beta}{N} \mathrm{Re}\,\mathrm{Tr}\,(U_l S_l) + \check{S}(\check{U}_l) \,. \tag{7.65}$$

The $N \otimes N$ matrix S_l is the sum of the products of link variables over the six 'staples', as illustrated by fig. 7.1. The expectation value $\langle A \rangle$ can be written as

$$\langle A \rangle = \frac{\int [\mathrm{d}\check{U}_l] \exp\left(-\check{S}(\check{U}_l)\right) \int \mathrm{d}U_l \exp\left[(\beta/N)\,\mathrm{Re}\,\mathrm{Tr}\,(U_l S_l)\right] \mathrm{Tr}\,\{ U_l R[\check{U}_l] \}}{\int [\mathrm{d}\check{U}_l] \exp\left(-\check{S}(\check{U}_l)\right) \int \mathrm{d}U_l \exp\left[(\beta/N)\,\mathrm{Re}\,\mathrm{Tr}\,(U_l S_l)\right]} \,. \tag{7.66}$$

In terms of the averaged link

$$\overline{U}_l \equiv \frac{\int \mathrm{d}U_l \exp\left[(\beta/N)\,\mathrm{Re}\,\mathrm{Tr}\,(U_l S_l)\right] U_l}{\int \mathrm{d}U_l \exp\left[(\beta/N)\,\mathrm{Re}\,\mathrm{Tr}\,(U_l S_l)\right]} \,, \tag{7.67}$$

this is given as

$$\langle A \rangle = \langle A_I \rangle \equiv \frac{\int [\mathrm{d}U] \mathrm{e}^{-S[U]} \mathrm{Tr}\,\{ \overline{U}_l R[\check{U}_l] \}}{\int [\mathrm{d}U] \mathrm{e}^{-S[U]}} \,. \tag{7.68}$$

It is intuitively clear that replacing the link U_l by the averaged link $\overline{U}_l$ gives a smaller variance. This expectation can be supported by direct numerical investigations [7.14].

Replacing in an expectation value the link variables by averaged links is allowed also on several links simultaneously, if no pair of the replaced links is on the same plaquette. In the case of SU(2) ($N = 2$) the group integration in (7.67) can be easily performed [7.15]. With $K_l \equiv |\det S_l|^{1/2}$ one obtains in terms of the Bessel functions $I_{1,2}$

$$\overline{U}_l = K_l S_l^{-1} \frac{I_2(\beta K_l)}{I_1(\beta K_l)} . \tag{7.69}$$

For SU(3) ($N = 3$) an explicit formula still exists, but it is rather lengthy [7.15], therefore it is in practice better to estimate $\overline{U}_l$ by numerical Monte Carlo integration. The procedure is called *multihit*, because it is the same as a repeated application of the heatbath updating on the link (see section 7.3).

A similar example of improved estimators can be constructed in the Ising model by replacing spins by averaged spins [7.16]. A very successful kind of improved estimators is obtained in spin models by using the clusters in the cluster algorithms (see section 7.7).

7.2   Metropolis algorithms

The general requirements for the updating (Markov) process were formulated in section 7.1, in particular by the relations (7.14)–(7.19). In practice the updating is in most cases built upon transition probability matrices $P([\varphi'] \leftarrow [\varphi])$, which satisfy the detailed balance condition (7.20) involving the normalized canonical (equilibrium) density $W_c[\varphi]$.

In order to illustrate some general classes of transition probability matrices let us consider a simple system with in total three possible configurations. Assume that in the canonical distribution the first configuration has a factor α smaller probability ($0 \leq \alpha \leq 1$) than the other two, which have equal probability. This means that the canonical distribution is given, as a column vector, by

$$W_c = \begin{pmatrix} \alpha(2 + \alpha)^{-1} \\ (2 + \alpha)^{-1} \\ (2 + \alpha)^{-1} \end{pmatrix} . \tag{7.70}$$

One can easily see that the most general transition probability matrix satisfying detailed balance in this case is

$$
P = \begin{pmatrix} 1 - P_{21} - P_{31} & \alpha P_{21} & \alpha P_{31} \\ P_{21} & 1 - \alpha P_{21} - P_{32} & P_{32} \\ P_{31} & P_{32} & 1 - \alpha P_{31} - P_{32} \end{pmatrix} . \tag{7.71}
$$

The three free parameters have to satisfy

$$
0 \le P_{21}, P_{31}, P_{32}, P_{21} + P_{31}, \alpha P_{21} + P_{32}, \alpha P_{31} + P_{32} \le 1 . \tag{7.72}
$$

A simple special case is the 'heatbath-like' transition matrix

$$
P = \begin{pmatrix} \alpha(2 + \alpha)^{-1} & \alpha(2 + \alpha)^{-1} & \alpha(2 + \alpha)^{-1} \\ (2 + \alpha)^{-1} & (2 + \alpha)^{-1} & (2 + \alpha)^{-1} \\ (2 + \alpha)^{-1} & (2 + \alpha)^{-1} & (2 + \alpha)^{-1} \end{pmatrix} , \tag{7.73}
$$

when every column is equal to W_c in (7.70). More details about this kind of choice will be discussed in section 7.3. Another two examples for P are

$$
P = \begin{pmatrix} 1/3 & \alpha/3 & \alpha/3 \\ 1/3 & (2 - \alpha)/3 & 1/3 \\ 1/3 & 1/3 & (2 - \alpha)/3 \end{pmatrix} ,
$$

$$
P' = \begin{pmatrix} 0 & \alpha/2 & \alpha/2 \\ 1/2 & (1 - \alpha)/2 & 1/2 \\ 1/2 & 1/2 & (1 - \alpha)/2 \end{pmatrix} . \tag{7.74}
$$

These are 'Metropolis-type' transition matrices which will be considered in this section. The difference between P and P' here is that, for instance, the first configuration goes according to P with equal probability to any of the three configurations, but according to P' only to configurations number two or three. Instead of the whole space (123) one can also consider the subspaces (12) or (13) and, similarly to P', define the transition matrices

$$
P_{(12)} = \begin{pmatrix} 0 & \alpha & 0 \\ 1 & 1 - \alpha & 0 \\ 0 & 0 & 1 \end{pmatrix} , \qquad P_{(13)} = \begin{pmatrix} 0 & 0 & \alpha \\ 0 & 1 & 0 \\ 1 & 0 & 1 - \alpha \end{pmatrix} . \tag{7.75}
$$

These are, however, not ergodic because they leave, respectively, the subspaces (12) and (13) invariant. Ergodicity is achieved if the subspaces (12) and (13) are updated in an alternating order. This corresponds to the transition matrix of the composite step $P \equiv P_{(13)} P_{(12)}$.

The transition probability matrix of the *Metropolis algorithm* [7.17, 7.1] is constructed in analogy with (7.74). For definiteness, let us consider a system with $\mathcal{N}$ discrete possible configurations, and define P for $[\varphi'] \ne [\varphi]$

by

$$P([\varphi'] \leftarrow [\varphi]) = \begin{cases} \mathcal{N}^{-1} & \text{if} \quad W_c[\varphi'] \geq W_c[\varphi] \\ \\ \mathcal{N}^{-1} W_c[\varphi']/W_c[\varphi] & \text{if} \quad W_c[\varphi'] < W_c[\varphi] \end{cases} , \quad (7.76)$$

that is

$$P([\varphi'] \leftarrow [\varphi]) = \mathcal{N}^{-1} \min\left\{1, \frac{W_c[\varphi']}{W_c[\varphi]}\right\} \equiv \mathcal{N}^{-1} \min\{1, R\} . \quad (7.77)$$

This transition matrix is realized by the following numerical procedure:

i) choose first a trial configuration randomly from $\mathcal{N}$ configurations, and then

ii) accept it as the next configuration in any case if the Boltzmann factor is increased (the action is decreased).

If the Boltzmann factor is decreased (the action is increased), then accept the change with probability equal to the ratio of the Boltzmann factors.

The accept–reject step can be implemented in a computer program by comparing the ratio of the Boltzmann factors to a pseudo-random number between 0 and 1.

The detailed balance condition (7.20), which is sufficient for the transition probability of an updating process, can be immediately verified for (7.77). Therefore, the Metropolis algorithm is a possible updating procedure. In the form defined above it also satisfies strong ergodicity (7.16).

In practice it is not convenient to consider all possible changes of a configuration simultaneously. In analogy to (7.75), one usually considers subspaces of the configurations which, for instance, can be reached by a local change of a single site or link variable. Even such subspaces are usually restricted by not allowing an arbitrary change, but only a subset of possible changes which can be generated, for instance, by choosing a random group element from a look-up table, and multiplying the variable by it. The look-up table can be refreshed from time to time, in order to improve ergodicity. All these variations are special cases of the following generalization [7.18]: generate P by a proposed change and an accept–reject step as $P = P_A P_C$, where

i) $P_C([\varphi'] \leftarrow [\varphi])$ is an arbitrary probability distribution for the proposed change of the configuration $[\varphi] \rightarrow [\varphi']$, and

ii) the acceptance probability $P_A([\varphi'] \leftarrow [\varphi])$ is defined in such a way that it compensates for P_C, namely

$$P_A([\varphi'] \leftarrow [\varphi]) \propto \min \left\{ 1, \frac{P_C([\varphi] \leftarrow [\varphi']) W_c[\varphi']}{P_C([\varphi'] \leftarrow [\varphi]) W_c[\varphi]} \right\} . \qquad (7.78)$$

Clearly, the simple Metropolis algorithm defined by (7.77) is a special case of (7.78), corresponding to a uniform P_C. The detailed balance condition (7.20) for $P = P_A P_C$ can also be easily proven by inspection.

Another kind of generalization of the Metropolis algorithm concerns the acceptance function $\min \{1, R\}$ in (7.77) or (7.78). For instance, one can also take $R/(1 + R)$ which gives, in the case of (7.77),

$$P([\varphi'] \leftarrow [\varphi]) = \mathcal{N}^{-1} \frac{W_c[\varphi']}{W_c[\varphi] + W_c[\varphi']} = \mathcal{N}^{-1} \frac{R}{1 + R} . \qquad (7.79)$$

Again, the proof of detailed balance goes by inspection.

Besides detailed balance the other important condition, which has to be satisfied by the updating process, is *ergodicity*. This means that the transition probability matrix has to be such that every configuration has to be connected to any other one by a finite probability. In other words, under the application of P there cannot be invariant subspaces other than the whole space of configurations. In practice the strong ergodicity condition in (7.16) is difficult to satisfy. The solution is analogous to taking the product of the two matrices in (7.75) as the transition probability matrix P. That is, a sequence of individually non-ergodic steps has to be considered as one large ergodic step. The transition probability of the individual steps is always assumed to be non-negative:

$$P_i ([\varphi'] \leftarrow [\varphi]) \geq 0 . \qquad (7.80)$$

In this case it is easy to see that if the transition probability matrices P_1 and P_2 both satisfy the conditions (7.15) and (7.19), then the product $P = P_2 P_1$ does the same, too.

A particular way to choose the trial element for updating goes under the name *overrelaxation*. This was first applied to multiquadratic actions [7.19], where it is similar to the technique of overrelaxation in algorithms for solving difference equations. The idea can also be generalized to gauge theories [7.20, 7.21]. For definiteness, let us consider here an SU(N) lattice gauge model. Let us assume that the action can be decomposed with respect to the link variable U_l as shown in (7.65). (Note that this is a more general case than the simple Wilson lattice action, if S_l is appropriately defined.) The aim of the overrelaxation is to speed up the updating process, especially in the vicinity of the continuum limit, where critical slowing down is dangerous (see (7.42)). This is achieved by choosing the trial link variable U_l' as far as possible from U_l. A convenient form for

U'_l is

$$U'_l = U_0 U_l^{-1} U_0 , \qquad (7.81)$$

with $U_0 \in SU(N)$. Since the inverse relation is $U_l = U_0 U_l'^{-1} U_0$, the factors P_C cancel in the acceptance probability (7.78). The choice of U_0 depends on the gauge group.

First let us consider the case of SU(2) ($N = 2$), when a natural choice is

$$U_0 = [\mathrm{Pr}\,(S_l)]^{-1} = S_l^{-1} [\det(S_l S_l^+)]^{1/4} = S_l^{-1} [\det(S_l)]^{1/2} . \qquad (7.82)$$

This is the inverse of the projection $\mathrm{Pr}\,(S_l)$ of the matrix S_l to the SU(2) group. Here we used a special feature of SU(2), namely that the sum of SU(2) matrices is always proportional to an SU(2) matrix. Using the reality of the trace of SU(2) elements, one realizes that

$$\mathrm{Tr}\,(U'_l S_l) = \mathrm{Tr}\,(U_l S_l) . \qquad (7.83)$$

Therefore, the action is unchanged and, according to the acceptance probability in (7.78), the proposed change of the link $U_l \to U'_l$ is always accepted. Another consequence of the constant action value is that this algorithm is non-ergodic. Instead of the canonical ensemble, it generates the *microcanonical ensemble* with constant action. (Remember that, according to the thermodynamic analogy, the action corresponds to the energy.) There are two ways to cure this difficulty: either the overrelaxed changes of the configuration are mixed with some ergodic procedure, like a normal Metropolis algorithm, or one uses the correspondence of the canonical and microcanonical ensembles, known from thermodynamics [A6]. Namely, for large volumes (in the 'thermodynamic limit') the canonical and microcanonical expectation values are very close to each other. In order to relate the two ensembles one has to match, for instance, the value of the average action.

In the case of gauge groups other than SU(2), the appropriate choice of the group element U_0 for the overrelaxed change in (7.81) is more involved, because in general it is not so straightforward to project the sum of group elements S_l onto the group. For a possible procedure in SU(3) see [7.22].

The main advantage of the overrelaxation algorithm is that it can be used to fight critical slowing down. The expected improvement can be expressed by a dynamical critical exponent in (7.42) of about $z \simeq 1$.

7.3   Heatbath algorithms

An efficient way of choosing the transition probability matrix for the
updating process would be to take the canonical (equilibrium) distribution
for the final configuration, irrespective of the initial one:

$$P([\varphi'] \leftarrow [\varphi]) = W_c[\varphi'] = Z^{-1}e^{-S[\varphi']} \ . \tag{7.84}$$

In the case of the simple model with three configurations, which was
discussed in section 7.2, P is given in (7.73). The transition matrix
in (7.84) can also be considered as a special case of the generalized
Metropolis algorithm in (7.78), if the trial configuration is distributed
according to $P_C = W_c$, and hence the acceptance probability is $P_A = 1$.
One can easily see, that P in (7.84) satisfies all requirements like (7.15),
(7.18) and (7.19), including the strong ergodicity condition (7.16), and also
detailed balance (7.20).

The direct numerical implementation of (7.84) is, however, in typical
cases practically impossible, because of the huge variety of configurations.
What can be done is to implement (7.84) locally, keeping all the other field
variables fixed, except for the ones at a single lattice site or link. In this
way one obtains the *heatbath algorithms* [7.23]. The name expresses the
fact that this is something like bringing the chosen site (or link) in contact
with an infinite 'heat bath', as one sometimes imagines in thermodynamics.

In order to formulate the transition probability matrix for heatbath
algorithms in mathematical terms, let us call the local variables on the
specific site (or link) by φ_x. All other field variables, which will be kept
fixed during the updating of φ_x, will be denoted by $\check{\phi}_x$. Let $W_c(\varphi_x; \check{\phi}_x)$ be
the conditional probability distribution of φ_x in the canonical ensemble,
for fixed $\check{\phi}_x$. The whole canonical distribution will be written as

$$W_c[\varphi] \equiv W_c(\varphi_x, \check{\phi}_x) \equiv W_c(\varphi_x; \check{\phi}_x)\check{W}_c(\check{\phi}_x) \ . \tag{7.85}$$

Similarly, the conditional transition probability for the updated variables
will be denoted by $P_x(\varphi'_x \leftarrow \varphi_x; \check{\phi}_x)$. The transition probability matrix for
the whole system is then

$$P_x([\varphi'] \leftarrow [\varphi]) = P_x(\varphi'_x \leftarrow \varphi_x; \check{\phi}_x)\delta(\check{\phi}'_x - \check{\phi}_x) \ . \tag{7.86}$$

It will be assumed that $P_x([\varphi'] \leftarrow [\varphi])$ satisfies the conditions (7.15) and
(7.19). A sufficient condition for (7.19) is *local detailed balance*:

$$P_x(\varphi'_x \leftarrow \varphi_x; \check{\phi}_x)W_c(\varphi_x; \check{\phi}_x) = P_x(\varphi_x \leftarrow \varphi'_x; \check{\phi}_x)W_c(\varphi'_x; \check{\phi}_x) \ . \tag{7.87}$$

The local updating step described by $P_x([\varphi'] \leftarrow [\varphi])$ is, of course, not
ergodic because it acts only on φ_x. One does, however, require *local
ergodicity*:

$$P_x(\varphi'_x \leftarrow \varphi_x; \check{\phi}_x) > 0 \ . \tag{7.88}$$

In order to achieve ergodicity for the whole configuration one can, for instance, perform a cicle of local updates, called *sweep*, over all sites (or links) consecutively:

$$P([\varphi'] \leftarrow [\varphi]) = \prod_x P_x([\varphi'] \leftarrow [\varphi]) . \tag{7.89}$$

Another possibility is to choose different sites randomly:

$$P([\varphi'] \leftarrow [\varphi]) = \sum_x p_x P_x([\varphi'] \leftarrow [\varphi]) , \tag{7.90}$$

with a probability distribution $\sum_x p_x = 1$ ($p_x > 0$ for all x).

The formulas (7.85)–(7.90) are valid in a more general context, that is they are applicable also to other kind of local updates (like e. g. Metropolis algorithms). The local heatbath algorithm corresponds to the conditional transition probability matrix

$$P_x(\varphi'_x \leftarrow \varphi_x; \check{\phi}_x) = W_c(\varphi'_x; \check{\phi}_x) . \tag{7.91}$$

The numerical task in a heatbath updating is to generate the distribution $W_c(\varphi'_x; \check{\phi}_x)$. This is in principle rather simple, if the integral of the distribution $W_c(\varphi_x; \check{\phi}_x)$ is known, that is

$$W_c(\varphi_x; \check{\phi}_x) \, \mathrm{d}\varphi_x = \mathrm{d}E_{\check{\phi}_x}(\varphi_x) . \tag{7.92}$$

Since the measure $\mathrm{d}E_{\check{\phi}_x}$ is translation invariant, one can generate the distribution of φ'_x from a random number $r \in [0, 1]$ by

$$\varphi'_x = E_{\check{\phi}_x}^{-1}\{E_{\check{\phi}_x}(a) + r[E_{\check{\phi}_x}(b) - E_{\check{\phi}_x}(a)]\} . \tag{7.93}$$

In these formulas $[a, b]$ is the range of field variables, and $E_{\check{\phi}_x}^{-1}$ is the inverse function of $E_{\check{\phi}_x}(\varphi_x)$. The generation of φ'_x is illustrated by fig. 7.2.

Using the exact integral $E_{\check{\phi}_x}$ and its inverse $E_{\check{\phi}_x}^{-1}$ can, however, be still impossible or too cumbersome. Let us assume, that there is a simple approximation $W_0(\varphi_x; \check{\phi}_x)$ with its integral $E_{0\check{\phi}_x}(\varphi_x)$, which are easier to handle. After generating the field variable φ'_x according to the distribution $W_0(\varphi'_x; \check{\phi}_x)$, one needs a correcting *accept–reject step*, in order to obtain the desired distribution $W_c(\varphi'_x; \check{\phi}_x)$. This can be achieved by a second random number $r' \in [0, 1]$. The proposed change to φ'_x is accepted, if r' satisfies

$$r' \leq \frac{W_c(\varphi'_x; \check{\phi}_x)}{W_0(\varphi'_x; \check{\phi}_x)} \min_{a \leq \varphi_x \leq b} \frac{W_0(\varphi_x; \check{\phi}_x)}{W_c(\varphi_x; \check{\phi}_x)} \leq 1 . \tag{7.94}$$

The transition probability matrix corresponding to this two-step process, with the random number pair (r, r'), is

$$P_x^{(1)} = \lambda_{\check{\phi}_x} P_x + (1 - \lambda_{\check{\phi}_x})\mathbf{1} . \tag{7.95}$$

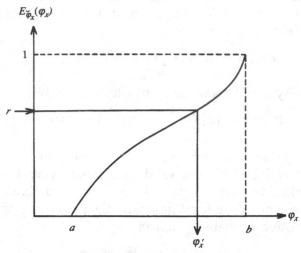

Fig. 7.2. The integral $E_{\breve{\varphi}_x}$ of the local heatbath distribution, and the generation of the field variable φ'_x from the random number r.

Here $\lambda_{\breve{\varphi}_x}$ is the average acceptance rate, which is equal to the area below the function in (7.94) divided by $(b - a)$. Repeating the trial step with the random number pair (r, r') as long as the proposed change is accepted, but maximally n-times, gives the transition probability matrix

$$P_x^{(n)} = \lambda_{\breve{\varphi}_x} P_x + (1 - \lambda_{\breve{\varphi}_x}) P_x^{(n-1)}$$

$$= [1 - (1 - \lambda_{\breve{\varphi}_x})^n] P_x + (1 - \lambda_{\breve{\varphi}_x})^n \mathbf{1} . \tag{7.96}$$

In the limit of $n \to \infty$, that is if new proposals are generated as long as one of them is accepted, we obtain the desired distribution

$$\lim_{n \to \infty} P_x^{(n)} = P_x . \tag{7.97}$$

From the point of view of numerical simulations on vectorizing or parallel computers, the $n \to \infty$ limit is not advantageous. Namely, the vectorization or parallelization goes over some mutually independent sets of sites (or links). This means that the accept–reject step cannot be finished, as long as there is some site where acceptance is not achieved. Near the end of this process there are only a few sites left over, where the proposed changes are still not accepted, and therefore the vectorization is ineffective (there is little parallelism). An important observation [7.24] is that the accept–reject process can be interrupted at any fixed number n of trials. This is because $P_x^{(n)}$ has the same stationary distribution as P_x, namely

$$\int d\varphi_x \, P_x^{(n)}(\varphi_x' \leftarrow \varphi_x; \breve{\varphi}_x) W_c(\varphi_x; \breve{\varphi}_x) = W_c(\varphi_x'; \breve{\varphi}_x) . \tag{7.98}$$

It is also true, that if P_x satisfies local detailed balance (7.87), then $P_x^{(n)}$ does it, too:

$$P_x^{(n)}(\varphi_x' \leftarrow \varphi_x; \breve{\varphi}_x) W_c(\varphi_x; \breve{\varphi}_x) = P_x^{(n)}(\varphi_x \leftarrow \varphi_x'; \breve{\varphi}_x) W_c(\varphi_x'; \breve{\varphi}_x) . \tag{7.99}$$

The optimal maximum number of trials n has to be found numerically, by requiring the smallest possible statistical errors of some important quantities, in a given computer time.

Another possibility is to iterate the transition $P_x^{(1)}$. This means that irrespective of whether the (r, r') proposal was accepted or not, a new trial is started from the actual configuration. Owing to the idempotency of the heatbath transition matrix ($P_x^2 = P_x$), we have

$$P_x^{(1)n} = P_x^{(n)} . \tag{7.100}$$

Therefore, in both ways the same distribution is generated.

7.3.1   Heatbath in lattice gauge theories

An important application of the heatbath updating is lattice gauge theory. Let us first consider the simple case of SU(2) gauge group [7.23].

It will be assumed that, with respect to the link variable U_l, the lattice action can be decomposed according to (7.65). This is the case for the Wilson action (7.6), but also in more general cases, when the action is built up linearly from some loop variables (not necessarily only loops around plaquettes). We shall use the projection of the sum of SU(2) elements S_l onto the group SU(2), which occurs also in (7.82). To be more specific, let us parametrize $U_l \in SU(2)$ and its adjoint as

$$U_l \equiv a_{l0} + \sum_{r=1}^{3} i\sigma_r a_{lr} , \quad U_l^+ = U_l^{-1} = a_{l0} - \sum_{r=1}^{3} i\sigma_r a_{lr} . \tag{7.101}$$

The unitarity condition implies

$$U_l^+ U_l = \sum_{r=0}^{3} a_{lr}^2 = 1 , \quad a_{l0} = z_l (1 - |a_l|^2)^{1/2} , \tag{7.102}$$

where $z_l = \pm 1$ is a Z_2 variable, and we defined

$$|a_l| \equiv \left[\sum_{r=1}^{3} a_{lr}^2 \right]^{1/2} . \tag{7.103}$$

The sum of SU(2)-variables S_l and its adjoint can be represented as

$$S_l \equiv s_{l0} - \sum_{r=1}^{3} i\sigma_r s_{lr} , \qquad S_l^+ = s_{l0} + \sum_{r=1}^{3} i\sigma_r s_{lr} . \qquad (7.104)$$

Therefore, we have

$$S_l S_l^+ = S_l^+ S_l = \sum_{r=0}^{3} s_{lr}^2 , \qquad S_l^{-1} = S_l^+ [\det (S_l S_l^+)]^{-1/2} = k^{-2} S_l^+ , \qquad (7.105)$$

where

$$k \equiv [\det (S_l S_l^+)]^{1/4} = \left[\sum_{r=0}^{3} s_{lr}^2 \right]^{1/2} = (\det (S_l))^{1/2} . \qquad (7.106)$$

It is clear from here that the projection of S_l on to SU(2) is $U_0 \equiv k S_l^{-1}$ in (7.82). The invariant Haar measure normalized to 1 is, in the variables $a_r \equiv a_{lr}$:

$$\int dU = \pi^{-2} \int d^4 a \, \delta \left(\sum_{r=0}^{3} a_r^2 - 1 \right)$$

$$= \frac{1}{2\pi^2} \int \frac{da^4}{(1 - |a|^2)^{1/2}} \left[\delta[a_0 - (1 - |a|^2)^{1/2}] \right.$$

$$\left. + \delta[a_0 + (1 - |a|^2)^{1/2}] \right] \theta(1 - |a|^2) . \qquad (7.107)$$

The three-dimensional integral over $\check{a} \equiv \{a_1, a_2, a_3\}$ can also be written as

$$\frac{d^3 \check{a}}{(1 - |a|^2)^{1/2}} = d^3 \check{a} \, \exp \left\{ \sum_{j=1}^{\infty} \frac{1}{2j} |a|^{2j} \right\}$$

$$= d|a| \frac{|a|^2}{(1 - |a|^2)^{1/2}} \, d^2 \Omega_a = da_0 (1 - a_0^2)^{1/2} \, d^2 \Omega_a . \qquad (7.108)$$

Here Ω_a is the solid angle belonging to $\check{a}$.

Using these relations the conditional probability distribution of the link variable U_l can be written as

$$W_c(U_l; \check{U}_l) \, dU_l \propto \exp \left\{ \frac{\beta}{2} \mathrm{Tr} \, (U_l S_l) \right\} dU_l$$

$$= \exp \left\{ \beta \sum_{r=0}^{3} a_{lr} s_{lr} \right\} dU_l . \qquad (7.109)$$

Since we have

$$U_l S_l = k U_l U_0^{-1} \equiv k U_{0l} \equiv k \left(a_0 + \sum_{r=1}^{3} i \sigma_r a_r \right), \qquad (7.110)$$

one can simplify the distribution by using the transformed link variable U_{0l}. Owing to the invariance of the Haar measure we have

$$\int \mathrm{d}U_l \, W_c(U_l; \check{U}_l) = \int \mathrm{d}U_{0l} \, W_c(U_{0l}; \check{U}_{0l})$$

$$\rightarrow \int_{-1}^{1} \mathrm{d}a_0 \, (1 - a_0^2)^{1/2} \, \mathrm{e}^{\beta k a_0} \int \mathrm{d}^2 \Omega_a . \qquad (7.111)$$

This is the heatbath distribution of a single link, which one has to generate numerically.

In order to simplify this distribution further, let us introduce the variable $y \equiv \exp(\beta k a_0)$. In terms of y one obtains from (7.111)

$$\rightarrow \int_{\mathrm{e}^{-\beta k}}^{\mathrm{e}^{+\beta k}} \mathrm{d}y \left[1 - \left(\frac{\log y}{\beta k} \right)^2 \right]^{1/2} \int \mathrm{d}^2 \Omega_a . \qquad (7.112)$$

According to this, one can produce the desired distribution in the following way:

i) Generate y uniformly in the interval $y \in [\exp(-\beta k), \exp(+\beta k)]$. This corresponds to the approximate distribution W_0 in (7.94).

ii) Correct for the factor $[\ldots]^{1/2}$ by an accept–reject step as discussed after (7.94).

iii) Generate the direction of $\mathring{a}$ randomly, for instance, according to $\mathrm{d}^2 \Omega_a = \mathrm{d}\phi \, \mathrm{d}(\cos \theta)$.

iv) Obtain the new link variable U_l' from U_{0l} by using $U_l' \equiv U_{0l} U_0$.

This procedure of generating the heatbath distribution of a link in SU(2) lattice gauge theory can also be extended to SU(3) [7.25]. For larger groups, however, it is difficult to find a direct generalization. In the case of SU(N) Cabibbo and Marinari [7.26] proposed to use the above SU(2) algorithm for appropriate subgroups. In this way one obtains the *pseudo-heatbath algorithm*.

For this one has to choose first a sufficiently large subset of SU(2) subgroups $\{SU(2)_k; \ k = 1, 2, \ldots, m\}$, in such a way that no subset of SU(N) is left invariant with respect to all the subgroups (except for the unit element and the whole group). As an example, one can take with $m = N - 1$

$$
a_k = \begin{pmatrix} 1 & & & & & & \\ & \ddots & & & & & \\ & & 1 & & & & \\ & & & (\,\alpha_k\,) & & & \\ & & & & 1 & & \\ & & & & & \ddots & \\ & & & & & & 1 \end{pmatrix} \in SU(2)_k \,, \tag{7.113}
$$

where α_k is an SU(2)-element located at the $(k, k+1)$ rows and columns. The new link variable is chosen according to

$$
U_l' = a_m a_{m-1} \cdots a_1 U_l \,. \tag{7.114}
$$

If we define

$$
U_l^{(k)} \equiv a_k a_{k-1} \cdots a_1 U_l \,, \qquad U_l^{(0)} \equiv U_l \,, \tag{7.115}
$$

then

$$
U_l^{(k)} = a_k U_l^{(k-1)} \,, \qquad U_l^{(m)} = U_l' \,. \tag{7.116}
$$

The lattice action $S[U] \equiv S(U_l, \check{U}_l)$ is assumed also here to have a decomposition as in (7.65).

The element of the $SU(2)_k$ subgroup a_k is chosen according to the distribution

$$
dP_l(a_k) = d^{(k)} a_k \, \exp\left\{ -S(a_k U_l^{(k-1)}, \check{U}_l) \right\} Z_k^{-1}(U_l^{(k-1)}) \,, \tag{7.117}
$$

where $d^{(k)} a_k$ is the Haar measure in $SU(2)_k$, and the normalization factor Z_k is defined by

$$
Z_k(U_l) \equiv \int_{SU(2)_k} da \, \exp\left\{ -S(a U_l, \check{U}_l) \right\} = Z_k(b U_l) \,. \tag{7.118}
$$

The last equality follows from the invariance of the integration measure. Performing the generation of a_k consecutively for $k = 1, 2, \ldots, m$, one obtains U_l' according to (7.116). In order to prove that the canonical distribution of configurations is left invariant by this procedure, let us assume that $U_l^{(k-1)}$ has the distribution

$$
W_c(U_l^{(k-1)}, \check{U}_l) \, dU_l^{(k-1)} = \exp\left\{ -S(U_l^{(k-1)}, \check{U}_l) \right\} Z^{-1} \, dU_l^{(k-1)} \,. \tag{7.119}
$$

Then $U_l^{(k)}$ is distributed according to

$$
W_c(U_l^{(k)}, \check{U}_l) \, dU_l^{(k)} =
$$

$$
\int_{SU(2)_k} d^{(k)} a \, \frac{\exp\left\{ -S(U_l^{(k)}, \check{U}_l) \right\} \exp\left\{ -S(a^{-1} U_l^{(k)}, \check{U}_l) \right\}}{Z_k(a^{-1} U_l^{(k)}) Z} d(a^{-1} U_l^{(k)})
$$

$$= \exp\left\{-S(U_l^{(k)}, \check{U}_l)\right\} Z^{-1} \, dU_l^{(k)} . \tag{7.120}$$

Therefore, if U_l has a canonical distribution, then $U_l^{(1)}, U_l^{(2)}, \ldots, U_l^{(m)} = U_l'$ have it, too.

The only task left is to generate the distributions in (7.117). Since according to (7.65) we have

$$S(a_k U_l, \check{U}_l) = -\frac{\beta}{N} \operatorname{Re} \operatorname{Tr}(a_k U_l S_l) + \check{S}(\check{U}_l)$$

$$= -\frac{\beta}{N} \operatorname{Re} \operatorname{Tr}(\alpha_k \rho_k) + \cdots . \tag{7.121}$$

Here the rest denoted by the dots is independent from α_k, and ρ_k stands for the $(k, k+1)$ block part of $U_l S_l$. Expanding α_k and ρ_k as usual:

$$\alpha_k \equiv \alpha_{k0} + \sum_{r=1}^{3} i\sigma_r \alpha_{kr} , \qquad \rho_k \equiv \rho_{k0} - \sum_{r=1}^{3} i\sigma_r \rho_{kr} , \tag{7.122}$$

the α_k-dependent part in (7.121) is

$$\frac{\beta}{N} \operatorname{Re} \operatorname{Tr}(\alpha_k \rho_k) = \frac{2\beta}{N} \sum_{r=0}^{3} \alpha_{kr} \operatorname{Re}(\rho_{kr}) . \tag{7.123}$$

Therefore the required distribution in (7.117) is

$$dP(\alpha_k) \propto d^4\alpha_k \, \delta\left(\sum_{r=0}^{3} \alpha_{kr}^2 - 1\right) \exp\left\{\frac{2\beta}{N} \sum_{r=0}^{3} \alpha_{kr} \operatorname{Re}(\rho_{kr})\right\} . \tag{7.124}$$

This is the same as the distribution in (7.109), needed for the SU(2) heatbath algorithm, therefore it can be generated in the same way as it was done above.

From the point of view of shorter autocorrelations, it can be advantageous to choose more SU(2) subgroups than the $(N-1)$ in (7.113). For instance, one can take any pairs of rows and columns, not only the neighbouring ones. The sequence of the chosen subgroups can also be changed randomly. In this way one obtains an efficient algorithm, in particular, also for the physically important case of the SU(3) gauge group.

7.4 Fermions in numerical simulations

As was shown in sections 4.1–4.3, in Euclidean quantum field theory the fermion fields are described by Grassmann variables. The direct evaluation of the Grassmannian path integrals is in principle possible in the polymer representation (see section 4.1), but the contributions have oscillating signs, due to the Pauli principle. This makes the numerical evaluation by importance sampling very difficult, except for some model quantum field theories in two dimensions [7.27]. At present the only practical possibility to simulate numerically a four-dimensional Euclidean quantum field theory with fermions is to perform the Grassmann integral analytically, and apply Monte Carlo simulation methods in the resulting effective bosonic theory. For definiteness, in the present and next two sections QCD with Wilson lattice fermions will be considered. The case of other fermion–boson theories is in most cases quite similar (for remarks concerning the numerical simulation of Yukawa models see chapter 6).

Let us consider the lattice action

$$S[U, \psi, \overline{\psi}] = S_g[U] + S_q[U, \psi, \overline{\psi}] . \tag{7.125}$$

S_g is the pure SU(3) gauge action (see chapter 3):

$$S_g[U] = \beta \sum_p \left(1 - \frac{1}{3} \operatorname{Re} \operatorname{Tr} U_p \right) . \tag{7.126}$$

S_q is the component depending on the quark fields $\psi_{qx} \equiv \{\psi_{qx\alpha c}, \alpha = 1, 2, 3, 4; c = 1, 2, 3\}$. The index $q = u, d, s, c, \dots$ stands for different quark flavours, x for the lattice points, α for the Dirac spinor and c for SU(3)-colour. Using the normalization defined in (4.88) we have

$$S_q[U, \psi, \overline{\psi}] = \sum_{q,x} \left\{ (\overline{\psi}_{qx} \psi_{qx}) - K_q \sum_{\mu=\pm 1}^{\pm 4} (\overline{\psi}_{qx+\hat{\mu}}[1 + \gamma_\mu] U_{x\mu} \psi_{qx}) \right\} . \tag{7.127}$$

Here the Wilson parameter is fixed to $r = 1$, $U_{x\mu}$ is the SU(3) link variable and K_q is the hopping parameter belonging to the flavour q. (Note that since the quark masses are different, the physical values of K_q are also different from each other.) In the following, for simplicity, often only a single flavour will be considered, or the flavour index will not be explicitly displayed. For instance, the fermionic component of the action will often be written as

$$S_q = \sum_{xy} (\overline{\psi}_y Q_{yx} \psi_x) , \tag{7.128}$$

where the *quark matrix* Q is

$$Q_{yx} = \delta_{yx} - K \sum_\mu \delta_{y,x+\hat{\mu}}(1 + \gamma_\mu) U_{x\mu} . \tag{7.129}$$

In matrix notation one can also write

$$Q \equiv 1 - KM \equiv 1 - K \sum_\mu M_\mu . \tag{7.130}$$

The adjoint of the quark matrix with respect to spinor and colour indices $Q^\dagger$ satisfies

$$Q_{yx} = \gamma_5 Q^\dagger_{xy} \gamma_5 \,, \tag{7.131}$$

therefore, taking the determinant on both sides, it follows that its determinant is real:

$$\det Q = \det Q^+ = (\det Q)^* \,. \tag{7.132}$$

The expectation value of a general function F of the field variables is given by

$$\langle F \rangle = \frac{\int [dU \, d\overline{\psi} \, d\psi] e^{-S_g - S_q} F[U, \psi, \overline{\psi}]}{\int [dU \, d\overline{\psi} \, d\psi] e^{-S_g - S_q}} \,. \tag{7.133}$$

After performing the Grassmann integral, for a function depending only on gauge variables, this becomes

$$\langle F \rangle = Z^{-1} \int [dU] e^{-S_g[U]} \det Q[U] F[U]$$

$$= Z^{-1} \int [dU] e^{-S_{eff}[U]} F[U] \,, \tag{7.134}$$

where the normalization is given by

$$Z = \int [dU] e^{-S_g[U]} \det Q[U] = \int [dU] e^{-S_{eff}[U]} \,, \tag{7.135}$$

and the *effective gauge action* S_{eff} is

$$S_{eff}[U] \equiv S_g[U] - \log \det Q[U] = S_g[U] - \text{Tr} \log Q[U] \,. \tag{7.136}$$

This shows how the fermion–boson theory is represented in the pure bosonic sector. The reality of the effective action is, however, only guaranteed if the determinant of the quark matrix is positive. According to (7.132) $\det Q$ is real, but the positivity is, in general, not satisfied. In fact, for small enough hopping parameter, namely for $|K_q| < \frac{1}{8}$, the matrix Q can be proven to be positive (similarly to the proof of positivity of B in (4.105)). This is, however, not enough because the critical values of K_q are, for finite gauge couplings, at $K_q > \frac{1}{8}$ (see chapter 5). An important case, where the positivity of $\det Q$ is true, is for a pair of exactly degenerate flavours (for instance, for u- and d-quarks), when

$$\det Q = \det Q_u \cdot \det Q_d = (\det Q_u)^2 \geq 0 \,. \tag{7.137}$$

In the case of a non-positive determinant one can, in principle, proceed by taking into account the absolute value of the determinant in the Monte Carlo integration measure, and include the phase $\det Q / |\det Q|$ in the measurable quantities. In practice, however, this procedure usually does not work, because the phase is strongly oscillating.

The formula for the expectation value in (7.134) can be easily generalized for quantities which depend also on the fermionic variables. The expectation value of a product of fermion bilinears with an arbitrary function of the gauge variables, according to (4.25), is

$$\langle \psi_{y_1} \overline{\psi}_{x_1} \psi_{y_2} \overline{\psi}_{x_2} \cdots \psi_{y_n} \overline{\psi}_{x_n} F[U] \rangle = Z^{-1} \int [dU] e^{-S_{eff}[U]} F[U]$$

$$\cdot \sum_{z_1 \cdots z_n} \epsilon_{y_1 y_2 \cdots y_n}^{z_1 z_2 \cdots z_n} Q[U]_{z_1 x_1}^{-1} Q[U]_{z_2 x_2}^{-1} \cdots Q[U]_{z_n x_n}^{-1} . \qquad (7.138)$$

The main difficulty of the numerical simulation of fermionic theories is, that the logarithm of the fermion determinant in the effective boson action is non-local. This makes the numerical evaluation of the transition probabilities in the Monte Carlo updating process rather slow. For instance, in a Metropolis algorithm the change of the effective action ΔS_{eff}, belonging to a change $\Delta U_{x\mu} \equiv U'_{x\mu} - U_{x\mu}$ of the link variable $U_{x\mu}$, is needed. This is given by

$$\Delta S_{eff} = \Delta S_g - \log \det \frac{1 - KM[U']}{1 - KM[U]}$$

$$= \Delta S_g - \log \det (1 - KD[U, U']) , \qquad (7.139)$$

where ΔS_g is the change of the pure gauge action and

$$D[U, U'] \equiv (1 - KM[U])^{-1} \Delta M[U, U'] , \qquad (7.140)$$

with

$$\Delta M_{x_2 x_1} \equiv (M[U'] - M[U])_{x_2 x_1} = \sum_{x,\mu} (1 + \gamma_\mu) \Delta U_{x\mu} \delta_{x_2, x+\hat{\mu}} \delta_{x, x_1} . \qquad (7.141)$$

Since for the evaluation of the change of the effective gauge action a large amount of arithmetics is required, one could try to perform a Monte Carlo updating with the pure gauge action, and include the quark determinant in the measurable quantities, as suggested by the first form of the expectation value in (7.134). This does not work, however (except for small hopping parameters [7.28]), because the value of the quark determinant fluctuates very strongly on an ensemble generated according to the pure gauge Boltzmann measure. The problem is similar to the one encountered at (7.22), where the possibility of obtaining the expectation values at neighbouring values of some action parameter was discussed.

A viable alternative is to include some approximation $\det Q^{(0)}$ of the quark determinant in the updating, and correct by $\det Q / \det Q^{(0)}$ in the measurements. A reasonably good approximation for large and intermediate quark masses, which correspond to hopping parameters not very close to the critical value, can be obtained by the hopping parameter expansion up to some maximum order l_{max} [7.29, 7.30]. According to (7.139) and (5.30) we have

$$\Delta S_{eff} \Rightarrow \Delta S_g + \sum_{l=1}^{l_{max}} \frac{K^l}{l} \mathrm{Tr} \left\{ M[U']^l - M[U]^l \right\} . \qquad (7.142)$$

Clearly, the crucial point in the updating process based on the effective boson action of fermionic theories is the evaluation of the matrix elements of Q^{-1}. With the help of an auxiliary complex scalar field $\chi_{q x \alpha c}$ we get

$$Q_{uv}^{-1} = \frac{\int [d\chi \, d\chi^+] \exp \left\{ -\sum_{xy} (\chi_y^+ [Q^+ Q]_{yx} \chi_x) \right\} \chi_u \chi_{v'}^+ Q_{v'v}^+}{\int [d\chi \, d\chi^+] \exp \left\{ -\sum_{xy} (\chi_y^+ [Q^+ Q]_{yx} \chi_x) \right\}} . \qquad (7.143)$$

Therefore, the matrix elements of the inverse quark matrix can, in principle, be calculated as expectation values of the *pseudofermion field* χ. The convergence of the Gaussian integral is guaranteed, since $Q^+ Q$ is a positive matrix. The action for the scalar pseudofermion field χ is local, since the matrix $Q^+ Q$ has non-vanishing elements only up to next-nearest-neighbour lattice sites. As suggested in [7.31], the evaluation of Q_{uv}^{-1} can be done by a heatbath Monte Carlo, and a small step updating of the link variables $U_{x\mu}$ is possible by the Langevin algorithm (see in section 7.5).

Another possibility of using an auxiliary complex scalar field $\phi_{q x \alpha c}$, which has the same number of components as the fermion field $\psi_{q x \alpha c}$, was proposed by Petcher and Weingarten [7.32] for the evaluation of $\det Q^+ \cdot \det Q = \det (Q^+ Q)$. We have, namely,

$$\det (Q^+ Q) \propto \int [d\phi \, d\phi^+] \exp \left\{ -\sum_{xy} (\phi_y^+ [Q^+ Q]_{yx}^{-1} \phi_x) \right\} . \qquad (7.144)$$

According to (7.137) and (7.132), this formula can be used in QCD with two degenerate flavours for the calculation of the quark determinant. We shall call the field ϕ also a *pseudofermion field*. (It is related to χ above by $\phi = (Q^+ Q)\chi$.) If $\phi_{q x \alpha c}$ and $\phi_{q x \alpha c}^+$ are considered to be, respectively, column and row vectors, then one can use short matrix notations like

$$\sum_{xy} (\phi_y^+ [Q^+ Q]_{yx}^{-1} \phi_x) \equiv \{\phi, [Q^+ Q]^{-1} \phi\} \equiv \{Q^{+-1} \phi, Q^{+-1} \phi\} . \qquad (7.145)$$

The expectation value of a pure gauge quantity $F[U]$ can be obtained as

$$\langle F \rangle = Z^{-1} \int [dU \, d\phi \, d\phi^+] \exp \left[-S_g[U] - \{\phi, [Q^+ Q]^{-1} \phi\} \right] F[U] . \qquad (7.146)$$

Here the normalization factor is defined by

$$Z \equiv \int [dU \, d\phi \, d\phi^+] \exp \left[-S_g[U] - \{\phi, [Q^+ Q]^{-1} \phi\} \right] . \qquad (7.147)$$

Other expectation values, which depend on the fermion fields too, can also be obtained from functions of the pseudofermion field ϕ. For instance, for the flavour with quark matrix Q, we have the relation

$$\langle \psi_u \overline{\psi}_v \rangle_Q = \left. \frac{\partial}{\partial \lambda_{vu}} \right|_{\lambda_{vu}=0} \log \int [dU \, d\overline{\psi} \, d\psi]$$

$$\cdot \exp \left\{ -S_g - \sum_{xy} (\overline{\psi}_y [Q + \lambda_{vu} I_{vu}]_{yx} \psi_x) \right\} , \qquad (7.148)$$

where the matrix I_{vu} is defined as

$$(I_{vu})_{yx} \equiv \delta_{vy} \delta_{ux} . \qquad (7.149)$$

Considering, instead of (7.144), $\det [Q^+(Q + \lambda_{vu} I_{vu})]$ one obtains from (7.148)

$$\langle \psi_u \overline{\psi}_v \rangle_Q = Z^{-1} \int [dU \, d\phi \, d\phi^+] \exp \left[-S_g[U] - \{\phi, [Q^+ Q]^{-1} \phi\} \right]$$

$$\cdot \{[Q^+ Q]^{-1} \phi, Q^+ I_{vu} [Q^+ Q]^{-1} \phi\} . \qquad (7.150)$$

This relation corresponds, in fact, to (7.143). Similarly, in the subspace of the quark flavour with fermion matrix Q^+, the corresponding formula is

$$\langle \psi_u \overline{\psi}_v \rangle_{Q^+} = Z^{-1} \int [dU \, d\phi \, d\phi^+] \exp \left[-S_g[U] - \{\phi, [Q^+ Q]^{-1} \phi\} \right]$$

$$\cdot \{[Q^+ Q]^{-1} \phi, I_{vu} Q [Q^+ Q]^{-1} \phi\} . \qquad (7.151)$$

In order to distinguish the formulas like (7.150) or (7.151) from the corresponding expressions in terms of the quark propagators (7.138), one can call them *pseudofermion estimators* or, due to the noise introduced by the pseudofermion field, *noisy estimators*.

The relations in (7.146), (7.150) and (7.151) show that QCD with two degenerate flavours (represented, respectively, by the quark matrices Q and Q^+) can be simulated by an updating process on gauge link variables $U_{x\mu}$ and pseudofermion field variables $\phi_{qx\alpha c}$. The difficulty in this formulation is that the action of the pseudofermion field ϕ is non-local, because $[Q^+ Q]^{-1}$ is non-local. Practical ways of overcoming this difficulty will be discussed in sections 7.5 and 7.6.

7.4.1   *Fermion matrix inversion*

A crucial point in every fermion simulation algorithm based on the effective bosonic action is the evaluation of the matrix elements of the inverse of the fermion matrix Q. Therefore, in this subsection some commonly used inversion algorithms will be briefly discussed.

The equation to be solved for the vector p is

$$Qp = v , \tag{7.152}$$

with some given vector v and the quark matrix $Q = 1 - KM$, for instance, in (7.130). It will be assumed here, that Q has no zero eigenvalues, therefore $p = Q^{-1}v$ is uniquely determined. There are many ways to solve (7.152) (see, for instance, [7.33, 7.34, 7.35]). A simple iterative method is the *Jacobi iteration*, defined by

$$p_0 \equiv v , \qquad p_{n+1} = v + KMp_n \quad (n \geq 0) . \tag{7.153}$$

The solution is $p = \lim_{n \to \infty} p_n$, if the limit exists. The convergence is guaranteed, if the absolute value of the largest eigenvalue of KM is less than 1. An improved version is the *Gauss–Seidel iteration* with some relaxation parameter $\lambda \neq 0$. This is based on the decomposition $M = M_l + M_u$, where

$$M_{l,ab} \equiv \begin{cases} M_{ab} & \text{if} \quad a > b \\ 0 & \text{if} \quad a \leq b \end{cases} , \tag{7.154}$$

and the iteration, instead of (7.153), is

$$p_{n+1} = (1 - \lambda)p_n + \lambda[v + K(M_l p_{n+1} + M_u p_n)] . \tag{7.155}$$

For $\lambda = 1$ this means a 'point-by-point' iteration of (7.153), that is the calculated new elements of the vector p_{n+1} are immediately stored in the array for p_n. λ is a *relaxation parameter*, which can be optimized in order to accelerate the convergence.

Another iterative scheme is the *minimal residue* iteration. In general the *residue* after n iteration steps is defined by

$$r_n \equiv v - Qp_n . \tag{7.156}$$

The iteration starts from some guess p_0 for the solution p, and then for $n = 0, 1, 2, \ldots$ one calculates

$$p_{n+1} = p_n + \frac{(Qr_n, r_n)}{|Qr_n|^2} r_n . \tag{7.157}$$

This implies

$$r_{n+1} = r_n - \frac{(Qr_n, r_n)}{|Qr_n|^2} Qr_n . \tag{7.158}$$

(Of course, the length of a vector v is defined by $|v| \equiv (v, v)^{1/2}$.) The iteration is stopped either if $r_n = 0$, when p_n is the exact solution, or if the absolute value of the residue $|r_n|$ is at least smaller than some small δ. Since the ratio of residues is given by

$$\frac{(r_{n+1}, r_{n+1})}{(r_n, r_n)} = 1 - \frac{|(Qr_n, r_n)|^2}{|Qr_n|^2 |r_n|^2} , \tag{7.159}$$

Simulation algorithms

and

$$|(Qr_n, r_n)|^2 = (r_n, Q_R r_n)^2 + (r_n, Q_I r_n)^2 , \qquad (7.160)$$

where $Q = Q_R + iQ_I$ is the decomposition of Q into Hermitean and anti-Hermitean parts (that is $Q_R = Q_R^+$ and $Q_I = Q_I^+$). If $\lambda_{min(max)}(A)$ denotes the eigenvalue of a Hermitean matrix A with minimal (maximal) absolute value, then (7.159) and (7.160) imply

$$\frac{(r_{n+1}, r_{n+1})}{(r_n, r_n)} \le 1 - \frac{\lambda_{min}^2(Q_R) + \lambda_{min}^2(Q_I)}{\lambda_{max}(Q^+ Q)} . \qquad (7.161)$$

Therefore, the minimal residue iteration is never divergent. The convergence speed is determined by the second term on the right hand side. This shows that a fast convergence is guaranteed if the eigenvalues of Q are not much different from each other.

The most popular iteration scheme for the inversion of the fermion matrix is the *conjugate gradient* iteration (see, for instance, [7.36]). Let us first consider the inversion of a positive definite Hermitean matrix A (for instance, $A \equiv Q^+ Q$):

$$Ap = v . \qquad (7.162)$$

If p_0 is a first guess for p, and the residue is defined as in (7.156) by $r_n \equiv v - Ap_n$, then we define

$$s_0 = r_0 = v - Ap_0 . \qquad (7.163)$$

For $n = 0, 1, 2, \ldots$ calculate

$$a_n = \frac{|r_n|^2}{(s_n, As_n)} , \quad p_{n+1} = p_n + a_n s_n , \quad r_{n+1} = r_n - a_n As_n . \qquad (7.164)$$

If the residue satisfies the stopping condition $|r_{n+1}|^2 < \delta$, then p_{n+1} is the solution with the required accuracy. Otherwise one continues by calculating

$$b_n = \frac{|r_{n+1}|^2}{|r_n|^2} , \qquad s_{n+1} = r_{n+1} + b_n s_n , \qquad (7.165)$$

and returns to (7.164) for the next iteration. One can prove by induction from the above definitions, that the residue vectors with different indices are orthogonal: $(r_n, r_m) = 0$ $(m \ne n)$, therefore the iteration has to converge to the exact solution in at most N steps, if the vector space is N-dimensional.

The conjugate gradient iteration can also be extended to non-positive matrices, if one finds another matrix B, such that $B^+ A$ is positive Hermitean. (In most cases one takes $A = B = Q$.) Then the iteration for the solution of (7.162) is defined by

$$r_0 = v - Ap_0 , \qquad s_0 = B^+ r_0 ,$$

$$a_n = \frac{|B^+ r_n|^2}{(s_n, B^+ A s_n)} , \qquad p_{n+1} = p_n + a_n s_n , \qquad r_{n+1} = r_n - a_n A s_n ,$$

$$b_n = \frac{|B^+ r_{n+1}|^2}{|B^+ r_n|^2} , \qquad s_{n+1} = B^+ r_{n+1} + b_n s_n . \tag{7.166}$$

Concerning convergence one can show, that if the eigenvalues of $A = B = Q$ are clustered around k different values $\{\mu_j; \ j = 1, 2, \dots, k\}$, in such a way that

$$\lambda_i(Q) = \mu_j + \epsilon(i, j) , \tag{7.167}$$

then the accuracy of the approximation after k iterations is governed by the largest $\epsilon(i, j)$. In particular, if there are only k different eigenvalues, then one obtains the exact solution in k steps.

The idea of *preconditioning* is to accelerate the convergence of the iteration in different inversion algorithms. For instance, if there are many different eigenvalues, then for the conjugate gradient iteration one tries to replace the matrix by another one with eigenvalues closer to each other. Considering again the equation in (7.152), one tries to find matrices A, B in such a way that $\tilde{Q} \equiv AQB$ is easier to invert. Solving for q the equation

$$\tilde{Q}q = Av , \tag{7.168}$$

one obtains the solution of the original equation as $p = Bq$. Popular preconditioning techniques are the *incomplete (LU)-decomposition* [7.37, 7.38] or the *Fourier acceleration* [7.39]. A simple example of preconditioning [7.40] uses the property of the hopping term KM in the quark matrix (7.129), that it only connects even sites to odd sites (according to $\sum_{\mu=1}^{4} x_\mu$). Decomposing the space into even–odd halfs, the quark matrix can be written as

$$Q = \begin{pmatrix} 1 & -KM_{eo} \\ -KM_{oe} & 1 \end{pmatrix} . \tag{7.169}$$

If we define

$$L = \begin{pmatrix} 1 & 0 \\ -KM_{oe} & 1 \end{pmatrix} , \qquad U = \begin{pmatrix} 1 & -KM_{eo} \\ 0 & 1 \end{pmatrix} , \tag{7.170}$$

then we have

$$L^{-1} = \begin{pmatrix} 1 & 0 \\ KM_{oe} & 1 \end{pmatrix} , \qquad U^{-1} = \begin{pmatrix} 1 & KM_{eo} \\ 0 & 1 \end{pmatrix} . \tag{7.171}$$

The preconditioned matrix $\tilde{Q}$ is defined by

$$\tilde{Q} \equiv L^{-1}QU^{-1} = \begin{pmatrix} 1 & 0 \\ 0 & 1 - K^2 M_{oe}M_{eo} \end{pmatrix} . \qquad (7.172)$$

The smallest eigenvalue of $\tilde{Q}$ is about twice as large as the smallest eigenvalue of Q, therefore in the critical region the inversion of $\tilde{Q}$ is easier. It is remarkable, that by such a simple transformation the quark matrix can be brought into a form, where the non-trivial piece acts only on the odd sites. Although the non-zero elements of M^2 connect next nearest neighbours, the multiplication by $\tilde{Q}$ can be done with the same amount of arithmetics as the multiplication by Q.

The preconditioning is a very useful tool for the acceleration of the iterative inversion of the quark matrix, but the achieved improvement is in most practical cases only moderate. A radical improvement can be obtained by *multigrid methods* [7.41, 7.42], which treat the low frequency dynamics of the system specially for an efficient acceleration of the convergence. The configuration on the original 'fine' lattice is projected on a 'coarse' lattice with, say, twice the lattice spacing. If an approximate solution $\bar{p}_y$ of the equation on the fine lattice

$$Q_{xy}p_y = v_x \qquad (7.173)$$

is known, then the 'residue' $r_x \equiv v_x - Q_{xy}\bar{p}_y$, and the 'error' $e_x \equiv p_x - \bar{p}_x$, satisfy

$$Q_{xy}e_y = r_x . \qquad (7.174)$$

In order to determine e_x, which is equivalent to the determination of p_x, one has to find an appropriate operator $\hat{Q}_{\hat{x}\hat{y}}$ on the coarse lattice in such a way, that the error e_x is well approximated by $\hat{e}_{\hat{x}}$ satisfying

$$\hat{Q}_{\hat{x}\hat{y}}\hat{e}_{\hat{y}} = \hat{r}_{\hat{x}} . \qquad (7.175)$$

(The indices on the fine lattice are denoted by $x, y, \ldots$, those on the coarse lattice by $\hat{x}, \hat{y}, \ldots$.) Besides $\hat{Q}_{\hat{x}\hat{y}}$ one has to define a projective mapping $r_x \to \hat{r}_{\hat{x}}$ and an interpolating mapping $\hat{e}_{\hat{x}} \to e_x$, too, which connect corresponding vectors on the fine and coarse lattices. Both $\hat{Q}$ and these mappings have to respect and exploit the detailed properties of the dynamics of the system (as gauge invariance etc.). In particular, the low frequency (low energy) modes of the fine lattice have to be treated with special care on the coarse lattice. The doubling of the lattice spacing can also be iterated. In this way the 'two-grid' procedure becomes a really 'multigrid' one.

7.5 Fermion algorithms based on differential equations

We have seen in section 7.4 that an interacting fermion–boson quantum field theory, like QCD with Wilson lattice fermions, can be simulated, after performing the Gaussian integrals over the fermion variables, in terms of the bosonic variables alone. The difficulty is that the effective bosonic action is non-local. For instance, in a straightforward Metropolis algorithm one has to evaluate the determinant of the fermion matrix many times. This makes the numerical simulation very slow. An improvement can be achieved by simulation algorithms based on difference equations (discretized differential equations).

7.5.1 *Classical dynamics algorithms*

A possibility to reduce the number of necessary fermion matrix inversions is to simulate, instead of the canonical ensemble of field configurations, the *microcanonical ensemble*, with fixed value of the 'energy' belonging to a suitably defined Hamiltonian. This was done in pure gauge field theory by Callaway and Rahman [7.43], who introduced conjugate momenta, and used the discretized version of the classical Hamiltonian equations. The idea to use this method for the simulation of fermionic quantum field theories was put forward by Polonyi and Wyld [7.44].

For definiteness, let us consider here lattice QCD with two degenerate flavours of quarks described by Wilson lattice fermions. After introducing the complex scalar pseudofermion field ϕ, the lattice action is, according to (7.146),

$$S[U, \phi] = S_g[U] + \sum_{xy}(\phi_y^+ [Q^+Q]_{yx}^{-1}\phi_x) \equiv S_g[U] + \{\phi, [Q^+Q]^{-1}\phi\} \; . \quad (7.176)$$

Here $S_g[U]$ is the pure gauge lattice action and $Q = Q[U]$ is the quark matrix in (7.129), which depend on the link variables $U_{x\mu} \in SU(3)$. Note that flavour, colour and spinor indices are not explicitly displayed here, for instance, $U_{x\mu} \equiv U_{x\mu,cd}$ and $\phi_x \equiv \phi_{qx\alpha c}$, where q is a flavour, α is a spinor, and c, d are colour indices. The conjugate momentum for ϕ_x is the complex valued $\pi_x \equiv \pi_{qx\alpha c}$, for $U_{x\mu}$ an element of the SU(3) Lie algebra $P_{x\mu} \in \mathscr{L}SU(3)$, namely

$$P_{x\mu} \equiv \sum_{j=1}^{8} i\lambda_j P_{x\mu j} \; , \quad (7.177)$$

with real $P_{x\mu j}$ and the Gell-Mann matrices λ_j ($j = 1, 2, \ldots, 8$) (see (8.14)–

(8.22)). The Hamiltonian H is defined as

$$H[P, \pi, U, \phi] = \frac{1}{2} \sum_{x\mu j} P_{x\mu j}^2 + \frac{1}{2} \sum_x \pi_x^+ \pi_x + S[U, \phi] , \qquad (7.178)$$

therefore the lattice action S plays the rôle of a potential. The expectation values, like (7.146) or (7.150) and (7.151), are given by a path integral over $[P, \pi, U, \phi]$ as

$$\langle F \rangle = Z^{-1} \int [\mathrm{d}P\, \mathrm{d}U\, \mathrm{d}\pi\, \mathrm{d}\pi^+\, \mathrm{d}\phi\, \mathrm{d}\phi^+] \exp\{-H[P, \pi, U, \phi]\}\, F[U, \phi] , \qquad (7.179)$$

where the normalization factor is, as usual,

$$Z = \int [\mathrm{d}P\, \mathrm{d}U\, \mathrm{d}\pi\, \mathrm{d}\pi^+\, \mathrm{d}\phi\, \mathrm{d}\phi^+] \exp\{-H[P, \pi, U, \phi]\} . \qquad (7.180)$$

This is obviously equivalent to the expressions in (7.146)–(7.151), because the Gaussian integrals over the canonical momenta P, π can be trivially performed, and the constants cancel by the normalization.

The above formulas show, how the expectation values can be calculated in an extended space of variables $[P, \pi, U, \phi]$, where the lattice action is replaced by the 'Hamiltonian' $H[P, \pi, U, \phi]$. The idea behind the classical dynamics algorithms is to perform the numerical simulation by solving some discretized version of the classical canonical equations with the Hamiltonian $H[P, \pi, U, \phi]$. Assuming ergodicity, this gives an approximation to the distributions in the *microcanonical ensemble* with constant 'energy': $H = $ const. In the limit of very large lattices the microcanonical ensemble is a good approximation to the canonical one required in (7.179), if the average potential energy $\langle S[U, \phi] \rangle$ is tuned appropriately. Therefore, the expectation values $\langle F \rangle$ can be obtained as time averages over sufficiently long classical trajectories.

In order to define a suitable discretization scheme, let us consider a classical Hamiltonian $H(p, q)$ of the general form

$$H(p, q) = \frac{p^2}{2} + V(q) . \qquad (7.181)$$

The corresponding canonical differential equations are

$$\frac{\partial H}{\partial p} = p = \dot{q} , \qquad -\frac{\partial H}{\partial q} = -\frac{\partial V}{\partial q} = \dot{p} = \ddot{q} . \qquad (7.182)$$

A suitable second order discretization in the time variable τ is

$$q(\tau_{n+1}) = q(\tau_n) + (\tau_{n+1} - \tau_n)p(\tau_n) - \frac{1}{2}(\tau_{n+1} - \tau_n)^2 \frac{\partial V(q(\tau_n))}{\partial q} ,$$

$$p(\tau_{n+1}) = p(\tau_n) - \frac{1}{2}(\tau_{n+1} - \tau_n)\left[\frac{\partial V(q(\tau_n))}{\partial q} + \frac{\partial V(q(\tau_{n+1}))}{\partial q}\right] . \qquad (7.183)$$

In the case of the pseudofermion field $\phi_x = \phi_{Rx} + i\phi_{Ix}$ ($\phi_{R,Ix} = real$), it follows from (7.176) that

$$\frac{\partial S}{\partial \phi_{Rx}} + i\frac{\partial S}{\partial \phi_{Ix}} = 2\sum_y [Q^+Q]_{xy}^{-1}(\phi_{Ry} + i\phi_{Iy}) \, , \qquad (7.184)$$

therefore, with $\Delta\tau \equiv \tau_{n+1} - \tau_n$, $(\phi(\tau_{n+1}), \pi(\tau_{n+1})) \to (\phi', \pi')$ and $(\phi(\tau_n), \pi(\tau_n)) \to (\phi, \pi)$ the equations corresponding to (7.183) are

$$\phi'_x = \phi_x + \Delta\tau\pi_x - (\Delta\tau)^2 \sum_y [Q^+Q]_{xy}^{-1}\phi_y \, ,$$

$$\pi'_x = \pi_x - \Delta\tau\sum_y \left\{ [Q^+Q]_{xy}^{-1}\phi_y + [Q'^+Q']_{xy}^{-1}\phi'_y \right\} . \qquad (7.185)$$

Here in the second equation the notation $Q' \equiv Q[U'] \equiv Q[U(\tau_{n+1})]$ was used.

In order to write down the equations for the gauge variables, let us introduce the derivatives $D_{x\mu j}$ ($j = 1, 2, \ldots, 8$) of a function of the gauge variable $U_{x\mu}$ by the definition

$$D_{x\mu j}f(U_{x\mu}) = \frac{\partial}{\partial\alpha}f\left[e^{i\alpha\lambda_j}U_{x\mu}\right]_{\alpha=0} . \qquad (7.186)$$

Note that this type of derivatives do not commute, but due to

$$D_{x\mu k}D_{x\mu j}f(U_{x\mu}) = \frac{\partial^2}{\partial\alpha_2\,\partial\alpha_1}f\left[e^{i\alpha_1\lambda_j}e^{i\alpha_2\lambda_k}U_{x\mu}\right]_{\alpha_1=\alpha_2=0} , \qquad (7.187)$$

and (8.16), satisfy the commutation relations

$$[D_{x\mu j}, D_{x\mu k}] = 2f_{jkl}D_{x\mu l} . \qquad (7.188)$$

From (7.176) follows that

$$D_{x\mu j}S[U, \phi] = D_{x\mu j}S_g[U]$$

$$- \left\{[Q^+Q]^{-1}\phi, [(D_{x\mu j}Q^+)Q + Q^+(D_{x\mu j}Q)][Q^+Q]^{-1}\phi\right\} . \qquad (7.189)$$

Using the same notations as in (7.185), we have

$$U'_{x\mu} = \exp\left\{\sum_{j=1}^8 i\lambda_j\left[\Delta\tau P_{x\mu j} - \frac{1}{2}(\Delta\tau)^2 D_{x\mu j}S[U, \phi]\right]\right\}U_{x\mu} \, ,$$

$$P'_{x\mu j} = P_{x\mu j} - \frac{1}{2}\Delta\tau\left[D_{x\mu j}S[U, \phi] + D'_{x\mu j}S[U', \phi']\right] . \qquad (7.190)$$

Here D' refers to the derivative with respect to U'.

According to (7.185) and (7.190), in one step of the classical dynamics algorithm for QCD, one has to evaluate the quark matrix inverse only twice, namely $[Q^+Q]^{-1}\phi$ and $[Q'^+Q']^{-1}\phi'$. In every step all the gauge and pseudofermion variables are changed. Therefore the number of required quark matrix inversions does not increase with the number of variables (that is, with the number of lattice points). This is a very important advantage of the classical dynamics algorithm in comparison to ordinary local updating algorithms. The problem is, however, ergodicity. In fact, since at weak gauge couplings QCD is equivalent to a system of weakly coupled oscillators, one expects a serious lack of ergodicity just in the vicinity of the continuum limit. A way of overcomming this difficulty is to consider, instead of the classical equations of motion, discretized versions of the Langevin stochastic differential equations.

7.5.2   *Langevin algorithms*

The conjugate momenta in the classical Hamiltonian system (7.178) have a Gaussian distribution, which is independent of the values of the field variables. This suggests that the problem of ergodicity in the classical dynamics algorithm can be solved, if instead of stepping along a single classical trajectory, one performs steps along different trajectories with Gaussian-distributed initial momenta. Going to the extreme, one can randomly refresh the momenta after every single step. This is the basis of the Langevin algorithms. The theoretical background of this way of simulating Euclidean quantum field theories is the so-called *stochastic quantization* [7.45]. (For reviews see [7.46].)

First let us consider a real scalar field ϕ_x with lattice action $S[\phi]$, and its conjugate momentum π_x. The discretized classical dynamics equation corresponding to the first equation in (7.185) is

$$\phi'_x = \phi_x + \Delta\tau\pi_x - \frac{1}{2}(\Delta\tau)^2 \frac{\partial S[\phi]}{\partial\phi_x} \; . \tag{7.191}$$

By making the substitutions

$$\sqrt{2}\pi_x \longrightarrow -\eta_x \; , \qquad \frac{1}{2}(\Delta\tau)^2 \longrightarrow \epsilon \; , \tag{7.192}$$

one obtains the *discretized Langevin equation*

$$\phi'_x = \phi_x - \epsilon\frac{\partial S[\phi]}{\partial\phi_x} - \sqrt{\epsilon}\eta_x \equiv \phi_x - f_x[\phi,\eta] \; . \tag{7.193}$$

Here η_x is a *Gaussian noise* satisfying

$$\langle\eta_x\rangle_\eta = 0 \; , \qquad \langle\eta_x\eta_y\rangle_\eta = 2\delta_{xy} \; . \tag{7.194}$$

The choice of the sign of η_x in (7.193) is just for convenience, an opposite sign convention is also possible. The expectation value $\langle\cdots\rangle_\eta$ is, of course, defined as

$$\langle F \rangle_\eta \equiv \frac{\int [\mathrm{d}\eta] \exp\{-\frac{1}{4}\sum_x \eta_x^2\} F[\eta]}{\int [\mathrm{d}\eta] \exp\{-\frac{1}{4}\sum_x \eta_x^2\}} . \qquad (7.195)$$

The development of the probability distribution of the field configurations $P[\phi]$ is described by the *Fokker–Planck equation*. This can be obtained from

$$P'[\phi'] = \left\langle \int [\mathrm{d}\phi] P[\phi] \prod_x \delta(\phi'_x - \phi_x + f_x[\phi, \eta]) \right\rangle_\eta . \qquad (7.196)$$

Expanding the δ-functions in powers of f_x gives

$$P'[\phi'] = \sum_{n=0}^{\infty} \frac{1}{n!} \sum_{x_1,\dots,x_n} \frac{\partial^n}{\partial \phi'_{x_1} \cdots \partial \phi'_{x_n}} \langle f_{x_1}[\phi', \eta] \cdots f_{x_n}[\phi', \eta] P[\phi'] \rangle_\eta .$$

$$(7.197)$$

The right hand side can be expanded according to the powers of the small ϵ. Up to first order one obtains

$$\frac{1}{\epsilon} (P'[\phi] - P[\phi]) = \sum_x \frac{\partial}{\partial \phi_x} \left(\frac{\partial P[\phi]}{\partial \phi_x} + \frac{\partial S[\phi]}{\partial \phi_x} P[\phi] \right) + O(\epsilon) . \qquad (7.198)$$

This is a discretized diffusion equation. For real action S one can prove, that the probability distribution P has a finite limit $\overline{P}$ at $\tau \to \infty$. According to (7.198), the limiting distribution

$$\overline{P}[\phi] \equiv e^{-\overline{S}[\phi]} \qquad (7.199)$$

satisfies

$$\sum_x \frac{\partial}{\partial \phi_x} \left(\frac{\partial \overline{P}[\phi]}{\partial \phi_x} + \frac{\partial S[\phi]}{\partial \phi_x} \overline{P}[\phi] \right) = O(\epsilon) . \qquad (7.200)$$

Therefore, in the limit of small ϵ

$$\overline{S}[\phi] = S[\phi] + O(\epsilon) . \qquad (7.201)$$

In other words, for $\epsilon \to 0$ the probability distribution of the field configurations tends to the Boltzmann distribution of the canonical ensemble. This shows that the Langevin equation (7.193) can indeed be used for the numerical simulation of Euclidean quantum field theories.

Nevertheless, since in pratice ϵ is always finite, (7.199)–(7.201) also show that the limiting distribution is not exactly the required one. In the effective action $\overline{S}$ there are $O(\epsilon)$ corrections. These can be determined by working out the higher order terms in (7.197). The order ϵ correction in

$$\overline{S}[\phi] = S[\phi] + \epsilon S_1[\phi] + O(\epsilon^2) \qquad (7.202)$$

turns out to be

$$S_1[\phi] = \frac{1}{2} \sum_x \left(\frac{\partial^2 S[\phi]}{\partial\phi_x \partial\phi_x} - \frac{1}{2} \frac{\partial S[\phi]}{\partial\phi_x} \frac{\partial S[\phi]}{\partial\phi_x} \right) . \tag{7.203}$$

This shows that the effect of a finite step size in the discretized Langevin equation is, in general, a small change in the effective lattice action. For instance, the correction terms in S_1 can shift the parameters of the original action a little bit, and/or can add some new higher dimensional terms to the action. These correction terms, however, usually do not change the universality class of the continuum limit. In this sense they are irrelevant.

The purpose of the *higher order Langevin algorithms* is to eliminate a part of the finite ϵ corrections from the effective action $\bar{S}$, by better approximating the *continuum Langevin equation*

$$\frac{\partial\phi(x,\tau)}{\partial\tau} = -\frac{\delta S[\phi]}{\delta\phi(x,\tau)} + \eta(x,\tau) . \tag{7.204}$$

This is a stochastic differential equation involving the continuum Gaussian noise $\eta(x,\tau)$ defined by

$$\langle\eta(x,\tau),\eta(x',\tau')\rangle_\eta = 2\delta(x-x')\delta(\tau-\tau') . \tag{7.205}$$

An example of an order ϵ^2 algorithm, where the first order correction S_1 in (7.202) vanishes, is the *Runge–Kutta algorithm* [7.47]:

$$\phi'_x = \phi_x - \frac{\epsilon}{2} \left[\frac{\partial S[\phi]}{\partial\phi_x} + \frac{\partial S[\bar{\phi}]}{\partial\bar{\phi}_x} \right] + \sqrt{\epsilon}\eta_x , \tag{7.206}$$

where $\bar{\phi}_x$ is a tentative update using the lowest order algorithm:

$$\bar{\phi}_x = \phi_x - \epsilon\frac{\partial S[\phi]}{\partial\phi_x} - \sqrt{\epsilon}\eta_x . \tag{7.207}$$

The Fokker–Planck equation corresponding to (7.197) implies that in this case the first order correction S_1 in the effective action $\bar{S}$ vanishes, indeed.

The discretized Langevin equation for QCD [7.48] can be obtained in the same way from the discretized classical dynamics equations (7.185) and (7.190), as (7.193) was obtained in the scalar case. Besides the real Gaussian noise $\eta_{x\mu j}^{(P)}$, which replaces the conjugate momenta $P_{x\mu j}$, and satisfies

$$\langle\eta_{x\mu j}^{(P)}\rangle_\eta = 0 , \qquad \langle\eta_{x\mu j}^{(P)}\eta_{y\nu k}^{(P)}\rangle_\eta = 2\delta_{xy}\delta_{\mu\nu}\delta_{jk} , \tag{7.208}$$

we also need a complex Gaussian noise $\eta_x^{(\pi)} \equiv \eta_{qx\alpha c}^{(\pi)}$ replacing the conjugate momentum π_x, and satisfying

$$\langle\eta_{qx\alpha c}^{(\pi)}\rangle_\eta = 0 , \qquad \langle\eta_{qx\alpha c}^{(\pi)*}\eta_{ry\beta d}^{(\pi)}\rangle_\eta = 4\delta_{qr}\delta_{xy}\delta_{\alpha\beta}\delta_{cd} . \tag{7.209}$$

The discretized Langevin equations in QCD with two degenerate quark flavours are then

$$\phi'_x = \phi_x - 2\epsilon \sum_y [Q^+Q]^{-1}_{xy} \phi_y - \sqrt{\epsilon} \eta^{(\pi)}_x \;,$$

$$U'_{x\mu} = \exp \left\{ -\sum_{j=1}^{8} i\lambda_j \left[\epsilon D_{x\mu j} S[U,\phi] + \sqrt{\epsilon} \eta^{(P)}_{x\mu j} \right] \right\} U_{x\mu} \;. \qquad (7.210)$$

The first of the equations in (7.210) can also be replaced by a direct generation of the Gaussian distribution of the pseudofermion field ϕ_x [7.47], because $S[U,\phi]$ is quadratic in ϕ_x. Generating the complex $\eta_x \equiv \eta_{qx\alpha c}$ field according to the distribution $\exp(-\sum_x \eta^+_x \eta_x)$, one can put

$$\phi_x = \sum_y (Q^+)_{xy} \eta_y \;. \qquad (7.211)$$

This has the required distribution $\exp(-\{\phi, [Q^+Q]^{-1}\phi\})$. The derivative appearing in the second equation in (7.210) is then replaced, according to (7.189), as

$$D_{x\mu j} S[U,\phi] \longrightarrow D_{x\mu j} S_g[U]$$

$$- \{Q^{-1}\eta, [(D_{x\mu j}Q^+)Q + Q^+(D_{x\mu j}Q)]Q^{-1}\eta\} \;. \qquad (7.212)$$

This variant of the Langevin algorithm has a close relationship to the pseudofermion algorithm [7.31], briefly discussed in section 7.4. It also offers a simple way for taking into account multiple (degenerate) quark flavours. Let us now consider N_f pairs of degenerate quarks, when according to (7.132) and (7.137) the fermion determinant is $[\det(Q^+Q)]^{N_f}$. The effective bosonic action corresponding to (7.136) is

$$S_{eff}[U] \equiv S_g[U] - N_f \, \text{Tr} \log (Q^+[U]Q[U]) \;. \qquad (7.213)$$

Therefore the derivative with respect to the gauge field is

$$D_{x\mu j} S_{eff}[U] = D_{x\mu j} S_g[U] - N_f \sum_{uv} (Q^+Q)^{-1}_{uv} \left[D_{x\mu j}(Q^+Q) \right]_{vu} \;. \qquad (7.214)$$

The required matrix element of $(Q^+Q)^{-1}$ can then be obtained by using (7.143). The simplest way to generate the χ-field there is to choose η according to the distribution $\exp(-\sum_x \eta^+_x \eta_x)$, and then put $\chi = Q^{-1}\eta$ [7.49]. This can also be repeated N_η times and averaged over $\eta = \eta_{(v)}$ ($v = 1, 2, \ldots, N_\eta$). As a result one obtains

$$D_{x\mu j} S_{eff}[U] = D_{x\mu j} S_g[U]$$

$$- \frac{N_f}{N_\eta} \sum_{v=1}^{N_\eta} \{Q^{-1}\eta_{(v)}, [(D_{x\mu j}Q^+)Q + Q^+(D_{x\mu j}Q)]Q^{-1}\eta_{(v)}\} \;, \qquad (7.215)$$

and the corresponding discrete Langevin equation for the change of the gauge field is

$$U'_{x\mu} = \exp\left\{-\sum_{j=1}^{8} i\lambda_j \left[\epsilon D_{x\mu j} S_{eff}[U] + \sqrt{\epsilon}\eta^{(P)}_{x\mu j}\right]\right\} U_{x\mu} . \qquad (7.216)$$

In this form it becomes clear that the pseudofermion fields $\eta_{(v)}$ are only auxiliary variables. Compared to (7.212) one can also see that, in the case of $N_\eta > 1$, the noise in $D_{x\mu j}S$ is diminished.

Before proceeding let us briefly remark that the Gaussian distributions needed in the Langevin algorithms can be easily obtained from a uniformly distributed x $(0 < x < 1)$. Putting $r = (-a^2 \log x)^{1/2}$ we have

$$P(r)\,dr = \left|\frac{dx}{dr}\right|\,dr = \frac{2r}{a^2}e^{-(r/a)^2}\,dr . \qquad (7.217)$$

With $x_1 = r\cos\varphi$ and $x_2 = r\sin\varphi$, and a uniformly distributed angle $0 < \varphi < 2\pi$, this gives

$$P(r)\,dr\,d\varphi \propto \exp\left[-\frac{1}{a^2}(x_1^2 + x_2^2)\right]\,dx_1\,dx_2 . \qquad (7.218)$$

Hence we have a pair of random numbers $-\infty < x_i < \infty$ $(i = 1, 2)$, which are distributed according to $\exp\left[-(x_i/a)^2\right]\,dx_i$.

The advantage of the classical dynamics algorithm is that it moves fast in phase space, and hence it generates statistically independent configurations in an effective way. It has, however, problems with ergodicity. The Langevin algorithms, on the other hand, are ergodic, but the succesive steps are in randomly changing directions. Therefore the distance in the configuration space only increases with the square root of the number of steps. The advantages of both algorithms are combined in the *hybrid classical–Langevin (HCL) algorithms* [7.50, 7.51]. In these algorithms at each step a random choice is made between a classical dynamics update, with probability α, and a Langevin update with probability $(1-\alpha)$. Therefore in the classical dynamics equations (7.185), (7.190) π_x and $P_{x\mu j}$ are replaced according to

$$\Delta\tau\,\pi_x \longrightarrow \alpha\,\Delta\tau\,\pi_x - (1-\alpha)\sqrt{\epsilon}\eta^{(\pi)}_x ,$$

$$\Delta\tau\,P_{x\mu j} \longrightarrow \alpha\,\Delta\tau\,P_{x\mu j} - (1-\alpha)\sqrt{\epsilon}\eta^{(P)}_{x\mu j} . \qquad (7.219)$$

A similar HCL procedure can also be formulated [7.52] in the variant of the Langevin algorithm, where the Gaussian distribution of the pseudofermion field ϕ_x is explicitly generated by (7.211). In this algorithm first ϕ_x and $P_{x\mu j} = \eta^{(P)}_{x\mu j}$ are generated according to the appropriate Gaussian distributions, and then the classical dynamics equations in (7.190) are used to evolve $U_{x\mu}$ and $P_{x\mu j}$ along a classical trajectory. Corresponding to

(7.219), after every time step of length $\Delta\tau$ it is decided whether the next step has to be performed or not. The probability of performing another step is α. Otherwise new ϕ_x and $\eta_{x\mu j}^{(P)}$ are generated, and the classical dynamics steps are restarted [7.53].

A potentially very important application of the Langevin algorithms would be to simulate Euclidean quantum field theories with complex (non-real) action [7.54, 7.55] (for further references see also [7.56]). Such actions occur in fermionic theories with non-positive fermion determinant, like QCD for non-zero chemical potential, chiral Yukawa models etc. The theory of complex Langevin algorithms is, however, not yet sufficiently developed. Important open questions are the convergence and ergodicity.

An inconvenience of general kind for the numerical simulation algorithms based on difference equations (discretized differential equations) is the dependence of the results on the finite step size in time. Even if these effects can be diminished by applying higher order discretization schemes, and can, in general, be expressed by a renormalization of bare parameters in the lattice action, still the appearance of a new parameter implies the increase of the dimensionality of the parameter space. The extrapolation of the results to zero time step size is in principle possible, but assumes the knowledge of the general behaviour of different quantities for small step sizes. In any case, the extrapolation is time consuming and introduces an additional source of systematic errors. These problems can be avoided if a suitable correction scheme is applied in the algorithm itself, as is done in the hybrid Monte Carlo algorithms discussed in the next section.

7.6   Hybrid Monte Carlo algorithms

In ordinary Metropolis updating processes, which were discussed in section 7.2, the bosonic field configuration is changed locally in every step: the proposed new trial configuration is typically obtained from the previous one by a random change of a few local field variables. In principle also non-local changes are possible. For instance, one could try to change randomly all (or a large part of) variables at once. This would, however, be very inefficient, because the new value of the action would be in most cases much larger than the old one, and hence the proposed change would practically never be accepted. (Remember that the updating process exploits only the vicinity of the minimum of the action with high probability.) The situation is different if conjugate momenta are introduced and the updating process is defined by a Hamiltonian (7.178), in the extended path integral representation (7.179). In this case the discretized classical dynamics equations can give new configurations, where every field variable is changed, but the value of the Hamiltonian is close to the starting value. One can consider the endpoint of a classical trajectory as a trial new configuration, and accept or reject it according to the general Metropolis acceptance probability (7.78). In this way the problem of the extrapolation of the classical dynamics difference equations to zero time step size is also avoided, because the accept–reject probability is formulated in terms of the exact Hamiltonian, and does not contain any approximation. (The only place where a poor approximation due to the discretization can hurt is a low acceptance probability.) This is the basic idea of the hybrid Monte Carlo algorithms [7.18, 7.57]. (For a more detailed discussion see also [7.58, 7.59].)

7.6.1   HMC for scalar fields

In order to introduce the basic ingredients of the *hybrid Monte Carlo (HMC)* algorithms, let us consider the simplest case of a real scalar field ϕ_x and its conjugate momentum π_x, with lattice action $S[\phi]$ and Hamiltonian

$$H[\pi, \phi] = \frac{1}{2} \sum_x \pi_x^2 + S[\phi] \; . \tag{7.220}$$

The HMC updating step is started by a random choice of the conjugate momenta according to the Gaussian distribution

$$P_G[\pi] \propto \exp\left\{ -\frac{1}{2} \sum_x \pi_x^2 \right\} \; . \tag{7.221}$$

Then, using the Hamiltonian $H[\pi, \phi]$, the configuration $[\pi, \phi]$ is changed along the discretized trajectory in phase space up to the end point $[\pi', \phi'] \equiv T_H[\pi, \phi]$. This is expressed by the change of the probability distribution of the configuration according to

$$P_H\left([\pi', \phi'] \leftarrow [\pi, \phi]\right) = \delta\left([\pi', \phi'] - T_H[\pi, \phi]\right) \; , \tag{7.222}$$

since the classical dynamics equation is deterministic. It will always be assumed that the discretized classical dynamics equations, which define the trajectory, are such that the mapping $[\pi, \phi] \rightarrow [\pi', \phi']$ is reversible in the sense

$$P_H\left([\pi', \phi'] \leftarrow [\pi, \phi]\right) = P_H\left([-\pi, \phi] \leftarrow [-\pi', \phi']\right) \; . \tag{7.223}$$

This is obviously satisfied by the continuous trajectories, given by the canonical differential equations, and we shall see below that there are also difference equations with finite step size, which obey (7.223) exactly. After the trajectory $[\pi, \phi] \rightarrow [\pi', \phi']$ is determined, the configuration $[\pi', \phi']$ is either accepted as a new one, or rejected and the old configuration $[\pi, \phi]$ is kept. The acceptance probability is, in analogy with the Metropolis algorithm discussed in section 7.2,

$$P_A \left([\pi', \phi'] \leftarrow [\pi, \phi]\right) = \min \left\{1, e^{-H[\pi', \phi'] + H[\pi, \phi]}\right\} . \qquad (7.224)$$

The total transition probability in ϕ-space is

$$P \left([\phi'] \leftarrow [\phi]\right) = \int [d\pi \, d\pi'] P_A \left([\pi', \phi'] \leftarrow [\pi, \phi]\right)$$

$$\cdot P_H \left([\pi', \phi'] \leftarrow [\pi, \phi]\right) P_G[\pi] . \qquad (7.225)$$

This has the canonical distribution $W_c[\phi] \propto \exp\{-S[\phi]\}$ as a unique fixed point, because it satisfies the detailed balance condition (7.20).

In order to prove this, one first uses the relations

$$W_c[\phi] P_G[\pi] = e^{-H[\pi, \phi]} , \qquad (7.226)$$

and

$$e^{-H[\pi, \phi]} \min \left\{1, e^{-H[\pi', \phi'] + H[\pi, \phi]}\right\}$$

$$= e^{-H[\pi', \phi']} \min \left\{1, e^{-H[\pi, \phi] + H[\pi', \phi']}\right\} , \qquad (7.227)$$

which imply

$$W_c[\phi] P_G[\pi] P_A \left([\pi', \phi'] \leftarrow [\pi, \phi]\right) = W_c[\phi'] P_G[\pi'] P_A \left([\pi, \phi] \leftarrow [\pi', \phi']\right)$$

$$= W_c[\phi'] P_G[-\pi'] P_A \left([-\pi, \phi] \leftarrow [-\pi', \phi']\right) . \qquad (7.228)$$

Therefore, due to reversibility (7.223), we have

$$W_c[\phi] \int [d\pi \, d\pi'] P_A \left([\pi', \phi'] \leftarrow [\pi, \phi]\right) P_H \left([\pi', \phi'] \leftarrow [\pi, \phi]\right) P_G[\pi]$$

$$= W_c[\phi'] \int [d\pi \, d\pi'] P_A \left([-\pi, \phi] \leftarrow [-\pi', \phi']\right)$$

$$\cdot P_H \left([-\pi, \phi] \leftarrow [-\pi', \phi']\right) P_G[-\pi'] . \qquad (7.229)$$

Taking into account that $[d\pi \, d\pi'] = [d(-\pi) \, d(-\pi')]$, this is just the detailed balance condition.

The next question is, how to define a finite difference approximation to the Hamiltonian equations, which fulfills the reversibility condition (7.223). One possibility is the *leapfrog integration*. Let us denote the starting point of the trajectory at $\tau = \tau_0 \equiv 0$ by $(\pi(0), \phi(0))$, and the end point at $\tau = \tau_n \equiv n \Delta \tau$ by $(\pi(\tau_n), \phi(\tau_n))$. At intermediate points of the trajectory the conjugate momenta are associated with the points at

$\tau = (j + \frac{1}{2}) \Delta\tau$ $(j = 0, 1, \ldots, n - 1)$, and the field variables to $\tau = (j + 1) \Delta\tau$. The first 'half step' is then

$$\pi\left(\frac{\Delta\tau}{2}\right) = \pi(0) - \frac{\partial S(\phi(0))}{\partial\phi}\frac{\Delta\tau}{2} ,$$

$$\phi(\Delta\tau) = \phi(0) + \pi\left(\frac{\Delta\tau}{2}\right)\Delta\tau . \tag{7.230}$$

The steps for $j = 1, 2, \ldots, n - 1$ are

$$\pi\left(j\Delta\tau + \frac{\Delta\tau}{2}\right) = \pi\left(j\Delta\tau - \frac{\Delta\tau}{2}\right) - \frac{\partial S(\phi(j\Delta\tau))}{\partial\phi}\Delta\tau ,$$

$$\phi(j\Delta\tau + \Delta\tau) = \phi(j\Delta\tau) + \pi\left(j\Delta\tau + \frac{\Delta\tau}{2}\right)\Delta\tau . \tag{7.231}$$

In the last half step the final momentum is obtained as

$$\pi(n\Delta\tau) = \pi\left(n\Delta\tau - \frac{\Delta\tau}{2}\right) - \frac{\partial S(\phi(n\Delta\tau))}{\partial\phi}\frac{\Delta\tau}{2} . \tag{7.232}$$

The half steps differ from the exact integration of the Hamiltonian equations by errors of order $O(\Delta\tau)^2$, whereas the intermediate steps have only errors of order $O(\Delta\tau)^3$.

The leapfrog integration is reversible, because starting at $\tau = \tau_n$ with $(-\pi(\tau_n), \phi(\tau_n))$, and performing the above steps with $-\Delta\tau$ instead of $\Delta\tau$, one arrives at $\tau = 0$ with $(-\pi(0), \phi(0))$. This is just the content of (7.223). In addition, the mapping $(\pi(0), \phi(0)) \to (\pi(\tau_n), \phi(\tau_n))$ is 'area preserving', because

$$[d\pi(0)\, d\phi(0))] = [d\pi(\tau_n)\, d\phi(\tau_n))] . \tag{7.233}$$

This can be seen by inspection. For instance, the Jacobian of the first half step is

$$\det\left(\frac{\partial(\pi(\Delta\tau/2), \phi(\Delta\tau))}{\partial(\pi(0), \phi(0))}\right) = 1 . \tag{7.234}$$

Note that for $n = 1$ the leapfrog equations (7.230), (7.232) imply

$$\phi(\Delta\tau) = \phi(0) + \pi(0)\,\Delta\tau - \frac{1}{2}(\Delta\tau)^2\frac{\partial S(\phi(0))}{\partial\phi} ,$$

$$\pi(\Delta\tau) = \pi(0) - \frac{\Delta\tau}{2}\left[\frac{\partial S(\phi(0))}{\partial\phi} + \frac{\partial S(\phi(\Delta\tau))}{\partial\phi}\right] . \tag{7.235}$$

This coincides with the discretized Hamiltonian equations used in section 7.5 (see (7.183)). Since this is a special case of the leapfrog integration, it is also reversible and area preserving. (It has an error of order $O(\Delta\tau)^3$ compared to the continuum Hamiltonian equations.) Therefore, one can also iterate (7.235), say, n times and use it to calculate the trajectory of length $n \Delta \tau$.

A question of utmost importance for the practicability of HMC updating is the acceptance rate as a function of step size $\Delta \tau$, number of steps per trajectory n, and the number of field variables (which is proportional to the number of lattice points). This question is investigated, for instance, in references [7.58]–[7.62]. The acceptance rate obviously depends on the distribution of the energy difference between the end point and starting point of the trajectory

$$\delta H \equiv H[\pi(\tau_n), \phi(\tau_n)] - H[\pi(0), \phi(0)] \, . \qquad (7.236)$$

Owing to the area preserving property (7.233) of the mapping $(\pi(0), \phi(0)) \to (\pi(\tau_n), \phi(\tau_n))$, we have for the HMC partition function

$$Z = \int [d\pi' \, d\phi'] e^{-H[\pi',\phi']} = \int [d\pi \, d\phi] e^{-H[\pi,\phi] - \delta H[\pi,\phi]} \, . \qquad (7.237)$$

This implies

$$1 = \langle e^{-\delta H} \rangle \geq e^{-\langle \delta H \rangle} \, . \qquad (7.238)$$

The first equality provides a useful tool for checking the correctness of HMC codes. The inequality follows from the convexity of the exponential function. Therefore the expectation value of δH is always non-negative: $\langle \delta H \rangle \geq 0$, and for small δH we have

$$\langle \delta H \rangle = \frac{1}{2} \langle (\delta H)^2 \rangle + O(\delta H)^3 = \frac{1}{2} \langle (\delta H - \langle \delta H \rangle)^2 \rangle + O(\delta H)^3 \, . \qquad (7.239)$$

This shows that in order to achieve a finite $\langle \delta H \rangle$ in the infinite lattice size limit, one has to keep the variance $\langle (\delta H - \langle \delta H \rangle)^2 \rangle$ finite. In this limit the higher cumulants of the distribution are negligible, and the expectation value of the acceptance probability is

$$\langle P_A \rangle = \frac{1}{(4\pi \langle \delta H \rangle)^{1/2}} \int_{-\infty}^{+\infty} dx \min \{1, e^{-x}\} \exp \left\{ -\frac{(x - \langle \delta H \rangle)^2}{4 \langle \delta H \rangle} \right\}$$

$$= \mathrm{erfc} \left(\frac{1}{2} (\langle \delta H \rangle)^{1/2} \right) \, . \qquad (7.240)$$

In the case of the above leapfrog integration one can argue, [7.58]–[7.62], that for large lattice sizes $\Omega \to \infty$

$$\langle \delta H \rangle \propto \Omega (\Delta \tau)^4 \, . \qquad (7.241)$$

Hence, in order to keep a constant acceptance rate, for large Ω one has to decrease the time step size as $\Delta \tau \propto \Omega^{1/4}$. The dependence of the acceptance rate on bare parameters, like the quark mass or gauge coupling, can be described by some empirical formulas (see, for instance, [7.61]).

7.6.2   HMC for gauge and fermion fields

The extension of the HMC algorithm to gauge fields is straightforward. The only thing one needs is a suitable definition of the leapfrog integration. For definiteness, let us consider again an SU(3) gauge group, since the generalization to SU(N) is trivial. One possibility is to iterate (7.190) n times, similarly to (7.235) in the scalar case. The leapfrog integration analogous to (7.230)–(7.232) is the following: in the first half step

$$P_{x\mu j}\left(\frac{\Delta\tau}{2}\right) = P_{x\mu j}(0) - D_{x\mu j}S[U(0)]\frac{\Delta\tau}{2} , \qquad (7.242)$$

and then, for $k = 0, 1, \ldots, n-1$

$$U_{x\mu}(k\,\Delta\tau + \Delta\tau) = \exp\left\{ \sum_{j=1}^{8} i\lambda_j P_{x\mu j}\left(k\,\Delta\tau + \frac{\Delta\tau}{2}\right)\Delta\tau \right\} U_{x\mu}(k\,\Delta\tau) , \quad (7.243)$$

and for $k = 1, 2, \ldots, n-1$

$$P_{x\mu j}\left(k\,\Delta\tau + \frac{\Delta\tau}{2}\right) = P_{x\mu j}\left(k\,\Delta\tau - \frac{\Delta\tau}{2}\right) - D_{x\mu j}S[U(k\,\Delta\tau)]\,\Delta\tau , \quad (7.244)$$

and finally

$$P_{x\mu j}(n\,\Delta\tau) = P_{x\mu j}\left(n\,\Delta\tau - \frac{\Delta\tau}{2}\right) - D_{x\mu j}S[U(n\,\Delta\tau)]\frac{\Delta\tau}{2} . \qquad (7.245)$$

The HMC algorithm for fermionic theories, like for instance QCD, can be formulated on the basis of the action $S[U, \phi]$ in (7.176) and the Hamiltonian in (7.178). Before starting the leapfrog integration steps for the gauge field (7.242)–(7.245), the Gaussian distribution for the pseudo-fermion field ϕ_x is generated according to (7.211). During the leapfrog steps the pseudofermion field ϕ_x is kept constant. The initial values of the conjugate momenta $P_{x\mu j}$ are also randomly generated with a distribution

$$P_G[P] \propto \exp\left\{ -\frac{1}{2}\sum_{x\mu j} P_{x\mu j}^2 \right\} . \qquad (7.246)$$

Of course, the action $S[U]$ in (7.242)–(7.245) is replaced by $S[U, \phi]$, and its derivative is given by (7.189).

7.7   Cluster algorithms

The autocorrelation of subsequent configurations in an updating process, which is discussed in section 7.1, reduces the effectiveness, because the amount of computational work required to reach statistically independent configurations is increased. This is particularly disturbing near the continuum limit, where the correlation length in lattice units ξ diverges and, according to (7.42), the autocorrelation behaves as ξ^z, with some dynamical critical exponent z. As argued in section 7.1, for an updating process based on local changes of the configuration, a typical value is $z = 2$. Some improvement, for instance in the case of the Metropolis algorithm, can be achieved by a clever choice of the local changes (see the 'overrelaxation' discussed in section 7.2), but the gain reached in this way corresponds only to $z \simeq 1$. Therefore some critical slowing down is still there.

A dramatic reduction of critical slowing down, corresponding to $z \simeq 0$, can be achieved in some simple scalar field models by the cluster algorithms (for a review see [7.63]). Unfortunately, at present it is not known, whether this method can effectively be implemented also for physically more important cases like, for instance, QCD. But the improvement in simple spin models is very impressive.

The cluster updating algorithm was introduced by Swendsen and Wang [7.64] in the Potts model (for discussions see also [7.65, 7.66]). The q-state Potts model is a statistical system of spins $[\sigma] \equiv \{\sigma_i, i = 1, 2, \ldots, \Omega\}$, which can have q different values $\{1, 2, \ldots, q\}$. Its partition function is defined as

$$Z_P \equiv \int [d\sigma] \exp\left\{ \sum_{(ij)} J_{ij}(\delta_{\sigma_i \sigma_j} - 1) \right\} . \tag{7.247}$$

Here (ij) denotes an (unordered) pair of lattice sites, usually nearest neighbours, and $J_{ij} = J_{ji}$ is the coupling for the pair (ij), which is assumed to be non-negative (ferromagnetic): $J_{ij} \geq 0$. The notation $\int [d\sigma]$ means here a summation over the spin configurations. An alternative form of Z_P is:

$$Z_P \equiv \int [d\sigma] \prod_{(ij)} \left\{ (1 - p_{ij}) + p_{ij} \delta_{\sigma_i \sigma_j} \right\} , \tag{7.248}$$

where

$$p_{ij} \equiv 1 - e^{-J_{ij}} . \tag{7.249}$$

The Potts model can be related [7.67] to a random cluster model defined by the bond occupation variables $n_{ij} = n_{ji} = 0, 1$ on the lattice site pairs.

The partition function of the random cluster model is

$$Z_{RC} \equiv \int [dn] \left(\prod_{(ij),\, n_{ij}=1} p_{ij} \right) \left(\prod_{(ij),\, n_{ij}=0} (1 - p_{ij}) \right) q^{C[n]} , \qquad (7.250)$$

where $C[n]$ is the number of connected components (including one site components) in the graph, whose edges are the bonds having $n_{ij} = 1$. The connection between Z_P and Z_{RC} can be best seen on a joint model defined by both spin $[\sigma]$ and bond $[n]$ variables as

$$Z_{SB} \equiv \int [d\sigma\, dn] \prod_{(ij)} \left\{ (1 - p_{ij})\delta_{n_{ij},0} + p_{ij}\delta_{n_{ij},1}\delta_{\sigma_i\sigma_j} \right\} . \qquad (7.251)$$

One can easily see that summing here over the bond variables $[n]$ gives Z_P, and summing over the spins $[\sigma]$ gives Z_{RC}.

According to (7.251) the conditional distribution of $[n]$ for fixed $[\sigma]$ is as follows: for each bond (ij) with $\sigma_i \neq \sigma_j$ one sets $n_{ij} = 0$, and for bonds with equal spins at the ends $\sigma_i = \sigma_j$, respectively, $n_{ij} = 0, 1$ with probability $(1 - p_{ij})$, p_{ij}. The conditional distribution of $[\sigma]$ for fixed $[n]$, on the other hand, is the following: for each connected cluster of lattice sites one sets all spins σ_i in the cluster to the same value, chosen with equal probability from the set of possible values $\{1, 2, \ldots, q\}$. The proposal of Swendsen and Wang was to simulate the joint model with partition function Z_{SB} by alternatively applying these conditional probabilities. That is, one alternatively generates new bond occupation variables (independent of the old ones) from the given spins, and new spin variables (independent of the old ones) from the given bonds. The success of this procedure in reducing (or even completely removing) critical slowing down in the spin system relies on an intimate and delicate matching of the critical dynamics of both spin and bond systems. At the critical point of the spin system the spins are correlated to become dominantly parallel over large distances. In this way the corresponding bond system is characterized by large clusters, like in a percolation problem. Flipping all spins in a large cluster provides the global changes necessary for an effective updating of the physically relevant long distance (low frequency) spin variables.

An important calculational task in the cluster algorithms is to find the connected clusters. A cluster is defined as a maximal subset of lattice sites connected by the occupied bonds (with $n_{ij} = 1$). An important requirement for cluster finding algorithms is that the computational work should grow only linearly with the number of lattice points. This requirement can be fulfilled [7.68, 7.69], but due to the irregularity of the cluster configurations it is difficult to write an effectively vectorizing code. (For proposals for vectorization see [7.70].) In the Hoshen–Kopelman cluster finding algorithm [7.68] the cluster structure is identified by a pointer

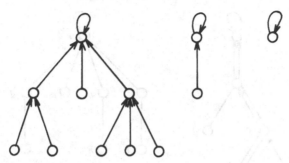

Fig. 7.3. Tree graph representation of clusters. A circle and a connected graph correspond to a site and a cluster, respectively. Each cluster has a unique root pointing to itself. By following pointers (arrows) one can always find the root for a given site.

array $POINTER(1, 2, \ldots, \Omega)$. The value of $POINTER(a)$ is the index of a site in the same cluster (not necessarily connected to a by an occupied bond). It is assumed that, representing the pointers by arrows, a graphical representation of a cluster can be given by a tree graph without closed loops and with a unique 'root' (see fig. 7.3.). The root is a site r pointing to itself: $POINTER(r) = r$. The construction of clusters is done by looping over the bonds. At an occupied bond $a \rightarrow b$ one looks for the corresponding roots r_a, r_b. If $r_a = r_b$, then a and b belong already to the same cluster. Otherwise one has to unify the two incomplete clusters by setting, for instance, $POINTER(r_a) = r_b$. The looping over the occupied bonds is stopped if all clusters are complete. In order to minimize the work, one can store the size of a cluster at its root: $SIZE(r)$, and when two incomplete clusters are unified, choose the root of the larger one to be the root of the unified cluster. Another important measure of optimization is 'path compression': in the course of root search remember all visited sites and, after finding the root, make them point directly to the root (see fig. 7.4.). This path compression assures that the required work for finding all clusters increases only linearly with the number of lattice sites.

The other cluster finding algorithm [7.69] starts from a site, and finds first all sites connected to it by an occupied bond. The new sites belong to the 'first generation' of cluster points. Looping over the first generation one finds new sites in the cluster, which are connected to first generation sites by an occupied bond, and calls these 'second generation'. Continuing to loop over the second generation etc., one finds all generations of sites, completing the cluster.

Once the cluster information is available from the updating, it can be used to obtain improved (reduced variance) estimators for the spin

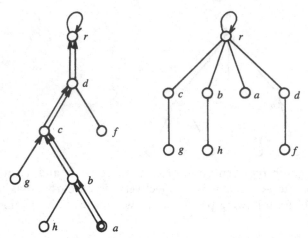

Fig. 7.4. Path compression when the root corresponding to site a is searched. Sites a, b and c, which are visited in the course of the search, will be directly connected to the root.

observables. For an arbitrary function of spins $F[\sigma]$ we have

$$\langle F[\sigma] \rangle = Z_{SB}^{-1} \int [\mathrm{d}\sigma \, \mathrm{d}n] \prod_{(ij)} \left\{ (1 - p_{ij}) \delta_{n_{ij},0} + p_{ij} \delta_{n_{ij},1} \delta_{\sigma_i \sigma_j} \right\} F[\sigma] \, . \quad (7.252)$$

Let us now consider, for definiteness, the Ising model with spin variables $\sigma_x = \pm 1$, which is equivalent to a single component scalar ϕ^4 model in the limit of infinite bare quartic self-coupling (see chapter 2). The lattice action is

$$S = -\kappa \sum_{x,\mu} \sigma_x \sigma_{x+\hat{\mu}} = -2\kappa \sum_x \sum_{\mu=1}^{4} \sigma_x \sigma_{x+\hat{\mu}} \, . \quad (7.253)$$

This is also equivalent to the two-state Potts model with couplings $J_{x,x+\hat{\mu}} = 4\kappa$. Owing to the global Z_2 symmetry $\sigma_x \rightarrow \pm\sigma_x$, the expectation value of an odd number of spins is zero:

$$\langle \sigma_{x_1} \sigma_{x_2} \ldots \sigma_{x_{2k+1}} \rangle = 0 \, . \quad (7.254)$$

The product of an even number of spins can be expressed by the cluster observables. For instance, according to (7.252), the two-point function is given by

$$\langle \sigma_x \sigma_y \rangle = \langle C_{xy} \rangle' \, , \quad (7.255)$$

where C_{xy} is defined to be 1 if x and y belong to the same cluster, and 0 otherwise. The prime denotes expectation values in the cluster configurations. This implies for the susceptibility

$$\chi_2 \equiv \frac{1}{\Omega} \sum_{xy} \langle \sigma_x \sigma_y \rangle = \frac{1}{\Omega} \left\langle \sum_c n_c^2 \right\rangle' , \qquad (7.256)$$

where $\sum_c$ means a summation over the clusters, and n_c denotes the number of points in cluster c. Similarly, the four-point function is

$$\langle \sigma_{x_1} \sigma_{x_2} \sigma_{x_3} \sigma_{x_4} \rangle = -2 \langle C_{x_1 x_2 x_3 x_4} \rangle'$$

$$+ \langle C_{x_1 x_2} C_{x_3 x_4} + C_{x_1 x_3} C_{x_2 x_4} + C_{x_1 x_4} C_{x_2 x_3} \rangle' , \qquad (7.257)$$

and the four-point (generalized) susceptibility

$$\chi_4 \equiv \frac{1}{\Omega} \sum_{x_1 x_2 x_3 x_4} \langle \sigma_{x_1} \sigma_{x_2} \sigma_{x_3} \sigma_{x_4} \rangle_c \equiv \frac{1}{\Omega} \sum_{x_1 x_2 x_3 x_4} [\langle \sigma_{x_1} \sigma_{x_2} \sigma_{x_3} \sigma_{x_4} \rangle$$

$$- \langle \sigma_{x_1} \sigma_{x_2} \rangle \langle \sigma_{x_3} \sigma_{x_4} \rangle - \langle \sigma_{x_1} \sigma_{x_3} \rangle \langle \sigma_{x_2} \sigma_{x_4} \rangle - \langle \sigma_{x_1} \sigma_{x_4} \rangle \langle \sigma_{x_2} \sigma_{x_3} \rangle]$$

$$= -\frac{2}{\Omega} \left\langle \sum_c n_c^4 \right\rangle' + \frac{3}{\Omega} \left[\left\langle \sum_{c_1 c_2} n_{c_1}^2 n_{c_2}^2 \right\rangle' - \left\langle \sum_c n_c^2 \right\rangle'^2 \right] . \qquad (7.258)$$

The remarkable property of these representations is that in terms of the cluster variables there are much fewer cancellations. In fact, in the case of the two-point function and susceptibility χ_2 there are no cancellations at all, because they are given by a sum of positive terms. This makes it plausible that the statistical errors are greatly reduced. The gain due to the improved estimators is particularly important in four-dimensional theories, where the present simulations are restricted to moderate correlation lengths (usually below $\xi \leq 10$), and hence the critical slowing down is not serious. An example is the four-dimensional Ising model studied in [7.69, 7.71].

Besides the Ising model, the other four-dimensional quantum field theories, which can be efficiently simulated by cluster algorithms, are the $O(N)$-symmetric σ-models. The first attempts to generalize the Swendsen–Wang algorithm to such cases were based on the idea of applying general $O(N)$-rotations on suitably constructed clusters [7.65, 7.66, 7.72]. The really successful generalization due to Wolff [7.73], however, exploits the stochastic embedding of Ising variables [7.74] into the $O(N)$ model.

The $O(N)$-symmetric non-linear σ-model is formulated in terms of the N-component real $O(N)$-vector field ϕ_{ax}, $(a = 1, 2, \ldots, N)$ of unit length: $\sum_{a=1}^N \phi_{ax}^2 \equiv \phi_{ax} \phi_{ax} = 1$. With a summation convention over $O(N)$-indices the lattice action is

$$S = -\kappa \sum_{x,\mu} \phi_{ax} \phi_{a,x+\hat{\mu}} = -2\kappa \sum_x \sum_{\mu=1}^4 \phi_{ax} \phi_{a,x+\hat{\mu}} . \qquad (7.259)$$

The embedding of the Ising variables $\sigma_x = \pm 1$ with respect to the random direction ρ_a ($\rho_a \rho_a = 1$) is defined by $\phi_{ax}^{\perp} \rho_a = 0$ and

$$\phi_{ax} = \phi_{ax}^{\perp} + \rho_a |\phi_{bx} \rho_b| \sigma_x . \tag{7.260}$$

That is, the Ising field σ_x gives the sign of the ρ-component of ϕ_x. The Ising spins are updated with the effective action

$$S_{WO} \equiv -2\kappa \sum_x \sum_{\mu=1}^{4} |\phi_{ax}\rho_a \phi_{bx}\rho_b| \sigma_x \sigma_{x+\hat{\mu}} . \tag{7.261}$$

Therefore in the Potts model notation the coupling is $J_{x,x+\hat{\mu}} = 4\kappa |\phi_{ax}\rho_a \phi_{bx}\rho_b|$.

An updating step in the Wolff algorithm consists of the following:

- choose a random direction ρ_a in the O(N) vector space, and a random lattice site x_C,

- grow one cluster C starting from x_C with bond activation probability $[1 - \exp(-4\kappa\phi_{ax}\rho_a\phi_{by}\rho_b)]\Theta(\phi_{cx}\rho_c\phi_{dy}\rho_d)$,

- reflect for all $x \in C$ the spins according to $\phi_{ax} \rightarrow \phi_{ax} - 2\rho_a\phi_{bx}\rho_b$.

The cluster search starting from the point x_C can be performed by the cluster growing algorithm of [7.69].

Ergodicity and the detailed balance relation (7.20) can be easily proven [7.73]. Therefore, this is a valid numerical simulation algorithm for the O(N) σ-model. In fact, the random choice of the site x_C implies that, compared to the original Swendsen–Wang algorithm, the probability to change a cluster is increased proportionally to its size. This favours large scale changes of the spin configuration, which is advantageous for decreasing the statistical errors of long range physical quantities. (This way of single cluster updating can, of course, also be implemented in the discrete Ising or Potts models.)

The improved estimators of spin expectation values based on the cluster variables can also be extended to the Wolff algorithm. For instance, the two-point function can be obtained from a single cluster (1C) expectation value as

$$\langle \phi_{ax}\phi_{ay} \rangle = N \left\langle \frac{\Omega}{n_C} C_{xy} \phi_{bx}\rho_b \phi_{cy}\rho_c \right\rangle^{1C} . \tag{7.262}$$

This corresponds to (7.255) in the case of the Swendsen–Wang algorithm, but here also the cluster sum is done stochastically (like a Monte Carlo integration). The factor Ω/n_C (where Ω is, as usual, the number of lattice sites, and n_C is the number of sites in the cluster C) is needed in order to compensate for the probability of picking out a single cluster.

Cluster algorithms for continuous $O(N)$ spin models were successfully applied both in four-dimensional ϕ^4 models [7.75], and in two-dimensional σ-models [7.76].

8

Appendix

8.1 Notation conventions and basic formulas

8.1.1 Pauli matrices

The Pauli matrices are

$$\sigma_1 = \begin{pmatrix} 0 & 1 \\ 1 & 0 \end{pmatrix}, \quad \sigma_2 = \begin{pmatrix} 0 & -i \\ i & 0 \end{pmatrix}, \quad \sigma_3 = \begin{pmatrix} 1 & 0 \\ 0 & -1 \end{pmatrix}. \tag{8.1}$$

These satisfy for $j, k, l = 1, 2, 3$

$$\sigma_j \sigma_k = \delta_{jk} + i\epsilon_{jkl}\sigma_l, \qquad \sigma_1\sigma_2\sigma_3 = i. \tag{8.2}$$

Defining $\sigma_{\pm} \equiv \frac{1}{2}(\sigma_1 \pm i\sigma_2)$ we have

$$\sigma_+ = \begin{pmatrix} 0 & 1 \\ 0 & 0 \end{pmatrix}, \quad \sigma_- = \begin{pmatrix} 0 & 0 \\ 1 & 0 \end{pmatrix}. \tag{8.3}$$

8.1.2 Dirac matrices

The connection between the Dirac matrices in Euclidean and Minkowski space is

$$\gamma_{1,2,3}^{\text{Euclidean}} \equiv -i\gamma_{1,2,3}^{\text{Minkowski}},$$

$$\gamma_4^{\text{Euclidean}} \equiv -i\gamma_4^{\text{Minkowski}} \equiv \gamma_0^{\text{Minkowski}}. \tag{8.4}$$

The Euclidean Dirac matrices are chosen to be Hermitean, and are simply denoted by γ_μ ($\mu = 1, 2, 3, 4$):

$$\gamma_\mu \equiv \gamma_\mu^{\text{Euclidean}} = \gamma_\mu^+. \tag{8.5}$$

These satisfy the anticommutation relations

$$\{\gamma_\mu, \gamma_\nu\} = 2\delta_{\mu\nu}. \tag{8.6}$$

434

The Dirac matrices in Minkowski space are then

$$\gamma_\mu^{(M)} \equiv \gamma_\mu^{\text{Minkowski}} = i\gamma_\mu \, . \tag{8.7}$$

An often used explicit representation of the Euclidean Dirac matrices is, in a $2 \otimes 2$ block notation, for $\mu = 1, 2, 3$:

$$\gamma_{1,2,3} = \begin{pmatrix} 0 & -i\sigma_{1,2,3} \\ i\sigma_{1,2,3} & 0 \end{pmatrix} \, . \tag{8.8}$$

For $\mu = 4$ and $\gamma_5 \equiv \gamma_1\gamma_2\gamma_3\gamma_4 = \gamma_5^+$ one can take either the 'non-relativistic' representation

$$\gamma_4 = \begin{pmatrix} 1 & 0 \\ 0 & -1 \end{pmatrix} \, , \qquad \gamma_5 = \begin{pmatrix} 0 & -1 \\ -1 & 0 \end{pmatrix} \, , \tag{8.9}$$

or the 'chiral' representation

$$\gamma_4 = \begin{pmatrix} 0 & 1 \\ 1 & 0 \end{pmatrix} \, , \qquad \gamma_5 = \begin{pmatrix} 1 & 0 \\ 0 & -1 \end{pmatrix} \, . \tag{8.10}$$

The projection operators on left- and right-handed chirality are, respectively

$$P_L \equiv \frac{1}{2}(1 - \gamma_5) \, , \qquad P_R \equiv \frac{1}{2}(1 + \gamma_5) \, . \tag{8.11}$$

The left-handed, respectively, right-handed chiral components of the fermion fields $\psi_x, \overline{\psi}_x$ are defined as

$$\psi_{Lx} \equiv P_L\psi_x \, , \quad \psi_{Rx} \equiv P_R\psi_x \, , \quad \overline{\psi}_{Lx} \equiv \overline{\psi}_x P_R \, , \quad \overline{\psi}_{Rx} \equiv \overline{\psi}_x P_L \, . \tag{8.12}$$

8.1.3 Lie algebra generators

Every element of SU(N) can be represented in the form

$$\Lambda = \exp\left\{ \sum_{a=1}^{N^2-1} iT_a\omega^a \right\} \, , \qquad \omega^a \in \mathbf{R}. \tag{8.13}$$

The matrices T_a are the generators of the Lie algebra su(N). They are traceless, Hermitean $N \otimes N$ matrices:

$$\text{Tr}\, T_a = 0, \qquad T_a^+ = T_a, \tag{8.14}$$

and are normalized according to

$$\text{Tr}\,(T_a T_b) = \frac{1}{2}\delta_{ab} \, . \tag{8.15}$$

They satisfy the commutation relations

$$[T_a, T_b] = i f_{abc} T_c, \tag{8.16}$$

where the structure constants f_{abc} are completely antisymmetric and real, and a summation convention over repeated indices is implied.

For SU(2) we have

$$T_a = \frac{\tau_a}{2}, \qquad a = 1, 2, 3 \tag{8.17}$$

with the Pauli matrices τ_a and

$$f_{abc} = \epsilon_{abc}. \tag{8.18}$$

Generalizing the Pauli matrices to the SU(N) Lie algebra one writes

$$T_j = \frac{\lambda_j}{2}, \qquad j = 1, 2, \ldots, N^2 - 1, \tag{8.19}$$

where λ_j are the SU(N) Gell-Mann matrices. The completeness relation is

$$\sum_{j=1}^{N^2-1} (\lambda_j)_{cd}(\lambda_j)_{ef} = 2\left(\delta_{cf}\delta_{de} - \frac{1}{N}\delta_{cd}\delta_{ef}\right). \tag{8.20}$$

The anticommutators are

$$\{\lambda_j, \lambda_k\} = \frac{4}{N}\delta_{jk} + 2d_{jkl}\lambda_l. \tag{8.21}$$

In case of SU(3) the conventional representation of the Gell-Mann matrices is

$$\lambda_1 = \begin{pmatrix} 0 & 1 & 0 \\ 1 & 0 & 0 \\ 0 & 0 & 0 \end{pmatrix}, \quad \lambda_2 = \begin{pmatrix} 0 & -i & 0 \\ i & 0 & 0 \\ 0 & 0 & 0 \end{pmatrix}, \quad \lambda_3 = \begin{pmatrix} 1 & 0 & 0 \\ 0 & -1 & 0 \\ 0 & 0 & 0 \end{pmatrix},$$

$$\lambda_4 = \begin{pmatrix} 0 & 0 & 1 \\ 0 & 0 & 0 \\ 1 & 0 & 0 \end{pmatrix}, \quad \lambda_5 = \begin{pmatrix} 0 & 0 & -i \\ 0 & 0 & 0 \\ i & 0 & 0 \end{pmatrix}, \quad \lambda_6 = \begin{pmatrix} 0 & 0 & 0 \\ 0 & 0 & 1 \\ 0 & 1 & 0 \end{pmatrix},$$

$$\lambda_7 = \begin{pmatrix} 0 & 0 & 0 \\ 0 & 0 & -i \\ 0 & i & 0 \end{pmatrix}, \quad \lambda_8 = \begin{pmatrix} 1/\sqrt{3} & 0 & 0 \\ 0 & 1/\sqrt{3} & 0 \\ 0 & 0 & -2/\sqrt{3} \end{pmatrix}. \tag{8.22}$$

8.1.4 Continuum gauge fields

The gauge covariant derivative is written

$$D_\mu = \partial_\mu + A_\mu(x), \tag{8.23}$$

where

$$A_\mu(x) = -ig A_\mu^a(x) T_a \tag{8.24}$$

is the gauge field. The gauge transformation law of the gauge field is

$$A'_\mu(x) = \Lambda^{-1}(x)(\partial_\mu + A_\mu(x))\Lambda(x) . \tag{8.25}$$

The field strength is

$$F_{\mu\nu}(x) = [D_\mu, D_\nu] = \partial_\mu A_\nu(x) - \partial_\nu A_\mu(x) + [A_\mu(x), A_\nu(x)] . \tag{8.26}$$

In component notation one has

$$F_{\mu\nu}(x) = -igF^a_{\mu\nu}(x)T_a , \tag{8.27}$$

$$F^a_{\mu\nu} = \partial_\mu A^a_\nu - \partial_\nu A^a_\mu + gf_{abc}A^b_\mu A^c_\nu . \tag{8.28}$$

8.1.5 Lattice notations

Most of the time a $L^3 \cdot T$ lattice is considered with an L^3 spatial lattice and T Euclidean timeslices. The spatial volume in lattice units is denoted by $V = L^3$. If an asymmetric spatial lattice is used, its extensions are denoted by L_1, L_2, L_3, therefore we have in general $V = L_1 L_2 L_3$. Eventually one can also use $L_4 \equiv T$. The total number of lattice points is

$$\Omega \equiv V \cdot T = L_1 L_2 L_3 L_4 . \tag{8.29}$$

The lattice spacing is denoted by a. If the spatial and timelike lattice spacings are different, then they are denoted by a_s and a_t, respectively.

The lattice points are usually denoted by $x, y, \ldots$. The integer coordinates of the point x on the lattice are (x_1, x_2, x_3, x_4). These are usually chosen in the intervals $0 \le x_\mu \le L_\mu - 1$ ($\mu = 1, 2, 3, 4$). The Euclidean time is the fourth coordinate: $t \equiv x_4$. The three-vector of the spatial coordinates is denoted by boldface letters, as $\mathbf{x} \equiv (x_1, x_2, x_3)$. The four orthogonal directions are $\mu, \nu, \ldots \in \{1, 2, 3, 4\}$. The unit vector in the direction μ is $\hat{\mu}$. Sometimes both positive and negative directions are considered: $\mu, \nu, \ldots \in \{\pm 1, \pm 2, \pm 3, \pm 4\}$. The summation over positive directions is

$$\sum_{\mu=1}^{4} . \tag{8.30}$$

If a summation over both positive and negative directions is meant, then we write it as

$$\sum_\mu \equiv \sum_{\mu=\pm 1}^{\pm 4} \equiv \sum_{\mu=\pm 1, \pm 2, \pm 3, \pm 4} . \tag{8.31}$$

In lattice gauge field theories the summation over positively oriented plaquettes is denoted by

$$\sum_p \equiv \sum_x \sum_{1 \le \mu < \nu \le 4} . \tag{8.32}$$

The summation over the space-like plaquettes is

$$\sum_{sp} \equiv \sum_{x} \sum_{1 \le \mu < \nu \le 3} , \qquad (8.33)$$

and over the time-like ones

$$\sum_{tp} \equiv \sum_{x} \sum_{1 \le \mu \le 3, \, \nu = 4} . \qquad (8.34)$$

Lattice site or momentum variables of the fields are denoted sometimes by an index, sometimes by a function argument: $\varphi_x \equiv \varphi(x)$, $\widetilde{\varphi}_k \equiv \widetilde{\varphi}(k)$. The Fourier transformation of the scalar lattice field $\varphi(x)$ to the discrete four-momenta $k \equiv (k_1, k_2, k_3, k_4)$ is

$$\widetilde{\varphi}(k) = \sum_{x} e^{-ik \cdot x} \varphi(x) , \qquad (8.35)$$

where the Euclidean scalar product is denoted by

$$k \cdot x \equiv k_\mu x_\mu \equiv \sum_{\mu=1}^{4} k_\mu x_\mu . \qquad (8.36)$$

The inverse Fourier transform is

$$\varphi(x) = \frac{1}{\Omega} \sum_{k} e^{ix \cdot k} \widetilde{\varphi}(k) . \qquad (8.37)$$

In the case of fermion fields $\psi_x, \overline{\psi}_x$ the corresponding relations are

$$\widetilde{\psi}_k = \sum_{x} e^{-ik \cdot x} \psi_x , \qquad \widetilde{\overline{\psi}}_k = \sum_{x} e^{ik \cdot x} \overline{\psi}_x ,$$

$$\psi_x = \frac{1}{\Omega} \sum_{k} e^{ix \cdot k} \widetilde{\psi}_k , \qquad \overline{\psi}_x = \frac{1}{\Omega} \sum_{k} e^{-ix \cdot k} \widetilde{\overline{\psi}}_k . \qquad (8.38)$$

The gauge fields are naturally associated with the middle point of the link $(x, x + \hat{\mu})$, therefore their Fourier transformation is defined by

$$\widetilde{A}^r_\mu(k) = \sum_{x} e^{-ik \cdot (x + \hat{\mu}/2)} A^r_\mu(x) = \sum_{x} e^{-ik \cdot x} e^{-ik_\mu/2} A^r_\mu(x) ,$$

$$A^r_\mu(x) = \frac{1}{\Omega} \sum_{k} e^{i(x + \hat{\mu}/2) \cdot k} \widetilde{A}^r_\mu(k) = \frac{1}{\Omega} \sum_{k} e^{ix \cdot k} e^{ik_\mu/2} \widetilde{A}^r_\mu(k) . \qquad (8.39)$$

Abbreviations of some functions occuring in lattice perturbation theory are

$$\hat{k}_\mu \equiv 2 \sin \frac{k_\mu}{2} , \quad \overline{k}_\mu \equiv \sin k_\mu , \quad \overline{\overline{k}}_\mu \equiv \frac{1}{2} \sin (2k_\mu) . \qquad (8.40)$$

The momentum sums or integrals in lattice Feynman graphs are briefly denoted by

$$\int_k \equiv \frac{1}{\Omega} \sum_k \implies \frac{1}{(2\pi)^4} \int_0^{2\pi} d^4k = \frac{1}{(2\pi)^4} \int_{-\pi}^{\pi} d^4k \ . \tag{8.41}$$

The sum approaches the integral for $L_1, L_2, L_3, L_4 \to \infty$.

The lattice forward derivative Δ^f and backward derivative Δ^b of the function $\varphi(x) \equiv \varphi_x$ are defined as

$$\Delta_\mu^f \varphi(x) \equiv \Delta_\mu^{(x)f} \varphi(x) \equiv \varphi(x + \hat{\mu}) - \varphi(x) \ ,$$

$$\Delta_\mu^b \varphi(x) \equiv \Delta_\mu^{(x)b} \varphi(x) \equiv \varphi(x) - \varphi(x - \hat{\mu}) \ . \tag{8.42}$$

The relation between them is

$$\Delta_\mu^b = -(\Delta_\mu^f)^+ \ . \tag{8.43}$$

The Laplace operator and d'Alembert operator on the lattice are defined by

$$\Delta\varphi(x) \equiv -\Box\varphi(x) \equiv \Delta_\mu^f \Delta_\mu^b \varphi(x) = \Delta_\mu^b \Delta_\mu^f \varphi(x)$$

$$= \sum_{\mu=1}^4 (\varphi(x + \hat{\mu}) + \varphi(x - \hat{\mu}) - 2\varphi(x)) \ . \tag{8.44}$$

8.1.6   Free fields

In this subsection positions and momenta are always in Minkowski space. A conventional normalization of the creation and annihilation operators of a scalar particle is defined by the commutation relation

$$[a(\mathbf{p}_1), a^+(\mathbf{p}_2)] = 2p_{10}(2\pi)^3 \delta^3(\mathbf{p}_1 - \mathbf{p}_2) \ . \tag{8.45}$$

$\mathbf{p}_1$ and $\mathbf{p}_2$ are the three-momenta. The zeroth component of the on-shell energy momentum four-vector is determined by the mass m as

$$p_0 \equiv (m^2 + \mathbf{p}^2)^{1/2} \ . \tag{8.46}$$

As a consequence, the resolution of the identity in the single particle Hilbert space, built on the vacuum state $|0\rangle$, is

$$I = \int \frac{d^3\mathbf{p}}{2p_0(2\pi)^3} a^+(\mathbf{p})|0\rangle\langle0|a(\mathbf{p})$$

$$= (2\pi)^{-3} \int d^4p \ \theta(p_0)\delta(p^2 - m^2)a^+(\mathbf{p})|0\rangle\langle0|a(\mathbf{p}) \ . \tag{8.47}$$

The free real scalar field is then

$$\phi(x) = \int \frac{d^3\mathbf{p}}{2p_0(2\pi)^3} \left\{ a(\mathbf{p})e^{-ip*x} + a^+(\mathbf{p})e^{ip*x} \right\} . \tag{8.48}$$

The scalar product in Minkowski space is distinguished from its analogue in Euclidean space (8.36) by the notation

$$p * x \equiv p_0 x_0 - \sum_{n=1}^{3} p_n x_n \equiv \sum_{\mu=0}^{3} p^\mu x_\mu \equiv \sum_{\mu=0}^{3} p_\mu x^\mu \equiv \sum_{\mu,\nu=0}^{3} p_\mu x_\nu g_{\mu\nu} . \tag{8.49}$$

The scalar product of the energy momentum four-vector with itself is, due to (8.46), $p * p = p_0^2 - \mathbf{p}^2 = m^2$. The single-particle to vacuum matrix element of the field is then, by using translation invariance,

$$\langle 0|\phi(x)a^+(\mathbf{p})|0\rangle = e^{-ix*p}\langle 0|\phi(0)a^+(\mathbf{p})|0\rangle = e^{-ix*p}$$

$$\longrightarrow Z^{1/2}e^{-ix*p} . \tag{8.50}$$

The value of $Z^{1/2}$ gives the normalization factor for a general *in*- or *out*-field $\phi_{in}(x)$ or $\phi_{out}(x)$.

The corresponding conventions for a massive ($m \neq 0$) spin-$\frac{1}{2}$ fermion field start by the anticommutation relations of the creation and annihilation operators of fermions (a) and antifermions (b):

$$\{a(\mathbf{p}_1,\sigma_1), a^+(\mathbf{p}_2,\sigma_2)\} = \{b(\mathbf{p}_1,\sigma_1), b^+(\mathbf{p}_2,\sigma_2)\}$$

$$= \frac{p_{10}}{m}(2\pi)^3\delta^3(\mathbf{p}_1 - \mathbf{p}_2)\delta_{\sigma_1\sigma_2} . \tag{8.51}$$

Besides the three-momenta $\mathbf{p}_{1,2}$ the states also depend on the spin projection quantum numbers $\sigma_{1,2} = \pm 1/2$. The resolution of the identity in the single fermion Hilbert space is now

$$I = \int \frac{d^3\mathbf{p}}{(2\pi)^3} \frac{m}{p_0} \sum_\sigma a^+(\mathbf{p},\sigma)|0\rangle\langle 0|a(\mathbf{p},\sigma) . \tag{8.52}$$

The free fermion fields are

$$\psi(x) = \int \frac{d^3\mathbf{p}}{(2\pi)^3} \frac{m}{p_0} \sum_\sigma \left\{ a(\mathbf{p},\sigma)e^{-ip*x}u(\mathbf{p},\sigma) + b^+(\mathbf{p},\sigma)e^{ip*x}v(\mathbf{p},\sigma) \right\} ,$$

$$\overline{\psi}(x) \equiv \psi(x)^+\gamma_0 =$$

$$\int \frac{d^3\mathbf{p}}{(2\pi)^3} \frac{m}{p_0} \sum_\sigma \left\{ b(\mathbf{p},\sigma)e^{-ip*x}\overline{v}(\mathbf{p},\sigma) + a^+(\mathbf{p},\sigma)e^{ip*x}\overline{u}(\mathbf{p},\sigma) \right\} . \tag{8.53}$$

The positive- and negative-energy solutions of the momentum space Dirac equation satisfy, respectively,

$$(\gamma * p - m)u(\mathbf{p},\sigma) = 0 , \quad (\gamma * p + m)v(\mathbf{p},\sigma) = 0 . \tag{8.54}$$

The normalization is fixed by

$$\bar{u}(\mathbf{p}, \sigma)u(\mathbf{p}, \sigma') = -\bar{v}(\mathbf{p}, \sigma)v(\mathbf{p}, \sigma') = \delta_{\sigma\sigma'} , \qquad (8.55)$$

or by

$$u^+(\mathbf{p}, \sigma)u(\mathbf{p}, \sigma') = v^+(\mathbf{p}, \sigma)v(\mathbf{p}, \sigma') = \frac{p_0}{m}\delta_{\sigma\sigma'} . \qquad (8.56)$$

The single-particle to vacuum matrix elements of the fields are

$$\langle 0|\psi(x)a^+(\mathbf{p}, \sigma)|0\rangle = u(\mathbf{p}, \sigma)e^{-ix\bullet p} \longrightarrow Z_2^{1/2}u(\mathbf{p}, \sigma)e^{-ix\bullet p} ,$$

$$\langle 0|\bar{\psi}(x)b^+(\mathbf{p}, \sigma)|0\rangle = \bar{v}(\mathbf{p}, \sigma)e^{-ix\bullet p} \longrightarrow Z_2^{1/2}\bar{v}(\mathbf{p}, \sigma)e^{-ix\bullet p} . \qquad (8.57)$$

8.1.7   Reduction formulas

The Lehmann–Symanzik–Zimmermann (LSZ) reduction formulas express the on mass shell matrix elements (the S-matrix elements) by residua of the vacuum expectation values of local fields. We consider here only simple special cases (for the general formulas see, for instance, [A3]).

Let us define the Fourier-transformed vacuum expectation value of a time ordered product

$$M(p) \equiv \int d^4x\, e^{ip\bullet x}\langle 0|T\{A(x)B(y)C(z)\ldots\}|0\rangle , \qquad (8.58)$$

where $A(x), B(y), C(z),\ldots$ are real scalar field operators. T denotes the time ordered product, in which the operators $A(x), B(y), C(z),\ldots$ are arranged from left to right in order of descending time coordinates $(x_0, y_0, z_0,\ldots)$. Let us suppose that $A(x)$ has the correct quantum numbers to annihilate the single particle state $|\mathbf{p}_A\rangle$ with four-momentum $p_A = \{p_{A0} = (m_A^2 + \mathbf{p}_A^2)^{1/2}, \mathbf{p}_A\}$ and mass m_A. According to (8.50) this means

$$\langle 0|A(x)|\mathbf{p}_A\rangle = Z_A^{1/2}e^{-ix\bullet p_A} . \qquad (8.59)$$

Let us investigate the behaviour of $M(p)$ for $p_0 \to +(m_A^2 + \mathbf{p}^2)^{1/2}$. In this limit the contribution of the above single particle state will dominate, because due to (8.47) it can be written as

$$M(p) = \int d^4x\, e^{ip\bullet x} \int \frac{d^3\mathbf{p}_A}{2p_{A0}(2\pi)^3}\theta(x_0 - \max\{y_0, z_0,\ldots\})$$

$$\cdot \langle 0|A(x)|\mathbf{p}_A\rangle\langle\mathbf{p}_A|T\{B(y)C(z)\ldots\}|0\rangle + \cdots . \qquad (8.60)$$

This comes from the contribution of the time ordering with $x_0 > \max\{y_0, z_0,\ldots\}$. The dots stand for all other contributions (other time orderings and intermediate states different from the single particle state).

Using (8.59) and

$$\int d^4x\, e^{ix*(p-p_A)}\theta(x_0 - t_0) = (2\pi)^3\delta^3(\mathbf{p} - \mathbf{p}_A) \lim_{\epsilon \searrow 0} \frac{ie^{it_0(p_0-p_{A0})}}{p_0 - p_{A0} + i\epsilon}, \qquad (8.61)$$

one obtains

$$M(p) = \frac{iZ_A^{1/2}\, \exp\{i[p_0 - (m_A^2 + \mathbf{p}_A^2)^{1/2}]\cdot \max(y_0, z_0, \ldots)\}}{2(m_A^2 + \mathbf{p}_A^2)^{1/2}[p_0 - (m_A^2 + \mathbf{p}_A^2)^{1/2} + i\epsilon]}$$

$$\cdot \langle \mathbf{p}_A = \mathbf{p}|T\{B(y)C(z)\ldots\}|0\rangle + \cdots . \qquad (8.62)$$

Therefore, in the on mass shell limit $p_0 \to +(m_A^2 + \mathbf{p}_A^2)^{1/2}$ we have

$$M(p) \xrightarrow{p_0 \to +(m_A^2 + \mathbf{p}_A^2)^{1/2}} \frac{iZ_A^{1/2}}{(p*p - m_A^2 + i\epsilon)}$$

$$\cdot \langle \mathbf{p}_A = \mathbf{p}|T\{B(y)C(z)\ldots\}|0\rangle + \cdots . \qquad (8.63)$$

This shows that $M(p)$ has a pole at $p*p = m_A^2$ (in the energy variable at $p_0 = +(m_A^2 + \mathbf{p}_A^2)^{1/2}$). None of the omitted terms contributes to this pole, and therefore the residue is given by the displayed on mass shell matrix element.

Similarly, in the limit $p_0 \to -(m_A^2 + \mathbf{p}_A^2)^{1/2}$ one can derive from the one particle contribution in the time ordering $x_0 < \min(y_0, z_0, \ldots)$

$$M(p) \xrightarrow{p_0 \to -(m_A^2 + \mathbf{p}_A^2)^{1/2}} \frac{iZ_A^{1/2}}{(p*p - m_A^2 + i\epsilon)}$$

$$\cdot \langle 0|T\{B(y)C(z)\ldots\}|\mathbf{p}_A = -\mathbf{p}\rangle + \cdots . \qquad (8.64)$$

By repeating the steps that led to (8.63) and (8.64), one can build up any physical on mass shell matrix elements out of the residua of the poles in the Fourier transforms of vacuum expectation values of time ordered products [A3].

References

Chapter 1

[1.1] Darrigol, O. (1986). The origin of quantized matter waves, *Historical studies in the physical and biological sciences* **16**, 197.

[1.2] Schweber, S.S. (1984). Some chapters for a history of quantum field theory: 1938–1952, in *Relativity, groups and topology II, Les Houches, Session XL, 1983*, ed. B.S. De Witt and R. Stora (Elsevier Science Publishers, Amsterdam).

[1.3] Pais, A. (1986). Inward bound: of matter and forces in the physical world (Oxford University Press, New York).

[1.4] Salam, A. and Wigner, E.P. (eds.) (1972). Aspects of quantum theory (Cambridge University Press, Cambridge).

[1.5] Weinberg, S. (1977). The search for unity: notes for a history of quantum field theory, *Daedalus* **106(4)**, 17.

[1.6] Heisenberg, W. (1925). Über quantentheoretische Umdeutung kinematischer und mechanischer Beziehungen, *Z. Phys.* **33**, 877.

[1.7] Born, M. and Jordan, P. (1925). Zur Quantenmechanik, *Z. Phys.* **34**, 858.

[1.8] Born, M., Heisenberg, W. and Jordan, P. (1926). Zur Quantenmechanik, II, *Z. Phys.* **35**, 557.

[1.9] Jordan, P. and Klein, O. (1927). Zum Mehrkörperproblem der Quantentheorie, *Z. Phys.* **45**, 751.

[1.10] Jordan, P. and Wigner, E.P. (1928). Über das Paulische Äquivalenzverbot, *Z. Phys.* **47**, 631.

[1.11] Dirac, P.A.M. (1927). The quantum theory of the emission and absorption of radiation, *Proc. Roy. Soc. London* **A 114**, 243.

[1.12] Jordan, P. and Pauli, W. (1928). Zur Quantenelektrodynamik ladungsfreier Felder, *Z. Phys.* **47**, 151.

[1.13] Heisenberg, W. and Pauli, W. (1929). Zur Quantendynamik der Wellenfelder, *Z. Phys.* **56**, 1.

[1.14] Heisenberg, W. and Pauli, W. (1930). Zur Quantentheorie der Wellenfelder, II, *Z. Phys.* **59**, 168.

[1.15] Fermi, E. (1929) Sopra l'elettrodinamica quantistica, *Rendiconti d.R. Accademia dei Lincei* **9**, 881.

[1.16] Pauli, W. and Weisskopf, V. (1934). Zur Quantisierung der skalaren relativistischen Wellengleichung, *Helv. Phys. Acta* **7**, 709.

[1.17] Stöckler, M. (1987). Philosophische Probleme der Elementarteilchenphysik, Habilitationsschrift (Universität Gießen).

[1.18] Feynman, R.P. (1948). Space-time approach to non-relativistic quantum mechanics, *Rev. Mod. Phys.* **20**, 267.

[1.19] Feynman, R.P. and Hibbs, A.R. (1965). *Quantum Mechanics and Path Integrals* (McGraw-Hill, New York).

[1.20] Dyson, F.J. (1949). The S-Matrix in quantum electrodynamics, *Phys. Rev.* **75**, 1736.

[1.21] Wick, G.C. (1954). Properties of Bethe-Salpeter wave functions, *Phys. Rev.* **96**, 1124.

[1.22] Schwinger, J. (1958). On the Euclidean structure of relativistic field theory, *Proc. Natl. Acad. Sci. U.S.A.* **44**, 956.

[1.23] Symanzik, K. (1966). Euclidean quantum field theory, I. Equations for a scalar model, *J. Math. Phys.* **7**, 510;
Symanzik, K. (1969). Euclidean quantum field theory, in *Local Quantum Field Theory*, ed. R. Jost (Academic Press, New York).

[1.24] Wentzel, G. (1940). Zum Problem des statischen Mesonfeldes, *Helv. Phys. Acta* **13**, 269.

[1.25] Schiff, L.I. (1953). Lattice-space quantization of a nonlinear field theory, *Phys. Rev.* **92**, 766.

[1.26] Wilson, K.G. (1983). The renormalization group and critical phenomena, *Rev. Mod. Phys.* **55**, 583.

[1.27] Dirac, P.A.M. (1933). The Lagrangian in quantum mechanics, *Phys. Z. Sowjetunion* **3**, 64.

[1.28] Reed, M. and Simon, B. (1972). *Methods of Modern Mathematical Physics, Vol. 1* (Academic Press, New York).

[1.29] Kac, M. (1949). On a distribution of certain Wiener functionals, *Trans. Am. Math. Soc.* **65**, 1.

[1.30] Gel'fand, I.M. and Yaglom, A.M. (1960). Integration in functional spaces, *J. Math. Phys.* **1**, 48.

[1.31] Reed, M. and Simon, B. (1975). *Methods of Modern Mathematical Physics, Vol. 2* (Academic Press, New York).

[1.32]   Wiener, N. (1923). Differential space, *J. Math. and Phys.* **58**, 131.

[1.33]   Streater, R.F. and Wightman, A.S. (1964). *PCT, Spin & Statistics, and All That* (Benjamin, New York).

[1.34]   Osterwalder, K. and Schrader, R. (1973). Axioms for Euclidean Green's functions, *Comm. Math. Phys.* **31**, 83;
Osterwalder, K. and Schrader, R. (1975). Axioms for Euclidean Green's functions. 2, *Comm. Math. Phys.* **42**, 281;
Osterwalder, K. (1973), Euclidean Green's functions and Wightman distributions, in *Constructive Quantum Field Theory* ed. G. Velo and A.S. Wightman, Lecture Notes in Physics 25 (Springer, Berlin).

[1.35]   Nelson, E. (1973). The construction of quantum fields from Markov fields, *J. Funct. Anal.* **12**, 97.

[1.36]   Coleman, S. (1985). Secret Symmetry (Erice lectures 1973), in *Aspects of Symmetry* (Cambridge University Press, Cambridge).

[1.37]   Guerra, F., Rosen, L. and Simon, B. (1975). The $P(\phi)_2$ Euclidean quantum field theory as classical statistical mechanics, *Ann. Math.* **101**, 111.

[1.38]   Wilson, K.G. (1976). Quantum chromodynamics on a lattice, in *New Developments in Quantum Field Theory and Statistical Mechanics, Cargèse 1976*, ed. M. Lévy and P.K. Mitter, (Plenum Press, New York).

[1.39]   Fröhlich, J. and Lieb, E. (1978). Phase transitions in anisotropic spin systems, *Comm. Math. Phys.* **60**, 233.

[1.40]   Osterwalder, K. and Seiler, E. (1978). Gauge field theories on the lattice, *Ann. Phys. (N.Y.)* **110**, 440.

[1.41]   Fisher, M.E. (1967). The theory of equilibrium critical phenomena, *Rep. Prog. Phys.* **30**, 615.

[1.42]   Stanley, H.E. (1971). *Introduction to Phase Transitions and Critical Phenomena* (Oxford University Press, Oxford).

[1.43]   Brézin, E., Le Guillou, J.C. and Zinn-Justin, J. (1976). Field theoretical approach to critical phenomena, in *Phase Transitions and Critical Phenomena, Vol. VI*, ed. C. Domb and M.S. Green (Academic Press, New York).

[1.44]   Bogoliubov, N.N. and Shirkov, D.V. (1959). *Introduction to the Theory of Quantized Fields* (Interscience, New York).

[1.45]   Hepp, K. (1969). *Théorie de la Renormalisation* (Springer, Berlin).

[1.46]   Zimmermann, W. (1970). Local operator products and renormalization in quantum field theory, in *Lectures on Elementary Particles and Quantum Field Theory, 1970 Brandeis University Summer Institute in Theoretical Physics, Vol. 1*, ed. S. Deser, M. Grisaru and H. Pendleton (M.I.T. Press, Cambridge, Mass.).

[1.47] Symanzik, K. (1970). Renormalization of models with broken
symmetry, in *Fundamental Interactions at High Energy II, Proceedings
of the Coral Gables Conference 1970*, ed. A. Perlmutter, G.J. Iverson
and R.M. Williams (Gordon and Breach, New York).

[1.48] Reisz, T. (1988). A power counting theorem for Feynman integrals on
the lattice, *Comm. Math. Phys.* **116**, 81;
A convergence theorem for lattice Feynman integrals with massless
propagators, *Comm. Math. Phys.* **116**, 573;
Renormalization of Feynman integrals on the lattice, *Comm. Math.
Phys.* **117**, 79;
Renormalization of lattice Feynman integrals with massless
propagators, *Comm. Math. Phys.* **117**, 639.

[1.49] Kadanoff, L.P. (1966). Scaling laws for Ising models near T_c, *Physics*
2, 263;
Kadanoff, L.P., Götze, W., Hamblen, D., Hecht, R., Lewis, E.A.S.,
Palciauskas, V.V., Rayl, M. and Swift, J. (1967). Static phenomena
near critical points: theory and experiment, *Rev. Mod. Phys.* **39**, 395.

[1.50] Wilson, K.G. (1970). Model of coupling-constant renormalization,
Phys. Rev. **D2**, 1438;
Wilson, K.G. (1971). Renormalization group and critical phenomena,
I. Renormalization group and the Kadanoff scaling picture, *Phys.
Rev.* **B4**, 3174; II. Phase-space cell analysis of critical behaviour, *Phys.
Rev.* **B4**, 3184.

[1.51] Wilson, K.G. and Kogut, J. (1974). The renormalization group and
the ϵ expansion, *Phys. Reports* **12C**, 75.

[1.52] Symanzik, K. (1980). Cutoff dependence in lattice ϕ_4^4 theory, in
Recent Developments in Gauge Theories, Cargèse lectures 1979, ed.
G. 't Hooft, C. Itzykson, A. Jaffe, H. Lehmann, P.K. Mitter,
I.M. Singer and R. Stora (Plenum Press, New York).

[1.53] Callan, C.G. (1970). Broken scale invariance in scalar field theory,
Phys. Rev. **D2**, 1541.

[1.54] Symanzik, K. (1970). Small distance behavior in field theory and
power counting, *Comm. Math. Phys.* **18**, 227.

[1.55] Brézin, E., Le Guillou, J.C. and Zinn-Justin, J. (1974). Addendum to
Wilson's theory of critical phenomena and Callan-Symanzik
equations in $4 - \varepsilon$ dimensions, *Phys. Rev.* **D9**, 1121.

[1.56] 't Hooft, G. (1973). Dimensional regularization and the
renormalization group, *Nucl. Phys.* **B 61**, 455.

[1.57] Weinberg, S. (1973). New approach to the renormalization group,
Phys. Rev. **D8**, 3497.

[1.58] Feynman, R.P. (1953). Atomic theory of the λ transition in helium,
Phys. Rev. **91**, 1291;

Kubo, R. (1957). Statistical mechanical theory of irreversible processes, *J. Phys. Soc. Japan* **12**, 570;
Martin, P.C. and Schwinger, J. (1959). Theory of many particle systems. I, *Phys. Rev.* **115**, 1342;
Abrikosov, A.A., Gorkov, L.P. and Dzyaloshinskii, I.E. (1959). *Sov. Phys. JETP* **9**, 636.

Chapter 2

[2.1] Münster, G. (1985). The size of finite size effects in lattice gauge theories, *Nucl. Phys.* **B 249**, 659.

[2.2] Reisz, T. (1988). A power counting theorem for Feynman integrals on the lattice, *Comm. Math. Phys.* **116**, 81;
A convergence theorem for lattice Feynman integrals with massless propagators, *Comm. Math. Phys.* **116**, 573;
Renormalization of Feynman integrals on the lattice, *Comm. Math. Phys.* **117**, 79;
Renormalization of lattice Feynman integrals with massless propagators, *Comm. Math. Phys.* **117**, 639.

[2.3] Lüscher, M. and Weisz, P. (1988). Scaling laws and triviality bounds in the lattice ϕ^4 theory, II. One-component model in the phase with spontaneous symmetry breaking, *Nucl. Phys.* **B 295**, 65.

[2.4] Jansen, K., Montvay, I., Münster, G., Trappenberg, T. and Wolff, U. (1989). Broken phase of the 4-dimensional Ising model in a finite volume, *Nucl. Phys.* **B 322**, 698.

[2.5] Wortis, M. (1974). Linked cluster expansion, in *Phase Transitions and Critical Phenomena, Vol. III*, ed. C. Domb and M.S. Green (Academic Press, New York).

[2.6] Lüscher, M. and Weisz, P. (1988). Application of the linked cluster expansion to the n-component ϕ^4 theory, *Nucl. Phys.* **B 300**, 325.

[2.7] Baker, G.A. and Kincaid, J.M. (1981). The continuous-spin Ising model, $g_0 : \phi^4 :_d$ field theory, and the renormalization group, *J. Stat. Phys.* **24**, 469.

[2.8] Lüscher, M. and Weisz, P. (1987). Scaling laws and triviality bounds in the lattice ϕ^4 theory, I. One-component model in the symmetric phase, *Nucl. Phys.* **B 290**, 25.

[2.9] Lüscher, M. and Weisz, P. (1989). Scaling laws and triviality bounds in the lattice ϕ^4 theory, III. n-component model, *Nucl. Phys.* **B 318**, 705.

[2.10] Wilson, K.G. and Kogut, J. (1974). The renormalization group and the ϵ expansion, *Phys. Reports* **12C**, 75.

[2.11] Aizenmann, M. and Graham, R. (1983). On the renormalized coupling constant and the susceptibility in ϕ^4_4 field theory and the Ising model in four dimensions, *Nucl. Phys.* **B 225**, 261.

[2.12]   Vohwinkel, C. and Weisz, P. (1992). Low-temperature expansion in the $d = 4$ Ising model, *Nucl. Phys.* **B 374**, 647.

[2.13]   Bhanot, G., Bitar, K., Heller, U.M. and Neuberger, H. (1990). ϕ^4 on F_4: analytical results, *Nucl. Phys.* **B 343**, 467;
Bhanot, G., Bitar, K., Heller, U.M. and Neuberger, H. (1991). ϕ^4 on F_4: numerical results, *Nucl. Phys.* **B 353**, 551.

[2.14]   Montvay, I. and Weisz, P. (1987). Numerical study of finite volume effects in the 4-dimensional Ising model, *Nucl. Phys.* **B 290**, 327.

[2.15]   Montvay, I., Münster, G. and Wolff, U. (1988). Percolation cluster algorithm and scaling behaviour in the 4-dimensional Ising model, *Nucl. Phys.* **B 305**, 143.

[2.16]   Jansen, K., Jersák, J., Montvay, I., Münster, G., Trappenberg, T. and Wolff, U. (1988). Vacuum tunneling in the 4-dimensional Ising model, *Phys. Lett.* **213B**, 203.

[2.17]   Münster, G. (1982). On the statistical mechanics of dense instanton gases, *Z. Phys.* **C12**, 43.

[2.18]   Lüscher, M. (1982). A semiclassical formula for the topological susceptibility in a finite space–time volume, *Nucl. Phys.* **B 205**, 483.

[2.19]   Lüscher, M. (1983). Some analytic results concerning the mass spectrum of Yang–Mills gauge theories on a torus, *Nucl. Phys.* **B 219**, 233;
Lüscher, M. and Münster, G. (1984). Weak coupling expansion of the low-lying energy values in the SU(2) gauge theory on a torus, *Nucl. Phys.* **B 232**, 445.

[2.20]   Lüscher, M. (1984). On a relation between finite size effects and elastic scattering processes, in *Progress in Gauge Field Theory, Cargèse lectures 1983*, ed. G. 't Hooft, A. Jaffe, H. Lehmann, P.K. Mitter, I.M. Singer and R. Stora (Plenum Publishing Corporation, New York);
Lüscher, M. (1986). Volume dependence of the energy spectrum in massive quantum field theories (I). Stable particle states, *Comm. Math. Phys.* **104**, 177.

[2.21]   Domb, C. (1960). On the theory of cooperative phenomena in crystals, *Adv. Phys.* **9**, 149.

[2.22]   Fisher, M.E. (1969). Aspects of equilibrium critical phenomena, *J. Phys. Soc. Japan Suppl.* **26**, 87.

[2.23]   Privman, V. and Fisher, M.E. (1983). Finite–size effects at first-order transitions, *J. Stat. Phys.* **33**, 385.

[2.24]   Münster, G. (1989). Tunneling amplitude and surface tension in ϕ^4 theory, *Nucl. Phys.* **B 324**, 630.

[2.25]   Brézin, E. and Zinn-Justin, J. (1985). Finite size effects in phase transitions, *Nucl. Phys.* **B 257**, 867.

[2.26] Münster, G. (1990). Interface tension in three-dimensional systems from field theory, *Nucl. Phys.* **B 340**, 559.

[2.27] Polyakov, A.M. (1977). Quark confinement and topology of gauge theories, *Nucl. Phys.* **B 120**, 429.

[2.28] Coleman, S. (1979). The use of instantons, in *The Whys of Subnuclear Physics (Erice lectures 1977)*, ed. A. Zichichi (Plenum Publishing Corporation, New York); reprinted in: S. Coleman, Aspects of Symmetry, (Cambridge University Press, 1985).

[2.29] Lüscher, M. and Weisz, P. (1988). Is there a strong interaction sector in the standard lattice Higgs model? *Phys. Lett.* **212B**, 472.

[2.30] Frick, Ch., Jansen, K., Jersák, J., Montvay, I., Münster, G. and Seuferling, P. (1990). Numerical simulation of the O(4)-symmetric ϕ^4-model in the symmetric phase, *Nucl. Phys.* **B 331**, 515.

[2.31] Hasenfratz, A., Jansen, K., Lang, C.B., Neuhaus, T. and Yoneyama, H. (1987). The triviality bound of the four-component Φ^4 model, *Phys. Lett.* **199B**, 531;
Hasenfratz, A., Jansen, K., Jersák, J., Lang, C.B., Neuhaus, T. and Yoneyama, H. (1989). Study of the 4-component Φ^4 model, *Nucl. Phys.* **B 317**, 81;
Neuhaus, T. (1989). Upgrade of results from Monte Carlo study of the O(4) invariant $\lambda\phi^4$ model in the broken phase, *Nucl. Phys. B (Proc. Suppl.)* **9**, 21.

[2.32] Kuti, J., Lin, L. and Shen, Y. (1988). Upper bound on the Higgs-boson mass in the standard model, *Phys. Rev. Lett.* **61**, 678;
Kuti, J., Lin, L. and Shen, Y. (1988). Fate of the standard model with a heavy Higgs particle, *Nucl. Phys. B (Proc. Suppl.)* **4**, 397;
Kuti, J., Lin, L. and Shen, Y. (1988). Non-perturbative lattice study of the Higgs sector in the standard model, in *Lattice Higgs Workshop*, ed. B. Berg, G. Bhanot, M. Burbank, M. Creutz and J. Owens (World Scientific, Singapore) p. 140.

[2.33] Hasenfratz, P. and Nager, J. (1988). The cut-off dependence of the Higgs meson mass and the onset of new physics in the standard model, *Z. Phys.* **C37**, 477.

[2.34] Jersák, J. (1990). Lattice Higgs systems at HLRZ, in *Physics at HLRZ*, ed. M. Kremer (Höchstleistungsrechenzentrum Jülich).

[2.35] Flyvbjerg, H., Larsen, F. and Kristjansen, C. (1991). $1/N$-expansion of σ-models in 2 and 4 dimensions: taking them to their technical limits, *Nucl. Phys. B (Proc. Suppl.)* **20**, 44.

Chapter 3

[3.1] Mack, G. (1981). Physical principles, geometrical aspects, and locality properties of gauge field theories, *Fortschr. Phys.* **29**, 135.

[3.2]   Yang, C.N. (1974). Integral formalism for gauge fields, *Phys. Rev. Lett.* **33**, 445;
Wu, T.T. and Yang, C.N. (1975). Concept of nonintegrable phase factors and global formulation of gauge fields, *Phys. Rev.* **D12**, 3845.

[3.3]   Yang, C.N. and Mills, R.L. (1954). Conservation of isotopic spin and isotopic gauge invariance, *Phys. Rev.* **96**, 191.

[3.4]   Wilson, K.G. (1974). Confinement of quarks, *Phys. Rev.* **D10**, 2445.

[3.5]   Wegner, F.J. (1971). Duality in generalized Ising models and phase transitions without local order parameters, *J. Math. Phys.* **10**, 2259.

[3.6]   Cornwell, J.F. (1984). *Group Theory in Physics, Vols.1,2* (Academic Press, London).

[3.7]   Elitzur, S. (1975). Impossibility of spontaneously breaking local symmetries, *Phys. Rev.* **D12**, 3978.

[3.8]   De Angelis, G.F., de Falco, D., Guerra, F. and Marra, R. (1978). Gauge fields on a lattice (selected topics), *Acta Physica Austriaca* **Suppl. XIX**, 205.

[3.9]   Seiler, E. (1982). *Gauge Theories as a Problem of Constructive Quantum Field Theory and Statistical Mechanics*, Lecture Notes in Physics 159 (Springer, Berlin).

[3.10] Durhuus, B. (1980). On the structure of gauge invariant classical observables in lattice gauge theories, *Lett. Math. Phys.* **4**, 515.

[3.11] Creutz, M. (1977). Gauge fixing, the transfer matrix, and confinement on a lattice, *Phys. Rev.* **D15**, 1128.

[3.12] Lüscher, M. (1977). Construction of a selfadjoint, strictly positive transfer matrix for Euclidean lattice gauge theories, *Comm. Math. Phys.* **54**, 283.

[3.13] Mack, G. and Petkova, V.B. (1979). Comparison of lattice gauge theories with gauge groups Z_2 and SU(2), *Ann. Phys. (N.Y.)* **123**, 442.

[3.14] Hartle, J.B. and Hawking, S.W. (1983). Wave function of the universe, *Phys. Rev.* **D28**, 2960.

[3.15] Gilmore, R. (1974). *Lie Groups, Lie Algebras, and Some of Their Applications* (John Wiley & Sons, New York);
Helgason, S. (1962). *Differential Geometry and Symmetric Spaces* (Academic Press, New York);
Wallach, N.R. (1973). *Harmonic Analysis on Homogeneous Spaces* (Dekker,New York);
Warner, G. (1972). *Harmonic Analysis on Semi-Simple Lie Groups, Vols.1,2* (Springer, Berlin);
Želobenko, D.P. (1973). *Compact Lie Groups and Their Representations* (Am. Math. Soc., Providence, Rhode Island).

[3.16] Bars, I. and Green, F. (1979). Complete integration of U(N) lattice gauge theory in a large-N limit, *Phys. Rev.* **D20**, 3311;

Bars, I. (1980). U(N) integral for the generating functional in lattice gauge theory, *J. Math. Phys.* **21**, 2678;

Samuel, S. (1980). U(N) integrals, $1/N$, and the De Wit–'t Hooft anomalies, *J. Math. Phys.* **21**, 2695;

Eriksson, K.E., Svartholm, N. and Skagerstam, B.S. (1981). On invariant group integrals in lattice QCD, *J. Math. Phys.* **22**, 2276.

[3.17] Osterwalder, K. and Seiler, E. (1978). Gauge field theories on the lattice, *Ann. Phys. (N.Y.)* **110**, 440.

[3.18] Manton, N.S. (1980). An alternative action for lattice gauge theories, *Phys. Lett.* **96B**, 328.

[3.19] Grosse, H. and Kühnelt, H. (1981). Phase structure of lattice models with and without SU(3) subgroup as gauge group, *Nucl. Phys.* **B 205**, 273.

[3.20] Drouffe, J.M. (1978). Transitions and duality in gauge lattice systems, *Phys. Rev.* **D18**, 1174;

Menotti, P. and Onofri, E. (1981). The action of SU(N) lattice gauge theory in terms of the heat kernel on the group manifold, *Nucl. Phys.* **B 190**, 288.

[3.21] Villain, J. (1975). Theory of one- and two-dimensional magnets with an easy magnetization plane, II. The planar, classical, two-dimensional magnet, *J. Physique* **36**, 581.

[3.22] Lang, C.B., Rebbi, C., Salomonson, P. and Skagerstam, B.-S. (1981). The transition from strong coupling to weak coupling in the SU(2) lattice gauge theory, *Phys. Lett.* **101B**, 173;

Mütter, K.H. and Schilling, K. (1983). Glueball mass from variant actions in lattice Monte Carlo simulations, *Phys. Lett.* **121B**, 267.

[3.23] Bhanot, G. and Creutz, M. (1981). Variant actions and phase structure in lattice gauge theory, *Phys. Rev.* **D24**, 3212;

Bhanot, G. (1982). SU(3) lattice gauge theory in 4 dimensions with a modified Wilson action, *Phys. Lett.* **108B**, 337;

Mack, G. and Pietarinen, E. (1982). Monopoles, vortices and confinement, *Nucl. Phys.* **B 205**, 141;

Fox, I.A. (1982). A study of the SU(2)–SO(3) phase diagram: analyticity domains, *Nucl. Phys.* **B 205**, 527;

Creutz, M. and Moriarty, K.J.M. (1982). Monte Carlo studies of SU(N)/Z_N lattice gauge theories in four dimensions, *Nucl. Phys.* **B 210**, 50;

Bhanot, G. and Dashen, R. (1982). Do numbers extracted near crossovers represent continuum physics? *Phys. Lett.* **113B**, 299;

Gonzales-Arroyo, A., Korthals-Altes, C.P., Peiro, J. and Perrottet, M. (1982). More structure in the string tension? *Phys. Lett.* **116B**, 414;

Bitar, K.M., Gottlieb, S. and Zachos, C.K. (1982). Phase structure and renormalization trajectories of lattice SU(2) gauge theory, *Phys. Rev.* **D26**, 2853;

Grossmann, B. and Samuel, S. (1983). Universality in SU(2) lattice gauge theory, *Phys. Lett.* **120B**, 383;
Dashen, R., Heller, U. and Neuberger, H. (1983). Analytical results for mixed action SU(2) lattice gauge theory, *Nucl. Phys.* **B 215**, 360;
Makeenko, Yu.M., Polikarpov, M.J. and Zhelonkin, A.V. (1983). Universality in the mixed SU(2) lattice gauge theory: nonperturbative approach to the ratio of Λ-parameters, *Phys. Lett.* **126B**, 82;
Flyvbjerg, H. and Marinari, E. (1983). Complex analytic structure of variant action SU(2) lattice gauge theory in 4d, *Phys. Lett.* **132B**, 385;
Jurkiewicz, J., Korthals-Altes, C.P. and Dash, J.W. (1984). Large-N universality of variant actions, *Nucl. Phys.* **B 233**, 457;
Decker, K. (1984). Strong coupling expansion for the mass gap in SU(2) lattice gauge theory with mixed action, *Nucl. Phys.* **B 240**, 543.

[3.24] Symanzik, K. (1983). Continuum limit and improved action in lattice theories (I). Principles and ϕ^4 theory, *Nucl. Phys.* **B 226**, 187.

[3.25] Weisz, P. (1983). Continuum limit improved lattice action for pure Yang–Mills theory (I), *Nucl. Phys.* **B 212**, 1.

[3.26] Curci, G., Menotti, P. and Paffuti,G. (1983). Symanzik's improved Lagrangian for lattice gauge theory, *Phys. Lett.* **130B**, 205; Erratum: *Phys. Lett.* **135B** (1984) 516.

[3.27] Weisz, P. and Wohlert, R. (1984). Continuum limit improved lattice action for pure Yang–Mills theory (II), *Nucl. Phys.* **B 236**, 397; Erratum: *Nucl. Phys.* **B 247**, 544.

[3.28] Lüscher, M. and Weisz, P. (1985). On-shell improved lattice gauge theories, *Comm. Math. Phys.* **97**, 59.

[3.29] Lüscher, M. and Weisz, P. (1985). Computation of the action for on-shell improved lattice gauge theories at weak coupling, *Phys. Lett.* **158B**, 250.

[3.30] Baaquie, B. (1977). Gauge fixing and mass renormalization in the lattice gauge theory, *Phys. Rev.* **D16**, 2612.

[3.31] Sharatchandra, H.S. (1978). Continuum limit of lattice gauge theories in the context of renormalized perturbation theory, *Phys. Rev.* **D18**, 2042.

[3.32] Hasenfratz, A. and Hasenfratz, P. (1980). The connection between the Λ parameters of lattice and continuum QCD, *Phys. Lett.* **93B**, 165.

[3.33] Kawai, H., Nakayama, R. and Seo, K. (1981). Comparison of the lattice Λ parameter with the continuum Λ parameter in massless QCD, *Nucl. Phys.* **B 189**, 40.

[3.34] Reisz, T. (1989). Lattice gauge theory: renormalization to all orders of the loop expansion, *Nucl. Phys.* **B 318**, 417.

[3.35] Hausner, M. and Schwartz, J.T. (1968). *Lie groups; Lie algebras* (Gordon and Breach, New York).

[3.36] Caracciolo, S., Menotti, P. and Pelissetto, A. (1992). One-loop analytic computation of the energy–momentum tensor for lattice gauge theories, *Nucl. Phys.* **B 375**, 195.

[3.37] Lüscher, M. and Weisz, P. (1986). Efficient numerical techniques for perturbative lattice gauge theory computations, *Nucl. Phys.* **B 266**, 309.

[3.38] Müller, V.F. and Rühl, W. (1981). Small coupling (low temperature) expansions of non-Abelian Yang–Mills fields on a lattice in temporal gauge, *Ann. Phys. (N.Y.)* **133**, 240;
Di Giacomo, A. and Rossi, G.C. (1981). Extracting $< (\alpha/\pi) \sum_{a,\mu\nu} G^a_{\mu\nu} G^a_{\mu\nu} >$ from gauge theories on a lattice, *Phys. Lett.* **100B**, 481;
Hattori, T. and Kawai, H. (1981). Weak coupling perturbative calculations of the Wilson loop in lattice gauge theory, *Phys. Lett.* **105B**, 43;
Horsley, R. and Wolff, U. (1981). Weak coupling expansion of Wilson loops in compact QED, *Phys. Lett.* **105B**, 290;
Curci, G., Paffuti,G. and Tripiccione, R. (1984). Perturbative background to Monte Carlo calculations in lattice gauge theories, *Nucl. Phys.* **B 240**, 91;
Heller, U. and Karsch, F. (1985). One-loop perturbative calculations of Wilson loops on finite lattices, *Nucl. Phys.* **B 251**, 254;

[3.39] Celmaster, W. and Gonzalves, R.J. (1979). Renormalization-prescription dependence of the quantum-chromodynamic coupling constant, *Phys. Rev.* **D20**, 1420.

[3.40] Jones, D.R.T. (1974). Two-loop diagrams in Yang–Mills theory, *Nucl. Phys.* **B 75**, 531.

[3.41] Caswell, W. (1974). Asymptotic behavior of non-Abelian gauge theories to two-loop order, *Phys. Rev. Lett.* **33**, 244.

[3.42] Weisz, P. (1981). On the connection between the Λ-parameters of Euclidean lattice and continuum QCD, *Phys. Lett.* **100B**, 330.

[3.43] Bardeen, W.A., Buras, A.J., Duke, D.W. and Muta, T. (1978). Deep inelastic scattering beyond the leading order in asymptotically free gauge theories, *Phys. Rev.* **D18**, 3998.

[3.44] Dashen, R. and Gross, D.J. (1981). Relationship between lattice and continuum definitions of the gauge-theory coupling, *Phys. Rev.* **D23**, 2340.

[3.45] Hasenfratz, A. and Hasenfratz, P. (1981). The scales of Euclidean and Hamiltonian lattice QCD, *Nucl. Phys.* **B 193**, 210.

[3.46] Gonzales-Arroyo, A. and Korthals-Altes, C.P. (1982). Asymptotic freedom scales for any single plaquette action, *Nucl. Phys.* **B 205**, 46;
Di Giacomo, A. and Paffuti, G. (1982) Some results related to the continuum limit of lattice gauge theories, *Nucl. Phys.* **B 205**, 313;

Lang, C.B., Rebbi, C., Salomonson, P. and Skagerstam, B.-S. (1982). Definitions of the gauge-theory coupling in lattice and continuum quantum chromodynamics: Implications of change in the lattice action, *Phys. Rev.* **D26**, 2028;
Iwasaki, Y. and Sakai, S. (1984). The Λ parameter for improved lattice gauge theory, *Nucl. Phys.* **B 248**, 441;
Bernreuther, W. and Wetzel, W. (1983). The relation between the Λ parameters of the standard and of the continuum limit improved lattice action for pure Yang–Mills theory, *Phys. Lett.* **132B**, 382;
Ukawa, A. and Yang, S. (1984). Asymptotic freedom scales for SU(N) lattice gauge theory with improved action, *Phys. Lett.* **137B**, 201;
Ellis, R.K. and Martinelli, G. (1984). Two-loop corrections to the Λ parameters of one-plaquette actions, *Nucl. Phys.* **B 235**, 93; Erratum: *Nucl. Phys.* **B 249** (1985) 750.

[3.47]   Ruelle, D. (1969). *Statistical Mechanics: Rigorous Results* (Benjamin, New York);
Rushbrooke, G.S., Baker, G.A. and Wood, P.J. (1974). Heisenberg model, in *Phase Transitions and Critical Phenomena, Vol. III*, ed. C. Domb and M.S. Green (Academic Press, New York).

[3.48]   Münster, G. (1981). High-temperature expansions for the free energy of vortices and the string tension in lattice gauge theories, *Nucl. Phys.* **B 180**, 23.

[3.49]   Balian, R., Drouffe, J.-M. and Itzykson, C. (1975). Gauge fields on a lattice, III. Strong-coupling expansions and transition points, *Phys. Rev.* **D11**, 2104; Erratum: *Phys. Rev.* **D19** (1979) 2514.

[3.50]   Drouffe, J.-M. (1980). Series analysis in four-dimensional Z_n lattice gauge systems, *Nucl. Phys.* **B 170**, 91.

[3.51]   Drouffe, J.-M. and Moriarty, K.J.M. (1981). The high-temperature expansion and Monte Carlo simulation of SU(5) gauge theory on a four-dimensional lattice: a comparison, *Phys. Lett.* **105B**, 449.

[3.52]   Drouffe, J.-M. and Moriarty, K.J.M. (1982). A comparison of the high-temperature expansion and Monte Carlo simulation of four-dimensional SU(4) and SU(6) gauge theory, *Phys. Lett.* **108B**, 333.

[3.53]   Drouffe, J.-M. and Moriarty, K.J.M. and Münster, G. (1982). Numerical studies of phase transitions in SU(N)/Z(N) lattice gauge theories, *Phys. Lett.* **115B**, 301.

[3.54]   Dashen, R., Heller, U. and Neuberger, H. (1983). Analytical results for mixed action SU(2) lattice gauge theory, *Nucl. Phys.* **B 215**, 360.

[3.55]   Drouffe, J.-M. and Zuber, J.-B. (1983). Strong coupling and mean field methods in lattice gauge theories, *Phys. Reports* **102**, 1.

[3.56]   Wilson, K.G., unpublished, cited in [3.55].

[3.57] Duncan, A. and Vaidya, H. (1979). Extrapolation of lattice gauge theories to the continuum limit, *Phys. Rev.* **D20**, 903.

[3.58] Falcioni, M., Marinari, E., Paciello, M.L., Parisi, G. and Taglienti, B. (1981). Phase transition analysis in Z_2 and U(1) lattice gauge theories, *Phys. Lett.* **105B**, 51.

[3.59] Falcioni, M., Marinari, E., Paciello, M.L., Parisi, G. and Taglienti, B. (1981). Complex singularities in the specific heat of the SU(2) lattice gauge model, *Phys. Lett.* **102B**, 270.

[3.60] Falcioni, M., Marinari, E., Paciello, M.L., Parisi, G. and Taglienti, B. (1981). On the link between strong and weak coupling expansion for the SU(2) lattice gauge theory, *Nucl. Phys.* **B190**, 782.

[3.61] Huang, K. (1982). *Quarks, Leptons & Gauge Fields* (World Scientific, Singapore).

[3.62] Kogut, J. and Susskind, L. (1975). Hamiltonian formulation of Wilson's lattice gauge theories, *Phys. Rev.* **D11**, 395.

[3.63] Kogut, J.B. (1980). Progress in lattice gauge theory, *Phys. Reports* **67**, 67.

[3.64] Marchesini, G. and Onofri, E. (1981). An elementary derivation of Wilson's and Polyakov's confinement tests from the Hamiltonian formulation, *Nuovo Cim.* **65A**, 298.

[3.65] Seiler, E. (1978). Upper bound on the color-confining potential, *Phys. Rev.* **D18**, 482.

[3.66] Drouffe, J.-M. (1978). Numerical analysis of transitions in 3-dimensional lattice gauge systems, Stony Brook preprint ITP-SB-78-35 (unpublished);
Kogut, J., Pearson, R. and Shigemitsu, J. (1979). Quantum-chromodynamic β function at intermediate and strong coupling, *Phys. Rev. Lett.* **43**, 484;
Münster, G. (1980). Vortex free energy and string tension at strong and intermediate coupling, *Phys. Lett.* **95B**, 59;
Kimura, N. (1980). Critical properties of Z(2) lattice gauge systems from strong-coupling expansions, *Prog. Theor. Phys.* **64**, 310.

[3.67] Münster, G. and Weisz, P. (1980). Estimate of the relation between scale parameters and the string tension by strong coupling methods, *Phys. Lett.* **96B**, 119.

[3.68] Münster, G. and Weisz, P. (1981). On the roughening transition in abelian lattice gauge theories, *Nucl. Phys.* **B180**, 13;
Münster, G. and Weisz, P. (1981). On the roughening transition in non-abelian lattice gauge theories, *Nucl. Phys.* **B180**, 330;
Drouffe, J.-M. and Zuber, J.-B. (1981). Roughening transition in lattice gauge theories in arbitrary dimension (I). The Z_2 case, *Nucl. Phys.* **B180**, 253;

Drouffe, J.-M. and Zuber, J.-B. (1981). Roughening transition in lattice gauge theories in arbitrary dimension (II). The groups Z_3, U(1), SU(2), SU(3), *Nucl. Phys.* **B 180**, 264.

[3.69]   Creutz, M. (1980). Monte Carlo study of quantized SU(2) gauge theory, *Phys. Rev.* **D21**, 2308; Asymptotic freedom scales, *Phys. Rev. Lett.* **45**, 313.

[3.70]   Kogut, J., Sinclair, D.K. and Susskind, L. (1976). A quantitative approach to low-energy quantum chromodynanics, *Nucl. Phys.* **B 114**, 199;
Kogut, J., and Shigemitsu, J. (1980). Crossover from weak to strong coupling in SU(*N*) lattice gauge theories, *Phys. Rev. Lett.* **45**, 410.

[3.71]   Lüscher, M., Münster, G. and Weisz, P. (1981). How thick are chromo-electric flux tubes? *Nucl. Phys.* **B 180**, 1.

[3.72]   van Beijeren, H. and Nolden, I. (1987). The roughening transition, in *Topics in Current Physics, Vol. 43: Structure and Dynamics of Surfaces II*, ed. W. Schommers and P. von Blanckenhagen (Springer, Berlin).

[3.73]   Parisi, G. (1980). On the structure of the phases in lattice gauge theories, in *Recent Developments in Gauge Theories, Cargèse lectures 1979*, ed. G. 't Hooft, C. Itzykson, A. Jaffe, H. Lehmann, P.K. Mitter, I.M. Singer and R. Stora (Plenum Publishing Corporation, New York).

[3.74]   Itzykson, C., Peskin, M.E. and Zuber, J.-B. (1980). Roughening of Wilson's surface, *Phys. Lett.* **95B**, 259.

[3.75]   Hasenfratz, A., Hasenfratz, E. and Hasenfratz, P. (1981). Generalized roughening transition and its effect on the string tension, *Nucl. Phys.* **B 180**, 353.

[3.76]   Lüscher, M. (1981). Symmetry-breaking aspects of the roughening transition in gauge theories, *Nucl. Phys.* **B 180**, 317.

[3.77]   Michael, C. (1990). Pure gauge: glueballs, potentials and the vacuum, *Nucl. Phys. B (Proc. Suppl.)* 17, 59.

[3.78]   Teper, M. (1991). Pure gauge theories, *Nucl. Phys. B (Proc. Suppl.)* 20, 159.

[3.79]   Bhanot, G. and Rebbi, C. (1981). SU(2) string tension, glueball mass and interquark potential by Monte Carlo computations, *Nucl. Phys.* **B 180**, 469;
De Grand, T.A. and Toussaint, D. (1981). Potential in lattice U(1) gauge theory, *Phys. Rev.* **D24**, 466;
Stack, J.D. (1983). Heavy quark potential in SU(2) lattice gauge theory, *Phys. Rev.* **D27**, 412; Heavy-quark potential in SU(3) lattice gauge theory, *Phys. Rev.* **D29** (1984) 1213;
Otto, S.W. and Stack, J.D. (1984). The SU(3) heavy quark potential with high statistics, *Phys. Rev. Lett.* **52**, 2328;

de Forcrand, P. (1986). QCD from Chippewa Falls, *J. Stat. Phys.* **43**, 1077;

Gutbrod, F., Hasenfratz, P., Kunszt, Z. and Montvay, I. (1983). String tension in SU(3) gauge theory on a 16^4 lattice, *Phys. Lett.* **128B**, 415;

Gutbrod, F. and Montvay, I. (1984). Scaling of the quark-antiquark potential and improved actions in SU(2) lattice gauge theory, *Phys. Lett.* **136B**, 411.

[3.80]   Kogut, J.B., Sinclair, D.K., Pearson, R.B., Richardson, J.L. and Shigemitsu, J. (1981). Fluctuating string of lattice gauge theory: the heavy-quark potential, the restoration of rotational symmetry, and roughening, *Phys. Rev.* **D23**, 2945;

Lang. C.B. and Rebbi, C. (1982). Potential and restoration of rotational symmetry in SU(2) lattice gauge theory, *Phys. Lett.* **115B**, 137;

Hasenfratz, A., Hasenfratz, P., Heller, U. and Karsch, F. (1984). Rotational symmetry of the SU(3) potential, *Z. Phys.* **C25**, 191.

[3.81]   Bacilieri, P., Fonti, L., Remiddi, E., Bernaschi, M., Cabasino, S., Cabbibo, N., Fernandez, L.A., Marinari, E., Paolucci, P., Parisi, G., Salina, G., Fiorentini, G., Galeotti, S., Lombardo, M.P., Passuello, D., Tripiccione, R., Marchesini, P., Marzano, F., Rapuano, F. and Rusack, R. (APE Collaboration) (1988). Scaling in lattice QCD: glueball masses and string tension, *Phys. Lett.* **205B**, 535;

Ford, I.J., Dalitz, R.H. and Hoek, J. (1988). Potentials in pure QCD on 32^4 lattices, *Phys. Lett.* **208B**, 286;

Michael, C. and Teper, M. (1989). The glueball spectrum in SU(3), *Nucl. Phys.* **B 314**, 347;

Vohwinkel, C. and Berg, B. (1989). Pure lattice gauge theory in intermediate volumes, *Phys. Rev.* **D40**, 584;

Born, K.D., Altmeyer, R., Ibes, W., Laermann, E., Sommer, R., Walsh, T.F. and Zerwas, P.M. (1991). The interquark potential, *Nucl. Phys. B (Proc. Suppl.)* **20**, 394;

Bali, G.S. and Schilling, K. (1992). Static quark–antiquark potential: scaling behaviour and finite-size effects in SU(3) lattice gauge theory, *Phys. Rev.* **D46**, 2636.

[3.82]   Fritzsch, H. and Gell-Mann, M. (1972). Current algebra: quarks and what else?, Proceedings of the XVIth International Conference on High Energy Physics, Vol.2 (Chicago).

[3.83]   Lüscher, M. (1983). Some analytic results concerning the mass spectrum of Yang–Mills gauge theories on a torus, *Nucl. Phys.* **B 219**, 233;

Lüscher, M. and Münster, G. (1984). Weak coupling expansion of the low-lying energy values in the SU(2) gauge theory on a torus, *Nucl. Phys.* **B 232**, 445;

Weisz, P. and Ziemann, V. (1987). Weak coupling expansion of the low-lying energy values in the SU(3) gauge theory on a torus, *Nucl.*

Phys. **B 284**, 157;
Koller, J. and van Baal, P. (1988). A non-perturbative analysis in
finite volume gauge theory, *Nucl. Phys.* **B 302**, 1;
van Baal, P. (1989). Gauge theory in a finite volume, *Acta Phys. Pol.*
B 20, 295.

[3.84] Bradley, C.J. and Cracknell, A.P. (1972). *The Mathematical Theory of
Symmetry in Solids* (Oxford University Press, London);
Janssen, T. (1973). *Crystallographic Groups* (North-Holland,
Amsterdam);
Madelung, O. (1972). *Festkörpertheorie II* (Springer-Verlag, Berlin).

[3.85] Berg, B. and Billoire, A. (1983). Glueball spectroscopy in 4d SU(3)
lattice gauge theory (I), *Nucl. Phys.* **B 221**, 109.

[3.86] Kogut, J., Sinclair, D.K. and Susskind, L. (1976). A quantitative
approach to low-energy quantum chromodynamics, *Nucl. Phys.* **B 114**,
199.

[3.87] Münster, G. (1981). Strong coupling expansions for the mass gap in
lattice gauge theories, *Nucl. Phys.* **B 190**, 439; Erratum: *Nucl. Phys.*
B 205 (1982) 648.

[3.88] Seo, K. (1982). Glueball mass estimates by strong coupling
expansions in lattice gauge theories, *Nucl. Phys.* **B 209**, 200.

[3.89] Decker, K. (1984). Strong coupling expansion for the mass gap in
SU(2) lattice gauge theory with mixed action, *Nucl. Phys.* **B 240**, 543.

[3.90] Decker, K. (1985). Critical behaviour of the 3d Ising model from an
extended low-temperature expansion of the inverse correlation length,
Nucl. Phys. **B 257**, 419;
Decker, K. (1986). An algorithm for high-order strong-coupling
expansions: The mass gap in 3d pure Z_2 lattice gauge theory, *Nucl.
Phys.* **B 270**, 292.

[3.91] Arisue, H. and Fujiwara, T. (1987). High order calculation of the
strong coupling expansion for the mass gap in lattice gauge theory,
Nucl. Phys. **B 285**, 253.

[3.92] Kimura, N. and Ukawa, A. (1982). Energy–momentum dispersion of
glueballs and the restoration of Lorentz invariance in lattice gauge
theories, *Nucl. Phys.* **B 205**, 637.

[3.93] Münster, G. (1985). Effective transfer matrix for low-lying glueball
states in lattice gauge theory, *Nucl. Phys.* **B 256**, 67.

[3.94] Smit, J. (1982). Estimate of glueball masses from their strong
coupling series in lattice QCD, *Nucl. Phys.* **B 206**, 309.

[3.95] Münster, G. (1983). Physical strong coupling expansion parameters
and glueball mass ratios, *Phys. Lett.* **121B**, 53.

[3.96] Münster, G. (1985). The size of finite size effects in lattice gauge
theories, *Nucl. Phys.* **B 249**, 659.

[3.97] Lüscher, M. (1984). On a relation between finite size effects and elastic scattering processes, in *Progress in Gauge Field Theory, Cargèse lectures 1983*, ed. G. 't Hooft, A. Jaffe, H. Lehmann, P.K. Mitter, I.M. Singer and R. Stora (Plenum Publishing Corporation, New York);
Lüscher, M. (1986). Volume dependence of the energy spectrum in massive quantum field theories (I). Stable particle states, *Comm. Math. Phys.* **104**, 177.

[3.98] Berg, B. (1980). Plaquette–plaquette correlations in the SU(2) lattice gauge theory, *Phys. Lett.* **97B**, 401.

[3.99] Falcioni, M., Marinari, E., Paciello, M.L., Parisi, G., Rapuano, F., Taglienti, B. and Zhang, Y.C. (1982). On the masses of the glueballs in pure SU(2) lattice gauge theory, *Phys. Lett.* **110B**, 295;
Ishikawa, K., Teper, M. and Schierholz, G. (1982). The glueball mass spectrum in QCD: first results of a lattice Monte Carlo calculation, *Phys. Lett.* **110B**, 399;
Berg, B., Billoire, A. and Rebbi, C. (1982). Monte Carlo estimates of the SU(2) mass gap, *Ann. Phys. (N.Y.)* **142**, 185; Addendum: *Ann. Phys. (N.Y.)* **146** (1983) 470.

[3.100] Berg, B. (1984). The spectrum in lattice gauge theories, in *Progress in Gauge Field Theory, Cargèse lectures 1983*, ed. G. 't Hooft, A. Jaffe, H. Lehmann, P.K. Mitter, I.M. Singer and R. Stora (Plenum Publishing Corporation, New York).

[3.101] Schierholz, G. (1985). Lattice QCD, in *Fundamental Forces*, 27th Scottish Universities Summer Schol in Physics, St. Andrews, 1984.

[3.102] Michael, C. and Perantonis, S. (1990). Potentials and glueballs at large beta, *Nucl. Phys. B (Proc. Suppl.)* **20**, 177.

[3.103] Schierholz, G. (1989). Status of lattice glueball mass calculations, in *Glueballs, Hybrids, and Exotic Hadrons (Proceedings of the BNL Workshop, Upton, New York, 1988)*, ed. Suh-Urk Chung (Brookhaven National Laboratory, Upton).

[3.104] Michael, C. and Teper, M. (1989). The glueball spectrum in SU(3), *Nucl. Phys.* **B 314**, 347.

[3.105] van Baal, P. and Kronfeld, A.S. (1989). Spectrum of the pure glue theory, *Nucl. Phys. B (Proc. Suppl.)* **9**, 227;
Kronfeld, A.S. (1992). Lattice calculations of glueball properties, in *Intersections between Particle and Nuclear Physics (Proceedings of the 4th International Conference, Tucson, 1991)*, ed. W.T.H. van Oers (American Institute of Physics, New York).

[3.106] Schierholz, G. (1989). The glueball mass spectrum in SU(3) lattice gauge theory, *Nucl. Phys. B (Proc. Suppl.)* **9**, 244.

[3.107] Simon, B. and Yaffe, L.G. (1982). Rigorous perimeter law upper bound on Wilson loops, *Phys. Lett.* **115B**, 145.

[3.108] 't Hooft, G. (1978). On the phase transition towards permanent quark confinement, *Nucl. Phys.* **B 138**, 1.

[3.109] Mack, G. (1980). Properties of lattice gauge theory models at low temperatures, in *Recent Developments in Gauge Theories, Cargèse lectures 1979*, ed. G. 't Hooft, C. Itzykson, A. Jaffe, H. Lehmann, P.K. Mitter, I.M. Singer and R. Stora (Plenum Publishing Corporation, New York).

[3.110] Yaffe, L.G. (1980). Confinement in SU(N) lattice gauge theories, *Phys. Rev.* **D21**, 1574.

[3.111] Samuel, S. (1983). A study of 't Hooft and Wilson loops, *Nucl. Phys.* **B 214**, 532.

[3.112] 't Hooft, G. (1979). A property of electric and magnetic flux in non-abelian gauge theories, *Nucl. Phys.* **B 153**, 141.

[3.113] Mack, G. and Petkova, V.B. (1980). Sufficient condition for confinement of static quarks by a vortex condensation mechanism, *Ann. Phys. (N.Y.)* **125**, 117.

[3.114] Münster, G. (1980). On the characterization of the Higgs phase in lattice gauge theories, *Z. Phys.* **C6**, 175.

[3.115] Balian, R., Drouffe, J.-M. and Itzykson, C. (1974). Gauge fields on a lattice, I. General outlook, *Phys. Rev.* **D10**, 3376.

[3.116] Dosch, H.G. and Müller, V.F. (1979). Lattice gauge theory in two spacetime dimensions, *Fortschr. Phys.* **27**, 547.

[3.117] Mack, G. (1979). Confinement of static quarks in two-dimensional lattice gauge theories, *Comm. Math. Phys.* **65**, 91.

[3.118] Petcher, D. and Weingarten, D.H. (1980). Monte Carlo calculations and a model of the phase structure for gauge theories on discrete subgroups of SU(2), *Phys. Rev.* **D22**, 2465.

[3.119] Göpfert, M. (1980). Peierls argument for Z(2) lattice gauge theory, Diplomarbeit (Universität Hamburg).

[3.120] Guth, A.H. (1980). Existence proof of a non-confining phase in four-dimensional U(1) lattice gauge theory, *Phys. Rev.* **D21**, 2291.

[3.121] Fröhlich, J. and Spencer, T. (1982). Massless phases and symmetry restoration in abelian gauge theories and spin systems, *Comm. Math. Phys.* **83**, 411.

[3.122] Göpfert, M. and Mack, G. (1981). Proof of confinement of static quarks in three-dimensional U(1) lattice gauge theory for all values of the coupling constant, *Comm. Math. Phys.* **82**, 545.

[3.123] Karliner, M. and Mack, G. (1983). Mass gap and string tension in QED$_3$: comparison of theory with Monte Carlo simulation, *Nucl. Phys.* **B 225**, 371.

[3.124]   Elitzur, S., Pearson, R.B. and Shigemitsu, J. (1979). The phase
          structure of discrete abelian spin and gauge systems, *Phys. Rev.* **D19**,
          3698;
          Horn, D., Weinstein, M. and Yankielowicz, S. (1979). Hamiltonian
          approach to Z(N) lattice gauge theories, *Phys. Rev.* **D19**, 3715;
          Ukawa, A., Windey, P. and Guth, A.H. (1980). Dual variables for
          lattice gauge theories and the phase structure of Z(N) systems, *Phys.
          Rev.* **D21**, 1013.

[3.125]   Polyakov, A.M. (1975). Compact gauge fields and the infrared
          catastrophe, *Phys. Lett.* **59B**, 82;
          Polyakov, A.M. (1977). Quark confinement and topology of gauge
          theories, *Nucl. Phys.* **B 120**, 429.

[3.126]   Banks, T., Myerson, R. and Kogut, J. (1977). Phase transitions in
          abelian lattice gauge theories, *Nucl. Phys.* **B 129**, 493.

[3.127]   Bellissard, J. and De Angelis, G.F. (1979). Gaussian limit of compact
          spin systems, in *Random Fields: Rigorous Results in Statistical
          Mechanics and Quantum Field Theory, Vol. 1*, Conf. Estergom,
          Hungary.

[3.128]   Creutz, M., Jacobs, L. and Rebbi, C. (1983). Monte Carlo
          computations in lattice gauge theories, *Phys. Reports* **95**, 201.

[3.129]   Berg, B. and Panagiotakopoulos, C. (1984). The photon in U(1) lattice
          gauge theory, *Phys. Rev. Lett.* **52**, 94;
          Jersák, J., Neuhaus, T. and Zerwas, P.M. (1985). Charge
          renormalization in compact lattice QED, *Nucl. Phys.* **B 251**, 299.

[3.130]   Creutz, M. (1979). Confinement and the critical dimensionality of
          space–time, *Phys. Rev. Lett.* **43**, 553;
          D'Hoker, E. (1981). Monte Carlo study of three-dimensional QCD,
          Nucl. Phys. **B 180**, 341;
          Creutz, M. and Moriarty, K.J.M. (1982). Phase transition in SU(6)
          lattice gauge theory, *Phys. Rev.* **D25**, 1724.

Chapter 4

[4.1]   Jordan, P. and Wigner, E. (1928). Über das Paulische
        Äquivalenzverbot, *Z. Phys.* **47**, 631.

[4.2]   Berezin, F.A. (1966). *The Method of Second Quantization* (Academic
        Press, New York).

[4.3]   Kerler, W. (1984). On the handling of fermion integrations in lattice
        gauge theory, *Z. Phys.* **C22**, 185.

[4.4]   Wilson, K.G. (1975). Quarks and strings on a lattice, in *New
        Phenomena in Subnuclear Physics*, ed. A. Zichichi (Plenum Press, New
        York), Part A, p. 69.

[4.5]   Kogut, J. and Susskind, L. (1975). Hamiltonian formulation of
        Wilson's lattice gauge theories, *Phys. Rev.* **D11**, 395;
        Banks, T., Kogut, J. and Susskind, L. (1976). Strong coupling
        calculations of lattice gauge theories: (1+1) dimensional exercises,
        Phys. Rev. **D13**, 1043;
        Susskind, L. (1977). Lattice fermions, *Phys. Rev.* **D16**, 3031.

[4.6]   Creutz, M. (1977). Gauge fixing, the transfer matrix, and confinement
        on a lattice, *Phys. Rev.* **D15**, 1128.

[4.7]   Lüscher, M. (1977). Construction of a self-adjoint, strictly positive
        transfer matrix for Euclidean lattice gauge theories, *Comm. Math.
        Phys.* **54**, 283.

[4.8]   Osterwalder, K. and Schrader, R. (1973). Axioms for Euclidean
        Green's functions, *Comm. Math. Phys.* **31**, 83;
        Osterwalder, K. and Schrader, R. (1975). Axioms for Euclidean
        Green's functions. 2, *Comm. Math. Phys.* **42**, 281.

[4.9]   Osterwalder, K. and Seiler, E. (1978). Gauge field theories on the
        lattice, *Ann. Phys. (N. Y.)* **110**, 440.

[4.10] Menotti, P. and Pelissetto, A. (1987). General proof of
        Osterwalder–Schrader positivity for the Wilson-action, *Comm. Math.
        Phys.* **113**, 369;
        Menotti, P. and Pelissetto, A. (1988). Osterwalder–Schrader positivity
        for the Wilson-action, *Nucl. Phys. B (Proc. Suppl.)* **4**, 644.

[4.11] Carpenter, D.B. and Baillie, C.F. (1985). Free fermion propagators
        and lattice finite size effects, *Nucl. Phys.* **B260**, 103.

[4.12] Sharatchandra, H.S., Thun, H.J. and Weisz, P. (1981). Susskind
        fermions on a Euclidean lattice, *Nucl. Phys.* **B192**, 205.

[4.13] Blairon, J.M., Brout, R., Englert, F. and Greensite, J. (1981). Chiral
        symmetry breaking in the action formulation of lattice gauge theory,
        Nucl. Phys. **B180 [FS2]**, 439.

[4.14] Kawamoto, N. and Smit, J. (1981). Effective Lagrangian and
        dynamical symmetry breaking in strongly coupled lattice QCD, *Nucl.
        Phys.* **B192**, 100.

[4.15] Kluberg-Stern, H., Morel, A., Napoly, O. and Petersson, B. (1981).
        Spontaneous symmetry breaking for a $U(N)$ gauge theory on a
        lattice, *Nucl. Phys.* **B190 [FS3]**, 504;
        Kluberg-Stern, H., Morel, A., Napoly, O. and Petersson, B. (1983).
        Flavours of Lagrangian Susskind fermions, *Nucl. Phys.* **B220 [FS8]**,
        447.

[4.16] Kluberg-Stern, H., Morel, A. and Petersson, B. (1982). The strong
        coupling limit of gauge theories with fermions on a lattice, *Phys. Lett.*
        114B, 152;

Kluberg-Stern, H., Morel, A. and Petersson, B. (1983). Spectrum of lattice gauge theories with fermions from a $1/d$ expansion at strong coupling, *Nucl. Phys.* **B215 [FS7]**, 527.

[4.17] Gliozzi, F. (1982). Spinor algebra of the one component lattice fermions, *Nucl. Phys.* **B204**, 419.

[4.18] Banks, T. and Zaks, A. (1982). Chiral analog gauge theories on the lattice, *Nucl. Phys.* **B206**, 23.

[4.19] Kitazoe, T., Ishihara, M. and Nakatani, H. (1978). Dirac equation on a lattice, *Lett. Nuovo Cim.* **21**, 59.

[4.20] Banks, T. and Windey, P. (1982). Supersymmetric lattice theories, *Nucl. Phys.* **B198**, 226.

[4.21] Jolicoeur, T., Morel, A. and Petersson, B. (1986). Continuum symmetries of lattice models with staggered fermions, *Nucl. Phys.* **B274**, 225.

[4.22] van den Doel, C. and Smit, J. (1983). Dynamical symmetry breaking in the flavor SU(*N*) and SO(*N*) lattice gauge theories, *Nucl. Phys.* **B228**, 122.

[4.23] Golterman, M. and Smit, J. (1984). Selfenergy and flavour interpretation of staggered fermions, *Nucl. Phys.* **B245**, 61.

[4.24] Becher, P. and Joos, H. (1982). The Dirac–Kähler equation and fermions on the lattice, *Z. Phys.* **C15**, 343.

[4.25] Rabin, J.M. (1982). Homology theory of lattice fermion doubling, *Nucl. Phys.* **B201**, 315.

[4.26] Kähler, E. (1962). Der innere Differentialkalkül, *Rend. Math. Ser. V* **21**, 425.

[4.27] Göckeler, M. and Joos, H. (1984). On Kähler's geometric description of Dirac fields, in *Progress in Gauge Field Theory*, Proceedings of the 1983 Cargèse Summer School, ed. G. 't Hooft *et al.* (Plenum Press, New York), p. 247.

[4.28] Joos, H. (1986). On geometry and physics of staggered lattice fermions, in *Clifford Algebras and Their Applications in Mathematical Physics*, ed. J.S.R. Chisholm and A.K. Common (Reidel Publishing Company, Dordrecht), p. 399.

[4.29] Singer, I.M. and Thorpe, J.A. (1967). *Lecture Notes on Elementary Topology and Geometry* (Springer, New York).

[4.30] Chodos, A. and Healy, J.B. (1977). Spectral degeneracy of the lattice Dirac-equation as a function of lattice shape, *Nucl. Phys.* **B127**, 426.

[4.31] Karsten, L.H. and Smit, J. (1981). Lattice fermions: species doubling, chiral invariance, and the triangle anomaly, *Nucl. Phys.* **B183**, 103.

[4.32] Pelissetto, A. (1988). Lattice nonlocal chiral fermions, *Ann. Phys. (N. Y.)* **182**, 177;

Inconsistency of nonlocal chiral fermions on the lattice, *Nucl. Phys. B (Proc. Suppl.)* **4**, 515.

[4.33]   Karsten, L.H. (1981). Lattice fermions in Euclidean space–time, *Phys. Lett.* **104B**, 315.

[4.34]   Wilczek, F. (1987). On lattice fermions, *Phys. Rev. Lett.* **59**, 2397.

[4.35]   Nielsen, H.B. and Ninomiya, M. (1981). Absence of neutrinos on a lattice. 1. Proof by homotopy theory, *Nucl. Phys.* **B185**, 20; erratum: **B195**, 541; Absence of neutrinos on a lattice. 2. Intuitive topological proof, *Nucl. Phys.* **B193**, 173.

[4.36]   Friedan, D. (1982). A proof of the Nielsen–Ninomiya theorem, *Comm. Math. Phys.* **85**, 481.

[4.37]   Golterman, M.F.L. and Petcher, D. (1989). The decoupling of righthanded neutrinos in chiral lattice gauge theories, *Phys. Lett.* **225B**, 159.

[4.38]   Schwinger, J. (1951). On gauge invariance and vacuum polarization, *Phys. Rev.* **82**, 664.

[4.39]   Adler, S. (1969). Axial-vector vertex in spinor electrodynamics, *Phys. Rev.* **177**, 2426.

[4.40]   Bell, J.S. and Jackiw, R. (1969). A PCAC-puzzle: $\pi^0 \to \gamma\gamma$ in the sigma-model, *Nuovo Cim.* **60A**, 47.

[4.41]   Seiler, E. and Stamatescu, I.O. (1982). Lattice fermions and theta vacua, *Phys. Rev.* **D25**, 2177; erratum: **D26**, 534.

[4.42]   Fujikawa, K. (1984). Chiral and conformal anomalies in lattice gauge theory, *Z. Phys.* **C25**, 179.

[4.43]   Coste, A., Korthals-Altes, C. and Napoly, O. (1986). Nonabelian anomaly from lattice fermions, *Phys. Lett.* **179B**, 125; Coste, A., Korthals-Altes, C. and Napoly, O. (1987). Calculation of the nonabelian chiral anomaly on the lattice, *Nucl. Phys.* **B289**, 645.

[4.44]   Jolicoeur, T., Lacaze, R. and Napoly, O. (1987). Anomalies for compact symmetries and the nonabelian anomaly on the lattice, *Nucl. Phys.* **B293**, 215.

[4.45]   Ambjørn, J., Greensite, J. and Peterson, C. (1983). The axial anomaly and the lattice Dirac sea, *Nucl. Phys.* **B221**, 381.

[4.46]   Landau, L.D. and Pomeranchuk, I.Y. (1955). On point interactions in quantum electrodynamics, *Dokl. Akad. Nauk SSSR* **102**, 489; Fradkin, E.S. (1955). The asymptote of Green's function in quantum electrodynamics, *JETP* **28**, 750; *Soviet Physics JETP* **1**, 604.

[4.47]   Gorishny, S.G., Kataev, A.L., Larin, S.A. and Surguladze, L.R. (1991). The analytic four-loop corrections to the QED β-function in the MS scheme and to the QED ψ-function. Total reevaluation, *Phys. Lett.* **256B**, 81.

[4.48] Miranskii, V.A. (1985). Dynamics of spontaneous chiral symmetry breaking and continuum limit in quantum electrodynamics, *Nouvo Cim.* **90A**, 149;
Fomin, P.I., Gusynin, V.P., Miranskii, V.A. and Sitenko, Yu.A. (1983). Dynamical symmetry breaking and particle mass generation in gauge field theories, *Riv. Nouvo Cim.* **6**, 1.

[4.49] Mitra, P. and Weisz, P. (1983). On bare and induced masses of Susskind fermions, *Phys. Lett.* **126B**, 355.

[4.50] Wilson, K.G. (1974). Confinement of quarks, *Phys. Rev.* **D10**, 2445.

[4.51] Kogut, J.B., Dagotto, E. and Kocic, A. (1988). A new phase of quantum electrodynamics: a nonperturbative fixed point in four-dimensions, *Phys. Rev. Lett.* **60**, 772; On the existence of quantum electrodynamics, *Phys. Rev. Lett.* **61**, 2416;
Kogut, J.B., Dagotto, E. and Kocic, A. (1989). Strongly coupled quenched QED, *Nucl. Phys.* **B317**, 253; A supercomputer study of strongly coupled QED, *Nucl. Phys.* **B317**, 271;
Kogut, J.B., Dagotto, E. and Kocic, A. (1990). Finite size, fermion mass and $N(f)$ systematics in computer simulations of quantum electrodynamics, *Nucl. Phys.* **B331**, 500.

[4.52] Nakamura, A. and Sinclair, R. (1990). Fermion propagators in U(1) lattice gauge theory, *Phys. Lett.* **243B**, 396.

[4.53] Sharatchandra, H.S. (1978). The continuum limit of lattice gauge theories in the context of renormalized perturbation theory, *Phys. Rev.* **D18**, 2042.

[4.54] Rossi, P. and Wolff, U. (1984). Lattice QCD with fermions at strong coupling: a dimer system, *Nucl. Phys.* **B248**, 105.

[4.55] Salmhofer, M. and Seiler, E. (1991). Proof of chiral symmetry breaking in lattice gauge theory, *Lett. Math. Phys.* **21**, 13; Proof of chiral symmetry breaking in strongly coupled lattice gauge theory, *Comm. Math. Phys.* **139**, 395.

[4.56] Lüscher, M. (1990). Charge screening and an upper bound on the renormalized charge in lattice QED, *Nucl. Phys.* **B341**, 341.

[4.57] Dagotto, E. and Kogut, J.B. (1987). First order phase transition in compact lattice QED with light fermions, *Phys. Rev. Lett.* **59**, 617;
Dagotto, E. and Kogut, J.B. (1988). Study of compact QED with light fermions, *Nucl. Phys.* **B295**, 123.

[4.58] Booth, S.P., Kenway, R.D. and Pendleton, B.J. (1989). The phase diagram of the gauge invariant Nambu–Jona-Lasinio model, *Phys. Lett.* **228B**, 115.

[4.59] Göckeler, M., Horsley, R., Laermann, E., Rakow, P., Schierholz, G., Sommer, R. and Wiese, U.-J. (1990). QED: a lattice investigation of the chiral phase transition and the nature of the continuum limit, *Nucl. Phys.* **B334**, 527; erratum: **B356**, 562; The continuum limit of

QED: renormalization group analysis and the question of triviality, *Phys. Lett.* **251B**, 567.

[4.60]   Dagotto, E. and Wyld, H.W. (1988). Nonperturbative study of QED in a strong Coulomb field, *Phys. Lett.* **205B**, 73.

[4.61]   Ng, Y.J. and Kikuchi, Y. (1990). Are there background fields that can induce QED phase transitions at weak coupling?, in *Vacuum Structure in Intense Fields,* Proceedings of the NATO Advanced Summer School, Cargèse, 1990.

[4.62]   Rakow, P. (1991). Renormalization group flow in QED: an investigation of the Schwinger–Dyson equation, *Nucl. Phys.* **B356**, 27.

[4.63]   Leung, C.N., Love, S.T. and Bardeen, W.A. (1986). Spontaneous symmetry breaking in scale invariant quantum electrodynamics, *Nucl. Phys.* **B273**, 649.

[4.64]   Horowitz, A.M. (1991). The renormalized charge and effective Yukawa-interactions from lattice QED, *Phys. Rev.* **D43**, 2461.

[4.65]   Nambu, Y. and Jona-Lasinio G. (1961). Dynamical model of elementary particles based on an analogy with superconductivity. I, *Phys. Rev.* **122**, 345.

Chapter 5

[5.1]   Wilson, K.G. (1974). Confinement of quarks, *Phys. Rev.* **D10**, 2445.

[5.2]   Wilson, K.G. (1975). Quarks and strings on a lattice, in *New Phenomena in Subnuclear Physics*, ed. A. Zichichi (Plenum Press, New York), Part A, p. 69.

[5.3]   Golterman, M. and Smit, J. (1984). Selfenergy and flavour interpretation of staggered fermions, *Nucl. Phys.* **B245**, 61.

[5.4]   Mitra, P. and Weisz, P. (1983). On bare and induced masses of Susskind fermions, *Phys. Lett.* **126B**, 355.

[5.5]   Napoly, O. (1983). Absence of Goldstone boson for a U(N) gauge theory with Dirac–Kähler fermions on the lattice, *Phys. Lett.* **132B**, 145.

[5.6]   Okubo, S. (1963). φ-meson and unitary symmetry model, *Phys. Lett.* **5**, 165;
Zweig, G. (1964). An SU(3) model for strong interaction symmetry and its breaking, CERN preprint TH 401;
Iizuka, J. (1966). A systematics and phenomenology of meson family, *Progr. Theor. Phys. Suppl.* **37–38**, 21.

[5.7]   Veneziano, G. (1968). Construction of a crossing-symmetric, Regge behaved amplitude for linearly rising trajectories, *Nuovo Cim.* **57A**, 190.

[5.8]   Hamber, H. and Parisi, G. (1981). Numerical estimates of hadronic masses in a pure SU(3) gauge theory, *Phys. Rev. Lett.* **47**, 1792; Marinari, E., Parisi, G. and Rebbi, C. (1981). Computer estimates of meson masses in SU(2) lattice gauge theory, *Phys. Rev. Lett.* **47**, 1795.

[5.9]   Weingarten, D. (1982). Monte Carlo evaluation of hadron masses in lattice gauge theories with fermions, *Phys. Lett.* **109B**, 57.

[5.10] Joos, H. and Montvay, I. (1983). The screening of colour charge in numerical hopping parameter expansion, *Nucl. Phys.* **B255 [FS9]**, 565.

[5.11] Berg, B., Billoire, A. and Foerster, D. (1982). A note on γ traces for the Wilson-action in the continuum limit, *Lett. Math. Phys.* **6**, 293.

[5.12] Hasenfratz, A., Hasenfratz, P., Kunszt, Z. and Lang, C.B. (1982). Hopping parameter expansion for the meson spectrum in SU(3) lattice QCD, *Phys. Lett.* **110B**, 289.

[5.13] Hasenfratz, P. and Montvay, I. (1983). Hadron spectrum: evidence for size problems on the lattice, *Phys. Rev. Lett.* **50**, 309.

[5.14] Hasenfratz, P. and Montvay, I. (1984). Meson spectrum in quenched QCD on a 16^4 lattice, *Nucl. Phys.* **B237**, 237.

[5.15] Kunszt, Z. and Montvay, I. (1984). Baryon spectrum in quenched QCD on a 16^4 lattice, *Phys. Lett.* **139B**, 195.

[5.16] Montvay, I. (1987). Numerical calculation of hadron masses in lattice quantum chromodynamics, *Rev. Mod. Phys.* **59**, 263.

[5.17] Kawamoto, N. and Smit, J. (1981). Effective Lagrangian and dynamical symmetry breaking in strongly coupled lattice QCD, *Nucl. Phys.* **B192**, 100.

[5.18] Hoek, J., Kawamoto, N. and Smit, J. (1982). Baryons in the effective Lagrangian of strongly coupled lattice QCD, *Nucl. Phys.* **B199**, 495.

[5.19] Kluberg-Stern, H., Morel, A. and Petersson, B. (1983). Spectrum of lattice gauge theories with fermions from a $1/d$ expansion at strong coupling, *Nucl. Phys.* **B215 [FS7]**, 527.

[5.20] Kawamoto, N. (1981). Towards the phase structure of Euclidean lattice gauge theories with fermions, *Nucl. Phys.* **B190 [FS3]**, 617.

[5.21] Jolicoeur, T., Kluberg-Stern, H., Lev, M., Morel, A. and Petersson, B. (1984). The strong coupling expansion of lattice gauge theories with Susskind fermions, *Nucl. Phys.* **B235 [FS11]**, 455.

[5.22] Hoek, J. and Smit, J. (1986). On the $1/g^2$ corrections to hadron masses, *Nucl. Phys.* **B263**, 129.

[5.23] Rossi, P. and Wolff, U. (1984). Lattice QCD with fermions at strong coupling: a dimer system, *Nucl. Phys.* **B248**, 105; Wolff, U. (1985). Baryons in lattice QCD at strong coupling, *Phys. Lett.* **153B**, 92.

[5.24]   Karsch, F. and Mütter, K.-H. (1989). Strong coupling QCD at finite
         baryon number density, *Nucl. Phys.* **B313**, 541;
         Karsch, F. (1990). The monomer dimer algorithm and QCD at finite
         density, in *Probabilistic Methods in Quantum Field Theory and
         Quantum Gravity*, ed. P.H. Damgaard, H. Hüffel and A. Rosenblum
         (Plenum Press, New York), p. 199.

[5.25]   Reisz, T. (1989). Lattice gauge theory: renormalization to all orders of
         the loop expansion, *Nucl. Phys.* **B318**, 417.

[5.26]   Symanzik, K. (1983). Continuum limit and improved action in lattice
         theories. I. Principles and ϕ^4 theory, *Nucl. Phys.* **B226**, 187.

[5.27]   Sheikholeslami, B. and Wohlert, R. (1985). Improved continuum limit
         lattice action for QCD with Wilson-fermions, *Nucl. Phys.* **B259**, 572.

[5.28]   Kawai, H., Nakayama, R. and Seo, K. (1981). Comparison of the
         lattice lambda parameter with the continuum lambda parameter in
         massless QCD, *Nucl. Phys.* **B189**, 40.

[5.29]   Karsten, L.H. and Smit, J. (1981). Lattice fermions: species doubling,
         chiral invariance, and the triangle anomaly, *Nucl. Phys.* **B183**, 103.

[5.30]   Gonzalez Arroyo, A., Yndurain, F.J. and Martinelli, G. (1982).
         Computation of the relation between the quark masses in lattice
         gauge theories and on the continuum, *Phys. Lett.* **117B**, 437; erratum:
         122B, 486.

[5.31]   Aoki, S. (1990). Chiral (gauge) symmetry and regularizations, *Phys.
         Lett.* **247B**, 357.

[5.32]   Stehr, J. and Weisz, P. (1983). Note on gauge fixing in lattice QCD,
         Lett. Nuovo Cim. **37**, 173.

[5.33]   Lepage, G.P. and Mackenzie, P.B. (1991). Renormalized lattice
         perturbation theory, *Nucl. Phys. B (Proc. Suppl.)* **20**, 173;
         Lepage, G.P. and Mackenzie, P.B. (1992). On the viability of lattice
         perturbation theory, FNAL preprint (1992).

[5.34]   Celmaster, W. and Gonsalves, R.J. (1979). The renormalization
         prescription dependence of the QCD coupling constant, *Phys. Rev.*
         D20, 1420.

[5.35]   Billoire, A. (1981). Another connection between the Λ parameters of
         the Euclidean lattice and continuum QCD, *Phys. Lett.* **104B**, 472;
         Kovacs, E. (1982). Lattice predictions for low-Q^2 phenomenology,
         Phys. Rev. **D25**, 871.

[5.36]   Weisz, P. (1981). On the connection between the Λ parameters of
         Euclidean lattice and continuum QCD, *Phys. Lett.* **100B**, 331.

[5.37]   Dashen, R. and Gross, D.J. (1981). The relationship between lattice
         and continuum definitions of the gauge theory coupling, *Phys. Rev.*
         D23, 2340.

[5.38]   Heller, U. and Karsch, F. (1985). One-loop perturbative calculation of Wilson loops on finite lattices, *Nucl. Phys.* **B251 [FS13]**, 254.

[5.39]   Hamber, H.W. and Wu, C.M. (1983). Some predictions for an improved fermion action on the lattice, *Phys. Lett.* **133B**, 351.

[5.40]   Pais, A. (1966). Dynamical symmetry in particle physics, *Rev. Mod. Phys.* **38**, 215.

[5.41]   Fucito, F., Martinelli, G., Omero, C., Parisi, G., Petronzio, R. and Rapuano, F. (1982). Hadron spectroscopy in lattice QCD, *Nucl. Phys.* **B210 [FS6]**, 407;
Joffe, B.L. (1981). Calculation of baryon masses in quantum chromodynamics, *Nucl. Phys.* **B188**, 317.

[5.42]   Billoire, A., Marinari, E. and Petronzio, R. (1985). Kogut–Susskind and Wilson-fermions in the quenched approximation: a Monte Carlo simulation, *Nucl. Phys.* **B251 [FS13]**, 141.

[5.43]   Weingarten, D.H. (1983). Mass inequalities for QCD, *Phys. Rev. Lett.* **51**, 1830;
Nussinov, S. (1983). Baryon meson mass inequalities, *Phys. Rev. Lett.* **51**, 2081;
Witten, E. (1983). Some inequalities among hadron masses, *Phys. Rev. Lett.* **51**, 2351.

[5.44]   Iwasaki, Y. and Yoshié, T. (1989). Hadron spectrum in quenched lattice QCD and quark potential models, *Phys. Lett.* **216B**, 387.

[5.45]   Wiese, U. (1989). Identification of resonance parameters from the finite volume energy spectrum, *Nucl. Phys. B (Proc. Suppl.)* **9**, 609.

[5.46]   Michael, C. (1989). Particle decay in lattice gauge theory, *Nucl. Phys.* **B327**, 515.

[5.47]   Lüscher, M. (1989). Selected topics in lattice field theory, in *Fields, strings and critical phenomena*, ed. E. Brézin and J. Zinn-Justin (North Holland, Amsterdam);
Lüscher, M. (1991). Two particle states on a torus and their relation to the scattering matrix, *Nucl. Phys.* **B354**, 531; Signatures of metastable particles in finite volume, *Nucl. Phys.* **B364**, 237.

[5.48]   Bowler, K.C., Chalmers, D.L., Kenway, A., Kenway, R.D., Pawley, G.S. and Wallace, D.J. (1985). A critique of quenched hadron mass calculations, *Phys. Lett.* **162B**, 354.

[5.49]   Bacilieri P. *et al.*, APE Collaboration, (1988). The hadronic mass spectrum in quenched lattice QCD: results at $\beta = 5.7$ and $\beta = 6.0$, *Phys. Lett.* **214B**, 115;
Bacilieri P. *et al.*, APE Collaboration, (1989). The hadronic mass spectrum in quenched lattice QCD: $\beta = 5.7$, *Nucl. Phys.* **B317**, 509;
Cabasino S. *et al.*, APE Collaboration, (1990). The APE with a small mass, *Nucl. Phys. B (Proc. Suppl.)* **17**, 431.

[5.50] Gupta, R. (1990). Hadron spectrum from the lattice, *Nucl. Phys. B (Proc. Suppl.)* **17**, 70.

[5.51] DeGrand, T.A. and Loft, R.D. (1990). Wave function tests for lattice QCD spectroscopy, *Comput. Phys. Commun.* **65**, 84; DeGrand, T.A. and Loft, R.D. (1991). Gaussian shell model trial wave functions for lattice QCD spectroscopy, Boulder preprint COLO-HEP-249.

[5.52] Güsken, S., Löw, U., Mütter, K.-H., Sommer, R., Patel, A. and Schilling, K. (1990). A study of smearing techniques for hadron correlation functions, *Nucl. Phys. B (Proc. Suppl.)* **17**, 361.

[5.53] Güsken, S., Löw, U., Mütter, K.-H., Sommer, R., Patel, A. and Schilling, K. (1989). Nonsinglet axial vector couplings of the baryon octet in lattice QCD, *Phys. Lett.* **227B**, 266.

[5.54] Eichten, E. (1988). Heavy quarks on the lattice, *Nucl. Phys. B (Proc. Suppl.)* **4**, 170.

[5.55] Lepage, G.P. and Thacker, B.A. (1988). Effective Lagrangians for simulating of heavy quark systems, *Nucl. Phys. B (Proc. Suppl.)* **4**, 199; Lepage, G.P. and Thacker, B.A. (1991). Heavy quark bound states in lattice QCD, *Phys. Rev.* **D43**, 196.

[5.56] Caswell, W.E. and Lepage, G.P. (1986). Effective Lagrangians for bound state problems in QED, QCD, and other field theories, *Phys. Lett.* **167B**, 437.

[5.57] Grinstein, B. (1990). The static quark effective theory, *Nucl. Phys.* **B339**, 253.

[5.58] Eichten, E. and Hill, B. (1990). An effective field theory for the calculation of matrix elements involving heavy quarks, *Phys. Lett.* **234B**, 511; Renormalization of heavy-light bilinears and f_B for Wilson-fermions, *Phys. Lett.* **240B**, 193; Static effective field theory: $1/m$ corrections, *Phys. Lett.* **243B**, 427.

[5.59] Eichten, E. (1991). B physics on the lattice, *Nucl. Phys. B (Proc. Suppl.)* **20**, 475.

[5.60] Goldstone, J. (1961). Field theories with 'superconductor' solutions, *Nuovo Cim.* **19**, 154; Goldstone, J., Salam, A. and Weinberg, S. (1962). Broken symmetries, *Phys. Rev.* **127**, 965.

[5.61] Bernstein, J. (1974). Spontaneous symmetry breaking, gauge theories, the Higgs mechanism and all that, *Rev. Mod. Phys.* **46**, 7.

[5.62] Gell-Mann, M. (1964). The symmetry group of vector and axialvector currents, *Physics* **1**, 63.

[5.63] Adler, S.L. and Dashen, R.F. (1968). *Current Algebras and Applications to Particle Physics* (Benjamin Inc., New York).

[5.64] Gell-Mann, M. and Lévy, M. (1960). The axialvector current in beta decay, *Nuovo Cim.* **16**, 705.

[5.65] Weinberg, S. (1970). Dynamic and algebraic symmetries, in *Lectures on Elementary Particles and Quantum Field Theory*, ed. S. Deser, M. Grisaru and H. Pendleton (MIT Press, Cambridge), Vol. 1 p. 285.

[5.66] Bochicchio, M., Maiani, L., Martinelli, G., Rossi, G. and Testa, M. (1985). Chiral symmetry on the lattice with Wilson-fermions, *Nucl. Phys.* **B262**, 331.

[5.67] Maiani, L., Martinelli, G., Paciello, M.L. and Taglienti, B. (1987). Scalar densities and baryon mass differences in lattice QCD with Wilson-fermions, *Nucl. Phys.* **B293**, 420.

[5.68] Curci, G. (1986). Infrared finiteness of normalization constants of currents in lattice gauge theories, *Phys. Lett.* **167B**, 425.

[5.69] Maiani, L. and Martinelli, G. (1986). Current algebra and quark masses from a Monte Carlo simulation with Wilson-fermions, *Phys. Lett.* **178B**, 265.

[5.70] Martinelli, G. and Zhang, I.-C. (1983). The connection between local operators on the lattice and in the continuum and its relation to meson decay constants, *Phys. Lett.* **123B**, 433; One loop corrections to extended operators on the lattice, *Phys. Lett.* **125B**, 77.

[5.71] Coste, A., Korthals-Altes, C. and Napoly, O. (1986). Nonabelian anomaly from lattice fermions, *Phys. Lett.* **179B**, 125; Coste, A., Korthals-Altes, C. and Napoly, O. (1987). Calculation of the nonabelian chiral anomaly on the lattice, *Nucl. Phys.* **B289**, 645.

[5.72] Atiyah, M. and Singer, I.M. (1971). The index of elliptic operators: V, *Ann. Math.* **93**, 139.

[5.73] Witten, E. (1979). Current algebra theorems for the U(1) 'Goldstone-boson', *Nucl. Phys.* **B156**, 269; Veneziano, G. (1979). U(1) without instantons, *Nucl. Phys.* **B159**, 213.

[5.74] Kronfeld, A.S. (1988). Topological aspects of lattice gauge theories, *Nucl. Phys. B (Proc. Suppl.)* **4**, 329.

[5.75] Campostrini, M., Di Giacomo, A. and Panagopoulos, H. (1988). The topological susceptibility on the lattice, *Phys. Lett.* **212B**, 206; Campostrini, M., Di Giacomo, A. and Panagopoulos, H. (1990). Topological charge, renormalization and cooling on the lattice, *Nucl. Phys. B (Proc. Suppl.)* **17**, 634.

[5.76] Lüscher, M. (1982). Topology of lattice gauge fields, *Comm. Math. Phys.* **85**, 39; Phillips, A. and Stone, D. (1986). Lattice gauge fields, principal bundles and the calculation of topological charge, *Comm. Math. Phys.* **103**, 599.

[5.77] Pugh, D.J.R. and Teper, M. (1989). Topological dislocations in the continuum limit of SU(2) lattice gauge theory, *Phys. Lett.* **224B**, 159.

[5.78] Hoek, J., Teper, M. and Waterhouse, J. (1986). Topology and the η' mass in SU(3) lattice gauge theory, *Phys. Lett.* **180B**, 112; Hoek, J., Teper, M. and Waterhouse, J. (1987). Topological fluctuations and susceptibility in SU(3) lattice gauge theory, *Nucl. Phys.* **B288**, 589.

[5.79] Smit, J. and Vink, J.C. (1988). Renormalized Ward–Takahashi relations and topological susceptibility with staggered fermions, *Nucl. Phys.* **B298**, 557; Laursen, M.L., Smit, J. and Vink, J.C. (1990). Small-scale instantons, staggered fermions and the topological susceptibility, *Nucl. Phys.* **B343**, 522.

[5.80] Itoh, S., Iwasaki, Y. and Yoshié, T. (1987). The U(1) problem and instantons on a lattice, *Phys. Lett.* **184B**, 375; The U(1) problem and topological excitations on a lattice, *Phys. Rev.* **D36**, 527.

[5.81] Bernard, C. and Soni, A. (1989). Review of weak matrix elements on the lattice, *Nucl. Phys. B (Proc. Suppl.)* **9**, 155.

[5.82] Sharpe, S.R. (1990). Lattice calculations of electroweak decay amplitudes, *Nucl. Phys. B (Proc. Suppl.)* **17**, 146.

[5.83] Kilcup, G. (1991). Electroweak matrix elements: 1990 update, *Nucl. Phys. B (Proc. Suppl.)* **20**, 417.

[5.84] Sommer, R. (1991). Hadron structure from lattice quantum chromodynamics, in *Computational Physics*, ed. A. Tenner (World Scientific, Singapore), p. 187.

[5.85] Maiani, L., Martinelli, G., Rossi, G. and Testa, M. (1987). The octet nonleptonic Hamiltonian and current algebra on the lattice with Wilson-fermions, *Nucl. Phys.* **B289**, 505.

[5.86] Bernard, C. (1990). Weak matrix elements on and off the lattice, in *From Actions to Answers*, ed. T. DeGrand and D. Toussaint, (World Scientific, Singapore), p. 233.

[5.87] Sachrajda, C.T. (1990). Lattice perturbation theory, in *From Actions to Answers*, ed. T. DeGrand and D. Toussaint, (World Scientific, Singapore), p. 293.

[5.88] Gaillard, M.K. and Lee, B.W. (1974). $\Delta I = \frac{1}{2}$ rule for nonleptonic decays in asymptotically free field theories, *Phys. Rev. Lett.* **33**, 108; Altarelli, G. and Maiani, L. (1974). Octet enhancement of nonleptonic weak interactions in asymptotically free gauge theories, *Phys. Lett.* **52B**, 351.

[5.89] Heatlie, G., Sachrajda, C.T., Martinelli, G., Pittori, C. and Rossi, G.C. (1991). The improvement of hadronic matrix elements in lattice QCD, *Nucl. Phys.* **B352**, 266.

[5.90] Hagedorn, R. (1965). Statistical thermodynamics of strong interactions at high energies, *Nuovo Cim.* **35**, 395.

[5.91] Collins, J.C. and Perry, M.J. (1975). Superdense matter: neutrons or asymptotically free quarks? *Phys. Rev. Lett.* **34**, 1353.

[5.92] Polyakov, A.M. (1978). Thermal properties of gauge fields and quark liberation, *Phys. Lett.* **72B**, 477;
Susskind, L. (1979). Lattice models of quark confinement at high temperature, *Phys. Rev.* **D20**, 2610.

[5.93] Linde, A.D. (1980). Infrared problem in thermodynamics of the Yang–Mills gas, *Phys. Lett.* **96B**, 289.

[5.94] Karsch, F. (1990). Simulating the quark–gluon plasma on the lattice, in *Quark–Gluon Plasma*, ed. R.C. Hwa (World Scientific, Singapore), p. 61.

[5.95] Fukugita, M. (1989). QCD thermodynamics, *Nucl. Phys. B (Proc. Suppl.)* **9**, 291.

[5.96] Ukawa, A. (1990). QCD phase transitions at finite temperatures, *Nucl. Phys. B (Proc. Suppl.)* **17**, 118.

[5.97] Gottlieb, S. (1991). Finite temperature QCD with dynamical fermions, *Nucl. Phys. B (Proc. Suppl.)* **20**, 247.

[5.98] Engels, J., Karsch, F., Montvay, I. and Satz, H. (1981). High temperature SU(2) gluon matter on the lattice, *Phys. Lett.* **101B**, 89;
Engels, J., Karsch, F., Montvay, I. and Satz, H. (1982). Gauge field thermodynamics for the SU(2) Yang–Mills system, *Nucl. Phys.* **B205 [FS5]**, 545.

[5.99] Montvay, I. and Pietarinen, E. (1982). Stefan–Boltzmann law at high temperature for the gluon gas, *Phys. Lett.* **110B**, 148;
Thermodynamical properties of the gluon matter, *Phys. Lett.* **115B**, 151.

[5.100] Karsch, F. (1982). SU(N) gauge theory couplings on asymmetric lattices, *Nucl. Phys.* **B205 [FS5]**, 285.

[5.101] Hasenfratz, A. and Hasenfratz, P. (1981). The scales of Euclidean and Hamiltonian lattice QCD, *Nucl. Phys.* **B193**, 210.

[5.102] Trinchero, R.C. (1983). One loop fermion contribution in an asymmetric lattice regularization of SU(N) gauge theories, *Nucl. Phys.* **B227**, 61.

[5.103] Karsch, F. and Stamatescu I.O. (1989). QCD thermodynamics with light quarks: quantum corrections to the fermionic anisotropy parameter, *Phys. Lett.* **227B**, 153.

[5.104] Burgers, G., Karsch, F., Nakamura, A. and Stamatescu, I.O. (1988). QCD on anisotropic lattices, *Nucl. Phys.* **B304**, 587.

[5.105]   Engels, J., Fingberg, J., Karsch, F., Miller, D. and Weber, M. (1990). Nonperturbative thermodynamics of SU(N) gauge theories, *Phys. Lett.* **252B**, 625.

[5.106]   Matsui, T. and Satz, H. (1986). J/ψ suppression by quark gluon plasma formation, *Phys. Lett.* **178B**, 416.

[5.107]   McLerran, L.D. and Svetitsky, B. (1981). A Monte Carlo study of SU(2) Yang–Mills theory at finite temperature, *Phys. Lett.* **98B**, 195; Kuti, J., Polonyi, J. and Szlachányi, K. (1981). Monte Carlo study of SU(2) gauge theory at finite temperature, *Phys. Lett.* **98B**, 199.

[5.108]   Svetitsky, B. and Yaffe, L.G. (1982). Critical behaviour at finite temperature confinement transitions, *Nucl. Phys.* **B210 [FS6]**, 423.

[5.109]   Kogut, J., Stone, M., Wyld, H.W., Gibbs, W.R., Shigemitsu, J., Shenker, S.H. and Sinclair, D.K. (1983). Deconfinement and chiral symmetry restoration at finite temperatures in SU(2) and SU(3) gauge theories, *Phys. Rev. Lett.* **50**, 393.

[5.110]   Pisarski, R.D. and Wilczek, F. (1984). Remarks on the chiral phase transition in chromodynamics, *Phys. Rev.* **D29**, 338.

[5.111]   Iwasaki, Y., Kanaya, K., Sakai, S. and Yoshié, T. (1991). Chiral properties of dynamical Wilson-quarks at finite temperature, *Phys. Rev. Lett.* **67**, 1494.

[5.112]   Petersson, B. and Reisz, T. (1991). Polyakov-loop correlations at finite temperature, *Nucl. Phys.* **B353**, 757.

[5.113]   Oleszczuk, M. and Polonyi, J. (1991). Dynamical symmetry breaking at high temperature, MIT preprint CTP#1984.

[5.114]   DeTar, C. (1985). A conjecture concerning the modes of excitation of the quark–gluon plasma, *Phys. Rev.* **D32**, 276.

[5.115]   DeTar, C. and Kogut, J.B. (1987). The hadronic spectrum of the quark plasma, *Phys. Rev. Lett.* **59**, 399; Measuring the hadronic spectrum of the quark plasma, *Phys. Rev.* **D36**, 2828.

[5.116]   Hasenfratz, P. and Karsch, F. (1983). Chemical potential on the lattice, *Phys. Lett.* **125B**, 308.

[5.117]   Kogut, J., Matsuoka, H., Stone, M., Wyld, H.W., Shenker, S., Shigemitsu, J. and Sinclair, D.K. (1983). Chiral symmetry restoration in baryon rich environments, *Nucl. Phys.* **B225**, 93.

[5.118]   Bilic, N. and Gavai, R.V. (1984). On the thermodynamics of an ideal Fermi gas on the lattice at finite density, *Z. Phys.* **C23**, 77; Gavai, R.V. (1985). Chemical potential on the lattice revisited, *Phys. Rev.* **D32**, 519.

[5.119]   Toussaint, D. (1990). Simulating QCD at finite density, *Nucl. Phys. B (Proc. Suppl.)* **17**, 248.

[5.120]   Barbour, I.M. (1990). Lattice QCD at finite density, *Nucl. Phys. B (Proc. Suppl.)* **17**, 243.

[5.121] Gottlieb, S., Liu, W., Shenker, R.L., Sugar, R.L. and Toussaint, D. (1987). Fermion number susceptibility in lattice gauge theory, *Phys. Rev. Lett.* **59**, 2247;
Gottlieb, S., Liu, W., Shenker, R.L., Sugar, R.L. and Toussaint, D. (1988). The quark number susceptibility of high temperature QCD, *Phys. Rev.* **D38**, 2888.

Chapter 6

[6.1] Montvay, I. (1989). Non-perturbative aspects of the Higgs sector in the standard electroweak theory, in *Heavy Flavours and High-Energy Collisions in the 1–100 TeV Range*, ed. A. Ali and L. Cifarelli (Plenum Press, New York), p. 469.

[6.2] Jersak, J. (1990). Lattice studies of the Higgs system, in *Higgs Particle(s) – Physics Issues and Searches in High Energy Collisions*, ed. A. Ali (Plenum Press, New York), p. 39.

[6.3] Shigemitsu, J. (1991). Higgs–Yukawa-chiral models. I, *Nucl. Phys. B (Proc. Suppl.)* **20**, 515.

[6.4] Golterman, M.F.L. (1991). Lattice chiral gauge theories: results and problems, *Nucl. Phys. B (Proc. Suppl.)* **20**, 528.

[6.5] Montvay, I. (1992). Higgs- and Yukawa-theories on the lattice, *Nucl. Phys. B (Proc. Suppl.)* **26**, 57.

[6.6] Kühnelt, H., Lang, C.B. and Vones, G. (1984). SU(2) gauge Higgs theory with radial degrees of freedom, *Nucl. Phys.* **B230 [FS10]**, 16.

[6.7] Montvay, I. (1985). Correlations in the SU(2) fundamental Higgs-model, *Phys. Lett.* **150B**, 441;
Montvay, I. (1986). Correlations and static energies in the standard Higgs-model, *Nucl. Phys.* **B269**, 170.

[6.8] Langguth, W., Montvay, I. and Weisz, P. (1986). Monte Carlo study of the standard SU(2) Higgs model, *Nucl. Phys.* **B277**, 11.

[6.9] Shrock, R. (1986). The phase structure of SU(2) $\otimes$ U(1)$_Y$ lattice gauge theory, *Nucl. Phys.* **B267**, 301.

[6.10] Montvay, I. (1987). Weak gauge coupling expansion near the critical line in the standard SU(2) Higgs model, *Nucl. Phys.* **B293**, 479.

[6.11] 't Hooft, G. (1971). Renormalization of massless Yang–Mills fields, *Nucl. Phys.* **B33**, 173; Renormalizable Lagrangians for massive Yang–Mills fields, *Nucl. Phys.* **B35**, 167.

[6.12] Elitzur, S. (1975). Impossibility of spontaneously breaking local symmetries, *Phys. Rev.* **D12**, 3978.

[6.13] Jersák, J., Lang, C.B., Neuhaus, T. and Vones, G. (1985). Properties of phase transitions of the lattice SU(2) Higgs model, *Phys. Rev.* **D32**, 2761.

[6.14]   Langguth, W. and Montvay, I. (1985). Two-state signal of the confinement-Higgs phase transition in the standard SU(2) Higgs model, *Phys. Lett.* **165B**, 135.

[6.15]   Evertz, H.G., Jersák, J., Lang, C.B. and Neuhaus, T. (1986). SU(2) Higgs boson and vector boson masses on the lattice, *Phys. Lett.* **171B**, 271.

[6.16]   Evertz, H.G., Katznelson, E., Lauwers, P. and Marcu, M. (1989). Towards a better quantitative understanding of the SU(2) Higgs model, *Phys. Lett.* **221B**, 143.

[6.17]   t'Hooft, G. (1980). Confinement and topology in nonabelian gauge theories, *Proceedings of the 19th Schladming Winter School*, p. 531.

[6.18]   Osterwalder, K. and Seiler, E. (1978). Gauge field theories on the lattice, *Ann. Phys.* **110**, 440;
Fradkin, E. and Shenker, S. (1979). Phase diagrams of lattice gauge theories with Higgs fields, *Phys. Rev.* **D19**, 3682.

[6.19]   Abbott, L.F. and Farhi, E. (1981). Are the weak interactions strong? *Phys. Lett.* **101B**, 69; A confining model of the weak interactions, *Nucl. Phys.* **B189**, 547.

[6.20]   Fredenhagen, K. and Marcu, M. (1988). Dual interpretation of order parameters for lattice gauge theories with matter fields, *Nucl. Phys. B (Proc. Suppl.)* **4**, 352.

[6.21]   Kirzhnits, D.A. and Linde, A.D. (1972). Macroscopic consequences of the Weinberg model, *Phys. Lett.* **42B**, 471.

[6.22]   Dolan, L. and Jackiw, R. (1974). Symmetry behavior at finite temperature, *Phys. Rev.* **D9**, 3320;
Weinberg, S. (1974). Gauge and global symmetries at high temperature, *Phys. Rev.* **D9**, 3357.

[6.23]   Damgaard, P.H. and Heller, U.M. (1987). The fundamental SU(2) Higgs model at finite temperature, *Nucl. Phys.* **B294**, 253;
Damgaard, P.H. and Heller, U.M. (1988). Search for symmetry restoration in the fundamental SU(2) Higgs model, *Nucl. Phys.* **B304**, 63.

[6.24]   Evertz, H.G., Jersák, J. and Kanaya, K. (1987). Finite temperature SU(2) Higgs model on a lattice, *Nucl. Phys.* **B285 [FS19]**, 229.

[6.25]   Bender, I., Hashimoto, T., Karsch, F., Linke, V., Nakamura, A., Schiestl, M. and Stamatescu, I.O. (1991). Results from finite temperature calculations on anisotropic lattices, *Nucl. Phys. B (Proc. Suppl.)* **20**, 329.

[6.26]   Langguth, W. and Montvay, I. (1987). A numerical estimate of the upper limit for the Higgs boson mass, *Z. Phys.* **C36**, 725.

[6.27] Hasenfratz, A. and Neuhaus, T. Upper bound estimate for the Higgs mass from the lattice regularized Weinberg–Salam model, *Nucl. Phys.* **B297**, 205.

[6.28] Dashen, R. and Neuberger, H. (1983). How to get an upper bound on the Higgs mass? *Phys. Rev. Lett.* **50**, 1897.

[6.29] Symanzik, K. (1970). Renormalizable models with simple symmetry breaking, *Comm. Math. Phys.* **16**, 48.

[6.30] Coleman, S. and Weinberg, E. (1973). Radiative corrections as the origin of spontaneous symmetry breaking, *Phys. Rev.* **D7**, 1888.

[6.31] Hasenfratz, A. and Hasenfratz, P. (1986). The continuum limit of an SU(2) gauge theory with a scalar doublet, *Phys. Rev.* **D34**, 3160.

[6.32] Weinberg, S. (1976). Mass of the Higgs boson, *Phys. Rev. Lett.* **36**, 294;
Linde, A.D. (1976). Dynamical symmetry restoration and constraints on masses and coupling constants in gauge theories, *JETP* **23**, 73.

[6.33] Montvay, I. (1986). Weak gauge coupling expansion in the lattice regularized standard SU(2) Higgs model, *Phys. Lett.* **172B**, 71.

[6.34] Smit, J. (1980). Chiral symmetry breaking in QCD: mesons as spin waves, *Nucl. Phys.* **B175**, 307;
Smit, J. (1986). Fermions on a lattice, *Acta Phys. Polonica* **B17**, 531;
Swift, P.V.D. (1984). The electroweak theory on the lattice, *Phys. Lett.* **145B**, 256.

[6.35] Hands, S. and Carpenter, D.B. (1986). Lattice sigma model and fermion doubling, *Nucl. Phys.* **B266**, 285.

[6.36] Montvay, I. (1987). A chiral SU(2)$_L$ ⊗ SU(2)$_R$ gauge model on the lattice, *Phys. Lett.* **199B**, 89;
Montvay, I. (1988). Nonperturbative approach to scalar–fermion theories, *Nucl. Phys. B (Proc. Suppl.)* **4**, 443.

[6.37] Lin, L., Montvay, I., Münster, G. and Wittig, H. (1991). A U(1)$_L$ ⊗ U(1)$_R$ symmetric Yukawa-model in the phase with spontaneously broken symmetry, *Nucl. Phys.* **B355**, 511.

[6.38] Lin, L., Ma, J.P. and Montvay, I. (1990). A scalar fermion model in the limit of infinitely heavy fermions, *Z. Phys.* **C48**, 355.

[6.39] Farakos, K., Koutsoumbas, G., Lin, L., Ma, J.P., Montvay, I. and Münster, G. (1991). A U(1)$_L$ ⊗ U(1)$_R$ symmetric Yukawa-model in the symmetric phase, *Nucl. Phys.* **B350**, 474.

[6.40] Montvay, I. (1988). The sigma-model with Wilson lattice fermions, *Nucl. Phys.* **B307**, 389.

[6.41] De, A.K. (1991). Smit–Swift model in the symmetric phase: investigations in the global symmetry limit, *Nucl. Phys. B (Proc. Suppl.)* **20**, 572.

[6.42] Lin, L. and Wittig, H. (1992). An $SU(2)_L \otimes SU(2)_R$ symmetric Yukawa-model in the symmetric phase, *Z. Phys.* **C54**, 331.

[6.43] Farakos, K., Koutsoumbas, G. and Montvay, I. (1990). Random walk approximation in a chiral Yukawa-model and global symmetries, *Z. Phys.* **C47**, 641;
Farakos, K. and Koutsoumbas, G. (1991). Hopping parameter expansion investigation of a $U(1)_L \otimes U(1)_R$ lattice Yukawa-model in the symmetric phase, *Nucl. Phys.* **B366**, 665;
Farakos, K. and Koutsoumbas, G. (1992). Hopping expansion for a $SU(2)_L \otimes SU(2)_R$ chiral Yukawa model, *Phys. Lett.* **B288**, 161.

[6.44] Montvay, I. (1988). Three mirror pairs of fermion families, *Phys. Lett.* **205B**, 315.

[6.45] Borrelli, A., Maiani, L., Rossi, G.C., Sisto, R. and Testa, M. (1990). Neutrinos on the lattice: the regularization of a chiral gauge theory, *Nucl. Phys.* **B333**, 335;
Maiani, L., Rossi, G.C. and Testa, M. (1991). On lattice chiral gauge theories, *Phys. Lett.* **261B**, 479.

[6.46] Golterman, M.F.L. and Petcher, D. (1989). The decoupling of righthanded neutrinos in chiral lattice gauge theories, *Phys. Lett.* **225B**, 159.

[6.47] Montvay, I. (1992). Mirror fermions in chiral gauge theories, to appear in the Proceedings of the Roma Workshop on Chiral Gauge Theories, *Nucl. Phys. B (Proc. Suppl.)*.

[6.48] Stephenson, D. and Thornton, A. (1988). Nonperturbative Yukawa couplings, *Phys. Lett.* **212B**, 479.

[6.49] Lee, I.-H., Shigemitsu, J. and Shrock, R.E. (1990). Lattice study of a Yukawa-theory with a real scalar field, *Nucl. Phys.* **B330**, 225; Study of different lattice formulations of a Yukawa-model with a real scalar field, *Nucl. Phys.* **B334**, 265.

[6.50] Hasenfratz, A., Liu, W. and Neuhaus, T. (1990). Phase structure and critical points in a scalar fermion model, *Phys. Lett.* **236B**, 339.

[6.51] Stephanov, M.A. and Tsypin, M.M. (1990). Phase diagram for the lattice scalar fermion model, *Phys. Lett.* **236B**, 344.

[6.52] Bock, W., De, A.K., Jansen, K., Jersák, J., Neuhaus, T. and Smit, J. (1990). Phase diagram of a lattice $SU(2) \otimes SU(2)$ scalar fermion model with naive and Wilson-fermions, *Nucl. Phys.* **B344**, 207.

[6.53] Lin, L., Montvay, I. and Wittig, H. (1991). Phase structure of a $U(1)_L \otimes U(1)_R$ symmetric Yukawa-model, *Phys. Lett.* **264B**, 407.

[6.54] Chanowitz, M., Furman, M. and Hinchliffe, I. (1979). Weak interactions of ultraheavy fermions. 2, *Nucl. Phys.* **B153**, 402.

[6.55] Duncan, M.J., Philippe, R. and Sher, M. (1985). Theoretical ceiling on quark masses in the standard model, *Phys. Lett.* **153B**, 155;

Sher, M. (1989). Electroweak Higgs potentials and vacuum stability, *Phys. Rept.* **179**, 273.

Chapter 7

[7.1] Bhanot, G. (1988). The Metropolis algorithm, *Rep. Prog. Phys.* **51**, 429.

[7.2] Ferrenberg, A.M. and Swendsen, R.H. (1988). New Monte Carlo technique for studying phase transitions, *Phys. Rev. Lett.* **61**, 2635.

[7.3] Huang, S., Moriarty, K.J.M., Myers, E. and Potvin, J. (1990). Using histograms in the study of the thermodynamics of QCD at finite temperature, *Nucl. Phys. B (Proc. Suppl.)* **17**, 281;
Huang, S., Moriarty, K.J.M., Myers, E. and Potvin, J. (1991). The density of states method and the velocity of sound in hot QCD, *Z. Phys.* **C50**, 221.

[7.4] Bhanot, G., Black, S., Carter, P. and Salvador, R. (1987). A new method for the partition function of discrete systems with application to the 2-d Ising model, *Phys. Lett.* **183B**, 331;
Bhanot, G., Bitar, K. and Salvador, R. (1987). The partition function of Z_2 and Z_3 lattice gauge theory in four dimensions: a novel approach to simulations of lattice systems, *Phys. Lett.* **187B**, 381; On solving four dimensional SU(2) gauge theory by numerically finding its partition function, *Phys. Lett.* **188B**, 246.

[7.5] Karliner, M., Sharpe, S.R. and Chang, Y.F. (1988). Zeroing in on SU(3), *Nucl. Phys.* **B302**, 204.

[7.6] O'Raifeartaigh, L., Wipf, A. and Yoneyama, H. (1986). The constraint effective potential, *Nucl. Phys.* **B271**, 653.

[7.7] Kuti, J. and Shen, Y. (1988). Supercomputing the effective action, *Phys. Rev. Lett.* **60**, 85.

[7.8] Montvay, I. and Weisz, P. (1987). Numerical study of finite volume effects in the four-dimensional Ising model, *Nucl. Phys.* **B290 [FS20]**, 327.

[7.9] Young, H.D. (1962). *Statistical Treatment of Experimental Data* (McGraw Hill, New York);
Frodesen, A.G., Skjaggerstad, O. and Tøfte, H. (1979). *Probability and Statistics in Particle Physics* (Universitetsvorlaget, Bergen);
Brandt, S. (1983). *Statistical and Computational Methods in Data Analysis* (North Holland, Amsterdam);
van der Waerden, B.L. (1969). *Mathematical Statistics* (Springer, New York).

[7.10] Toussaint, D. (1990). Error analysis of simulation results: a sample problem, in *From Actions to Answers*, ed. T. DeGrand and D. Toussaint, (World Scientific, Singapore), p. 121.

[7.11]   Efron, B. (1979). Computers and the theory of statistics: thinking the unthinkable, *SIAM Review* **21**, 460.

[7.12]   Gottlieb, S., Mackenzie, P.B., Thacker, H.B. and Weingarten, D. (1986). Hadronic coupling constants in lattice gauge theory, *Nucl. Phys.* **B263**, 704;
Gupta, R., Guralnik, G., Kilcup, G., Sharpe, S.R. and Warnock, T. (1987). The hadron spectrum on a $18^3 \cdot 42$ lattice, *Phys. Rev.* **D36**, 2813.

[7.13]   Berg, B.A. (1990). Double jackknife bias corrected estimators, FSU-SCRI preprint 90-100.

[7.14]   Parisi, G. (1984). The strategy for computing the hadronic mass spectrum, *Phys. Rept.* **103**, 203;
Parisi, G., Petronzio, R. and Rapuano, F. (1983). A measurement of the string tension near the continuum limit, *Phys. Lett.* **128B**, 418.

[7.15]   Brower, R., Rossi, P. and Tan, C.-I. (1981). The external field problem for QCD, *Nucl. Phys.* **B190 [FS3]**, 699.

[7.16]   Swendsen, R.H. (1984). Monte Carlo calculation of renormalized coupling parameters, *Phys. Rev. Lett.* **52**, 1165.

[7.17]   Metropolis, N., Rosenbluth, A.W., Rosenbluth, M.N., Teller, A.H. and Teller, E. (1953). *J. Chem. Phys.* **21**, 1087.

[7.18]   Scalettar, R.T., Scalapino, D.J. and Sugar, R.L. (1986). New algorithm for the numerical simulation of fermions, *Phys. Rev.* **B34**, 7911.

[7.19]   Adler, S.L. (1981). Overrelaxation method for the Monte Carlo evaluation of the partition function for multiquadratic actions, *Phys. Rev.* **D23**, 2901;
Adler, S.L. (1988). Overrelaxation algorithm for lattice field theories, *Phys. Rev.* **D37**, 458.

[7.20]   Brown, F.R. and Woch, T.J. (1987). Overrelaxation heat bath and Metropolis algorithms for accelerating pure gauge Monte Carlo calculations, *Phys. Rev. Lett.* **58**, 2394.

[7.21]   Creutz, M. (1987). Overrelaxation and Monte Carlo simulation, *Phys. Rev.* **D36**, 515.

[7.22]   Petronzio, R. and Vicari, E. (1990). An overrelaxed Monte Carlo algorithm for SU(3) lattice gauge theories, *Phys. Lett.* **248B**, 159.

[7.23]   Creutz, M. (1980). Monte Carlo study of quantized SU(2) gauge theory, *Phys. Rev.* **D21**, 2308.

[7.24]   Fredenhagen, K. and Marcu, M. (1987). A modified heat bath method suitable for Monte Carlo simulations on vector and parallel processors, *Phys. Lett.* **193B**, 486.

[7.25]   Pietarinen, E. (1981). String tension in SU(3) lattice gauge theory, *Nucl. Phys.* **B190 [FS3]**, 349.

[7.26] Cabibbo, N. and Marinari, E. (1982). A new method for updating SU(*N*) matrices in computer simulations of gauge theories, *Phys. Lett.* **119B**, 387.

[7.27] Montvay, I. (1990). Simulation of staggered fermions by polimer algorithms, in *Probabilistic Methods in Quantum Field Theory and Quantum Gravity*, ed. P.H. Damgaard, H. Hüffel and A. Rosenblum (Plenum Press, New York), p. 87.

[7.28] Joos, H. and Montvay, I. (1983). The screening of colour charge in numerical hopping parameter expansion, *Nucl. Phys.* **B255 [FS9]**, 565.

[7.29] Montvay, I. (1984). Monte Carlo simulation with unquenched Wilson-fermions, *Phys. Lett.* **139B**, 70;
Langguth, W. and Montvay, I. (1984). Monte Carlo calculation of hadron masses with light dynamical quarks, *Phys. Lett.* **145B**, 261.

[7.30] Montvay, I. (1987). Numerical calculation of hadron masses in lattice quantum chromodynamics, *Rev. Mod. Phys.* **59**, 263.

[7.31] Fucito, F., Marinari, E., Parisi, G. and Rebbi, C. (1981). A proposal for Monte Carlo simulations of fermionic systems, *Nucl. Phys.* **B180 [FS2]**, 369.

[7.32] Weingarten, D.H. and Petcher, D.N. (1981). Monte Carlo integration for lattice gauge theories with fermions, *Phys. Lett.* **99B**, 333.

[7.33] Rossi, P., Davies, C.T.H. and Lepage, G.P. (1988). A comparison of a variety of matrix inversion algorithms for Wilson-fermions on the lattice, *Nucl. Phys.* **B297**, 287.

[7.34] Householder, A. (1964). *The Theory of Matrices in Numerical Analysis* (Blaisdell, New York);
Varga, R.S. (1965). *Matrix Iterative Analysis* (Prentice-Hall, Englewood Cliffs).

[7.35] Aho, A.V., Hopcroft, J.E. and Ullman, J.D. (1974). *The Design and Analysis of Computer Algorithms* (Addison-Wesley, Reading);
Stoer, J. and Bulirsch, R. (1980). *Introduction to Numerical Analysis* (Springer, Berlin).

[7.36] Hestenes, M. (1980). *Conjugate Direction Method in Optimization* (Springer, New York).

[7.37] Fukugita, M., Oyanagi, Y. and Ukawa, A. (1986). Hadron spectroscopy in lattice QCD with dynamical quark loops, *Phys. Rev. Lett.* **57**, 953.

[7.38] Oyanagi, Y. (1986). An incomplete LDU decomposition of lattice fermions and its application to conjugate residual methods, *Comp. Phys. Commun.* **42**, 333.

[7.39] Katz, G., Batrouni, G., Davies, C., Kronfeld, A., Lepage, P., Rossi, P., Svetitsky, B. and Wilson, K. (1988). Fourier acceleration in lattice gauge theories. 1. Landau gauge fixing, *Phys. Rev.* **D37**, 1581; Fourier acceleration. 2. Matrix inversion and the quark propagator, *Phys. Rev.* **D37**, 1589.

[7.40] DeGrand, T.A. (1988). A conditioning technique for matrix inversion for Wilson-fermions, *Comp. Phys. Commun.* **52**, 161; DeGrand, T.A. and Rossi, P. (1990). Conditioning technique for dynamical fermions, *Comp. Phys. Commun.* **60**, 211.

[7.41] Ben-Av, R., Brandt, A. and Solomon, S. (1990). The fermion matrix, instantons, zero modes and multigrid, *Nucl. Phys.* **B329**, 193; Ben-Av, R., Brandt, A., Harmatz, M., Katznelson, E., Lauwers, P.G., Solomon, S. and Wolowesky, K. (1991). Fermion simulations using parallel transported multigrid, *Phys. Lett.* **253B**, 185; Parallel transported multigrid and its application to the Schwinger model, *Nucl. Phys. B (Proc. Suppl.)* **20**, 102.

[7.42] Brower, R.C., Rebbi, C. and Vicari, E. (1991). Projective multigrid for propagators in lattice gauge theory, *Phys. Rev. Lett.* **66**, 1263; Brower, R., Moriarty, K., Rebbi, C. and Vicari, E. (1991). Variational multigrid for nonabelian gauge theory, *Nucl. Phys. B (Proc. Suppl.)* **20**, 89.

[7.43] Callaway, D. and Rahman, A. (1982). The microcanonical ensemble: a new formulation of lattice gauge theory, *Phys. Rev. Lett.* **49**, 613.

[7.44] Polonyi, J. and Wyld, H.W. (1983). Microcanonical formulation of fermionic systems, *Phys. Rev. Lett.* **51**, 2257; erratum: **52**, 401.

[7.45] Parisi, G. and Wu, Y.-S. (1981). Perturbation theory without gauge fixing, *Sci. Sin.* **24**, 483.

[7.46] Damgaard, P.H. and Hüffel, H. (1987). *Phys. Rept.* **152**, 227; *Stochastic Quantization*, ed. P.H. Damgaard and H. Hüffel (World Scientific, Singapore, 1988).

[7.47] Batrouni, G.G., Katz, G.R., Kronfeld, A.S., Lepage, G.P., Svetitsky, B. and Wilson, K.G. (1985). Langevin simulations of lattice field theories, *Phys. Rev.* **D32**, 2736.

[7.48] Ukawa, A. and Fukugita, M. (1985). Langevin simulation including dynamical quark loops, *Phys. Rev. Lett.* **55**, 1854.

[7.49] Creutz, M. and Gavai, R. (1987). Monte Carlo simulation of fermionic fields, *Nucl. Phys.* **B280**, 181.

[7.50] Duane, S. (1985). Stochastic quantization vs. the microcanonical ensemble: getting the best of both worlds, *Nucl. Phys.* **B257 [FS14]**, 652.

[7.51] Duane, S. and Kogut, J.B. (1986). The theory of hybrid stochastic algorithms, *Nucl. Phys.* **B275 [FS17]**, 398.

[7.52] Gottlieb, S., Liu, W., Toussaint, D., Renken, R.L. and Sugar, R.L. (1987). Hybrid molecular dynamics algorithms for the numerical simulation of quantum chromodynamics, *Phys. Rev.* **D35**, 2531.

[7.53]   Weingarten, D. (1989). Monte Carlo algorithms for QCD, *Nucl. Phys. B (Proc. Suppl.)* **9**, 447.

[7.54]   Klauder, J.R. (1983). Stochastic quantization, in *Recent Developments in High Energy Physics*, ed. H. Mitter and C.B. Lang (Springer, Berlin), p. 251;
Klauder, J.R. (1984). Coherent state Langevin equations for canonical quantum systems with applications to the quantized Hall effect, *Phys. Rev.* **A29**, 2036.

[7.55]   Parisi, G. (1983). On complex probabilities, *Phys. Lett.* **131B**, 393.

[7.56]   Okano, K., Schülke, L. and Zheng, B. (1991). Kernel controlled complex Langevin simulation: Field dependent kernel, *Phys. Lett.* **258B**, 421.

[7.57]   Duane, S., Kennedy, A.D., Pendleton, B.J. and Roweth, D. (1987). Hybrid Monte Carlo, *Phys. Lett.* **195B**, 216.

[7.58]   Creutz, M. (1988). Global Monte Carlo algorithms for many fermion systems, *Phys. Rev.* **D38**, 1228.

[7.59]   Kennedy, A.D. (1990). The theory of hybrid stochastic algorithms, in *Probabilistic Methods in Quantum Field Theory and Quantum Gravity*, ed. P.H. Damgaard, H. Hüffel and A. Rosenblum (Plenum Press, New York), p. 209.

[7.60]   Gupta, R., Kilcup, G.W. and Sharpe, S.R. (1988). Tuning the Hybrid Monte Carlo algorithm, *Phys. Rev.* **D38**, 1278;
Gupta, R., Patel, A., Baillie, C.F., Guralnik, G., Kilcup, G.W. and Sharpe, S.R. (1989). QCD with dynamical Wilson-fermions, *Phys. Rev.* **D40**, 2072.

[7.61]   Gupta, S., Irbäck, A., Karsch, F. and Petersson, B. (1990). The acceptance probability in the Hybrid Monte Carlo method, *Phys. Lett.* **242B**, 437.

[7.62]   Kennedy, A.D. and Pendleton, B. (1991). Acceptances and autocorrelations in Hybrid Monte Carlo, *Nucl. Phys. B (Proc. Suppl.)* **20**, 118.

[7.63]   Wolff, U. (1990). Critical slowing down, *Nucl. Phys. B (Proc. Suppl.)* **17**, 93.

[7.64]   Swendsen, R.H. and Wang, J.-S. (1987). Nonuniversal critical dynamics in Monte Carlo simulations, *Phys. Rev. Lett.* **58**, 86.

[7.65]   Wolff, U. (1988). Monte Carlo simulation of a lattice field theory as correlated percolation, *Nucl. Phys.* **B300 [FS22]**, 501.

[7.66]   Edwards, R.G. and Sokal, A.D. (1988). Generalization of the Fortuin–Kasteleyn–Swendsen–Wang representation and Monte Carlo algorithm, *Phys. Rev.* **D38**, 2009.

[7.67]   Kasteleyn, P.W. and Fortuin, C.M. (1969). *J. Phys. Soc. Jpn.* **26** (Suppl.), 11;

Fortuin, C.M. and Kasteleyn, P.W. (1972). On the random cluster model. I. Introduction and relation to other models, *Physica (Utrecht)* **57**, 536.

[7.68] Hoshen, J. and Kopelman, R. (1976). Percolation and cluster distribution. I. Cluster multiple labeling technique and critical concentration algorithm, *Phys. Rev.* **B14**, 3438.

[7.69] Montvay, I., Münster, G. and Wolff, U. (1988). Percolation cluster algorithm and scaling behaviour in the 4-dimensional Ising model, *Nucl. Phys.* **B305 [FS23]**, 143.

[7.70] Mino, H. (1991). A vectorized algorithm for cluster formation in the Swendsen–Wang dynamics, *Comp. Phys. Commun.* **66**, 25;
Evertz, H.G. (1991). Vectorized cluster search, Tallahassee preprint FSU-SCRI-91-183.

[7.71] Jansen, K., Jersák, J., Montvay, I., Münster, G., Trappenberg, T. and Wolff, U. (1988). Vacuum tunneling in the 4-dimensional Ising model, *Phys. Lett.* **213B**, 203;
Jansen, K., Montvay, I., Münster, G., Trappenberg, T. and Wolff, U. (1989). Broken phase of the 4-dimensional Ising model in a finite volume, *Nucl. Phys.* **B322**, 698.

[7.72] Niedermayer, F. (1988). A general cluster updating method for Monte Carlo simulations, *Phys. Rev. Lett.* **61**, 2026.

[7.73] Wolff, U. (1989). Collective Monte Carlo updating for spin systems, *Phys. Rev. Lett.* **62**, 361.

[7.74] Edwards, R.G. and Sokal, A.D. (1989). Dynamical critical behavior of Wolff's collective Monte Carlo algorithm for the two-dimensional $O(N)$ nonlinear sigma model, *Phys. Rev.* **D40**, 1374.

[7.75] Frick, C., Jansen, K., Jersák, J., Montvay, I., Münster, G. Seuferling, P. (1990). Numerical simulation of the $O(4)$ symmetric ϕ^4 model in the symmetric phase, *Nucl. Phys.* **B331**, 515;
Frick, C., Jansen, K. and Seuferling, P. (1989). Cluster algorithms in the $O(4)$ ϕ^4 theory in four dimensions, *Phys. Rev. Lett.* **63**, 2613.

[7.76] Wolff, U. (1989). Continuum behavior in the lattice $O(3)$ nonlinear sigma model, *Phys. Lett.* **222B**, 473;
Lüscher, M. and Wolff, U. (1990). How to calculate the elastic scattering matrix in two-dimensional quantum field theories by numerical simulation, *Nucl. Phys.* **B339**, 222.

Chapter 8

[A1] Bjorken, J.D. and Drell, S.D. (1964). *Relativistic Quantum Machanics* (McGraw-Hill, New York).

[A2] Bjorken, J.D. and Drell, S.D. (1965). *Relativistic Quantum Fields* (McGraw-Hill, New York).

[A3]   Itzykson, C. and Zuber, J.-B. (1980). *Quantum Field Theory* (McGraw-Hill, New York).

[A4]   Becher, P., Böhm, M. and Joos, H. (1984). *Gauge Theories of Strong and Electroweak Interactions* (Wiley, New York).

[A5]   Zinn-Justin, J. (1989). *Quantum Field Theory and Critical Phenomena* (Clarendon, Oxford).

[A6]   Huang, K. (1963). *Statistical Mechanics* (John Wiley & Sons, New York).

[A7]   Creutz, M. (1983). *Quarks, Gluons and Lattices* (Cambridge University Press, Cambridge).

[A8]   Collins, J. (1984). *Renormalization* (Cambridge University Press, Cambridge).

Index

Printed in the United States
By Bookmasters